Materials for Smart Systems II

MATERIALS RESEARCH SOCIETY
SYMPOSIUM PROCEEDINGS VOLUME 459

Materials for Smart Systems II

Symposium held December 2-5, 1996, Boston, Massachusetts, U.S.A.

EDITORS:

Easo P. George
Oak Ridge National Laboratory
Oak Ridge, Tennessee, U.S.A.

Rolf Gotthardt
École Polytechnique Fédérale de Lausanne
Lausanne, Switzerland

Kazuhiro Otsuka
University of Tsukuba
Tsukuba, Japan

Susan Trolier-McKinstry
Pennsylvania State University
University Park, Pennsylvania, U.S.A.

Marilyn Wun-Fogle
Naval Surface Warfare Center
Silver Spring, Maryland, U.S.A.

MATERIALS
RESEARCH
SOCIETY

PITTSBURGH, PENNSYLVANIA

This work was supported in part by the Army Research Office under Grant Number ARO: DAAG55-97-1-0005. The views, opinions, and/or findings contained in this report are those of the author(s) and should not be construed as an official Department of the Army position, policy, or decision, unless so designated by other documentation.

Single article reprints from this publication are available through University Microfilms Inc., 300 North Zeeb Road, Ann Arbor, Michigan 48106

CODEN: MRSPDH

Published by:

Materials Research Society
9800 McKnight Road
Pittsburgh, Pennsylvania   15237
Telephone (412) 367-3003
Fax (412) 367-4373
Website: http://www.mrs.org/

Library of Congress Cataloging in Publication Data

Materials for smart systems II : symposium held December 2–5, 1996, Boston, Massachusetts, U.S.A. / editors, Easo P. George, Rolf Gotthardt, Kazuhiro Otsuka, Susan Trolier-McKinstry, Marilyn Wun-Fogle
   p. cm—(Materials Research Society symposium proceedings ; v. 459)
   Includes bibliographical references and index.
   ISBN 1-55899-363-0
   1. Smart materials—Congresses. 2. Detectors—Materials—Congresses. 3. Ferroelectric devices—Materials—Congresses. 4. Magnetostrictive transducers—Materials—Congresses. 5. Thin film devices—Materials—Congresses. 6. Shape memory alloys—Congresses. 7. Piezoelectric devices—Materials—Congresses. I. George, Easo P. II. Gotthardt, Rolf III. Otsuka, Kazuhiro IV. Trolier-McKinstry, Susan, V. Wun-Fogle, Marilyn VI. Series: Materials Research Society symposium proceedings ; v. 459.
TA418.9.S62M377 1997   97-6489
620.1'1—dc21   CIP

Manufactured in the United States of America

CONTENTS

*Invited Paper

*Invited Paper

PART IV: POSTER SESSION I

*Invited Paper

PART VI: SHAPE-MEMORY ALLOYS II

*Invited Paper

*Invited Paper

PREFACE

This volume represents a record of the proceedings of the symposium on "Materials for Smart Systems II," which was held in conjunction with the 1996 MRS Fall Meeting in Boston, Massachusetts. The symposium revolved around the use of ferroelectrics, magnetostrictors, and shape-memory materials in smart systems. Joint sessions were also held on actuator materials for microsystems. Seventy-five oral presentations were scheduled over the course of four days, including nine in a joint session with the symposium on "Materials in Microsystems." In addition, about 60 posters were presented during two evening sessions. Of the 79 papers in this volume, approximately 40 (or 51%) came from outside the U.S.

In the piezoelectrics section, several major themes emerged. First, the presence of multiple phase transformations is a common thread found in all families of actuator materials. The manipulation of these transformations to develop highly active ferroelectric actuator materials, including materials with high electromechanical coupling coefficients, was discussed in several papers. A second focus revolved around the influence of mobile twin domain walls in affecting the electrical and electromechanical response of ferroelectric perovskites under a variety of loading conditions. Several papers dealt with recent developments in thin-film actuators for integrated silicon micromachined devices and for integrated optical-fiber devices. Finally, the development of polymeric actuators was also discussed.

The session on magnetostrictive materials and applications covered both high-strain materials and amorphous magnetostrictive materials. Several papers focused on bulk Terfenol-D, a high-magnetostriction room temperature material. The discussions on Terfenol-D centered around its use in high-power transducers and actuators as well as the development of Terfenol-D/epoxy composite materials. Several papers covered the recent progress in the development of giant magnetostrictive thin films with large strains for microsystem applications. Papers in the area of amorphous magnetostrictive materials ranged from their uses in remote sensors to utilizing the strong nonlinear behavior of these materials to manipulate chaos. Finally, the giant magnetic strain in magnetic shape memory alloys was discussed.

The new trends in shape-memory alloys (SMA) may be summarized as follows. The first relates to composites, which consist of SMA-polymer, SMA-TiC or SMA-metal. They may be used to change elastic constants over a wide range by adjusting temperature, or to strengthen metals, and are thus useful in space technology. The second is TiNi thin films, which have potential applications in micromachines. It is now fairly easy to make such fine-grained films by sputter deposition. However, several problems remain, e.g., uniformity in composition and control of the composition since the transformation temperatures are very sensitive to composition. The constraint from the substrate and the difference in thermal expansion are other problems, which may become serious during operation, when the films are made on Si chips. New developments in SMAs that were discussed in detail include computer simulation of microstructure evolution, a new thermodynamic approach to matensitic transformations, incommensurability, rubber-like behavior, and application of R-phase, which enables the linear control in thermostatic valves.

In addition to the work on shape memory, magnetostrictive, and piezoelectric actuators, papers were presented on alternative sensing and actuating mechanisms. Materials for gas sensors, electrochromic devices, magnetorheological fluids, and thermoplastic elastomers were described.

Easo P. George
Rolf Gotthardt
Kazuhiro Otsuka
Susan Trolier-McKinstry
Marilyn Wun-Fogle

January 1997

ACKNOWLEDGMENTS

Financial support for the symposium on "Materials for Smart Systems II" was provided by:

Army Research Office
Furukawa Electric Co., Ltd.
Oak Ridge National Laboratory
PIOLAX, Inc.

MATERIALS RESEARCH SOCIETY SYMPOSIUM PROCEEDINGS

MATERIALS RESEARCH SOCIETY SYMPOSIUM PROCEEDINGS

Prior Materials Research Society Symposium Proceedings available by contacting Materials Research Society

Part I
Piezoelectrics

HIGH ELECTROMECHANICAL COUPLING PIEZOELECTRICS - HOW HIGH ENERGY CONVERSION RATE IS POSSIBLE

KENJI UCHINO
International Center for Actuators and Transducers, Materials Research Laboratory
The Pennsylvania State University, University Park, PA 16802

ABSTRACT

A new category of piezoelectric ceramics with very high electromechanical coupling was discovered in a lead zinc niobate:lead titanate solid solution in a single crystal form. The maximum coupling factor k_{33} reaches 95%, which corresponds to the energy conversion rate twice as high as the conventional lead zirconate titanate ceramics. This paper reviews the previous studies on superior piezoelectricity in relaxor ferroelectric: lead titanate solid solutions and on the possible mechanisms of this high electromechanical coupling.

KEY WORDS: electromechanical coupling, piezoelectric, relaxor ferroelectric, domain motion

INTRODUCTION

Lead zirconate-titanate (PZT) ceramics are well known piezoelectrics widely used in many transducers. Their applications include gas igniters, force/acceleration sensors, microphones, buzzers, speakers, surface acoustic wave filters, piezoelectric transformers, actuators, ultrasonic motors, ultrasonic underwater transducers, and acoustic scanners [1,2]. Particularly in recent medical acoustic imaging, higher electromechanical coupling materials are eagerly required to improve the image resolution. Under these circumstances, relaxor ferroelectric: lead titanate solid solution systems with superior electromechanical coupling factors to the conventional PZT have been refocused, which were initially discovered in a lead zinc niobate:lead titanate system by the author's group in 1981 [3].

This paper reviews the previous studies on superior piezoelectricity in relaxor ferroelectric: lead titanate solid solutions, then on peculiar domain motions in these materials, finally possible mechanisms are considered for this extremely high electromechanical coupling.

ELECTROMECHANICAL COUPLING FACTORS

The terminologies, electromechanical coupling factor and efficiency are sometimes confused. Let us consider them at first. The electromechanical coupling factor k is defined as

$$k^2 = \text{(Stored mechanical energy / Input electrical energy)} \quad (1)$$

or

$$= \text{(Stored electrical energy / Input mechanical energy)}. \quad (2)$$

When an electric field E is applied to a piezoelectric actuator, since the input electrical energy is $(1/2) \, \varepsilon_0 \varepsilon \, E^2$ per unit volume and the stored mechanical energy per unit volume under zero external stress is given by $(1/2) \, x^2 / s = (1/2) \, (d \, E)^2 / s$, k^2 can be calculated as

$$k^2 = [(1/2) \, (d \, E)^2 / s] / [(1/2) \, \varepsilon_0 \varepsilon \, E^2]$$
$$= d^2 / \varepsilon_0 \varepsilon \cdot s. \quad (3)$$

On the other hand, the efficiency η is defined as

$$\eta = \text{(Output mechanical energy)} / \text{(Consumed electrical energy)} \quad (4)$$

or

$$= \text{(Output electrical energy)} / \text{(Consumed mechanical energy)}. \quad (5)$$

In a work cycle (e. g. an electric field cycle), the input electrical energy is transformed partially into mechanical energy and the remaining is stored as electrical energy (electrostatic energy like a capacitor) in an actuator. Then, this ineffective energy can be returned to the power source, leading to near 100 % efficiency, if the loss is small. Typical values of dielectric loss in PZT are about 1 - 3 %.

The electromechanical coupling factor is different according to the sample geometry. Figure 1 shows two sample geometries corresponding to k_{33} and k_{31}, which we will discuss later. In some particular applications such as Non-Destructive Testing, large piezoelectric anisotropy, i. e. a large value of the ratio k_{33}/k_{31}, is required to improve the image quality in addition to a large value of k_{33} itself. However, the empirical rule suggests that these two requirements are contradictory to each other. In Figure 2 we plotted the k_{33} versus k_{31} relation for various perovskite oxide piezoelectric polycrystal and single crystal samples such as Pb(Zr,Ti)O$_3$, PbTiO$_3$, Pb(Zn$_{1/3}$Nb$_{2/3}$)O$_3$ and Pb(Mg$_{1/3}$Nb$_{2/3}$)O$_3$ based compositions [4]. It is obvious from this convex tendency that the piezoelectricity becomes isotropic (i. e. the k_{33}/k_{31} ratio approaches to 1) with increasing the electromechanical coupling factor (i. e. the k_{33} value).

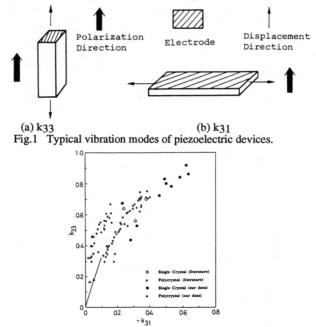

(a) k_{33} (b) k_{31}

Fig.1 Typical vibration modes of piezoelectric devices.

Fig.2 Relation between k_{33} and k_{31} for various perovskite oxide piezoelectrics.

HIGH ELECTROMECHANICAL COUPLING MATERIALS

Morphotropic Phase Boundary Composition

Conventionally, Pb(Zr,Ti)O$_3$ (PZT), PbTiO$_3$ (PT) and PZT based ternary ceramics with a small amount of a relaxor ferroelectric have been utilized for piezoelectric applications. Figure 3 shows the composition dependence of permittivity and electromechanical coupling factor k_p in the PZT system. It is notable that the morphotropic phase boundary (MPB) composition between the rhombohedral and tetragonal phases exhibits the maximum enhancement in dielectric and piezoelectric properties; this is explained in terms of a phenomenological theory [5]. The physical properties of a perovskite solid solution between A and B, (1-x) A - x B, can be estimated through the Gibbs elastic energy of a solid solution, if we suppose a linear combination of the Gibbs elastic energy of each component:

$$G_1(P,X,T) = (1/2)[(1 - x)\alpha_A + x\alpha_B] P^2 + (1/4)[(1 - x)\beta_A + x\beta_B] P^4$$
$$+ (1/6)[(1 - x)\gamma_A + x\gamma_B] P^6$$
$$- (1/2)[(1 - x)s_A + xs_B] X^2 - [(1 - x)Q_A + xQ_B] P^2 X, \qquad (6)$$

where $\alpha_A = (T - T_{0,A}) / \varepsilon_0\, C_A$ and $\alpha_B = (T - T_{0,B}) / \varepsilon_0\, C_B$.

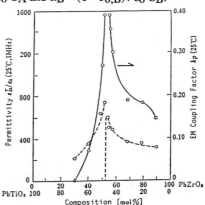

Fig.3 Composition dependence of permittivity and electromechanical coupling factor k_p in the PZT system.

The solution provides reasonable first-order estimates of the Curie temperature, spontaneous polarization and strain, as well as the enhancement of the permittivity, piezoelectric constant and electromechanical coupling at the MPB composition.

Note that in virgin samples of piezoelectric ceramics, the polarizations of grains (micro single crystals making up a polycrystalline sample) are randomly oriented (or domains are oriented randomly even in each grain, if the grain size is large enough for a multi-domain state) so as to cancel the net polarization in total. In a similar fashion, the net strain is negligibly small under an external electric field. Hence, before use, it is necessary to apply a relatively large electric field (> 3 kV/mm) to align the polarization direction of each grain as much as possible. Such a treatment is called electric poling.

Relaxor Ferroelectric Based Composition

On the other hand, relaxor ferroelectrics such as $Pb(Mg_{1/3}Nb_{2/3})O_3:PbTiO_3$ compositions have been focused due to their giant electrostriction [6]. Few work has been conducted on piezoelectric properties at the MPB region before the trial by the author's group.

We focused on single crystals of $(1-x)Pb(Zn_{1/3}Nb_{2/3})O_3- xPbTiO_3$ (PZN-PT) which are relatively easily grown by a flux method over the whole composition range, compared with the case in the $Pb(Zr,Ti)O_3$ (PZT) system. This system exhibits a drastic change from a diffuse phase transition to a sharp transition with an increase of the PT content, x, correlating to the existence of a morphotropic phase boundary from a rhombohedral to a tetragonal phase around $x = 0.1$ [7]. Figure 4 shows the phase diagram of this system near the MPB region.

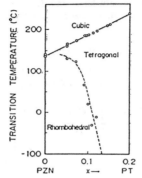

Fig.4   Phase diagram of $(1-x)Pb(Zn_{1/3}Nb_{2/3})O_3- xPbTiO_3$ (PZN-PT).

The most intriguing piezoelectric characteristics have been found in the MPB compositions with $0.05< x <0.143$, which exhibit two multiple phase transitions, changing the crystal symmetry from rhombohedral to tetragonal, then to cubic during heating [3]. Figure 5 shows the composition dependence of the pyroelectric coefficient λ_3^T, piezoelectric constants d_{ij}, electromechanial coupling factors k_{ij}, elastic compliances s_{ij}^E and dielectric constant ε_3^T measured at room temperature [3]. The superscript * is for the crystal with the rhombohedral symmetry at room temperature poled along the pseudocubic [001] direction. Figure 6 shows the temperature dependence of electromechanical coupling factors k_{33} and k_{31} measured with bar-shaped specimens of 0.91PZN-0.09PT [8]. Special notations have been introduced to describe the elastic and electromechanical constants for a sample poled in a certain direction. The $s^E_{[001]//}$ or $s^E_{[111]//}$ are defined from the resonance frequency of a bar sample elongated and poled in the pseudocubic [001] or [111] axes, respectively. The [001] and [111] axes are the principle axes of the tetragonal and rhombohedral phases and also the poling directions for each sample. The coupling coefficients $k_{[001]//}$ and $k_{[111]//}$ are consequently calculated from the resonance and antiresonance frequencies of the same sample. All the electromechanical components show a very large kink anomaly at the rhombohedral-tetragonal transition temperature, and a rapid decrease in k or an increase in d and s^E on approaching the Cuie point. These components vanish just above the Curie point.

Table I summarizes the elastic, piezoelectric, electromechanical and dielectric constants and the spontaneous polarization of 0.91PZN-0.09PT for the rhombohedral and tetragonal phases. It is important that the sample electrically poled along the pseudocubic [001] axis (not the principal axis in the rhombohedral phase!) reveals a very large piezoelectric constant $d_{[001]//} = 1.5x10^{-9}$

C/N and a high electromechanical coupling factor $k_{[001]//}$ = 0.92 at room temperature, both of which are much larger than $d_{[111]//}$ and $k_{[111]//}$, respectively. Also these are the highest values among all perovskite piezoelectric materials reported so far. It was found that such a high electromechanical coupling factor could not be explained consistently in terms of a mono-domain single crystal model without considering the complicated domain dynamical motion.

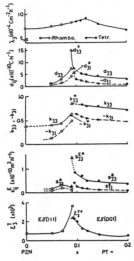

Fig.5   Composition dependence of the pyroelectric coefficient $\lambda_3{}^T$, piezoelectric constants d_{ij}, electromechanial coupling factors k_{ij}, elastic compliances $s_{ij}{}^E$ and dielectric constant $\varepsilon_3{}^T$ measured at room temperature.

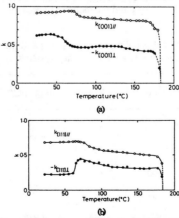

Fig.6   Temperature dependence of electromechanical coupling factors k_{33} and k_{31} measured with bar-shaped specimens of 0.91PZN-0.09PT.

Table I Elastic, piezoelectric, electromechanical and dielectric constants and the spontaneous polarization of 0.91PZN-0.09PT for the rhombohedral and tetragonal phases.

	Rhomb. phase	Tet. phase	Unit
Temp.	25	130	°C
$s^E_{[111]\perp}$	18	15.5	
$s^E_{[111]//}$	13.6	10.3	
$s^E_{[001]\perp}$	36.9	17.7	
$s^E_{[001]//}$	143	56	$(\text{TPa})^{-1}$
$s^D_{[111]\perp}$	17.1	13.9	
$s^D_{[111]//}$	7.6	7.3	
$s^D_{[001]\perp}$	22.6	13.6	
$s^D_{[001]//}$	21.8	17.6	
$-d_{[111]\perp}$	194	352	
$d_{[111]//}$	625	450	pC/N
$-d_{[001]\perp}$	493	266	
$d_{[001]//}$	1570	795	
$-k_{[111]\perp}$	0.23	0.32	
$k_{[111]//}$	0.68	0.53	—
$-k_{[001]\perp}$	0.62	0.48	
$k_{[001]//}$	0.92	0.83	
$\varepsilon^T_{[111]//}$	4100	8200	
$\varepsilon^T_{[001]//}$	2200	1880	—
$\varepsilon^S_{[111]//}$	2200	—	
$\varepsilon^S_{[001]//}$	295	570	
P_s	0.52	0.30	C/m²

Recently, two groups of Yamashita (Toshiba) and Shrout (Penn State) reconfirmed the author's original work, and extended the investigation to a wide range of relaxor ferroelectrics such as $Pb(Mg_{1/3}Nb_{2/3})O_3$-PT, $Pb(Sc_{1/2}Ta_{1/2})O_3$-PT and $Pb(Sc_{1/2}Nb_{1/2})O_3$-PT. Large electromechanical coupling factors k_p, k_{33} and piezoelectric constant d_{33} of these binary systems are listed in Table II.

Table II Large electromechanical coupling factors k_p, k_{33} and piezoelectric constant d_{33} of relaxor ferroelectric binary systems.

MATERIAL	FEATURE	k_p (%)	k_{33} (%)	d_{33} (pC/N)	REFERENCE
PZT 53/47	Polycrystal	52	67	220	[9]
		67	76	400	[10]
PZN:PT 91/9	Single crystal		92	1500	[8]
PMN:PT 67/33	Polycrystal	63	73	690	[11]
	Single crystal			1500	[12]
PST:PT 55/45	Polycrystal	61	73	655	[13]
PSN:PT 58/42	Polycrystal	71	77	450	[14]

Measurements on electric field-induced polarization and strain were carried out on a pure PZN single crystal by Shrout et al [15]. The polarization and strain curves are plotted for the <111> (the spontaneous polarization direction!) and <100> orientations in Fig. 7. Even though the [100] plate sample showed the "ideal" P vs. E or strain vs. E behaviors of a mono-domain crystal, notice that the absolute value of P is much larger in the [111] plate sample; this indicates again the spontaneous polarization along the <111> axis. He also reported the poling direction-dependent electromechanical coupling factors in PZN: $k_{[001]}// = 0.85$ was much larger than $k_{[111]}// = 0.38$, in a similar fashion to 0.91PZN-0.09PT. These results also suggest the importance of the domain contribution to dielectric and piezoelectic properties.

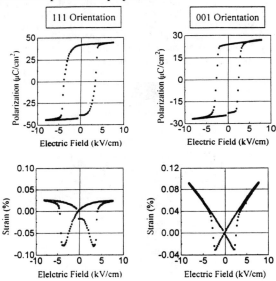

Fig.7 Polarization and strain curves plotted for the <111> and <100> orientations in pure PZN.

DOMAIN MOTION IN RELAXOR FERROELECTRICS

Historical Background of Domain-Controlled Piezoelectric Transducers

Historically, most of the studies on ferroelectric single crystals and polycrystalline materials have aimed to simulate the mono-domain state, desiring to derive the better characteristics from the specimens; ceramics as well as single crystals were poled electrically and/or mechanically to reorient the domains along one direction. Researches on controlling domains intentionally can be found in electrooptic devices and ferroelectric memory devices in particular. However, the intentional domain control has not been utilized or applied occasionally in the actuator and transducer areas. Recent requirements for the higher performance transducers encourages the investigation on the possibility of domain-related effect usage.

Developments in high resolution CCD optical microscope systems and in single crystal growth techniques are also accelerating these domain-controlled piezoelectric devices. A high resolution CCD (Charge Coupled Device) camera was attached to a Nikon Transmission Petrographic Microscope which was connected to a monitor and VCR (illustrated in Fig.8)[16]. The birefringence between the domains permits the observation of the domains with the polarizing light microscope. The microscope system also allows magnifications up to x1300 on the monitor. The temperature-controlled sample stage (Linkam Inc.) in conjunction with the deep focal point of the objective lenses allows an electric field to be safely applied across the sample between -185 and 600°C. The stationary and switching domains can be instantaneously recorded by the VCR and observed on the monitor.

Single crystal growth methods of PZN-PT are described here for example [8]. The powders used were Pb_3O_4, TiO_2, ZnO and Nb_2O_5. Excess ZnO was added in some cases to counteract the evaporation during crystal growth. PbO was used as the flux. The mole ratio of the flux to composition was varied from 1:1 to 3:2. The batch sizes were changed from 75 grams to 500 grams. The raw powders were loaded into a platinum crucible and charged several times at 900°C until the crucible was full. The crucible was then partially sealed with a Pt lid. The sealed Pt crucible was placed in an alumina crucible and sealed once again. The crucibles were placed in a box furnace with a temperature controller. The cooling rate was veried from 0.5 to 3°C/hr down to 900°C. After the temperature reaches 900°C, the furnace was fast cooled at 50°C/hr to room temperature. The single crystals were leached from from the flux with warm 25 vol% nitric acid. Single crystals with a dimension of 1 cm^3 could be obtained.

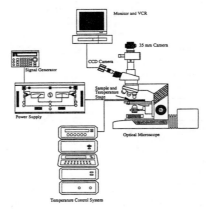

Fig.8   CCD optical microscope system.

Domain Configurations in PZN-PT

Let us review the relation between domain configurations and physical properties in relaxor ferroelectrics. Figure 9 shows the static domain configurations for the samples of x = 0, 0.07 and 0.095 of $(1-x)Pb(Zn_{1/3}Nb_{2/3})O_3$- $xPbTiO_3$ (PZN-PT) [17]. The pure PZN did not exhibit large clear domains in the whole temperature range when it was unpoled (i. e. micro-domains). Rhombohedral domains could be described as having an ambiguous spindle-like morphology. Ambiguity refers to the variation of domain widths and lengths which appear as an interpenetrating structure. With increasing the PT content, this small spindle-like domain was enlarged and the domain wall became sharp. Tetragonal domains appear to have a well defined lamellar morphology, and are either at right angles or antiparallel. Therefore, even though the widths and lengths of the tetragonal domains vary, the divisions between the domains are well defined and no interpenetrating structure is observable. Notice that the morphotropic phase boundary (MPB) composition (x = 0.095) shows the coexistence of both rhombohedral (spindle-like) and tetragonal domains (sharp straight line); the two-phase coexistence can not be found statically in normal ferroelectric materials such as $BaTiO_3$.

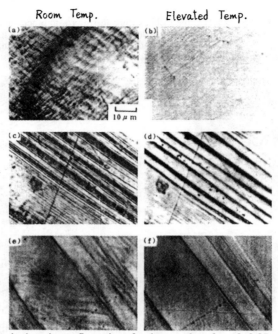

Room Temp.   Elevated Temp.

Fig.9   Static domain configurations for the samples of x = 0, 0.07 and 0.095
of $(1-x)Pb(Zn_{1/3}Nb_{2/3})O_3$- $xPbTiO_3$ (PZN-PT).

Figures 10 and 11 show the actual domain reversal processes and their schematical illustration under an applied electric field [18]. Sharp 90° domain walls corresponding to the tetragonal symmetry in the sample with x = 0.2 moved abruptly and rather independently each other above a coercive field of 1 kV/mm. The situation resembles to the case in normal ferroelectrics. On the contrary, pure PZN (x = 0.0) revealed very different domain motion. During electric field cycles, micro- to macro-domain growth occurred, and long narrow spindle-like domains (aspect ratio = 10) were arranged rather perpendicularly to the electric field (18 degree canted). When a field above 0.5 kV/mm was applied, the ambiguous domain walls rippled simultaneously in a certain size region, so that each domain should change synchronously like cooperative phenomena. The domain reversal front (180° domain wall) moved rather slower than in the sample of x = 0.2. It is noteworthy that the stripe period of the dark and bright domains (probably corresponding to up and down polarizations) was not changed by the domain reversal, and that each domain area changed under an AC external field with zero net polarization at zero field. Thus, the relaxor crystal is electrically-poled easily when an electric field is applied around the transition temperature, and depoled completely without any remanent polarization. This can explain large apparent secondary non-linear effects in physical properties such as electrostrictive and electrooptic phanomena, without exhibiting any hysteresis.

The relation between the dielectric property and the domain structure was clearly demonstrated in the permittivity measurement of the pure PZN. Figures 12(a) and 12(b) show the permittivity vs. temperature curves taken for the annealed (unpoled) and poled states of the PZN sample[19]. Large dielectric relaxation (frequency dependence of the permittivity) was observed in a wide temperature range below the Curie point for the unpoled state, while the dielectric dispersion was measured only in a narrow temperature range between 100°C and the Curie point for the poled state. Considering that below 100°C the PZN exhibits the micro- to macro-domain growth under a high electric field, we can conclude that the dielectric relaxation is attributed to the micro-domains. Thus, we learned how to control the micro- and macro-domains through temperature change and an external electric field, and how to change the dielectric dispersion, elastic and piezoelectric constants according to these phase transitions.

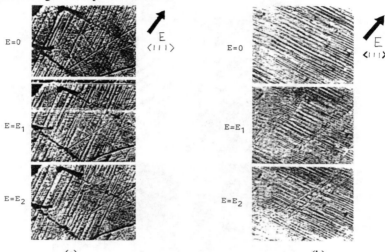

(a)   (b)

Fig.10   Actual domain reversal processes under an applied electric field for the samples of x = 0.2 (a) and 0 (b) of (1-x)Pb(Zn$_{1/3}$Nb$_{2/3}$)O$_3$- xPbTiO$_3$.

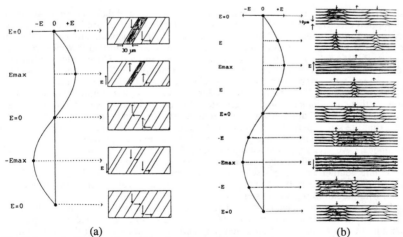

(a)   (b)

Fig.11   Schematical illustration of the domain reversal processes under an applied electric field for the samples of x = 0.2 (a) and 0 (b) of (1-x)Pb(Zn$_{1/3}$Nb$_{2/3}$)O$_3$- xPbTiO$_3$.

12

a.) A depoled 100%PZN single crystal measured on the <111>.

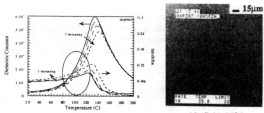

No field, 25°C

b.) A poled 100% PZN single crystal measured on the <111>.

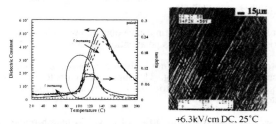

+6.3kV/cm DC, 25°C

Fig.12 Permittivity vs. temperature curves taken for the annealed (unpoled) state (a) and poled state (b) of the PZN sample.

<u>Hierarchical Domain Structures</u>

Figure 13 shows the domain structures observed in a 0.89PZN-0.11PT crystal, which exists on the morphotropic phase boundary. The typical tetragonal stripe domain pattern was observed without an external electric field at room temperature, while the spindle-like rhombohedral domain pattern appeared as overlapped on the stripe pattern, when the electric field was applied along the perovskite pseudo-cubic [111] directions. This domain hierarchy suggests that the morphotropic phase boundary composition may easily change the domain configuration and the crystal symmetry according to the applied electric field direction, much easily than in the normal ferroelectric PZT.

E = 0

E = E₁

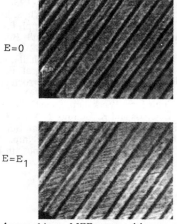

Fig.13 Domain structures observed in an MPB composition crystal 0.89PZN-0.11PT.

Future work will include the dynamic domain observation in the 0.91PZN-0.09PT sample (rhombohedral phase is stable at room temperature) poled along the perovskite [100] axis; this enhances remarkably the electromechanical coupling factor k up to 92 - 95 %. Also the possibility of the different poling direction which enhances the coupling factor more will be explored.

CONCLUSIONS

1. The electromechanical coupling factor k_{33} of more than 90 % can be obtained in the solid solutions between the relaxor and normal ferroelectrics.

2. Promising compositions include: $Pb(Zn_{1/3}Nb_{2/3})O_3$-$PbTiO_3$, $Pb(Mg_{1/3}Nb_{2/3})O_3$-$PbTiO_3$ and $Pb(Sc_{1/2}Nb_{1/2})O_3$-$PbTiO_3$.

3. The highest k in a single crystal form can be obtained when it is electrically-poled along a different axis from the spontaneous polarization direction.

Domain cotrolled single crystals (not in the monodomain state!) may be the key for obtaining the highest electromechanical coupling. The important factors to the domain reconstruction will be realized by changing external electric field, stress and temperature, as well as the sample preparation history.

This work was supported by the Office of Naval Research under Grant No. N00014-91-J-4145.

REFERENCES

1. K. Uchino, Proc. 4th Int'l Conf. on Electronic Ceramics & Appl. Vol.1, p.179, Aachen, Germany, Sept. 5-7 (1994).
2. K. Uchino, Piezoelectric Actuators and Ultrasonic Motors, Kluwer Academic Publ., Boston (1996).
3. J. Kuwata, K. Uchino and S. Nomura, Ferroelectrics 37, 579 (1981).
4. K. H. Hellwege et al.: Landolt - Bornstein, Group III, Vol.11, Springer-Verlag, N.Y. (1979).
5. K. Uchino and S. Nomura, Jpn. J. Appl. Phys. 18, 1493 (1979).
6. K. Uchino, Bull. Amer. Ceram. Soc. 65(4), 647 (1986).
7. S. Nomura, T. Takahashi and Y. Yokomizo, J. Phys. Soc., Jpn. 27, 262 (1969).
8. J. Kuwata, K. Uchino and S. Nomura, Jpn. J. Appl. Phys. 21(9), 1298 (1982).
9. B. Jaffe, R. S. Roth and S. Marzullo, J. Res. Nat'l. Bur. Stand. 55,, 239 (1955).
10. H. Igarashi, Mem. Nat'l. Def. Adad. 22, 27 (1982).
11. S. W. Choi, T. R. Shrout, S. J. Jang and A. S. Bhalla, Ferroelectrics 100, 29 (1989).
12. T. R. Shrout, Z. P. Chang, N. Kim and S. Markgraf, Ferroelectrics Lett. 12, 63 (1990).
13. J. F. Wang, J. R. Giniewicz and A. S. Bhalla, Ferroelectics Lett. 16, 113 (1993).
14. Y. Yamashita, Jpn. J. Appl. Phys. 33, 4652 (1994).
15. T. R. Shrout, ONR Transducer Workshop, State College (March, 1996)
16. M. L. Mulvihill, L. E. Cross and K. Uchino, J. Amer. Ceram. Soc. 78, 3345 (1996).
17. K. Kato, K. Suzuki and K. Uchino, J. Jpn. Ceram. Soc. 98(8), 840 (1990).
18. R. Ujiie and K. Uchino, Proc. IEEE Ultrasonic Symp. '90, Hawaii, 2, 725 (1991).
19. M. L. Mulvihill, L. E. Cross and K. Uchino, Proc. 8th Europian Mtg. on Ferroelectrics, Ferroelectrics 186, 325 (1996).

DOMAIN WALL CONTRIBUTIONS TO THE PIEZOELECTRIC PROPERTIES OF FERROELECTRIC CERAMICS AND THIN FILMS, AND THEIR SIGNIFICANCE IN SENSOR AND ACTUATOR APPLICATIONS

D. DAMJANOVIC, D.V. TAYLOR, A. L. KHOLKIN, M. DEMARTIN, K. G. BROOKS, and N. SETTER,
Laboratory of Ceramics, Department of Materials, Swiss Federal Institute of Technology--EPFL, 1015 Lausanne, Switzerland

ABSTRACT

The piezoelectric and dielectric properties of ferroelectric thin films and ceramics were investigated in detail as a function of the frequency and amplitude of the driving field. A description, which is based on the theories of domain wall pinning by randomly distributed imperfections in magnetic materials, is used to interpret the electromechanical behaviour of several ferroelectric bulk ceramic and thin film compositions. With this approach, it is possible to make quantitative estimates of the domain wall contributions to the electromechanical properties of ferroelectric sensors and actuators.

INTRODUCTION

The domain walls displacement is an important extrinsic contribution to the electro-mechanical (piezoelectric, dielectric and elastic) properties of ferroelectric ceramics and thin films [1]. The practical significance of the domain wall and other extrinsic contributions to the properties lays in the fact that they are often responsible for the field and frequency dependence of the electromechanical devices. Despite a significant progress in understanding the different mechanisms which contribute to the electromechanical properties of ferroelectric materials, a satisfactory and a general description of their behaviour under weak field (well below coercive field) conditions still does not exist [2,3,4].

Domain wall vibrations and their contributions to material properties have been a subject of intensive research for many years. The results of the theories of domain wall pinning by randomly distributed imperfections have been successfully applied to interpret the frequency and field dependence of the magnetic susceptibility and magnetization in (anti)ferromagnetic materials. Recently, it has been demonstrated experimentally that the main results of the theories may also be applicable for the pinning of ferroelectric-ferroelastic domain walls in ferroelectric systems [5,6]. It has been shown that the field dependence of the direct piezoelectric coefficient in many ferroelectric ceramics may be described by the so called Rayleigh law [5] and that the direct piezoelectric coefficient decreases linearly with the logarithm of the frequency of driving field [6].

In this paper, the recent experimental results of the field and frequency dependence of the piezoelectric coefficients in bulk ceramics are reviewed. New data on the frequency and field dependence of the dielectric permittivity and piezoelectric coefficients of lead zirconate titanate (PZT) thin films are also presented and discussed. The presented data suggest that the electromechanical behaviour of both ferroelectric thin films and ceramics may be satisfactorily interpreted within the framework of the theories of domain wall (or interface) pinning by imperfections. The obtained results may be useful for modeling the response of the piezoelectric sensors and actuators.

Mat. Res. Soc. Symp. Proc. Vol. 459 © 1997 Materials Research Society

FIELD DEPENDENCE OF THE DIELECTRIC PERMITTIVITY AND THE PIEZOELECTRIC COEFFICIENT

Ferroelectric ceramics

The experimental procedure for measurements of the direct piezoelectric effect in bulk ceramics and preparation of the ceramic samples are described in detail in Ref. [5,7]. Figure 1 shows the piezoelectric coefficient of $Bi_4Ti_3O_{12}$ doped with ~1.3at% Nb (BTO12-Nb) and PZT (63/37) doped with 4at%Nb as a function of the amplitude of ac pressure. In both materials the dependence of the piezoelectric coefficient on the field may be described by the Rayleigh equation [8]:

$$d_{33} = d_{init} + \alpha X_0 \qquad (1)$$

where d_{33} is the direct piezoelectric coefficient, X_0 is the amplitude of the ac pressure $X(t)$, d_{init} is the sum of the lattice (so called intrinsic) contribution, and the contribution from reversible displacement of the walls. In piezoelectric ceramics, d_{init} depends also on the poling conditions. The Rayleigh coefficient α is due to irreversible displacement of the walls. The Rayleigh law is valid in the limit of weak field signal, where domain walls do not interact and where restoring forces due to stray fields can be neglected. Therefore, displacement of the walls is controlled only by pinning of the walls on defects. One consequence of the domain wall pinning and depinning as the field is cycled is the appearance of piezoelectric hysteresis. For the direct piezoelectric effect, the charge density (Q) vs. pressure hysteresis may be written in terms of the Rayleigh parameters as:

$$Q(X) = (d_{init} + \alpha X_0)X \pm \frac{\alpha}{2}(X_0^2 - X^2) \qquad (2)$$

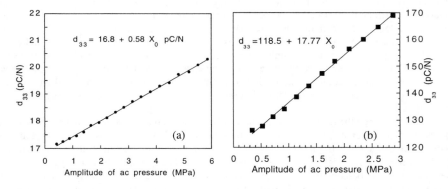

Fig. 1. The direct longitudinal piezoelectric coefficient of the BTO12-Nb (a) and PZT (63/37) -4%Nb (b) as a function of the amplitude of ac pressure. Full symbols represent experimental data and solid lines are linear fits with the Rayleigh equation (1). The frequency of the ac field was 1 Hz. [5,9]

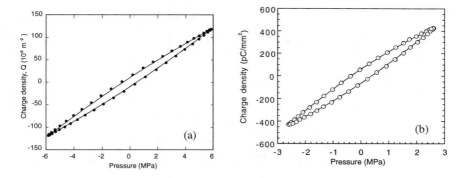

Fig. 2. Experimental loops (circles) and loops calculated from the Rayleigh law and Eq. (2) (solid lines) for BTO12-Nb (a) and PZT (63/37)-Nb (b). [5,9]

where "+" stands for decreasing and "—" for increasing field. Figure 2 compares calculated [Eq. (2)] and experimental loops for BTO12-Nb and PZT (63/37)-Nb.

The area of the hysteresis is related to the piezoelectric phase angle δ_p (or the piezoelectric "loss"). The phase angle δ_p may be calculated as a function of X_0 from the Rayleigh relationships as [7,10]:

$$\delta_p \approx \frac{4\alpha X_0}{3\pi d_{33}} \qquad (3)$$

The calculated and experimentally determined piezoelectric phase angle are shown in Figure 3 for BTO12-Nb. The measured phase angle is larger than the calculated value suggesting that dissipative processes other than those related to the irreversible displacement of the domain walls contribute to the hysteresis. [7]

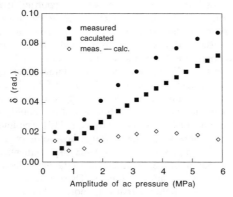

Fig. 3. The calculated, measured, and difference between the calculated and measured piezoelectric phase angle of BTO12-Nb as a function of the amplitude of ac pressure.

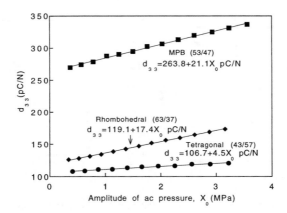

Fig. 4. The Rayleigh relationship (1) for the direct d_{33} piezoelectric coefficient in a rhombohedral $(Pb(Zr_{0.63}Ti_{0.37})_{0.96} Nb_{0.04}O_3)$, a MPB $(Pb(Zr_{0.53}Ti_{0.47})_{0.96}Nb_{0.04}O_3)$ and a tetragonal $(Pb(Zr_{0.43}Ti_{0.57})_{0.96}Nb_{0.04}O_3)$ composition of Nb-doped PZT. The external dc pressure is 15 MPa The frequency of the ac pressure is 1 Hz. [9]

The contribution of the irreversible displacement of the domain walls to the piezoelectric effect is strongly affected by various microstructural features, such as grain size and local variations in the composition, crystal structure and by external fields [11]. Figure 4 shows the effect of the crystal structure on the irreversible Rayleigh parameter α for three compositions of PZT ceramics around the morphotropic phase boundary (MPB). The irreversible component of the piezoelectric coefficient at a given field (αX_0) is, as expected, strongest in the MPB composition. It is more than three times higher in the rhombohedral composition than in the tetragonal, which may be explained by smaller spontaneous strains in the rhombohedral material. In general, it has been found that domain wall contributions from displacement of the rhombohedral walls (109° and 71°) are stronger than from the displacement of tetragonal (90°) walls. It is interesting that the fraction of the total piezoelectric coefficient which is due to the irreversible displacement of the walls [defined as $(\alpha X_0)/(d_{init} + \alpha X_0)$] is the largest in the rhombohedral ceramic. A practical consequence of this is that instability in the piezoelectric coefficient (i.e. a relative increase of d_{33} with increasing field) and the relative piezoelectric hysteresis are larger in the rhombohedral than in the MPB composition.

Figure 5 illustrates the effect of the grain size on the domain wall contributions to the piezoelectric properties of barium titanate ceramics ($BaTiO_3$ or BT). The direct longitudinal piezoelectric coefficient in coarse (average grain size ~ 26μm) and fine grained (average grain size ~ 0.7μm) barium titanate ceramics is shown as a function of the amplitude of ac pressure. A detailed description of experimental conditions may be found in [12]. The linear increase of d_{33} vs. X_0 in coarse grained ceramics shows that the Rayleigh law holds over the pressure range investigated. At $X_0 \approx 2$ MPa, 34% of d_{33} is due to irreversible displacements of domain walls. In fine grained ceramics, the contribution of the domain walls to the piezoelectric effect appears to be significantly smaller than in coarse grained ceramics. These results are in agreement with recent theoretical predictions which concluded that it is more difficult to move thin domains (fine grains) than thick ones (large grains) [4]. The much finer domain structure in fine grained ceramics implies that interaction of domain walls as well as restoring forces cannot be neglected.

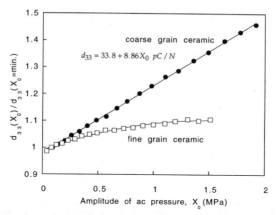

Fig. 5. The relative direct d_{33} piezoelectric coefficient for coarse (circles) and fine (squares) grained ceramics of barium titanate as a function of the amplitude of ac pressure. The external dc pressure is 1 MPa. The frequency of the ac force is 1 Hz. [12]

The Rayleigh relationship (1) is therefore not valid in fine grained ceramics even at low fields and, as shown in Fig. 5, the second order term must be included in the d_{33} vs. X_0 relationship. Another possible reason for the deviation from the linear behavior could be the presence of a strong internal pressure in the fine grain ceramics [12].

Ferroelectric thin films

It is next shown that the description proposed for the field dependence of the piezoelectric coefficient in ferroelectric ceramics [5] is also valid for the dielectric permittivity and the converse piezoelectric coefficient in PZT thin films. Figure 6 plots the dielectric permittivity of a sol-gel (111) oriented PZT (45/55) film (1.3 μm thick) as a function of the amplitude of ac electric field. The preparation of the film is described in more detail in Ref. [13]. The Rayleigh relationships for the weak field dielectric permittivity and polarization vs. electric field hysteresis are given by:

$$\varepsilon_{33} = \varepsilon_{init} + \alpha E_0 \qquad (4)$$

and

$$P(E) = (\varepsilon_{init} + \alpha E_0)E \pm \frac{\alpha}{2}(E_0^2 - E^2) \qquad (5)$$

The parameters in Eq. (4) and (5) correspond to those in Eq. (1) and (2). In a virgin state (as grown films), the dielectric permittivity vs. field relationship has a strong quadratic component in addition to the linear term. However, after exposing the film to subcoercive ac field for a sufficient time, the linear term becomes dominant up to relatively high fields, Fig 6 [14].

The calculated hysteresis loop overestimates the dielectric phase angle at higher ac fields and underestimates at low ac fields (not shown). The reasons for such behaviour are presently

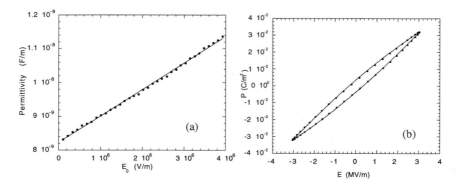

Fig. 6. (a) The dielectric permittivity as a function of the amplitude of ac electric field for sol-gel (111) oriented PZT (45/55) thin film. The solid line is obtained by fitting the experimental data (circles) with Eq. (4). (b) Weak field polarization vs. electric field hysteresis loop for the same film. The solid line is obtained by fitting the experimental data (circles) with Eq. (5). [14]

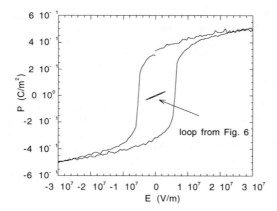

Fig. 7. Weak field polarization vs. electric field hysteresis loop from Fig. 6b and a loop obtained by cycling the field above the coercive field. [14]

being investigated. Figure 7 compares the weak field hysteresis loop from Fig. 6b with a strong field (E_0 > coercive field) loop.

The relative converse d_{33} piezoelectric coefficient for a 7.1 μm thick 53/47 PZT film with random orientation is shown in Fig. 8 as a function of the amplitude of ac electric field. The measurement procedure and preparation of the sample are described in detail in Ref. [15,16]. Above a certain threshold field (~ 0.3 kV/cm) the piezoelectric coefficient increases linearly with the amplitude of ac field. The Rayleigh relationship (1) is valid up to approximately 5 kV/cm. At stronger fields, the higher order nonlinear terms (above ~5 kV/cm) and domain wall switching effects (above ~8 kV/cm) start to dominate the response [15].

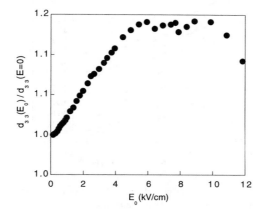

Fig. 8. The relative converse d_{33} piezoelectric coefficient of a 7.1 μm thick 53/47 PZT film with random orientation as a function of the amplitude of ac electric field. The linear region is observed below approximately 5 kV/cm. [15]

FREQUENCY DEPENDENCE OF THE DIELECTRIC PERMITTIVITY AND THE PIEZOELECTRIC COEFFICIENT

Ferroelectric ceramics

The theories of the interface pinning by imperfections in random systems predict logarithmic frequency dependence of the magnetic susceptibility χ:

$$\chi \sim \left[\frac{1}{t_0 \omega} \right]^{\Theta} \tag{6}$$

where exponent Θ is related to the roughness of interface [17]. The recent experimental study has shown [6] that the same type of the dependence is valid for the displacement of ferroelectric-ferroelastic domain walls in ferroelectric ceramics. Figure 9 shows the direct d_{33} piezoelectric coefficient of PZT (52/48)-2.5%Nb ceramics as a function of the frequency of ac pressure, for different amplitudes of the pressure. It can be easily shown [6] that the Rayleigh relationship (1) can be derived from the data presented in Fig. 9. In the case of the piezoelectric coefficient the exponent of the logarithmic term is ~1.

The same type of the frequency dependence has been observed, as shown in Fig. 10, for the converse d_{33} piezoelectric coefficient in PZT thin films [18], as well as for the dielectric permittivity of the thin films [14, 15].

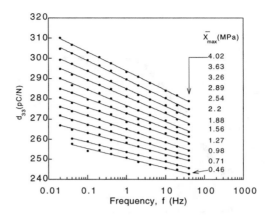

Fig. 9. The direct d_{33} piezoelectric coefficient of 52/48 PZT-2.5%Nb ceramic as a function of the frequency of ac pressure, for different amplitudes of the ac pressure [6].

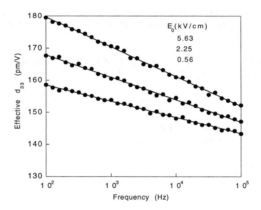

Fig. 10. The converse d_{33} piezoelectric coefficient of a 53/47 PZT thin film (thickness=7.1 μm) as a function of the frequency of ac electric field, for different amplitudes of the ac field [18].

CONCLUSIONS

In summary, the presented experimental results show that the linear relationship (the Rayleigh law) between the dielectric permittivity, the direct and converse piezoelectric coefficients and the driving field, is valid in several compositions of ferroelectric ceramics and thin films. These results, together with the observed logarithmic frequency dependence of the piezoelectric coefficients and the permittivity, suggest that the weak field dependence of the electromechanical properties of ferroelectric ceramics and thin films are controlled by the pinning of domain walls by imperfections. The validity of the very general relationships (the Rayleigh law and the logarithmic frequency dependence) indicate that the general theories of the interface pinning in random systems may be a useful starting point for interpretation of the electromechanical behaviour of the ferroelectric thin films and ceramics.

Acknowledgments: This work was supported in part by the Swiss Federal Office for Science and Education and performed within the European research program on ferroelectric thin films (COST 514); by the Swiss Committee for Encouragement of Scientific Research (CERS); and by the Swiss National Science Foundation.

REFERENCES

1. L.E. Cross, in *Ferroelectric Ceramics*, edited by N. Setter and E. Colla (Birkhäuser, Basel, Switzerland, 1993), p. 1.

2. S. Li, W. Cao, and L.E. Cross, *J. Appl. Phys.* **69**, 7219 (1991).

3. Q.M. Zhang, W.Y. Pan, S.J. Jang and L.E. Cross, *J. Appl. Phys.* **64**, 6445 (1988)

4. G. Arlt and N.A. Pertsev, *J. Appl. Phys.* **70**, 2283 (1991).

5. D. Damjanovic and M. Demartin, *J. Phys. D.: Appl. Phys.* **29**, 2057 (1996)..

6. D. Damjanovic, *Phys. Rev. B* **55**, (1997) (in press).

7. D. Damjanovic (unpublished).

8. D. Jiles, *Introduction to Magnetism and Magnetic Materials* (Chapman and Hall, London, 1991) p. 95-175.

9. M. Demartin, PhD Thesis, Swiss Federal Institute of Technology-EPFL, Lausanne, 1996.

10. Lord Rayleigh, *Phil. Mag.* **23**, 225 (1887).

11. D. Damjanovic, M. Demartin, F. Chu, and N. Setter in *Proceedings of the Tenth IEEE International Symposium on Applications of Ferroelectrics,* Rutgers Univ., East Brunswick, NJ, USA, Aug 18-21, 1996. (in press)

12. M. Demartin and D. Damjanovic, *Appl. Phys. Lett.* **68** 3046 (1996).

13. D. V. Taylor, K.G. Brooks, A.L. Kholkin, D. Damjanovic and N. Setter, Proceedings of the 5th International Conference on Electronic Ceramics & Applications ELECTROCERAMICS V, Univ. of Aveiro, Portugal, Sept. 2-4, 1996. Vol. 1, Ed. by J.L. Baptista, J.A. Labrincha and P.M. Vilarinho (Print: TIPAVE, Aveiro) pp. 341-344.

14. D.V. Taylor (unpublished)

15. A.L. Kholkin, A.K. Tagantsev, K.G. Brooks, D.V. Taylor, and N. Setter, in *Proceedings of the Tenth IEEE International Symposium on Applications of Ferroelectrics,* Rutgers Univ., East Brunswick, NJ, USA, Aug 18-21, 1996. (in press)

16. K. G. Brooks, R. Klissurska, P. Moeckli, and N. Setter, *Microelectron. Eng.* **29**, 293 (1995).

17. T. Nattermann, Y. Shapir, and I. Vilfan, *Phys. Rev. B* **42** 8577 (1990).

18. A.L. Kholkin (unpublished).

APPLICATIONS OF ACTIVE THIN FILM COATINGS ON OPTICAL FIBERS

G.R. Fox,* C.A.P. Muller,* C.R. Wüthrich,* A.L. Kholkin,* N. Setter,* D.M. Costantini,**
N.H. Ky,** and H.G. Limberger**
*Laboratory of Ceramics, Swiss Federal Institute of Technology, CH-1015 Lausanne,
Switzerland, Glen.Fox@lc.dmx.epfl.ch
**Institute of Applied Optics, Swiss Federal Institute of Technology , CH-1015 Lausanne,
Switzerland

ABSTRACT

Active thin film coatings on optical fibers provide a variety of functions that are being used to develop active all-fiber optical devices. Two types of active coatings that are of interest for device development include resistive and piezoelectric coatings. Resistive coatings can be used to heat an optical fiber, while piezoelectric coatings can be used to strain the fiber. Localized changes in the fiber waveguiding properties can be achieved by electrically activating the fiber coating. These coated fibers show promise for applications such as optical phase shifters and modulators.

Recent developments in the fabrication of diffraction gratings within the core of an optical fiber have provided the means for making a variety of intra-core reflection and band pass filters. By combining these passive intra-core fiber devices with active coatings, wavelength tunable devices have been demonstrated. Wavelength tunable devices are expected to have a variety of applications in telecommunications and sensing networks. A review of recent developments in fiber coating and analysis techniques, device fabrication, and applications of active all-fiber devices are presented along with a discussion of which coating materials are of interest in active devices.

INTRODUCTION

A wide range of opportunities exist for the use of active fiber coatings in the development of miniature electrically controlled all-fiber devices. Electrically tuned or modulated all-fiber devices are of interest for applications in telecommunications, fiber optic sensing networks, and a variety of other applications where optical fiber waveguides are used. The function of an active thin film optical fiber coating is to provide an electrical means for regulating the optical response of a section of fiber underlying the active coating. Both fiber temperature and strain can be adjusted to tune the optical properties of the fiber material and temporal variations of these two variables can be utilized for modulation of the optical properties. Three different classes of active fiber coatings have been used to provide an electrical means for tuning or modulation; these include resistive [1], piezoelectric [2], and magnetostrictive coatings [3]. Joule heating of resistive coatings is used to produce devices based on thermal regulation. The converse piezoelectric effect is used to achieve strain regulation with piezoelectric coatings. With magnetostrictive coatings, a current carrying coil is used to produce a magnetic field that induces a strain in the coating, which is coupled to the underlying fiber. Four different devices that utilize either resistive or piezoelectric fiber coatings have been demonstrated and a description of there fabrication and operation are presented below. The devices include a piezoelectric fiber optic phase modulator (PFOM), wavelength tunable thermal fiber Bragg grating (TFBG), wavelength modulated piezoelectric fiber Bragg grating (PFBG), and a flexural/bending mode piezoelectric fiber actuator. A description of the fabrication processes for active fiber coatings is reviewed and materials issues that influence device performance are also discussed.

FABRICATION OF PIEZOELECTRIC AND RESISTIVE FIBER COATINGS

A dual magnetron sputtering machine as shown schematically in FIG. 1 was used for depositing piezoelectric and resistive fiber coatings. The vacuum chamber contains two 100 mm diameter magnetron sources that are powered by dc power supplies. In order to produce concentric

25

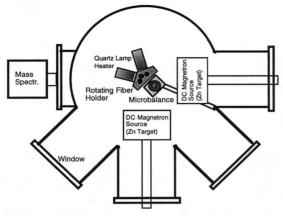

FIG. 1. Schematic of dual magnetron sputtering system used for deposition of ZnO optical fiber coatings.

fiber coatings, a sample holder was designed that allowed for rotation of the fiber substrates while they were held vertically in front of the magnetron sources. The lengths of the coatings deposited onto the fibers were determined by shadow masks.

Piezoelectric ZnO coatings were deposited by using either a single magnetron source or two magnetron sources that were positioned with their target normals at 90°. A single magnetron was used for the deposition of ZnO coatings up to 6 μm thick while a dual magnetron configuration was used to produce coatings up to 20 μm thick. Zn metal targets were sputtered in a reactive mixed atmosphere of oxygen and argon while maintaining an oxidized target surface layer. The total chamber pressure, power, and oxygen partial pressure were carefully maintained near the transition between an oxidized target surface and a metal target surface since these conditions resulted in a high deposition rate and ZnO films with preferred orientation and a piezoelectric wurtzite type structure. Since ZnO exhibits a fixed polarization that cannot be reoriented after growth, ZnO thin films with a preferred orientation are required in order to excite a macroscopic piezoelectric response. The ZnO fiber coatings consist of columnar grains that exhibit an [0001] preferred orientation along the column axis. A radial [0001] preferred orientation of the fiber coating results since the columnar ZnO grains grow along the radial direction with respect to the fiber axis. Debye-Scherrer X-ray diffraction was used to determine the radial texture of the ZnO film. The ZnO film exhibits an ∞m symmetry with respect to the radial vector, but the complete fiber structure has an ∞/mm symmetry. Concentric Cr/Au electrode coatings for ZnO based devices were deposited by thermal evaporation. A rotating fiber holder and shadow masks similar to those used for the ZnO deposition were employed. Details on the deposition process of the ZnO and Cr/Au electrode structures have been previously reported [4, 5].

Concentric fiber coatings of Ti and Pt were employed as a resistive coating for thermally tunable fiber devices. The Ti layer improves the adhesion between the Pt coating and the glass fiber, while the Pt layer provides the resistive coating that is heated by passing a current. This combination of metal thin films for the resistive coating was chosen because it has been demonstrated that the coating remains stable in oxidizing atmospheres to temperatures above 500°C [6]. The Ti and Pt were sequentially deposited using the same sputtering system that was used for the ZnO fiber coatings, but the geometry of the system was slightly altered. Instead of placing the two magnetrons at a 90° angle, the two magnetrons were placed at a 45° angle. The conditions for the Ti and Pt deposition have been reported elsewhere [7].

PIEZOELECTRIC FIBER OPTIC MODULATORS (PFOM)

PFOM devices consist of concentric electrode and piezoelectric coatings that form a cylindrical actuator around the fiber as shown in FIG. 2 [8, 9]. The devices described in this work consist of Cr/Au electrodes and a ZnO piezoelectric coating although other piezoelectric coatings have been investigated, including polyvinylidene fluoride [10], vinylidene fluoride/tetrafluoroethylene copolymer [11], vinylidene fluoride/trifluoroethylene copolymer [12], and $PbZr_xTi_{1-x}O_3$ (PZT) [13, 14]. When an electric field is applied to the piezoelectric coating, a strain results due to the converse piezoelectric effect. Since the piezoelectric and bottom electrode

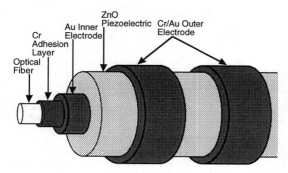

FIG. 2. Schematic drawing of ZnO based PFOM.

coatings are bonded to the fiber, the strain induced in the piezoelectric coating is elastically coupled into the fiber. Light waves propagating along the fiber experience a phase shift in the section of fiber that is strained by the piezoelectric coating. A simplified model that assumes plane strains and quasistatic stresses results in a relationship between the optical phase shift ($\Delta\phi$) and the strain within the fiber as given by Equation (1) [11]. The optical wave number in vacuum is designated as k, n is the refractive index, l is the length of the PFOM device, S_r is the radial strain, S_z is the axial strain, and P_{11} and P_{12} are the photoelastic coefficients of the fiber core. This relation clearly illustrates how both radial and axial strains contribute to the optical phase shift and provides a useful description of device operation at low frequencies. For ZnO coatings with an [0001] radial texture, radial strains are produced by a combination of the longitudinal and transverse piezoelectric effects, the magnitude of which is determined by the d_{33} (longitudinal) and d_{31} (transverse) piezoelectric coefficients. Axial strains result from the d_{31} effect as well as from the Poisson's ratio coupling of radial stresses applied to the fiber. According to the quasistatic model, large phase shifts are obtained by having large piezoelectric, elastic stiffness, and photoelastic coefficients. It should also be noted that the phase shift is proportional to the actuator length and inversely proportional to the wavelength of the propagated light wave.

$$\Delta\phi = knl\left\{S_z - \frac{n^2}{2}\left[(P_{11} + P_{12})S_r + P_{12}S_z\right]\right\}$$

(1)

At high frequencies, inertial effects must be taken into account and the plane strain approximation does not sufficiently explain the interaction between the acoustic wave introduced by the oscillating piezoelectric coated fiber and the optic wave propagating through the core. Recently, a model for a ZnO based PFOM device has been developed that takes into account the symmetry of the piezoelectric ZnO coating, the stiffened piezoelectric elastic response, inertial effects, and the interaction between non-uniform strain and the propagating optical mode [15]. This model provides a means for describing the phase shift amplitude under radial resonance conditions and can be used to determine the radial mode resonance frequency as well as the cut-off frequency above which the acoustic wave becomes ineffective for optical phase modulation. The three main contributions to the phase shift are the overlap of the fiber strains and the optical field in the fiber, the overlap of the acoustic mode and the electrical driving force in the piezoelectric layer, and damping, which can result from both mechanical and electrical losses. Equation (2) gives the relation describing the phase amplitude modulation where n_o and β_o are the unperturbed core index and propagation constant, respectively, S_θ is the azimuthal strain, Γ is the power of the guided wave, Ψ is the eigenmode of the optical light field in the fiber core, and R is the radius of the fiber.

$$\Delta\phi \approx -\frac{\pi k^2 n_o l}{2\Gamma\beta_o}\int_0^R (P_{11} + P_{12})(S_r + S_\theta)\Psi^2 rdr$$

(2)

27

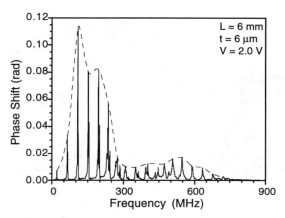

FIG. 3. Optical phase shift spectrum (solid line) resulting from radial mode resonances induced by a 6 μm thick ZnO coating. The dashed line indicates the envelope of the peak intensities.

PFOM devices consisting of the multilayer structure Cr(13 nm)/Au(130 nm)/ZnO(6 μm)/Cr(25 nm)/Au (400 nm) and lengths of 2 and 6 mm were analyzed using an all-fiber Mach-Zehnder interferometer and a optical wavelength of 1.56 μm [16]. Below 10 MHz the optical phase shift is constant with respect to frequency except for a few peaks that generally occur around the 10 kHz frequency range. The position and amplitude of these peaks are dependent on the sample holder and mounting of the PFOM and are believed to result from shear or axial mode resonances associated with fiber mounting. In general, the broadband phase shift response reflects the behavior described by the quasistatic model. The highest broadband phase shifts achieved are on the order of 10^{-4} to 10^{-3} rad with maximum driving potential amplitudes of 10 V.

FIG. 3 shows the phase shift as a function of frequency between 1 and 900 MHz. The peaks in phase shift occur at the radial mode resonances of the coated fiber structure. These resonances can also be observed in the spectrum of the reflected rf power. For the ZnO and electrode coating thicknesses given above, the fundamental radial mode resonance occurs at approximately 21 MHz and the higher order resonances are separated in frequency by 42 MHz. The position of the resonance peaks are dependent upon the thickness of the fiber coatings, but are insensitive to driving potential and mounting conditions unless the device is driven at potentials approaching the electrical breakdown strength of the ZnO [17]. In contrast, the amplitudes of the resonance peaks are strongly dependent on both fiber mounting and the driving potential. The maximum phase shift

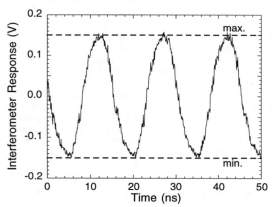

FIG. 4. PFOM modulation of interferometer output intensity. A π rad peak-to-peak phase shift has been achieved as indicated by modulation over the entire interfermeter response range of ±0.15 V.

can occur at resonances between 50 and 200 MHz and seems to be strongly influenced by impedance matching, heat dissipation, and mechanical damping provided by the sample holder. The model based on elastic vibrational theory can be used to accurately calculate the resonance frequencies, but agreement between measured and calculated amplitudes has not been achieved because the quality factor of the device is difficult to evaluate. Qualitatively, the envelope of the resonance amplitudes and the cut-off frequency agrees with the model's description of the overlap integral between acoustic and optic waves, but a better understanding of fiber mounting and impedance matching is required. A linear dependence of the phase shift amplitude on voltage is observed at low driving potentials, but a saturation of the phase shift occurs at sufficiently high potentials. The potential at which saturation occurs is dependent on both the device thickness and length. This saturation effect is related to heating and nonlinear conduction mechanisms within the ZnO coating [5]. The largest phase shift achieved with these all-fiber modulators is π rad peak-to-peak at a driving frequency of 67 MHz and driving potential amplitude of 6.3 V. The π rad phase shift is demonstrated in FIG. 4 which shows intensity modulation over the entire response range of the interferometer.

In comparison with polymer coatings, ZnO has several advantages for modulator design. Since ZnO is deposited from the vapor directly onto the fiber substrate, concentric coatings of uniform thickness can easily be produced. ZnO exhibits elastic stiffness coefficients that are more closely matched to the underlying glass substrate, resulting in more efficient strain coupling between the fiber and the ZnO coating. In addition, the driving potential of the ZnO can be significantly lower than for polymer coatings since ZnO coatings under 10 μm in thickness can be used for devices. The melt extrusion process used for piezoelectric polymer coatings produces films between 25 and 150 μm in thickness, which requires high PFOM driving potentials, and the process has lower accuracy in controlling the uniformity of concentric coatings [11, 12]. Polymers also have relatively low elastic stiffness coefficients and exhibit no advantage with respect to the piezoelectric response. The primary advantages of ZnO over PZT include the ease of fabrication at low temperatures, low dielectric constant, which is advantageous when operating at high frequencies, and the textured growth that makes poling unnecessary. Since PZT does have higher piezoelectric coefficients than ZnO and can be poled to provide various modes of piezoelectric response, work is continuing on PFOM devices based on PZT.

WAVELENGTH TUNABLE THERMAL FIBER BRAGG GRATINGS (TFBG)

A reflection filter can be made within the core of an optical fiber by forming an intra-core Bragg grating. The Bragg grating is fabricated by side irradiation of an optical fiber with a UV laser. An interference pattern is formed with the incident radiation to produce a periodic irradiation pattern along the length of the fiber. The exposure results in a periodic variation of the refractive index because the incident radiation induces changes in the glass structure [18, 19]. With proper control of the magnitude and spatial distribution of the photo-induced refractive index changes, a reflection filter is formed with a specific reflected wavelength, λ_B, which is defined by the effective refractive index of the guided mode, n_{eff}, and the grating period, Λ, as given by Equation (3). Bragg gratings commonly exhibit a reflection peak or transmission valley with a spectral width on the order of 0.1 to 1 nm [20].

$$\lambda_B = 2n_{eff}\Lambda \tag{3}$$

$$\Delta\lambda_B / \lambda_B = (\alpha + \xi)\Delta T \tag{4}$$

A shift of the Bragg wavelength, $\Delta\lambda_B$, results from a change in the fiber temperature, ΔT, due to thermal expansion and temperature induced changes of the refractive index. The temperature dependence of the Bragg wavelength is given by Equation (4) where α is the thermal expansion coefficient of the fiber and ξ is the thermooptic coefficient [21]. This principle of temperature induced changes in the Bragg wavelength was used to make a wavelength tunable filter.

A compact device based on an active fiber coating was fabricated by coating a 10 mm long Bragg grating with a Ti/Pt resistive coating as shown in FIG. 5. The Bragg grating, with a Bragg wavelength of 1552 nm, was formed in hydrogenated standard monomode telecommunications fiber (9 μm core and 125 μm outside diameter) using KrF excimer laser irradiation and a phase mask to produce the periodic exposure pattern [7]. Gold wires were connected at the ends of the coating with silver paint. A Ti/Pt bilayer, with the Ti acting as an adhesion layer and the Pt acting as the resistive layer, was chosen for the resistive fiber coating since this type of electrode coating has been shown to be stable in oxidizing atmospheres to temperatures as high as 500°C. In addition, the dimensions of sputter deposited Pt fiber coatings can easily be tailored to give device

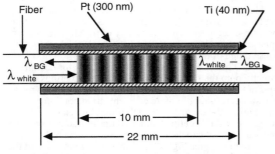

FIG. 5. Schematic diagram of wavelength tunable TFBG device.

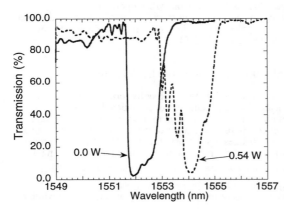

FIG. 6. TFBG transmission spectrum at room temperature and Joule heated with an applied DC electrical power of 0.54 W.

operating voltages and currents of less than 10 V and 1 A, respectively, which allows the use of inexpensive driving electronics. The TFBG device had a room temperature resistance of 38 Ω. Passing current through the coating resulted in Joule heating of the Ti/Pt coating and a subsequent heat transfer to the fiber. Transmission spectra were measured using a tunable laser and photodiode. FIG. 6 shows the TFBG transmission spectrum at room temperature with no applied power and with an applied power of 0.54 W. Heating of the coated fiber resulted in a transmission minimum shift of 2.2 nm. Measurements of the resistance change with temperature indicated that a temperature of approximately 100°C was attained. The wavelength shift was linear with applied electrical power resulting in a wavelength shift efficiency of 4.1 nm/W. Temperature and strain gradients are believed to be the cause of the broadening that accompanies the shift of the reflection peak. Modulation of the Bragg wavelength was also possible at frequencies as high as 500 Hz, but the wavelength modulation efficiency decreases strongly with increasing frequency due to limitations of the thermal conductivity [22].

WAVELENGTH MODULATED PIEZOELECTRIC FIBER BRAGG GRATINGS (PFBG)

Bragg gratings can also be wavelength tuned by straining the section of fiber that contains the grating. The relation between the shift of the Bragg wavelength and strain, assuming quasistatic and plain strain conditions, is given in Equation (5), where n is the refractive index of the core, P_{11} and P_{12} are the photoelastic coefficients and S_z and S_r are the axial and radial strains, respectively [20]. By comparing Equation (1) and Equation (5), it is clearly seen that the shift of the Bragg wavelength and the phase shift have the same dependence on strain. This suggests that optimization of the piezoelectric actuator to obtain large optical phase shifts will result in PFBG devices with a large Bragg wavelength shift.

$$\Delta\lambda_B / \lambda_B = S_z - \frac{n^2}{2}\left[(P_{11} + P_{12})S_r + P_{12}S_z\right]$$

(5)

A miniature device based on the principle of strain tuning was fabricated by coating a 10 mm long Bragg grating with a piezoelectric coating as shown in FIG. 7. The Bragg grating was

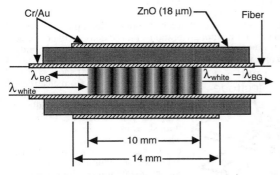

FIG. 7. Schematic diagram of wavelength modulating PFBG device.

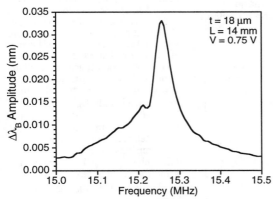

FIG. 8. PFBG modulated wavelength shift dependence for frequencies near the fundamental radial mode resonance and for a driving potential amplitude of 0.75 V.

produced by the same process used in the TFBG fabrication. The Cr/Au and piezoelectric ZnO coatings were deposited by the same methods used for PFOM fabrication, but the ZnO coating was deposited by dual magnetron sputtering in order to obtain a coating thickness of 18 μm as compared with a 6 μm thickness for the PFOM devices. ZnO was chosen as the piezoelectric coating material for the same reasons as those described above for the PFOM devices, but of particular importance is the low deposition temperature of the ZnO. Bragg gratings exhibit a significant decay in the reflectivity as the temperature is increase above 100°C [23]. Since the ZnO and Cr/Au coatings can be deposited with fiber temperatures of approximately 100°C, a large decay of the grating reflectivity can be avoided [7].

Modulation of the Bragg wavelength was measured with a tunable laser, photodiode, and spectrum analyzer. The wavelength of the tunable laser was fixed at half the height of the Bragg reflection peak, and the modulation of the transmitted intensity was recorded. For small amplitude modulations of the transmitted intensity, the intensity is proportional to the Bragg wavelength. The modulation amplitude of the Bragg wavelength was determined from the linear dependence of the intensity on Bragg wavelength. FIG. 8 shows a peak in the Bragg wavelength modulation amplitude for a PFBG device driven with a frequency of 15.25 MHz and an applied potential amplitude of 0.75 V. Measurements of the impedance and phase angle of the PFBG were used to confirm that the peak in $\Delta\lambda_B$ corresponded to the fundamental radial resonance. The fundamental radial resonance exhibited the highest $\Delta\lambda_B$ modulation efficiency, and the modulation efficiency decreased rapidly for higher order resonances. The maximum observed $\Delta\lambda_B$ modulation amplitude of 0.044 nm was obtained at the fundamental radial resonance using a driving potential amplitude of 1.26 V. Since the highest $\Delta\lambda_B$ modulation efficiencies are correlated with the radial mode resonances, it is apparent that the mechanisms controlling the $\Delta\lambda_B$ modulation efficiency are strongly linked to the optical phase shift response as observed for PFOM devices. This is expected due to the similarities of the quasistatic plane strain models for the two devices.

FLEXURAL/BENDING PIEZOELECTRIC FIBER ACTUATORS

Beside the ability to excite radial and axial mode vibrations in optical fiber, piezoelectric actuator coatings can also excite flexural or bending mode vibrations. Piezoelectric actuators that can provide a flexural motion of a fiber tip show promise for applications in scanning near field optical microscopy and endoscopy. Flexural vibrations in fibers with cylindrical actuators arise from a combination of shear vibrations that result from electrode edge effects, as well as asymmetrical mounting of the fiber, which allows coupling of radial and axial vibrations with shear mode vibrations. Thickness gradients in the fiber coatings can also be a source for the excitation of flexural vibrational modes. A flexural/bending fiber actuator was fabricated as shown in FIG. 9. The same process used for making PFOM devices was used to make a 2 mm long cylindrical actuator with a 6 μm thick ZnO coating having a thickness gradient of approximately 2%. The 2 mm long actuator segment was fixed at the end of a board using silver paint which acted as both the top electrode contact and the fixation point for the fiber segment. The free fiber tip hanging over the end of the board had a length (L_{FT}) of 29 mm.

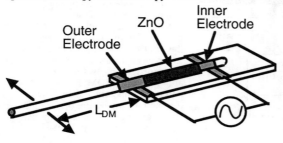

FIG. 9. Schematic diagram of flexural/bending mode actuator. The arrows indicate the direction of measured displacement.

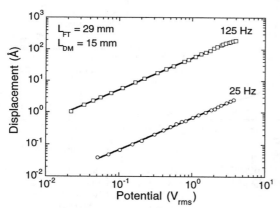

FIG. 10. Displacement dependence on driving potential for a flexural fiber actuator with a fiber tip length, L_{DM}, of 29 mm and piezoelectric exciting element of 2 mm.

The horizontal displacement of the fiber tip was measured using a Mach-Zehnder interferometer. Measurements were made by reflecting the focused incident laser beam from a gold coated section of fiber positioned at a distance (L_{DM}) of 15 mm away from the fixation point and the actuator segment. Several resonances were observed in the frequency range between 25 Hz and 100 kHz. The rms displacement as a function of driving potential for driving frequencies of 25 and 125 Hz is shown in FIG. 10. At 125 Hz the first flexural resonance is observed for the fiber tip. Driving the tip in resonance resulted in a linear dependence of displacement on driving potential and displacements as large as 190 Å were achieved with an rms driving potential of 3.9 V. Below the 125 Hz resonance, the displacement is constant as a function of frequency until 25 Hz, which was the lowest measured frequency. The non-resonant displacement at 25 Hz is approximately two orders of magnitude smaller than the displacements measured for equivalent driving potentials in resonance. Similar to the resonant behavior, a linear dependence of the displacement is observed at the 25 Hz driving frequency, but a displacement of only 2.5 Å was achieved with an rms driving potential of 3.7 V. Flexural fiber actuator devices with segmented electrode and ZnO coatings have also been fabricated by etching line trenches in the fiber coatings along the length of the fiber. The fabrication process and demonstration of the devices are presented elsewhere in this proceedings [24].

CONCLUSIONS

Four different devices based on piezoelectric and resistive active fiber coatings have been fabricated and their operation has been demonstrated. These four devices include a piezoelectric fiber optic phase modulator (PFOM), wavelength tunable thermal fiber Bragg grating (TFBG), wavelength modulating piezoelectric fiber Bragg grating (PFBG), and flexural/bending fiber actuator. The PFOM uses a concentric piezoelectric fiber coating to strain the optical fiber and produce a change of the optical path length resulting in an optical phase shift. A π rad optical phase shift has been achieved for ZnO based PFOM devices, which have applications in both telecommunications and local area fiber networks. TFBG combine a resistive Ti/Pt coating with an intra-core fiber Bragg grating that acts as an in-fiber wavelength filter. These Joule heated devices provide a compact high efficiency (4.1 nm/W) all-fiber tunable filter that can also provide wavelength modulation to frequencies as high as 500 Hz. PFBG combine a piezoelectric ZnO fiber coating and fiber Bragg grating to give a wavelength modulation device that can work at high frequencies. A Bragg wavelength modulation amplitude of 0.044 nm was achieved with a driving potential amplitude of 1.26 V while driving the device at the fundamental radial mode resonance frequency of 15.25 MHz. Both TFBG and PFBG are of interest for a variety of multiplexed telecommunication and sensing networks. The flexural/bending fiber actuator was also fabricated with a concentric ZnO coating, and fiber tip displacements of 190 Å were achieved with an rms driving potential of 3.7 V and a resonant driving frequency of 125 Hz. Applications of flexural fiber tip actuators may include scanning near field microscopy and endoscopy. Demonstration of these four different types of devices indicate that active fiber coatings show promise for a wide variety of applications in the field of optical fiber communications as well as micro-electro-mechanical systems.

ACKNOWLEDGMENTS

This work was supported by the optical Sciences, Applications, and Technology Priority Program of the Board of the Swiss Federal Institute of Technology.

REFERENCES

1. B.J. White, J.P. Davis, L.C. Bobb, H.D. Krumboltz, and D.C. Larson, *J. Lightwave Technol.*, **LT-5** (9), 1169 (1987).

2. D.S. Czaplak, J.F. Weller, L. Goldberg, F.S. Hickernell, H.D. Knuth, and S.R. Young, *IEEE Ultrason. Symp. Proc.*, 491 (1987).

3. A. Dandridge, A.B. Tveten, and T.G. Giallorenzi, *Electron. Lett.*, **17** (15) 523 (1981).

4. G.R. Fox, N. Setter, and H.G. Limberger, *J. Mater. Res.*, **11** (8), 2051 (1996).

5. G.R. Fox, C.R. Wüthrich, C.A.P. Muller, N. Setter, and H.G. Limberger, submitted to *Ferroelectrics*, (1996).

6. G.R. Fox, S. Trolier-McKinstry, S.B. Krupanidhi, and L.M. Casas, *J. Mater. Res.*, **10** (6), 1508 (1995).

7. G.R. Fox, C.A.P. Muller, N. Setter, D.M. Costantini, N.H. Ky, and H.G. Limberger, *J. Vac. Sci. Technol. A*, **15** (3), (1997).

8. G.R. Fox, D. Damjanovic, P.A. Danai, N. Setter, H.G. Limberger, and N.H. Ky, in *Materials for Smart Systems*, edited by E.P. George, S. Takahashi, S. Trolier-McKinstry, K. Uchino, and M. Wun-Fogle, (Mater. Res. Soc. Proc., Vol. 360, Materials Research Society, Pittsburg, PA, 1995) pp. 389-394.

9. N.H. Ky, H.G. Limberger, R.P. Salathé and G.R. Fox, *IEEE Photon. Technol. Lett.*, **8** (5), 629 (1996).

10. L.J. Donalds, W.G. French, W.C. Mitchell, R.M. Swinehart, and T. Wei, *Electron. Lett.*, **18** (8), 327 (1982).

11. J. Jarzynski, *J. Appl. Phys.*, **55** (9) 3243 (1984).

12. M. Imai, T. Shimizu, Y. Ohtsuka, and A. Odajima, *J. Lightwave Technol.*, **LT-5** (7), 926 (1987).

13. G.R. Fox, C.A.P Muller, N. Setter, N.H. Ky, and H.G. Limberger, *J. Vac. Sci. Technol. A*, **14** (3), 800 (1996).

14. G. Yi, M. Sayer, C.K. Jen, J.C.H. Yu, and E.L. Adler, *IEEE Ultranson. Symp. Proc.*, 1231 (1989).

15. A. Gusarov, N.H. Ky, H.G. Limberger, R.P. Salathé, and G.R. Fox, *J. Lightwave Technol.* **14** (12), (1996).

16. N.H. Ky, H.G. Limberger, R.P. Salathé, and G.R. Fox, *Optical Fiber Communication Conference*, Vol. 2 1996 OSA Technical Digest Series (Optical Society of America, Washington, D.C., 1996), pp. 244-245.

17. G.R. Fox, C.A.P. Muller, M. Kuhn, N. Setter, N.H. Ky, and H.G. Limberger, in *Microelectronic Structures and Microelectromechanical Devices for Optical Processing and Multimedia Applications*, edited by W. Bailey, M. Edward Motamedi, and F.C. Luo, (Proc. SPIE, Vol. 2641, SPIE, Bellingham, WA, 1995) pp. 55-61.

18. G. Meltz and W.W. Morey, in *International Workshop on Photoinduced Self-Organization Effects in Optical Fiber*, (Proc. SPIE, Vol. 1516, SPIE, Bellingham, WA, 1991) pp. 185-199.

19. H.G. Limberger, P.Y. Fonjallaz, R.P. Salathé, and F. Cochet, *Appl. Phys. Lett.*, **68** (22), 3069, (1996).

20. G. Meltz, W.W. Morey, and W.H. Glenn, *Optics Lett.*, **14** (15), 823 (1989).

21. W.W. Morey, G. Meltz, and W.H. Glenn, in *Fiber Optic and Laser Sensors VII*, (Proc. SPIE, Vol. 1169, SPIE, Bellingham, WA, 1989) pp. 98-107.

22. N. H. Ky, H.G. Limberger, D.M. Costantini, R.P. Salathé, C.A.P. Muller, and G.R. Fox, submitted to CLEO'97 (Conference on Lasers and Electro-Optics)

23. H. Patrick, S.L. Gilbert, A. Lidgard, and M.D. Gallagher, *J. Appl. Phys.*, **78** (5), 2940 (1996).

24. S. Trolier-McKinstry, G.R. Fox, A.L. Kholkin, C.A.P. Muller, and N. Setter, in *Materials for Smart Systems II*, (Mater. Res. Soc. Proc., this volume, Materials Research Society, Pittsburg, PA, 1997).

SMART PIEZOELECTRIC PZT MICROCANTILEVERS
WITH INHERENT SENSING AND ACTUATING ABILITIES
FOR AFM AND LFM

C. Lee[*,**,1], T. Itoh[*,2], J. Chu[*], T. Ohashi[*], R. Maeda[**], A. Schroth[**,3], and T. Suga[*]
*RCAST, The University of Tokyo, Komaba 4-6-1, Meguro-Ku, Tokyo 153, Japan
**Mechanical Eng. Lab., AIST, MITI, Namiki 1-2, Tsukuba, Ibaraki, 305 Japan

ABSTRACT

Novel designs of the force sensing components for an atomic force microscope (AFM) and lateral force microscope (LFM) have been proposed in this study. By using PZT thin layers, a smart structure that can perform force sensing and feedback actuation at the same time is applied to the AFM. Clear images can be derived by an AFM equipped with this smart structure. A structure of two parallel PZT bars integrated on a SiO_2 free standing cantilever has shown potential for operation in an LFM, because a difference in the piezoelectric charge outputs from these two beams will be induced by frictional force when the cantilever end quasi-staticly contacts with the sample surface in dynamic scanning across the surface.

INTRODUCTION

The atomic force microscope (AFM) has received much attention as a nanometer-scale characterization and micromachining tool. It is possible to investigate a sample surface by measuring the force between a sharp tip on the cantilever end and the sample surface. It also shows potential applications for observations of bio-samples and as a recording tool for high density data storage devices [1-4]. It has static and dynamic modes of operation. During static scanning, the tip is in close contact to the sample surface, and the tip drags along the surface. The deflection of the cantilever is measured and used to control a feedback loop to keep the trace force constant. In the dynamic mode the tip either periodically touches the surface in each vibration cycle, i.e., cyclic contact or tapping mode, or does not touch the surface during scanning, i.e., non-contact mode, while the tip is driven at its first resonance frequency. Although the contact mode can offer high resolution, it tends to hurt soft samples, e.g. biological specimens. In contrast the non-contact mode can avoid damaging the sample surface, but it can not be used for high resolution images in air due to the fact that the cantilever end is easily trapped by the surface layer that forms in air. The cyclic contact is becoming the prevailing operation mode in air, because this mode offers high resolution, like the contact mode, while not damaging the surface of soft samples. In the operation of commercial AFMs, an optical deflection component is used to measure the variation of the cantilever deflection resulting from the topographical changes of the sample. The optical component occupies a large part of the whole volume of the imaging unit of AFMs. It also requires precise alignment of a laser beam on the cantilever, which must be maintained during scanning. This characteristic makes observations of samples by AFM in liquid environments and UHV difficult.

The lateral force microscope (LFM) is a variation of the AFM [5]. It operates in a similar method as the contact AFM, but the cantilever scans a sample in the direction parallel to the sample surface and perpendicular to the longitudinal direction of the cantilever. The lateral or friction force between the tip and the sample surface induces the cantilever beam to twist. For commercial LFMs, the induced torsion angle is typically monitored by optical components which include a

1. Present address : Microsystems Lab., ITRI, 195, Sec. 4, Chung Hsing Rd., Chutung,
         Taiwan 310, Republic of China.
2. Author to whom correspondence should be addressed. e-mail: itoh@suga.rcast.u-tokyo.ac.jp
3. NEDO fellow on leave from Micromachine Center Japan.

laser diode and a 4 segment photo detector, via separating the reflected laser light signals into lever bending and torsion. Again, the necessity of optical components brings the same problems as in the case of the AFM.

Several force sensing cantilevers have been proposed and developed to get rid of the problems caused by the optical sensing components [6-8]. Among the sensing mechanisms of force sensing cantilevers, the piezoelectric scheme shows superiority in the operation of dynamic mode AFM, because a piezoelectric cantilever can be excited by an applied ac voltage without any disturbance of the sensing action [9]. This excellent characteristic also makes for simple operation of dynamic AFM in liquid [10] and vacuum [11]. Moreover, AFMs equipped with a cantilever array show potential applications in both nanolithography and imaging on large scales and high speed. These are important for applications in high density data storage and wafer inspection [4, 12-14]. Conventional AFMs use a single probe that can only scan an area of less than 10' μm'. The separate control of each cantilever in an array is needed, because 1) the resonance frequency of each cantilever will be slightly different due to the deviation of geometric structure in the microfabrication process, 2) the original tip-sample spacing of each cantilevers will be different, and 3) the different traces of each cantilever must be individually feedback controlled.

This paper describes a smart structure, PZT (Lead zirconate titanate) microcantilever with the inherent ability of force sensing and feedback-actuation. Based on this structure, the individual feedback control of each cantilever in a piezoelectric cantilever array for multiprobe AFM becomes possible. In addition a smart cantilever, which is a free standing SiO_2 beam integrated with two parallel PZT beams on one side, is proposed to measure the torsion due to the friction force for the LFM application.

FABRICATION

Fabrication starts from sol-gel deposition of the $Pb(Zr_{0.53}Ti_{0.47})O_3$ layer on a $Pt/Ti/SiO_2/Si$ substrate. Then a micromachining process is used to sculpt the free standing cantilever structure. The details of the fabrication procedure are similar to a previously reported process [15]. Fig. 1 shows a schematic drawing and an SEM micrograph of a smart structure for AFM. This structure includes a 125 μm long and 50 μm wide cantilever integrated with a PZT layer of 100 μm length , and a PZT reference pattern of the same surface area. The smart force sensing cantilever for LFM is shown in Fig. 2 (a). The tip on the end of cantilever shown in Fig. 2 (a) is fixed by glue via a micro-manipulator. Two structures were made, one contains a PZT layer with separated surface electrodes (Fig. 2 b), the other was integrated with two parallel beams of PZT/surface electrode

Fig. 1(a) SEM photo of the PZT smart structure for AFM application.

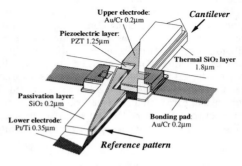

Fig. 1(b) Schematic drawing of the PZT smart structure shown in (a).

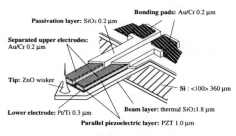

Fig. 2(a) SEM photo of a force sensing cantilever as shown schematically in (b).

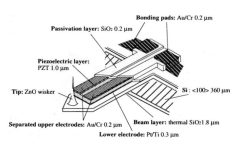

Bonding pads: Au/Cr 0.2 µm
Passivation layer: SiO₂ 0.2 µm
Piezoelectric layer: PZT 1.0 µm
Tip: ZnO wisker
Si : <100> 360 µm
Separated upper electrodes: Au/Cr 0.2 µm
Beam layer: thermal SiO₂1.8 µm
Lower electrode: Pt/Ti 0.3 µm

Fig. 2 (b) Schematic drawing of a 125 µm long and 50 µm wide cantilever integrated with one PZT layer and two separated upper electrodes (Type 1).

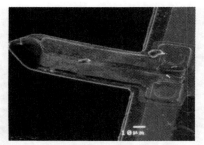

Bonding pads: Au/Cr 0.2 µm
Passivation layer: SiO₂ 0.2 µm
Separated upper electrodes: Au/Cr 0.2 µm
Tip: ZnO wisker
Si : <100> 360 µm
Lower electrode: Pt/Ti 0.3 µm
Beam layer: thermal SiO₂1.8 µm
Parallel piezoelectric layer: PZT 1.0 µm

Fig. 2 (c) Schematic drawing of a 125 µm long and 50 µm wide cantilever integrated with two parallel bars of electrode/PZT layer of 100 µm length and 15 µm width (Type 2).

layers (Fig. 2 c).

A SMART STRUCTURE FOR FORCE SENSING AND FEEDBACK ACTUATION IN AFM

Fig. 3 shows the control loop of a dynamic AFM using the smart PZT structure shown in Fig. 1. An ac voltage from the frequency synthesizer is applied to both of the free standing cantilever and the reference pattern. The cantilever is excited at its resonance frequency. The current output from the cantilever and the reference pattern are used as the input to the differential current amplifier. When the cantilever vibrates at the resonance frequency, a piezoelectric current output will result from the PZT layer due to the piezoelectric effect. The output is determined by the vibrational amplitude, i.e., the strain. Simultaneously, a capacitance current occurs in the PZT layer due to the applied ac voltage. The current output from the cantilever contains both signals. In contrast to the cantilever, the current output from the reference pattern only includes the capacitance current. The

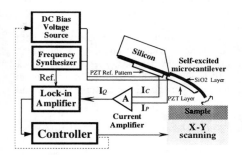

Fig. 3 Block diagram of a dynamic AFM using a PZT smart structure.

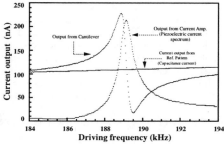

Fig. 4. Current outputs from the PZT cantilever and the PZT reference pattern. The piezoelectric current output spectrum can be recorded from the output of the differential current amplifier.

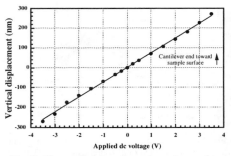

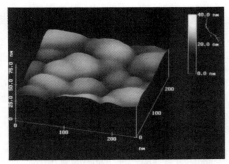

Fig. 5. The applied dc voltage dependence of the cyclic contact force curve displacement of a 125 μm cantilever. It shows that the actuating ability of the cantilever is about 75 nm/V.

Fig. 6 A cyclic contact AFM image of an evaporated Au film. It is taken by the AFM equipped with a PZT smart structure.

current outputs measured separately for the cantilever and the reference pattern are shown in Fig. 4. Fig. 4 also shows the curve of the driving frequency versus the current output from the differential amplifier, which is determined by the piezoelectric current caused by the vibration. This spectrum is similar to the one measured by optical methods. The current output from the differential amplifier can be used to represent the vibrational amplitude. This conversion can be done by external electronic circuits, however, it is difficult for a PZT cantilever [9, 16], because the PZT has a relatively large capacitance.

In the operation of a piezoelectric AFM, the force curve means the piezoelectric current outputs, i.e., the vibrational amplitude of cantilever, versus the displacement of sample surface, when the cantilever is excited at a fixed position by a constant energy, and the sample surface is moved upward and downward by a XYZ piezoelectric scanner. There is an intersecting point in a force curve. This point means an absolute position that the central point of a vibrational trace of the cantilever end meets the sample surface, i.e., the point of changing from cyclic contact region to static contact region. Once the cantilever end being bended upward or downward by dc voltages, the absolute position of this intersecting point will be moved upward or downward as well. When various superimposed dc voltages are applied to a vibrating cantilever, which is excited by a 25 mV ac voltage, the actuating ability for the cantilever end can be estimated. This is done by measuring the shift of intersecting point in the absolute position. The measured dc actuation ability of displacement for the cantilever end in the z-directional is illustrated in the Fig. 5. A linear depenence is observed in this displacement range and the actuating ability is 75 nm/V. Because the vertical actuated displacement of the cantilever end may reach 1 μm by applying a ± 6V dc voltage, the actuating ability of PZT cantilever is enough for satisfying most of the surface features for smooth samples. From this actuating ability the piezoelectric constant d_{31} is calculated as -58 pC/N, which is about 60% of the bulk PZT. Images were successfully obtained when the z-directional feedback signals were sent to the cantilever during scanning [17]. As shown in Fig. 6, an AFM image of an Au film evaporated on a smooth glass plate has been taken by using this smart force sensing and feedback-actuation structure via the loop of Fig. 3.

SMART CANTILEVERS FOR FRICTION FORCE SENSING

By using the cantilever structures shown in Fig. 2 (b) and (c) and a loop of Fig. 7, the piezoelectric charges from the two electrodes are simultaneously recorded by two charge amplifiers, while the cantilever is excited by an external oscillator at the first resonance frequency. The related spectra of piezoelectric charge outputs from type1, and 2 structures are illustrated in Fig. 8 (a), and (b),

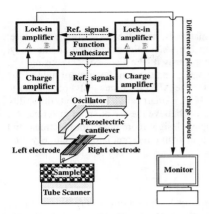

Fig. 7 Block diagram of an experimental setup to detect piezoelectric charge output difference between right and left electrode from the piezoelectric cantilever

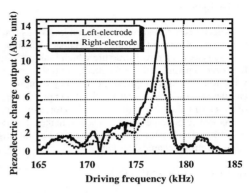

Fig. 8 (a) Piezoelectric charge outputs of the type 1 cantilever measured from left and right electrodes versus the driving frequency.

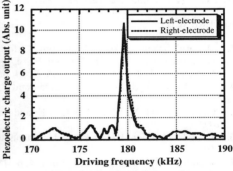

Fig. 8 (b) Piezoelectric charge outputs of the type 2 cantilever measured from left and right electrodes versus the driving frequency.

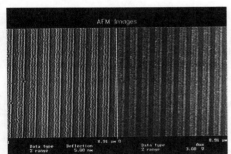

Fig. 9 Topographic image obtained from the piezoelectric charge output is shownon the left, while the image of change output difference is shown in right.

respectively. The difference in resonance frequency between the two spectra is possibly due to the deviation at the microfabrication process. We can conclude that the type1 cantilever exhibits a lower electrical interference corresponding to the geometric structure.

As shown in Fig. 9, an image related to the piezoelectric charge output differences can be recorded by the method depicted in Fig. 7, where the cantilever is excited at the 1st resonance frequency. The slight difference between the right and left images is considered to be an influence of the frictional force. The relationship between the piezoelectric charge output difference and the friction force is currently being investigated. Contact imaging in the dynamic AFM mode has been studied by exciting the cnatilever at the resonance frequency of the 2nd or 3rd flexural mode in stead of the 1st mode [18, 19]. Vibrating the tip at the 2nd or 3rd flexural modeallowss the cantilever to be kept in quasi-static contact with the sample surface during scanning. The output difference between the two electrodes will show the influence attributed to the torsion caused by the friction force in this quasi-static mode.

CONCLUSION

Piezoelectric materials have been known as smart materials, nevertheless, a simple structure, which shows inherent sensing and feedback actuation capabilities, is seldom reported. In this study, we have created a novel structure which is able to simultaneously do the displacement sensing, i.e., the force sensing, and the feedback control of displacement, i.e., the cantilever amplitude of the vibration, for AFM application. The key components for the AFM becomes just a piezoelectric cantilever. In addition, by using the direct piezoelectric effect, a smart cantilever exhibits the ability of measuring the torsion due to friction force influences. We believe the use of smart materials, eg. piezoelectric materials, can be used to make a less complex design, more powerful functions, and lead to an intelligent system in the field of intelligent instruments, precision machinery, and microelectromechanical systems.

ACKNOWLEDGEMENTS

The authors would like to thank the Japan Science and Technology Corporation (JST) for offering the National Institute Post Doctoral Fellowship to C. Lee. A part of the work was performed under the manegement of the Micromachine Center as the Industrial Science and Technology Forntier Program "Research and Development of Micromachine Technology" of MITI, supported by the New Energy and Industrial Technology Development Organization (NEDO).

REFERENCES

1. D. Sarid, <u>Scanning Force Microscopy</u>, Oxford Univ. Press, New York, 1991.
2. H. G. Hansma, J. Vac. Sci. Technol. B, 14, p.1390 (1996).
3. K. Yano, et al., J. Vac. Sci. Technol. B, 14, p.1353 (1996).
4. S. C. Minne, S. R. Manalis, A. Atalar, and C. F. Quate, J. Vac. Sci. Technol. B, 14, p.2456 (1996).
5. C. M. Mate, G. M. McClelland, R. Erlandsson, and S. Chiang, Phys. Rev. Lett., 59, p.1942 (1987).
6. T. Goddenhenrich, H. Lemke, U. Hartmann, and C. Heiden, J. Vac. Sci. Technol. A, 8, p.383 (1990).
7. T. Itoh and T. Suga, Nanotechnology, 4, p.218 (1993).
8. Tortonese, R. C. Barrett, and C. F. Quate, Appl. phys. Lett., 62, p.834 (1993).
9. T.Itoh, T.Ohashi, and T.Suga, IEICE Trans. on Electron., E78-C, p.146 (1995).
10. C. Lee, R. Maeda, T.Itoh, and T.Suga, Proc. 3rd France-Japan Congress & 1st Europe-Asia Congress on Mechatronics, Besançon, France, 1, p.285 (1996).
11. J. Chu, T. Itoh, C. Lee, and T.Suga, Abstracts of NANO IV, 4th Int. Conf. on Nanometer-scale Sci. & Technol., Beijing, P.R.China (the American Vacuum Soc., New York, NY 10005, 1996), p.27.
12. S. C. Minne, Ph. Flueckiger, H. T. Soh, and C. F. Quate, J. Vac. Sci. Technol. B, 13, p.1380 (1995).
13. T. Itoh, T. Ohashi, and T. Suga, Proc. IEEE MEMS '96, San Deigo, (IEEE, Piscataway, NJ 08855, 1996), p.451.
14. B. W. Chui, T. D. Stowe, T. W. Kenny, H. J. Mamin, B. D. Terris, and D. Rugar, Appl. Phys. Lett., 69, p.2767 (1996).
15. C. Lee, T.Itoh, and T.Suga, IEEE Trans. Ultras., Ferro., and Fre. control, 43, p.553 (1996).
16. C. Lee, Ph.D. Dissertation, The University of Tokyo, Tokyo, 1996, p.116.
17. T. Itoh, C. Lee, and T. Suga, Appl. Phys. Lett., 69, p.2036 (1996).
18. S. C. Minne, S. R. Manalis, A. Atalar, and C. F. Quate, Appl. Phys. Lett., 68, p.1427 (1996).
19. U. Rabe, K. Janser, and W. Arnold, Rev. Sci. Instrum., 67, p.3281 (1996).

IN-SITU CONTROL OF THE DIRECTION OF SPONTANEOUS POLARIZATION IN FERROELECTRIC THIN FILMS BY RF-MAGNETRON SPUTTERING

K. IIJIMA*, K. NIIHARA**
* Synergy Ceramics Laboratory, Fine Ceramics Research Association, Nagoya, Aichi 462, Japan, iijimak@ctmo.mei.co.jp
**Institute of Scientific and Industrial Research, Osaka Univ. Ibaraki, Osaka 567, Japan, niihara@sanken.osaka-u.ac.jp

ABSTRACT

La-modified $PbTiO_3$ thin films were grown on (100)MgO by rf-magnetron sputtering . Thin films were prepared by "two step method" in which the different Pb/Ti ratios during the growth were used for the first step(0-4nm of thickness) and the subsequent second step(>4nm). The Pb/Ti ratios of the first step were selected from 1.2 to 1.5. The Pb/Ti ratio of the second step, on the other hand, was fixed at 1.4. Strong dependency of the direction of the spontaneous polarization(Ps) to the Pb/Ti ratio of the first step was observed. It is concluded that the direction of the Ps in the films is determined by the atomic species of the first layer of the films.

INTRODUCTION

Ferroelectric materials have many useful properties, such as piezoelectricity, pyroelectricity, ferroelectricity, opto-electric property etc. and are widely used in various kinds of transducers. Recently, progress in the processing technology of thin films makes it possible to prepare high quality ferroelectric thin films and micro-devices using ferroelectrics [1].

However, usually ferroelectrics have a domain structure in as-grown state and we do not use most of the properties of ferroelectrics. To utilize them, a polarizing treatment applying a high voltage to ferroelectrics at high temperature is required to polarize the Ps in ferroelectrics. The polarizing treatment sometimes leads to a break-down of the thin film. It is difficult to control the direction of the Ps at any position in the film using the conventional method. Therefore, an in-situ process which can control the orientation of the Ps in the films is strongly required. The establishment of such in-situ process is essential to develop a smart material.

A "self-polarized" thin film was reported for the Pb-based ferroelectric thin film prepared by rf-magnetron sputtering[2,3]. However, the direction of Ps was not controlled. The domain orientation of the surface of $LiNbO_3$ is controlled by proton-exchange method[4]. This method only controls the direction of Ps at the surface.

In this paper, we report the in-situ control of the direction of Ps in Pb-based ferroelectric thin films by rf-magnetron sputtering. We found that the precise control of the composition at the initial stage of the film growth effectively controls the direction of the Ps in the film.

Mat. Res. Soc. Symp. Proc. Vol. 459 © 1997 Materials Research Society

EXPERIMENTAL

Lanthanum-modified PbTiO$_3$ (PLT) thin films were grown on the (100)MgO single crystal substrates by planar rf-magnetron sputtering using an epitaxial condition. Typical sputtering conditions are shown in Table I. In this study, we used a "two step growth" consisting of the first step (film thickness of 0-4nm) and the second step of the growth(>4nm). In each step, we use different Pb/Ti ratios: Pb/Ti ratios of the first step were chosen between 1.2 and 1.5. On the other hand, Pb/Ti ratio of the second step was fixed at 1.4. Pb/Ti ratio was adjusted by controlling the input power during the sputtering. The films thus obtained were examined by X-ray diffraction and high-resolution transmission electron microscopy to confirm the perovskite structure formation and epitaxial growth of the films.

After the PLT film deposition, 0.8×0.8mm Pt top electrodes were formed on the PLT thin film by rf-magnetron sputtering and the technique of photolithography. The substrate MgO under the top electrodes was etched off by phosphoric acid and then Pt bottom electrode was deposited. The orientations of the Ps in the films were determined by the direction of pyroelectric current. Pyroelectric current was measured by pico-ampere meter(HP4862) changing the temperature from -20℃ to 100℃ with the temperature gradient of 1℃ per minute. Pyroelectric coefficient, γ, was calculated using pyroelectric current and temperature gradient.

Table I. Summary of sputtering conditions

Target composition	0.8(Pb$_{1-x}$La$_x$Ti$_{1-x/4}$O$_3$)+0.2PbO, x=0.15
Substrate temperature	600℃
Input power density	about 3W/cm^2
Sputtering gases	Ar/O$_2$=90/10
Gas pressure	2mTorr
Deposition rate	2nm/min.

RESULTS

Figure 1 shows an X-ray diffraction pattern of the sputtered films. In this figure, mainly *00l* peaks of perovskite structure are observed, indicating that the strongly c-axis oriented PLT thin film was obtained. No difference was observed on the films deposited with different Pb/Ti ratios in the first step of the growth process.

Figure 2 shows the high-resolution TEM photograph of the PLT film near the film/substrate interface. It is seen that the sputtered film was epitaxially grown on (001)MgO. The {200} lattice planes are continuous across the interface, demonstrating that the film is epitaxially grown on the substrate.

Figure 3 shows the results of the "2 step growth". When the Pb/Ti ratio is 1.2, γ of the film is minus 6×10^{-8}C/cm^{2}K. The absolute value of γ decrease with increasing Pb/Ti ratio and becomes zero at the Pb/Ti of 1.3. Then γ changes its sign and increases with increasing Pb/Ti ratio. This results implies that the Ps in the film is oriented pointing upward when the Pb/Ti ratio is under 1.3 and the Ps is arranging upward and downward randomly at Pb/Ti=1.3. When Pb/Ti ratio is larger than 1.3, the Ps in the film is oriented pointing upward (arrow indicating the Ps is drown from substrate to film surface) as shown in Fig. 3.

We tried the "two-step growth" in different schemes: changing the Pb/Ti ratios at the top of the films and the middle of the films. In those cases, the direction of Ps was not controlled.

Some samples were annealed at elevated temperature for ten minutes above the Curie temperature. The orientation of Ps was kept even after the anneals above the Curie temperature of the film.

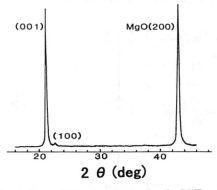

Fig. 1 X-ray diffraction pattern for the PLT thin film on MgO substrate.

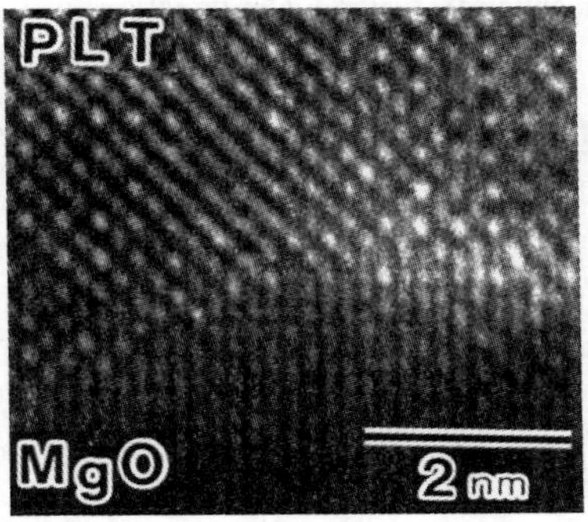

Fig. 2 Cross-sectional TEM image of the PLT thin film.

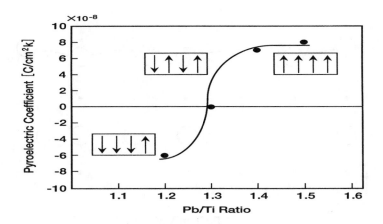

Fig. 3 Pyroelectric coefficient and Pb/Ti ratio of the first step of the growth. Inset diagrams represent schematically the orientation of the Ps in the films.

DISCUSSION

The direction of the Ps was changed only when the Pb/Ti ratios at the film/substrate interface were changed. This means that the origin controlling the Ps direction is hidden at the interface. The possibility of the stress or strain in the film is ruled out because the thermal stress in films is not essentially different from each other. The substrate temperature dependency of the direction of the Ps of PLT thin film has been reported[5]. This result also ruled out the possibility of a mechanical origin. The directions of the Ps were not controlled when the Pb/Ti ratios were

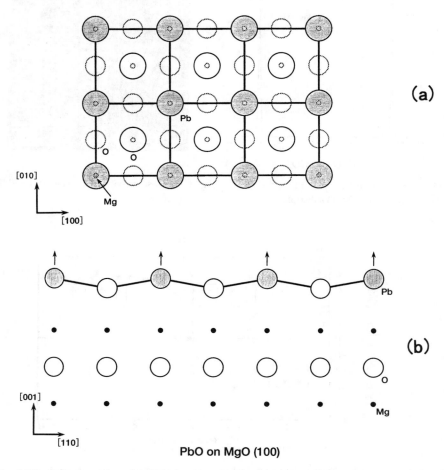

PbO on MgO (100)

Fig. 4 Plan (a) and cross-sectional(b) view of the PbO -layer/MgO-substrate interface explaining the small shift of ions. In the plan view, dotted circles indicate the ions in the substrate.

changed at the top or middle of the films. Therefore, the compositional gradient in the film is not the reason of the orientation of the Ps in the films. The possibility of the effect of space charge is also difficult to accept, because of the results of high temperature anneals and two step growth in different schemes.

We are considering that the direction of the Ps is determined by the small shift of the ions in the film at the interface. The (001) oriented PLT thin film is constructed by the alternate stacking of (PbLa)O or PbO layer and TiO_2 layer. So, there are two cases of the film growth; one is the growth starting from PbO layer and the other one is from TiO_2 layer. When the growth starts from PbO layer, Pb atoms may slightly shift upward due to the cation(Mg^{2+})-cation(Pb^{2+})

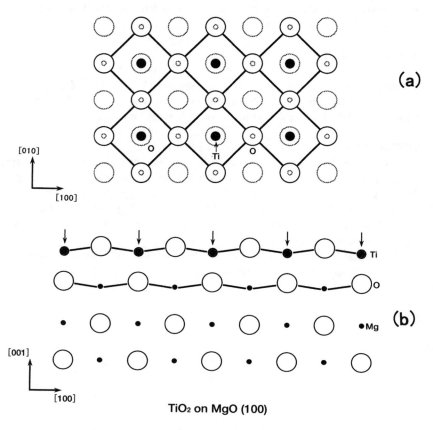

(a)

[010]

[100]

(b)

[001]

[100]

TiO_2 on MgO (100)

Fig. 5 Plan (a) and cross-sectional(b) view of the TiO_2 -layer/MgO-substrate interface explaining the small shift of ions. In the plan view, dotted circles indicate the ions in the substrate.

repulsion as shown in Fig. 4 and this small shift is a driving force to point the Ps upward below the Curie temperature. On the other hand, when the first layer of the film growth is TiO_2, Ti ions will be located on O ions of the MgO substrate and Ti ions are attracted to the substrate. Consequently, the films have the trigger directing the Ps downward at the interface of the film as shown in Fig.5. Therefore, the structural imprint at the interface in the films is considered to be controlling the direction of the Ps.

CONCLUSION

It is found that the direction of the Ps in Pb-based ferroelectric thin films can be controlled by the precise adjustment of the composition at the initial stage of the film growth by sputtering. We are conclude that the direction of the Ps in the films is determined by the atomic species of the first layer of the films.

ACKNOWLEDEMENT

Work supported by NEDO as part of the Synergy Ceramics Project under the Industrial Science and Technology Frontier(ISTF) Program promoted by AIST, MITI, Japan. The authors are members of the Joint Research Consortium of Synergy Ceramics.

REFERENCES
1. E.P.George, S.Takahashi, S.Trolier-McKinstry, K.Uchino, M.Wun-Fogle, Materials for Smart Systems, Materials Research Society Symposium Proceedings Vol. 360, 1995.
2. K.Iijima, Y.Tomita, R.Takayama and I.Ueda, J. Appl. Phys., 60, 361 (1986).
3. K.Iijima, R.Takayama, Y.Tomita, and I.Ueda, J. Appl. Phys., 60, 2914 (1986)
4. N.Ohnishi, Jpn. J. Appl. Phys. 16, 1069(1977), K. Nakamura, H.Ando and H.Shimizu, Appl. Phys. Lett. 50, 1413(1987)
5. T.Kamada, R.Takayama, A.Tomozawa, S.Fujii, K.Iijima and T.Hirao, Materials Research Society Symposium Proceedings (in press)

THE EFFECTS OF BIAXIAL STRESS ON THE FERROELECTRIC CHARACTERISTICS OF PZT THIN FILMS

JOSEPH F. SHEPARD JR.*[1], SUSAN TROLIER-McKINSTRY[1], MARY HENDRICKSON[2]
ROBERT ZETO[2]

[1] Intercollege Materials Research Laboratory, The Pennsylvania State University, University Park, Pennsylvania 16802
[2] Advanced MicroDevices Branch, Army Research Laboratory, Fort Monmouth, New Jersey, 07703

ABSTRACT

The design and implementation of microelectromechanical (MEMS) systems requires a sound understanding of the influence of film stress on both the ferroelectric and piezoelectric characteristics of thin films to be used for sensing and/or actuation. Experiments were conducted in which thin film samples of sol-gel derived PZT were subjected to applied biaxial stresses from -139 to 142 MPa. Films were characterized at known stress states (derived from known values of residual stress and large deflection plate theory) in terms of their ferroelectric polarization (saturated and remnant), dielectric constants, coercive field strengths, and tan δ. Results obtained indicate that domain wall motion in thin films contributes much less to the observed response than is typical for bulk PZT materials. Alternative mechanisms are proposed in an attempt to explain the discrepancies.

INTRODUCTION

The recent advances in electroceramic thin film deposition technology has resulted in a great deal of research and development in the area of microelectromechanical systems (MEMS) since the sensing and/or actuation function of MEMS devices can be accomplished though the use of thin film piezoelectric materials. Most piezoelectric-based devices are currently constructed with zinc oxide (ZnO) as their active material. The small piezoelectric coefficients of the ZnO when compared to those of the perovskite oxides (e.g. barium titanate and lead zirconate titanate) have led to inevitable attempts at implementation of thin film perovskites, most often lead zirconate titanate (PZT), in MEMS devices.

It has been well documented in the past that thin film materials, be it metals, ceramics or polymers, are in a state of residual stress upon completion of the deposition procedure [1]. The state of stress is dependent upon a number of factors, a major component of which is the mismatch of thermal expansion coefficients between the substrate and film. Electroceramic thin films are no different in that respect and are subjected to residual stresses, either tensile or compressive in nature (depending upon the substrate and process conditions), on the order of hundreds of MPa [2]. Previous experiments with bare PZT films (i.e. no top electrode) have shown residual stresses to be tensile with magnitudes on the order of 100 MPa [3]. That fact becomes important when one designs micromechanical systems, for if the active material is subjected to an in-plane stress, the effective piezoelectric coefficients of the film may change relative to the unstressed values. In particular, the mechanical stress applied to the material may act to restrict domain wall motion, thus significantly reducing the extrinsic component of the piezoelectric response and making the material less desirable from an application point of view [4].

---

* jfs12@email.psu.edu

Mat. Res. Soc. Symp. Proc. Vol. 459 © 1997 Materials Research Society

EXPERIMENTAL

Sol-gel PZT

PZT solutions were synthesized with compositions near the morphotropic phase boundary (52/48) using a variation of the procedure described earlier by Budd, Dey, and Payne [5]. Completed solutions were diluted to 0.5M concentrations, then spin coated onto 3" platinized Si wafers pre-annealed prior to deposition between 650 and 750°C for 60 sec. A thin coating of 0.5M lead-oxide (PbO) was applied after PZT deposition to aid in the minimization of lead volatilization during the subsequent crystallization step [6]. The amorphous films were rapid thermal annealed at 750°C for 60 sec. Final film thickness was on the order of ~2500-3000 Å. Pt top electrodes were sputtered through a 1.5 mm diameter shadow mask and post annealed between 650 and 750°C for 60 sec.

Uniform Pressure Rig

The biaxial stress rig used in this study was designed to yield a tensile or compressive stress over the surface of a 3" substrate and is in effect a combination of two earlier designs [7,8]. The base of the stress rig was constructed from an aluminum round bored out to a 2" internal diameter [3]. The open end of the chuck is sealed with a coated substrate clamped between two 2.5" Viton O-rings. Evacuation (with a 1 micron mechanical pump) or pressurization (with a common gas bottle/regulator combination) of the cavity behind the wafer then flexes the sample and places the PZT on the surface of the wafer in a controlled state of biaxial tension or compression. Applied stress was calculated at the center of the wafer using large deflection plate theory [9] and assuming a "fixed but not held" boundary condition.

Electrical Characterization

Polarization measurements were made using a Radiant Technologies RT66A ferroelectrics tester. Voltages for the cases reported herein were ±10 V, producing a maximum electric field on the order of 360 kV/cm (film thickness was assumed to be ~0.28 μm). Capacitance and tan δ studies were conducted using a Hewlett Packard 4275A Multi-Frequency LCR Meter at applied frequencies of 10 kHz and an electric field of 0.36 kV/cm. Polarization measurements were made at tensile pressures ranging from 5 to 15 psig. (producing stresses of ~52 and 142 MPa, respectively) and compressive pressures up to 14.6 psi. (-139 MPa). Capacitance and tan δ were measured at applied pressures from -14.6 psi. to 15 psig. (-139 to 142 MPa).

RESULTS

Polarization as a Function of Applied Stress

Changes in the polarization-electric field hysteresis loops were measured as a function of applied biaxial stress for test capacitors subjected to applied pressures between -14.6 psi. and +15 psig. (-139 and 142 MPa). Results of the experiments conducted showed that the remanent polarization decreased with applied tension and increased with applied compression. Figure 1 shows the changes in polarization for one sample subjected to maximum applied fields of ±360 kV/cm. The magnitudes of the changes in remanent polarization are on the order of 12% (+11.6% compression, -12.5% tension). Upon release of the gas pressure from the chuck hysteresis loops returned to their unstressed shapes, illustrating the reversible nature of the process.

Capacitance as a Function of Applied Stress

The capacitance of PZT test capacitors was monitored as a function of applied biaxial stress at applied fields of 0.36 kV/cm at frequencies of 10 kHz. Experiments were conducted in both the poled and unpoled conditions and revealed that the material response is comparable for the two cases. It was determined that the dielectric constant decreases between 2 and 5%, for poled and unpoled material respectively, for an applied tensile stress of approximately 142 MPa. Conversely, the capacitance of the material increased between 0.5 and 3% (for the poled and unpoled condition) with an applied compressive stress on the order of -139 MPa. The effects of both tensile and compressive stresses are illustrated in Figures 2 and 3. Starting values of capacitance in the unpoled condition were 56.9 nF, which increased to 62.3 nF after poling with a field of 320 kV/cm for 10 minutes. Data in the plots were normalized to the initial (i.e. prestressed) starting values of capacitance to more clearly illustrate the magnitude of the observed change.

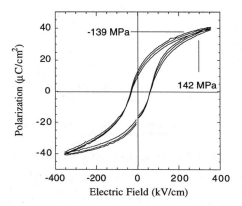

Figure 1. Hysteresis loops as measured under applied tension and compression (positive sign indicates tension, negative indicates compression).

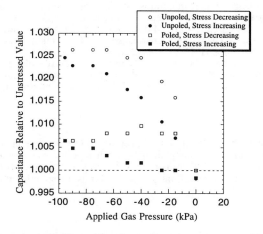

Figure 2. The change in capacitance of unpoled and poled samples with applied compressive stress over the range from 0 to - 139 MPa.

49

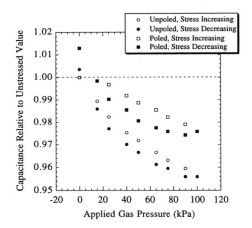

Figure 3. The change in capacitance of a poled and unpoled sample as applied stress was increased (becomes more tensile) from 0 to 142 MPa.

Dielectric Loss and Coercive Field Strength as a function of Applied Stress

The dielectric loss, tan δ, and coercive field strengths of the PZT films tested were found to be insensitive to applied mechanical loads in either the compressive or tensile states. Results indicate that tan δ remained constant at ~3%. Coercive field strengths were found to be on the order of +60 and -36 kV/cm (for hysteresis measured with ±10 V) as illustrated in Figure 1.

DISCUSSION

Comparison of the thin film stress response with earlier observations of bulk PZT materials indicates an obvious disagreement. In contrast to the reversible *increase* of dielectric constant shown here (Fig. 2), earlier studies have reported an irreversible *decrease* of capacitance over a similar compressive stress range [10,11]. Maximum changes observed were on the order of 50% (compared to the 3% observed here), with a permanent ~30% reduction of dielectric constant found to result for soft compositions (PZT-5). Reasons for the large changes were attributed to 90° domain reorientation caused by the application of a compressive stress normal to the poling direction. The obvious contradiction of thin film and bulk material response suggests that an alternative mechanism (i.e. alternative to non-180° twin wall motion) is responsible for the recorded behavior.

Grain Size Effects

Effects similar to those of Fig. 2 have been reported elsewhere [12] for fine grained (~1 μm) barium titanate ceramics. The anomalous behavior reported was thought to result from the reduced number (or mobility) of 90° domain walls in fine-grained materials and suggests that one explanation for the empirical observations presented here is the submicron grain structure of the PZT film.

Phase Transformations

An alternative explanation is that applied stress might result in a shift in either the tetragonal to rhombohedral phase content or a shift in the morphotropic phase boundary (MPB). The dielectric constants of the rhombohedral and tetragonal phases are different in both the poled and unpoled conditions [13] so that any change in the relative amount would be reflected in the film's capacitance. Alternatively, since the dielectric constant of PZT peaks at the MPB, any change in its position would lead to appreciable changes in the capacitance.

CONCLUSIONS

The reported disagreement with earlier studies on bulk materials suggests domain wall motion to be negligible in the stress response of PZT thin films. That realization has led to the proposal of grain size effects and tetragonal-rhombohedral phase transformations as alternative mechanisms for the observed behavior. Data collected thus far are insufficient to provide definitive support for either mechanism and more extensive work is therefore required before the extent to which each mechanism operates might be quantified.

ACKNOWLEDGMENTS

The authors would like to acknowledge Tao Su for his assistance in depositing the PZT films used in this study. This work is funded by DARPA under contract DABT63-95-C-0053.

REFERENCES

1. J. A. Thorton and D. W. Hoffman, Thin Solid Films **171**, p. 5-31 (1989).

2. B. A. Tuttle, J. A. Voigt, T. J. Garino, D. C. Goodnow, R. W. Schwartz, D. L. Lamppa, T. J. Headley, M. O. Eatough, Proceedings of 8th International Symposium on Applications of Ferroelectrics, 1992.

3. J. Shepard, S. Trolier-McKinstry, M. Hendrickson, R. Zeto, to be published in, Proceedings of 10th International Symposium on Applications of Ferroelectrics, New Brunswick, New Jersey, 1996.

4. M. Demartin and D. Damjanovic, Appl. Phys. Lett. **68**, p. 3046-3048 (1996).

5. K. D. Budd, S. K. Dey, D. A. Payne, Electrical Ceramics, British Ceramic Proceedings **36**, p. 107-121 (1985).

6. T. Tani and D. A. Payne, J. Am. Ceram. Soc. **77**, p. 1242-1248 (1994).

7. R. J. Jaccodine and W. A. Schlegel, J. Appl. Phys. **37**, p. 2429-2434 (1966).

8. T. J. Garino and M. Harrington, Ferroelectric Thin Films II, Mat. Res. Soc. Symp. Proc. **243**, A. I. Kingon, E. R. Meyers, and B. Tuttle, Eds. Pittsburgh: Materials Research Society, p. 341-347 (1992).

9. R. J. Roark, Formulas for Stress and Strain, McGraw-Hill, 1989.

10. R. F. Brown, Can. J. Phys. **39**, p. 741-753 (1961).

11. R. F. Brown and G. W. McMahon, Can. J. Phys. **40**, p. 672-674 (1962).

12. W. R. Buessem, L. E. Cross, A. K. Goswami, J. Am. Ceram. Soc. **49,** p. 36-39 (1966).

13. B. Jaffe, W. R. Cook, H. Jaffe, Piezoelectric Ceramics, R.A.N. Publishers, 1971.

DIELECTRIC PROPERTIES OF PIEZOELECTRIC POLYIMIDES

Z. OUNAIES[1], J. A. YOUNG[2], J. O. SIMPSON[3], B. L. FARMER[2]

[1]National Research Council, NASA Langley Research Center, Hampton, VA 23681
[2] University of Virginia, Department of Materials Science and Engineering, Charlottesville, VA 22903
[3]Composites and Polymers Branch, NASA Langley Research Center, Hampton, VA 23681

ABSTRACT

Molecular modeling and dielectric measurements are being used to identify mechanisms governing piezoelectric behavior in polyimides such as dipole orientation during poling, as well as degree of piezoelectricity achievable. Molecular modeling on polyimides containing pendant, polar nitrile (CN) groups has been completed to determine their remanent polarization. Experimental investigation of their dielectric properties evaluated as a function of temperature and frequency has substantiated numerical predictions. With this information in hand, we are then able to suggest changes in the molecular structures, which will then improve upon the piezoelectric response.

INTRODUCTION

This investigation is motivated by NASA's interest in developing high performance piezoelectric polymers for a variety of high temperature aerospace applications. Reported herein are numerical calculations and experimental results which are used to characterize and understand the poling of a piezoelectric, nitrile-substituted aromatic polyimide [1, 2]. Molecular modeling provides fundamental understanding of the polyimide's response to temperature and electric field. The experimental studies are used to evaluate the accuracy of the model.

Molecular structures of the polyimides investigated are given in Figure 1 below.

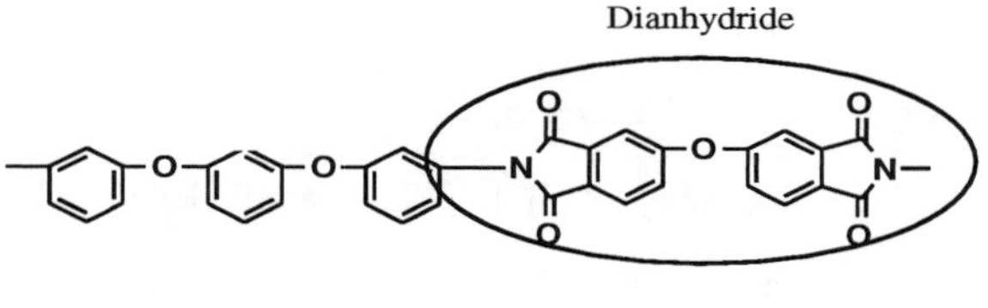

APB/ODPA
$T_g = 185°C$

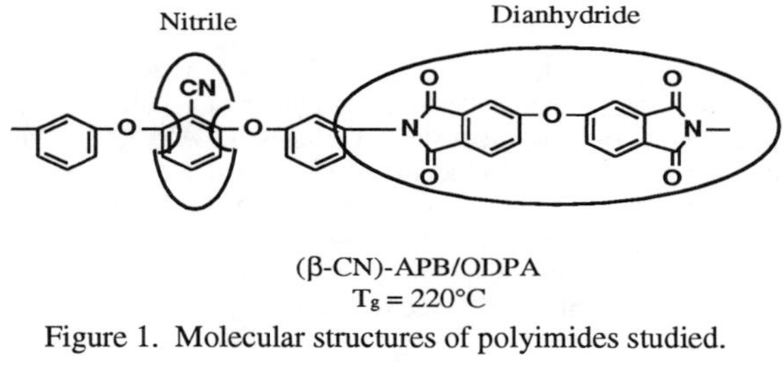

(β-CN)-APB/ODPA
$T_g = 220°C$

Figure 1. Molecular structures of polyimides studied.

53

METHODS

In order to induce a piezoelectric response in the polyimide systems shown above, they are poled by applying a strong electric field (E_p) at an elevated temperature $(T_p \geq T_g)$ which produces orientation of the molecular dipoles. Partial retention of this orientation is achieved by lowering the temperature below T_g in the presence of E_p. This is known as orientation polarization . In order to maximize piezoelectricity, the remanent polarization (P_r) must be maximized. P_r is related to the poling field by

$$P_r = \varepsilon_o \, \Delta\varepsilon \, E_p \qquad (1)$$

where ε_o is the permittivity of space, and $\Delta\varepsilon$ is the change in dielectric constant upon traversing the glass transition. As pointed out by Furukawa [3], $\Delta\varepsilon$ is the parameter of greatest interest in designing amorphous polymers with large piezoelectric activity. Due to dielectric breakdown of the polymeric materials, E_p is limited to approximately 100 MV/m. Hence, molecular design must be implemented to increase the $\Delta\varepsilon$ and consequently P_r.

Computational

Molecular modeling of high temperature polyimides is done within the BIOSYM molecular modeling package. Initial quantum mechanical calculations using MOPAC have been done on segments of the polyimide to assign force field parameters and partial atomic charges for the polymer. The computational model used in the molecular dynamic simulations is a chain of five repeat units within a three-dimensionally periodic cell. The cell is 'plated' with dummy atoms upon which partial charges are placed to emulate the applied electric field in the z-direction [4].

Molecular dynamics is employed to solve the classical equations of motion for all of the atoms comprising the polymer. A dynamic picture of the system is obtained as a function of temperature, applied electric field and time. Experimental polymer relaxation times tend to be quite long in the vicinity of T_g (seconds or minutes). By using an artificially high temperature in the simulation, however, the polarization is seen to reach an steady-state value in 50 ps. An analysis of the relaxation behavior will be reported elsewhere [4]. The time/temperature relationship is of the familiar Arrhenius form presented in detail by [5],

$$\tau = [\, 3\varepsilon_s \,/\, (2\,\varepsilon_s + \varepsilon_\infty)\,]\; [h\,/\,k\,T]\; [\exp(\Delta F\,/\,kT\,)\,] \qquad (2)$$

The magnitude of the electric field is also increased in order to maintain the population density of dipoles (i.e. m E / kT). Molecular dynamics is performed at 1700 °C for 50 ps in the absence of a poling field in order to obtain starting conformations. A poling field of 700 MV/m is then applied for 200 ps. The polarization is then frozen in by running molecular dynamics for 200 ps at 25 °C. This yields to the steady-state value of the remanent polarization.

Experimental

Experimental studies are made on films synthesized at NASA Langley Research Center. Silver electrodes are evaporated on both sides of the polymer. The dielectric constant and loss as a function of temperature and frequency are evaluated on unpoled APB/ODPA and (β-CN)-APB/ODPA samples using a Hewlett Packard 4192A Impedance Analyzer. The measurements are done at temperatures ranging from ambient to 300°C at 5 Hz, 10 Hz, 100 Hz and 1 kHz.

Poling of the polyimide samples is done in an oil bath at $T_p = T_g + 4°C$ at $E_p = 50, 80,$ and 100 MV/m for $t_p = 30$ min. To evaluate the P_r in the poled samples, a thermally stimulated current (TSC) analysis is used[6]. A Keithley 6517 electrometer is connected to the sample to record the short-circuit current while the sample is heated at a constant rate of 1°C/min from room temperature to T_g+20°C. The current is created as the material depolarizes with increasing temperature.

RESULTS

Using the molecular modeling outlined previously, three cells of (β-CN)-APB/ODPA are independently built and poled at 150 MV/m. The polarization of each cell is then calculated using the atomic charges and positions during the molecular dynamics run. As seen in Figure 2 the final polarization in the direction of the applied field, the z direction, is independent of the initial starting conformation. Δε is obtained using equation 1. The average value of Δε is 17.8 ± 1.1 for (β-CN)-APB/ODPA.

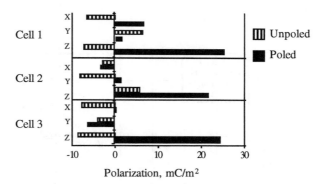

Figure 2. Polarization along principle axes for the three starting configurations.

The contributions of specific dipoles along the polymer chain (Figure 3) to the total polarization of the cell are examined. The ability of nitrile substituents to create large polarizations in polymeric systems has been previously demonstrated [7,8]. On average, the nitrile substituent constitutes only 48 % of the total polarization. The dianhydride also provides significant polarization, constituting about 39 % of the total.

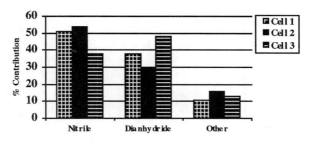

Figure 3. Segmental contribution to polarization.

The unsubstituted APB/ODPA is then modeled. Assuming that the polar segments of the polymer contribute to the polarization independently, APB/ODPA is expected to have a polarization of 48% less than that of (β-CN)-APB/ODPA, i.e. Δε = 9.2. Molecular simulations of APB/ODPA actually yield Δε = 7.7 ± 1.6, in excellent agreement with the expected value. Examination of the molecular dynamics trajectories shows that the majority of the polarization, i.e. 81%, is now due to the dianhydride portion of the polymer.

Figure 4 shows the dielectric constant and loss as a function of frequency and temperature for the (β-CN)-APB/ODPA polymer. In analyzing the data, care must be taken to separate the electrical conduction effects from dipolar reorientation. Inspection of Figure 4 reveal that the dielectric properties are reaching a plateau at temperatures around 300°C. It was observed that conductivity effects and cross-linking of the substituents occur at temperatures higher than 300°C as evidenced by the increase in the dielectric loss. As the frequency decreases, the onset of the dielectric relaxation occurs at a lower temperature. The value of Δε also increases with decreasing frequency because the molecular dipoles are more able to follow the low frequency signal. This indicates that the lowest measurement frequency (i.e. 5 Hz) most closely approximates the dc poling conditions. As shown in Table I, the Δε at 5 Hz is in excellent agreement with computational predictions.

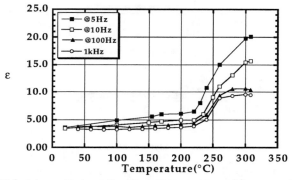

Figure 4a. Dielectric constant as a function of temperature and frequency for the (β-CN)-APB/ODPA polymer.

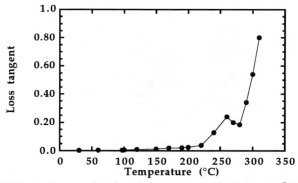

Figure 4b. Dielectric loss as a function of temperature at 1 kHz for the (β-CN)-APB/ODPA polymer.

Table I. Comparison between computed and measured $\Delta\varepsilon$

	$\Delta\varepsilon_{10Hz}$	$\Delta\varepsilon_{5Hz}$	$\Delta\varepsilon_{comp}$
(β-CN)-APB/ODPA	11.5 ± 0.6	16 ± 1.3	17.8 ± 1.1
APB/ODPA	4.2 ± 0.5	5.7 ± 1.5	7.7 ± 1.6

An example of the spectrum obtained from the TSC analysis is shown in Figure 5. It is described by a single current peak that correlates well with the glass transition temperature of the material as measured by DSC. By integrating the current/temperature curve as a function of time, the remanent polarization is obtained. Figure 6 shows the value of P_r as a function of applied field. As expected, there is a linear correlation between the poling field and the resultant P_r. Once the poling field is removed, a certain amount of dipole relaxation is expected, and this is seen by the difference between P_r computed from equation 1 and P_r obtained from TSC.

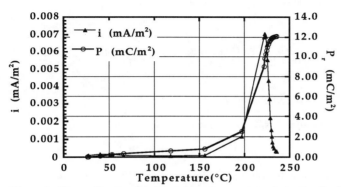

Figure 5. Thermally stimulated current and remanent polarization for (β-CN)-APB/ODPA. E_p = 100 MV/m, T_p = 223°C, t_p = 30 min

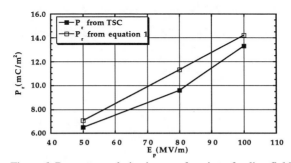

Figure 6. Remanent polarization as a function of poling field.

CONCLUSIONS

In this analysis, a high T_g, amorphous polymer is studied using both modeling and experimental methods. The goal of both methods is to evaluate and optimize the remanent polarization P_r and hence the piezoelectric behavior. The dipole moment of the nitrile substituent as well as the dianhydride contribute significantly to the polarization and piezoelectric response of the polyimide. Both the dielectric measurements and the TSC technique yield similar P_r.

The computational and experimental results for the APB/ODPA systems are in good agreement. Through the computational modeling, we have succesfully developed a tool to guide the development of polymeric materials with improved piezoelectric responses. Such tools could reduce the time and cost of material production.

ACKNOWLEDGEMENTS

The authors gratefully acknowledge the technical insight of Drs. Jeffrey Hinkley and Terry St. Clair of NASA Langley Research Center. Also acknowledged is Dr. Catherine Fay of the National Research Council for the synthesis of polyimide films.

REFERENCES

1. J. O. Simpson. S. S. Welch and T. L. St. Clair, in Electrical, Optical and Magnetic Properties of Organic Solid State Materials III, edited by A. K-Y. Jen, C. Y-C. Lee, L. R. Dalton, M. F. Rubner and G. E. Wnek (Mater. Res. Soc. Proc. 143, Boston, MA 1995), p.351-356.

2. J. A. Hinkley, High Performance Polymers 8, (1996).

3. T. Furukawa, IEEE Trans. Elec. Ins. 24, 375 (1989).

4. J. A. Young, (to be published).

5. B. Hilezer and J. Mlecki, Electrets, Polish Scientific Publishers, Warszawa, 1986.

6. J. P. Ibar, P. Denning, T. Thomas, A. Bernes, C. de Goys, J. R. Saffell, P. Jones and C. Lacabanne, in Polymer Characterization: Physical Property, Spectroscopy, and Chromotographic methods, edited by C. D. Craver and T. Provder (Adv. Chem. Series 227, 1990), p. 167-190.

7. S. Tasaka, T. Toyama and N. Inagaki, Jpn. J. Appl. Phys. 33, 5838 (1994).

8. S. Tasaka, N. Inagaki, T. Okutani and S. Miyata, Polymer 30, 1639 (1989).

POLARIZATION AND PIEZOELECTRIC PROPERTIES
OF A NITRILE SUBSTITUTED POLYIMIDE

Joycelyn Simpson[1], Zoubeida Ounaies[2], and Catharine Fay[2]
[1]Composites and Polymers Branch, NASA Langley Research Center, Hampton, VA 23681
[2]National Research Council, NASA Langley Research Center, Hampton, VA 23681

ABSTRACT

This research focuses on the synthesis and characterization of a piezoelectric (β-CN)-APB/ODPA polyimide. The remanent polarization and piezoelectric d_{31} and g_{33} coefficients are reported to assess the effect of synthesis variations. Each of the materials exhibits a level of piezoelectricity which increases with temperature. The remanent polarization is retained at temperatures close to the glass transition temperature of the polyimide.

INTRODUCTION

Fluoropolymers and copolymers such as polyvinylidene fluoride (PVDF) currently represent the state of the art in piezoelectric polymers. The development of other classes of piezoelectric polymers such as polyimides could provide enabling materials technology for a variety of aerospace and commercial applications. Due to their exceptional thermal, mechanical, and dielectric properties, polyimides are already widely utilized as matrix materials in aircraft and as dielectric materials in microelectronic devices. Particularly interesting is the potential use of piezoelectric polyimides in micro electro mechanical systems (MEMS) devices since fluoropolymers do not possess the chemical resistance or thermal stability necessary to withstand conventional MEMS processing.

The majority of literature on poled polymers presents semicrystalline systems such as fluoropolymers and nylons [1-3]. In this analysis a high glass transition temperature (T_g), amorphous polymer is studied. The piezoelectric phenomena in amorphous polymers differs significantly from semicrystalline polymers and inorganic crystals. In order to induce a piezoelectric response in amorphous systems the polymer is poled by the application of a strong electric field (E_p) at an elevated temperature (T_p) sufficient to allow mobility of the molecular dipoles in the polymer. The dipoles are polarized with the applied field and will partially retain this polarization when the temperature is lowered below T_g in the presence of E_p. The resulting remanent polarization (P_r) is key in developing materials with a useful level of piezoelectricity as it is directly proportional to the material's piezoelectric response.

In a previous report [4], computational predictions and experimental measurements of P_r are reported for a nitrile substituted polyimide. In this investigation the piezoelectric coefficients, g_{33} and d_{31}, are measured.

METHODS
Synthesis

Polyimide films are synthesized using two different diamines, (2,6-bis(3-aminophenoxy) benzonitrile ((β-CN)-APB) and 1,3-aminophenoxy benzene (APB), to determine the effect of the pendant nitrile group on the piezoelectric response. Each of the diamines is reacted with 4,4' oxidiphthalic anhydride (ODPA) resulting in the two polyimides shown in Figure 1. The influence of cure cycle and thermal versus chemical imidization on the piezoelectric properties is investigated for the nitrile substituted polyimide. Two different thermal imidization cycles are used: 1) *standard*-1 hour each at 100, 200, and 300°C and 2) *relative*- 1 hour each at 50, 150, 200, and T_g + 20°C. The chemically imidized films [5] are cast from a solution of polyimide powder in NMP solvent and the solvent is removed using the relative thermal cycle. As shown in Table 1, the chemically imidized polyimide has a slightly lower T_g and thermal stability than the thermally imidized films.

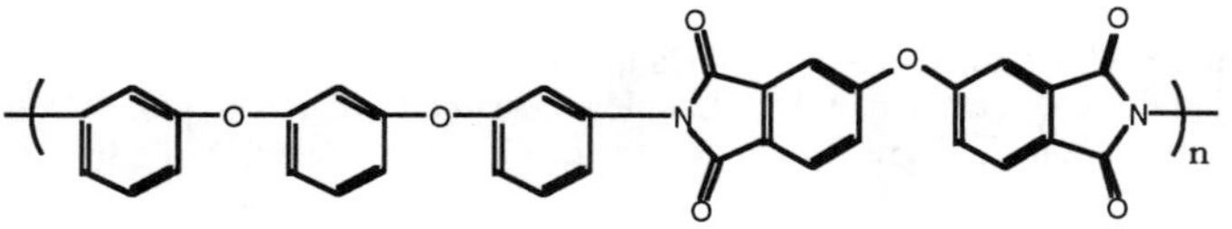

APB/ODPA

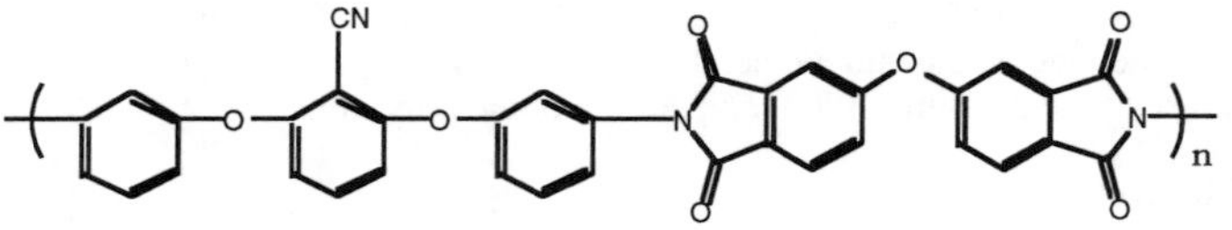

(β-CN)-APB/ODPA

Figure 1. Chemical structures of polyimides studied.

Table 1. Thermal properties of polyimide films.

(β-CN)-APB/ODPA	$T_g (°C)^1$	TGA, $(°C)^2$
Standard thermal cure	225	480
Relative thermal cure	222	484
Chemical imidization	199	459
APB/ODPA	185	431

^1T_g measured by DSC at 20°C/min.
^{2}Temperature of 5% weight loss by TGA in air at 2.5°C/min.

Poling

A silver electrode is evaporated onto each side of the film for electrical contact. For this study all polymers are poled at the polymer's T_g at a field strength of 80 MV/m for 30 minutes unless otherwise indicated. The 80 MV/m field strength is used because it yields films which are best suited for piezoelectric measurements.

Characterization

Three techniques are used to characterize the dielectric and piezoelectric properties of the polymers. To evaluate the P_r in the poled samples, a thermally stimulated current (TSC) method is used. A Keithley 6517 electrometer is used to measure the discharged current as temperature is ramped at 1°C/min from room temperature to T_g +20°C.

The d_{31} is measured by mounting the polymer sample on a load frame and applying a stress parallel to the electrodes. The charge (Q) is measured using an electrometer and d_{31} is given by:

$$d_{31} = Q/AX_1 \tag{1}$$

where A is the area of the electrodes and X_1 is the applied stress. A linear response is obtained for each of the polymers in the range from 2 to 10 newtons of force.

The g_{33} coefficient is given by:

$$g_{33} = V_o/ X_3 t , \tag{2}$$

where V_o is the open circuit voltage, X_3 is the applied stress normal to the surface, and t is the film thickness. A ceramic piezoelectric actuator applies a 10 Hz force to the polymer film sample normal to the surface. A calibrated piezoelectric load cell located beneath the sample measures the force applied to the sample. An electrometer is used to measure the voltage developed across the sample in response to the applied force.

RESULTS

A summary of the remanent polarization and piezoelectric properties for the polyimides studied is presented in Table 2. Each of the materials demonstrates a piezoelectric response although lower than previously reported [6]. The P_r values and room temperature piezoelectric coefficients are low relative to other known piezoelectric polymers.

Table 2. Polarization and piezoelectric properties of polyimide films.

$(\beta\text{-CN})\text{-APB/ODPA}$	$P_r (mC/m^2)$	$-d_{31} (pC/N)^1$	$-g_{33} (mVm/N)^1$
Standard thermal cure	9.6	0.12	16
Relative thermal cure	10.8	0.15	20
Chemical imidization	11.5	0.14	14
APB/ODPA	3.7	0.01	6

[1]Measured at 25°C.

The influence of cure temperature on the piezoelectric properties is investigated for the nitrile substituted polyimide. Previous work [7] indicates that optimum mechanical film properties are obtained by limiting the ultimate cure temperature to $T_g + 20°C$. Based on the measurements completed in this study, there is no significant difference in the piezoelectric response for the different (standard or relative) cure cycles. Likewise, the small differences in P_r, d_{31}, and g_{33} values for the chemically and thermally imidized polyimides are within the experimental uncertainty of each of the methods.

To determine the effect of the pendant nitrile group on the piezoelectric response the unsubstituted APB/ODPA was evaluated. From Table 2 it is seen that APB/ODPA exhibits an extremely low but measurable remanent polarization and piezoelectric response. Approximately 40% of the P_r in the nitrile substituted polyimide is attributed to the polarization of the dianhydride portion of the polyimide [4]. This indicates that incorporating more polar groups in the dianhydride may yield polyimides with improved remanent polarizations.

An objective of this work is to develop polyimides with useful, stable piezoelectric properties at elevated temperatures. Figure 2 is a plot of the calculated g_{33} coefficients for each of the polyimides as a function of temperature. The g_{33} values increase with temperature in the range from 25 to approximately 95°C.

In Table 3 the piezoelectric properties at 95°C for a chemically imidized polyimide poled at an E_p of 100 MV/m is compared to PVDF. At this temperature the g_{33} and d_{33} values of the polyimide are three times greater than those at room temperature but are still an order of magnitude lower than PVDF. For most polymers significant enhancement in piezoelectric activity is achieved by altering the morphology of the polymer through uniaxial stretching prior to poling [8,9]. Hence, with improvements in polymer design and processing it is conceivable that piezoelectric activity in polyimides will approach a range feasible for practical use in high temperature applications.

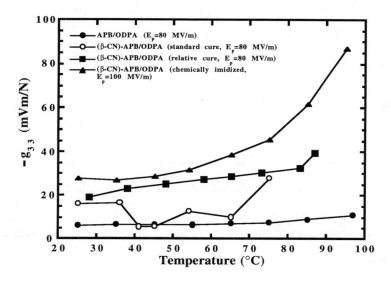

Figure 2. Temperature dependence of the piezoelectric coefficient.

Table 3. Comparison of piezoelectric properties of chemically imidized polyimide at 95°C and PVDF.

Polymer	$-g_{33}$ (mVm/N)	$-d_{33}$ (pC/N)
(β-CN)-APB/ODPA[1]	87	2.7
PVDF[2]	339	33

[1] g_{33} measured and d_{33} calculated by $d_{ij} = \varepsilon\, \varepsilon_0\, g_{ij}$ [9].
[2] Piezoelectric g_{33} and d_{33} coefficients are from [10].

To illustrate the thermal stability of the polyimides, Figure 3 shows the percentage of remanent polarization which is lost in (β-CN)-APB/ODPA as it is heated from 30°C to 240°C. At 150°C the polymer retains 94% of its remanent polarization. Over 80% of the P_r is retained at 200°C. This thermal stability indicates that polyimides can potentially be used for high temperature piezoelectric applications.

CONCLUSIONS

Polarization and piezoelectric properties have been measured for APB/ODPA and its nitrile substituted derivative. The effect of polyimide synthesis method was investigated and no significant difference in piezoelectric response was observed. The level of piezoelectricity in the (β-CN)-APB/ODPA is an order of magnitude lower than what is necessary for practical utility in devices. The high temperature stability of the remanent polarization has been verified. Future efforts will concentrate on increasing the piezoelectric response through processing such as mechanical stretching and by incorporating the use of computational tools to guide synthesis of other polyimides.

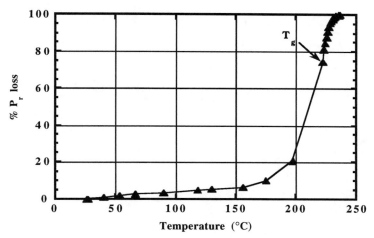

Figure 3. Percentage P_r loss as a function of temperature for (β-CN)-APB/ODPA.

ACKNOWLEDGEMENTS

The authors gratefully acknowledge the technical support of Dr. Terry St. Clair of NASA Langley Research Center. Also acknowledged is the support of Mr. James Clemmons of Vigyan, Inc. for innovative electronic design and measurement support; and Mitsui Toatsu for preparation of the (β-CN)-APB diamine.

REFERENCES

1. G. Thomas Davis, in Polymers for Electronic and Photonic Applications, edited by C.P. Wong (Academic Press, Inc., Boston, 1993) pp. 435-465.

2. T. Furukawa, IEEE Trans. Elect. Insul., 21, 543 (1986).

3. B. Newman, P. Chen, K. Pae, and J. Scheinbeim, J. Appl. Phys., 51, 5161 (1980).

4. Z. Ounaies, J. Young, J. Simpson and B. Farmer, Materials Research Society Symposium Proceedings, Boston, MA, (1996).

5. A.K. St. Clair and T.L. St. Clair, US Patent No. 5,338,826 (16 August 1994).

6. J.O. Simpson, S.S. Welch, and T.L. St. Clair, Materials Research Society Symposium Proceedings, Ed. Jen, A. K-Y., 413, 351 (1996).

7. J. F. Dezern, and C.I. Croall, NASA Technical Memorandum 104178 (1991).

8. H. Ueda and S.H. Carr, <u>Polymer J.</u>, 16, 661 (1984).

9. G. Thomas Davis, in <u>The Applications of Ferroelectric Polymers</u>, edited by T. Wang, J. Herbert, and A. Glass (Blackie, London, 1988) pp. 37-65.

10. Piezo Film Sensors Technical Manual, AMP, Inc., (1993).

INNOVATIVE PROCESS DEVELOPMENT OF $Sr_2(Nb_{0.5}Ta_{0.5})O_7$/PVDF HYBRID MATERIALS FOR SENSORS AND ACTUATORS

W. Kowbel, V. Chellappa, and J.C. Withers, MER Corp., Tucson, AZ and M.J. Crocker, Mechanical Engineering, Auburn University, Auburn, AL and B. Wada, JPL, Pasadena, CA

ABSTRACT

Over the past several years interest in adaptive 'smart'materials development has gained momentum. Smart structures utilize both polymeric sensors and lead-based piezo-ceramic actuators. This paper addresses the development of an alternative smart material which consists of a novel lead-free piezo-ceramic/PVDF hybrid composite, to be used as a single component system capable of performing multiple tasks. A lead-free controlled porosity perovskite ceramic of the type $A_2B_2O_7$ ($Sr_2(Nb_{0.5}Ta_{0.5})_2O_7$ was developed utilizing hot forging.
Long, oriented grains along the x-y plane, perpendicular to the forging direction were obtained in the ceramics. PVDF was subsequently infiltrated into the porous piezo-ceramic resulting in a three dimensional architecture in which the piezo-ceramic is oriented perpendicular to the PVDF. It is anticipated that manufacturability combined with the ease of functional tailorability of such a class of lead-free hybrid materials can be useful in a variety of smart structures applications.

INTRODUCTION

Over the past several years, piezoelectric ceramic/polymer composites have been extensively investigated in order to get improved piezo- and pyroelectric properties [1]. Several methods are employed to make these composites including injection molding [2], dice and fill [3] and reticulation [4].
These materials are designed so that the high piezoelectric coefficients and low dielectric constants of bulk ceramics are combined with low density and good flexibility of the polymers. This paper reports the fabrication of novel $Sr_2(Nb_{0.5}Ta_{0.5})_2O_7$/PVDF composites. The hot-forged $Sr_2(Nb_{0.5}Ta_{0.5})_2O_7$ (SNTO) ceramic is lead-free and was previously reported to exhibit high Curie temperature which can be adjusted by controlling the content of Ta [5].

EXPERIMENTAL

Figure 1 shows the fabrication flow chart for $Sr_2(Nb_{0.5}Ta_{0.5})_2O_7$/PVDF composites. Initially, appropriate amounts of $SrCO_3$, Nb_2O_5 and Ta_2O_5 powders were mixed with a 3% binder and an appropriate amount of a polystyrene pore-former. Subsequently, the ceramic powders were cold pressed at 1.4 MPa into rectangular composites. The green bodies were sintered at 1250°C and then hot-forged at 1400°C. The porous ceramic material was subsequently infiltrated with the PVDF solution and poled using the conventional DC poling method at 20 kV/cm. Density of the resulting composites was measured using the Archimedes principle. SEM was used to analyze the microstructure, while XRD was used to perform the phase identification and
determine the degree of the preferred orientation by the calculation of the orientation factor. The frequency spectra was obtained using an impedance analyzer and the electromechanical properties were determined by the resonance method.

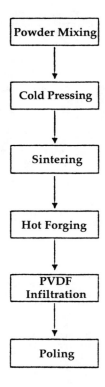

Figure 1.   Schematic of ceramic/polymer composite processing.

RESULTS AND DISCUSSION

Figure 2 shows an SEM micrograph of the as-sintered $Sr_2(Nb_{.5}Ta_{.5})_2O_7$ ceramics. Small grains (about 2μm) are observed combined with a significant amount of porosity.

Figure 2.   SEM micrograph of as-sintered SNTO ceramics.

Figures 3a and 3b show the SEM micrographs of the $Sr_2(Nb_{.5}Ta_{.5})_2O_7$ ceramic processed with 10% pore former after hot forging (pependicular to FA direction). No preferred grain orientation is observed in either case.

a)   b)

Figure 3.   SEM micrograph of SNTO with 10% pore former after HF (perpendicular direction): a) low magnification, b) higher magnification.

Figures 4a and 4b show the SEM micrographs of the $Sr_2(Nb_{.5}Ta_{.5})O_7$ ceramics with 10% pore former (parallel to FA direction). Elongated grains (about 20μm long) are seen to be aligned perpendicular to the forging axis (FA). In addition, small pores (about 2μm in diameter) are also observed.

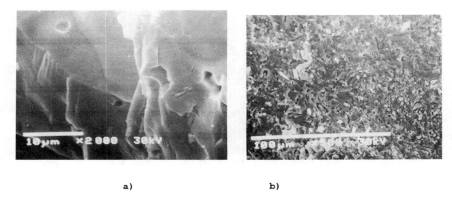

a)   b)

Figure 4. SEM micrograph of SNTO with 10% pore former after HF (parallel direction): a) low magnification, b) higher magnification.

Figures 5a and 5b show the SEM micrographs of the $Sr_2(Nb_{.5}Ta_{.5})_2O_7$ ceramics fabricated with 20% pore former after hot forging (perpendicular to FA). Small gains (about 1-2 μm) are observed combined with a significant porosity.

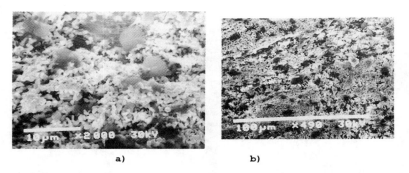

a)   b)

Figure 5. SEM micrograph of SNTO with 20% pore former after HF (perpendicular direction): a) low magnification, b) higher magnification.

Figures 6a and 6b show the SEM micrographs of the $Sr_2(Nb_{.5}Ta_{.5})_2O_7$ ceramics fabricated with 20% pore former after hot forging (parallel to FA). Slightly elongated, smaller grains (about 10 μm) as compared with 10% pore formers, are observed. In addition, significant porosity is observed. The degree of alignment is not nearly as good as in the case of 10% pore former. Thus, increased porosity appears to lower the amount of the preferred orentation.

68

a)   b)

Figure 6. SEM micrograph of SNTO ceramics with 20% pore former after HF (parallel direction): a) low magnification, b) higher magnification.

The porosity present in the ceramics made with 20% pore former shows some degree of preferred orientation. Thus, with subsequent polymer infiltration there is the potential to produce 3:3 or 3:1 connectivity.

Figures 7a and 7b show the X-ray spectra corresponding to the hot forged $Sr_2(Nb_{.5}Ta_{.5})_2O_7$ ceramics in the directions parallel and perpendicular to FA, respectively. The stoichiometric $Sr_2(Nb_{.5}Ta_{.5})_2O_7$ structure was identified. In addition, strongly preferred orientation in the parallel direction is observed.
Thus, the hot forging results in a high degree of grain orentation.

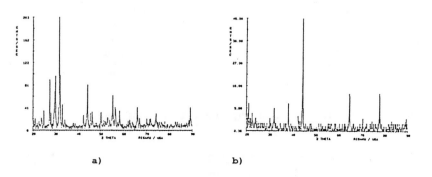

a)   b)

Figure 7. X-ray diffraction spectra of SNTO: a) parallel direction, b) perpendicular direction.

Figures 8a and 8b show SEM micrographs of the PVDF infiltrated $Sr_2(Nb_{.5}Ta_{.5})_2O_7$ ceramic fabricated with 20% pore former. Good pore infiltration is observed.
In addition, the ceramics/polymer composite exhibited improved machinability and handlibility critical to commercial applications. However, it is believed

that better handlibility will be achieved at a higher polymer volume fraction.

a)   b)

Figure 8. SEM of ceramic/polymer composite: a) parallel direction, b) perpendicular direction.

Table I summarized the properties of different ceramics/polymer composites. The use of pore former results in a significant density reduction, e.g. 33% open porosity in the case of 20% pore former and a corresponding density of just 4.47 g/cm^3. Thus the use of pore former has been demonstrated as a method of fabricating ceramic/polymer piezo-composite materials.

In addition, following hot forging of porous ceramics, good alignment perpendicular to the FA is observed. With respect to actuator applications, d_{33} is a critical parameter. The measured d_{33} value of 50 x 10^{-12} C/N in the case of a ceramic/polymer composite suggests that poling occurs in both the ceramics and polymer. However, the technical difficulty with the exact determining of the resonance spectrum peak positions possibly introduces an experimental error into the piezoeletric coefficient. Thus, the actual value could be lower. The actual magnitude of d_{33} is expected to be improved via better processing and control of defect chemistry in future work.

Table I. Properties of porous SNTO and ceramic/PVDF composite.

Compound	% Pore Former	% Deformed	% Open Porosity	Density (g/cm^3)	Dielectric Constant $(10^{-12}C/N)$
$Sr_2(Nb_{.5}Ta_{.5})_2O_7$	10	78	20	5.13	---
$Sr_2(Nb_{.5}Ta_{.5})_2O_7$	20	81	33	4.47	---
$Sr_2(Nb_{.5}Ta_{.5})_2O_7$ + PVDF	20	81	---	4.8	50

CONCLUSIONS

1.   Controlled porosity $Sr_2(Nb_{.5}Ta_{.5})_2O_7$ ceramics were fabricated via the addition of a pore former.

2.   Grain elongation can occur in the porous $SR_2(Nb_{.5}TA_{.5})_2O_7$ ceramics in the direction perpendicular to FA.

3.   Infiltration of porous $Sr_2(Nb_{.5}Ta_{.5})_2O_7$ ceramics with PVDf results in a ceramic/polymer piezo-composite.

4.   Simultaneous poling of the ceramic/polymer composite was demonstrated.

5.   The ceramic/polymer composite exhibited lower density and better machinability than pure $Sr_2(Nb_{.5}Ta_{.5})_2O_7$ ceramics.

6.   The $Sr_2(Nb_{.5}Ta_{.5})_2O_7$/PVDF composite exhibits a d_{33} value of 50×10^{-12} C/N.

ACKNOWLEDGEMENT.

This work was supported by NASA contract NAS8-40709

REFERENCES

1.   A Safari, J.Phys.III, 4, 1129 (1994)
2.   R.L. Gentilman, at el, Proc. 6th US-Japan Seminar on Dieletric Ceramics, 134 (1993)
3.   H.P. Savakus, K.A. Klucker and R.E. Newnham, Mat. Res. Bull. 16, 677 (1981)
4.   M.J. Credon and W.A. Schultze, Ferroelectrics, 153, 333 (1994)
5.   P.A. Fuierer, MRS Proc. 276, 51 (1992)

Part II

Other Mechanisms

Parameter Optimization of a Microfabricated Surface Acoustic Wave Sensor for Inert Gas Detection*

S. Ahuja,[†**] A. DiVenere,[‡] C. Ross,[†] H. T. Chien,[†] and A. C. Raptis,[†]

† Energy Technology Division, Argonne National Laboratory, Argonne, IL 60439
** sanjay@anl.gov
‡ Materials Research Center, Northwestern University, Evanston, IL 60208

ABSTRACT

This work is related to designing, fabricating, and testing a surface acoustic wave sensor to be used for detecting metastable inert gases, particularly helium. The assembly consists of two microsensor configurations: (a) a reference device with no deposition at the delay line and (b) a sensing device with an Au-activated TiO_2 e-beam-deposited thin film on the delay line. The interdigitated transducers and delay lines are fabricated by photolithography techniques on a single Y-cut $LiNbO_3$ substrate oriented for Z-propagation of the acoustic waves. Variation in electrical conductivity of the Au-activated TiO_2 film due to exposure to metastable He is translated as a frequency change in the assembly. Various characteristics of the surface acoustic microsensor have been studied to better understand and optimize the variation of acoustic wave velocity and the operating frequency of the microdevice. Methods for the TiO_2 thin-film deposition are discussed.

INTRODUCTION

Surface acoustic wave (SAW) sensors have been used in many fields because of their wide applicability. The technology of varying the SAW velocity by some physical variable is attractive due to the ease in measurement and quick sensor turnaround time during manufacturing. SAW sensors have been tested for robustness and have proven to provide repeatable results with good sensitivity and resolution.

Many SAW sensor applications, e.g., sensing of vapors, flow, optical orientation, force, pressure, mechanical positioning, voltage, temperature, acceleration, viscosity, density, and thermal conductivity have been thoroughly investigated [1-8]. Most of these applications rely on mass loading on films deposited on a delay line intermediate between the transmitting and receiving transducers. This mass loading is propagated into a frequency shift or phase change of the SAW either of which is easily detected.

The basic principle of the SAW sensor is that a frequency signal is applied to the transmitting interdigitated transducer (IDT). The SAW undergoes a variation in the delay line due to interaction with a physical variable most commonly realized by employing a sensing film. This variation in the SAW is transmitted to the receiving IDT. Applying signal amplification to the received wave allows accurate detection of a change in the operating frequency of the device.

Metastable He atoms ($1s2s\,^1S$- or ^3S-states) belong to electron-excited states that are spin-prohibited from returning to their ground state. A large number of metastable He atoms exist in ionized plasma because of the long lifetime of metastable He atoms. In the past, electrical

---

* Work supported by the U.S. Department of Energy, Energy Efficiency and Renewable Energy, Office of Industrial Technologies, under Contract W-31-109-ENG-38.

conductivity of ZnO has been found to have a direct relationship with the presence of metastable atoms. Kupriyanov et al. [9, 10] employed activating thin-film surfaces with microcrystals of metals that interact with metastable atoms (thus demonstrating electron coupling with semiconductors) but the results did not demonstrate the direct correlation of a parametric change with He pressure. Much other work has been done using metastable He but none has reported the use of a microsensor that was fabricated specifically for detection of metastable He [11, 12].

Surface interaction in the gaseous phase is due primarily to the high density of metastable atoms, and a robust method of inert gas detection is available through metastable He detection. This paper discusses a micromachined SAW sensor that detects metastable He. The sensing system used in this device consists of an Au-activated TiO_2 film on an $LiNbO_3$ piezoelectric substrate. A method to deposit the TiO_2 film has also been proposed. The unique aspect of the present research is the development of a microsensor with the potential for detecting a frequency change of the SAW with respect to a variation in the electrical conductivity of an Au-activated TiO_2 thin film deposited on the delay line of the SAW microsensor when exposed to metastable He atoms.

EXPERIMENT

Figure 1 shows the configuration of the He SAW microsensor, which consists of (a) a sensing device that uses a pair of transmitting and receiving IDTs with a delay line coated with an Au-activated TiO_2 film capable of sensing metastable He and (b) a reference device configured in the same manner but without the sensing film on the delay line [13]. By comparing the outputs from the sensing and reference devices, we investigated the exposure and reactivity of the sensing film on the delay line to metastable He atoms. An Au-activated TiO_2 thin film was deposited by using an Edwards Auto 306 reactive electron beam evaporator. Reactive electron-beam evaporation was utilized in thin-film deposition because the metal transforms directly from vapor to solid and the thin films are rapidly cooled ($\approx 10^{13}$ K/s), leading to ultrafine structures [14].

TiO_2 thin films are chemically resistant and have high temperature stability. Several researchers have studied different starting materials such as TiO_2, Ti_3O_5, and TiO [15]. The present investigation was oriented toward utilizing Ti_2O_3 as a source material due to its uniform evaporation rate. Ti_2O_3 has not been explored greatly in terms of surface roughness, and a successful attempt was made to deposit consistent-quality thin TiO_2 films by introducing O_2 into the chamber. Oxygen chamber pressure was altered until good-quality TiO_2-rich films were produced.

To begin the deposition, we placed the Ti_2O_3 powder in an Mo crucible inside the evaporator. The background chamber pressure of 5×10^{-6} torr was used with a deposition rate of 0.2 nm/s. When Ti_2O_3 began evaporating and a stable deposition rate was achieved, oxygen was leaked into the chamber such that the chamber pressure did not exceed the threshold of 3×10^{-4} torr. Electron dispersion spectroscopy (EDS) studies determined the film composition with a Hitachi S-4500 scanning electron microscope. Scans were performed with a working distance of 15 mm, an EDS distance of 50 mm, and an accelerating voltage of 5 kV to detect lower-energy Ti L-lines. An oxygen pressure of $1.2\text{-}2.9 \times 10^{-4}$ torr was necessary to produce stoichiometric TiO_2 thin films.

To pattern the IDTs, photolithography was employed by using a positive photoresist to coat the substrate, exposing the photoresist-covered substrate to UV light through the IDT pattern by using a photomask, and developing and rinsing the photoresist to expose the substrate

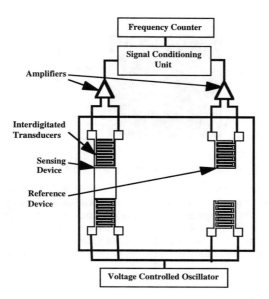

Fig. 1. SAW microsensor configuration.

Fig. 2. Optical photomicrograph of a portion of interdigitated transducer of microsensor showing consistent finger width (50x).

specifically patterning the IDTs. 1800 Å Cr and 200 Å Au films were deposited by electron beam evaporation on the substrate, patterning the IDTs of the reference and the sensing device. The Au layer reduced the contact resistance between the sensor external electronics. After deposition, unwanted metal was removed by liftoff. Figure 2 is a magnified phomicrograph of the IDT, showing the consistent thickness of the fingers. With the same deposition and liftoff procedure, the sensing film of Au/TiO_2 was applied to the delay line of the sensing device. The contact pads of the SAW microsensor were interfaced to the external electronics with gold wire. With conductive epoxy, 3-mil gold wires were adhered to the contact pads, which were then connected to a 24-pin integrated circuit socket. The pluggable sensor, shown in Fig. 3, was connected directly to external electronics for testing.

Fig. 3. SAW microsensor mounted on a 24-pin socket.

The IDTs of the sensors were designed to operate at 108.9 MHz with an impedance of 50 Ω at the transducers. A voltage-controlled oscillator was used to provide an output frequency of 108.9 MHz controlled by a tuning voltage applied externally to the circuit [15]. The integrated electronics were attached to the IC socket of the SAW microsensor, which was enclosed in a sensing chamber in the test setup. Because the sensing film detects metastable He, a plasma chamber was incorporated into the gas flow system to excite stable He to the higher-energy metastable state.

RESULTS

The as-fabricated SAW microsensor was used to detect metastable He and the mechanism was found stable due to the direct correspondence of metastable atoms exposed during surface interaction. Variation in the frequency at the receiving ends of the IDTs of both the sensing and the reference devices provided the frequency change with respect to exposure to metastable He. To ensure consistent film thickness, a Tencor P-10 Surface Profiler was employed to determine surface roughness and film thickness with profile lengths of 100 μm.

Figure 4(a) shows the effect of the oxygen chamber pressure to the root-mean-squared surface roughness (R_{rms}) of the thin film and Fig. 4(b) shows how the R_{rms} varies with film thickness. A consistent roughness of 2 nm was observed, with an uncertainty of $\pm$ 0.4 nm due to equipment limitations. These values are much lower than the film thickness employed for TiO_2 and therefore do not affect the results significantly. Figure 5 shows the variation of the frequency difference with increasing He pressure; on-set of plasma is at an He pressure of 10 kPa. Because both the sensing and the reference devices experience identical environments, the frequency shifts are independent of most parameters. Due to possible change in processing conditions, and also to dust trapped in the IDT fingers, small variations in the sensing behavior were found. These small changes were investigated and provided a baseline frequency difference of $\approx$4 kHz, noticeable when experiments were run in vacuum for long periods.

From Fig. 5, when production of metastable He atoms began, a rise in the frequency shift was observed due to activation of the Au atoms overlying the TiO_2 thin film. The metastable He collided with the sensing film, thereby imparting its decay energy to the film. With the bombardment of He atoms with Au, additional energy was imparted and the Au impurities in the semiconducting TiO_2 film were activated as deep traps, varying the electrical conductivity of the film. The change in electrical conductivity was translated as a change in the frequency of the sensing device and therefore as a change in the frequency difference between the sensing and the

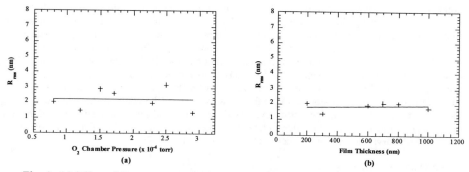

Fig. 4. (a) Effect of O$_2$ pressure on surface roughness (R$_{rms}$) of film; (b) surface roughness variation for different film thicknesses.

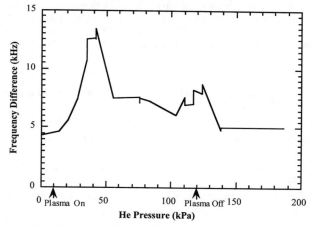

Fig. 5. He pressure vs. frequency difference between sensor and reference devices.

reference devices. Due to the limited number of Au atoms, no variation was observed in the curve where increased pressure of He did not increase the electrical conductivity of the sensing film. Due to stripping of the Au atoms by metastable He atoms, the variation becomes abrupt at certain He pressures. This abrupt variation may be attributed to sensor vibration in the chamber. When the plasma was shut off, no metastable He atoms were produced and, after a momentary time lapse for evacuation of the leftover metastable He, the change in frequency shift decreased dramatically, as expected, eventually becoming constant at a level equal to the baseline difference.

CONCLUSIONS

We have developed a microsensor for the detection of metastable He. The microsensor assembly consisted of two SAW configurations, one a reference with no deposition at the delay line and the other sensing with an Au-activated TiO$_2$ e-beam-deposited thin film (sensitive to electrical conductivity changes) on the delay line. The sensitivity of the device was limited to the number of Au atoms present in the Au-TiO$_2$ sensing film. As the Au atoms interact with the metastable He, a change in the electrical conductivity of TiO$_2$ was observed as a change in the frequency difference between the sensing and the reference devices of the microsensor. Due to bombardment of the metastable He atoms with Au atoms, some of the Au atoms were removed from the sensing film and therefore, even after saturation of the sensing film with metastable He atoms, some variation in frequency was observed due to varying numbers of Au atoms in the sensing Au/TiO$_2$ film. These results indicate the role of metastable He atoms in the detection of surface interaction with Au-atoms. In the future, we will study the width of the delay line, temperature exposure, operating frequency, IDT finger width and separation, and separation between IDTs and delay line to better understand and optimize the variation of acoustic wave velocity with varying operating frequency of the SAW device.

ACKNOWLEDGMENTS

This work was supported by the U.S. Department of Energy, Energy Efficiency and Renewable Energy, and by the Argonne National Laboratory Directed Research and Development project. The authors thank C. J. Sjoberg, an REU Program student, for his help in e-beam evaporation of TiO$_2$ using the MRSEC Central Facilities of Northwestern University which is supported by the National Science Foundation under Award No. DMR-9120521.

REFERENCES

[1]   S. J. Martin, K. S. Schweizer, S. S. Schwartz, and R. L. Gunshor, *IEEE Ultrasonics Symposium*, pp. 207-212 (1984).

[2]   S. G. Joshi, *Sensors and Actuators*, Vol. 44A, pp. 191-197 (1994).

[3]   F. Seifert, W.-E. Bulst, and C. Ruppel, *Sensors and Actuators*, Vol. 44A, pp. 231-239 (1994).

[4]   B. A. Martin, S. W. Wenzel, and R. M. White, *Sensors and Actuators*, Vol. A21-A23, pp. 704-708 (1990).

[5]   E. Gatti, A. Palma, and E. Verona, *Sensors and Actuators*, Vol. 4, pp. 45-54 (1983).

[6]   M. Hoummady and D. Hauden, *Sensors and Actuators*, Vol. 44A, pp. 177-182 (1994).

[7]   H. Wohltjen, *Sensors and Actuators*, Vol. 5, pp. 307-325 (1984).

[8]   G. W. Watson, R. D. Ketchpel, and E. J. Staples, *IEEE Ultrasonics Symposium*, pp. 253-256 (1992).

[9]   L. Yu. Kupriyanov, V. I. Tsivenko, and I. A. Myasnikov, *Zh. Fiz. Khim.*, Vol. 58, pp. 1156-1162 (1984).

[10] L. Y. Kupriyanov and V. I. Tsivenko, *Rus. J. Phy. Chem.*, Vol. 61(3), pp. 430-431 (1987).

[11] J. W. Sheldon, E. D. Anderson, and K. A. Hardy, *J. Appl. Phys.*, Vol. 56 (3), pp. 798-800 (1984).

[12] R. Muller-Fiedler, P. Schlemmer, K. Jung, H. Hotop, and H. Ehrhardt, *J. Phys. B: At. Mol. Phys.*, Vol. 17, pp. 259-268 (1984).

[13] S. Ahuja, M. Hersam, C. Ross, H.-T. Chien, and A. C. Raptis, *IEEE Ultrasonics Symposium* (1996).

[14] L. Bianchi, *JOM*, pp. 45-47 (1991).

[15] H. K. Pulkner, G. Paesold, and E. Ritter, *Appl. Opt.*, Vol. 15, pp. 2986-2991 (1976).

CHARACTERIZATION AND ELECTRICAL PROPERTIES OF WO$_3$ SENSING THIN FILMS

C. CANTALINI*, S. DI NARDO**, L. LOZZI**, M. PASSACANTANDO**, M. PELINO*, A. R. PHANI*, S. SANTUCCI**.
*Department of Chemistry and Materials - 67040 Monteluco di Roio - L'Aquila -Italy
**Department of Physics - 61010 - L'Aquila - Italy - canta@dsiaq1.ing.univaq.it

ABSTRACT

The microstructure and the electrical properties of thermally evaporated WO$_3$ thin films have been investigated by glancing angle XRD, atomic force microscopy AFM, X-ray photoelectron spectroscopy XPS and dc techniques. Thin films of 1500 Å thickness have been obtained by evaporating high purity WO$_3$ powders by an electrically heated crucible at 5×10^{-4} Pa on sapphire substrates. The as-deposited films have resulted to be amorphous. After annealing at 500 °C in dry air for 6, 12 and 24 hours the films have shown well crystallized structures with preferential orientations of WO$_3$ in the (200) direction. The increase of the annealing time has shown marked influence on the microstructural features of the films surface, as highlighted by AFM investigations. The binding energies of W 4f$_{7/2}$ have been close to that of WO$_3$, the 24 h annealed yielded an O/W ratio close to 2.9 which is in good agreement with the theoretical one. The gas sensitivity, selectivity and stability of the annealed films in presence of NO$_2$ gas (between 0.7 and 5 ppm) have been evaluated by measuring the electrical change of the film resistance in dry air and in gas atmosphere conditions. The influence of NO and Humidity interfering gases to the NO$_2$ electrical response has been also evaluated. The 500 °C annealed at 24 h has shown better electrical properties in terms of NO$_2$ sensitivity, stability and cross sensitivity effects.

INTRODUCTION

Among the transition metal oxides pure WO$_3$ [1] and metal tungstates [2] thick films have been reported to have outstanding sensitive properties towards nitrogen oxides at low and elevated operating temperatures respectively. In the case of air air-quality systems the TLV (Threshold Limit Value) limits for NO$_2$ and NO are respectively 3 and 25 ppm, being NO$_2$ the main gas to be detected, mainly because in ambient atmosphere NO easily oxidize to NO$_2$. Market needs for air-quality NO$_x$ sensor demand for high sensitivity to NO$_2$ at low concentration (between 0 and 5 ppm), reduced cross sensitivity effects and long term stability of the electrical response.

Thin WO$_3$ films, with respective advantages in quality and cost, have been prepared via physical and chemical routes, focusing on sol-gel coating [3], chemical vapor deposition [4], RF sputtering and thermal annealing [5-6]. Both thin and thick films are sensitive to NO$_x$ at elevated temperature, but thin film exhibits high performances including large gas sensitivity, fast response and low working temperature.

Our previous research [7-8] has established that WO$_3$ thin-film prepared by high vacuum thermal evaporation of pure WO$_3$ powder and subsequent annealing for 1 h at 500 °C in static air, posses high sensitivity to sub-ppm levels of NO$_2$ gas.

This paper outlines the results obtained for a series of WO$_3$ thin films, prepared by annealing the evaporated substrates at 500 °C for different times of 6, 12 and 24 hours. Aim of this work is to determine the influence of the annealing time on the NO$_2$ sensitivity, as well as the cross sensitivity effect played by H$_2$O, NO interfering gases. The long term stability of the electrical response is also investigated, in order to asses the WO$_3$ films as sensors for air-quality monitoring.

EXPERIMENT

Commercial WO_3 powder with 99.995% purity was electrically heated in a tungsten crucible at 5×10^{-4} Pa. Polished sapphire with platinum interdigital electrodes of 0.1 mm width was used as the substrate. The vapor phase was condensed on the substrate at room temperature and covered 4.0 mm × 4.0 mm area. The film was deposited at 6 nm/min rate up to 150 nm thickness, monitored by a quartz crystal microbalance. Annealing was made in air at 500° for 6, 12 and 24 hours with a heating and cooling rate of 1 °C/min.

Crystalline phases of the substrate and film were examined by a wide-angle XRD (PW1820, Philips) with thin film attachments. Cu Kα radiation ($\lambda = 0.154$ nm) and 0.02° angle step were used for the XRD analysis. The surface topography was observed by a large sample probe microscope (NanoScope III, Digital Instrument Inc.). The binding energy of chemical elements on the film surface was characterized by an XPS system containing a hemispherical analyzer (PHI 5600, Perkin-Elmer).The X-ray was generated by an non-monochromatic Mg source (Kα, 1253.6 eV) and the measurement was performed in an ultra high vacuum about 5×10^{-8} Pa. The spectrometer was calibrated from binding energy of Au $4f_{7/2}$ (84.0 eV) with respect to the Fermi level. Survey scans were up to 1000 eV at 1 eV/s and details were obtained at 0.1 eV/s.

The electrical response to NO_2 was measured by a computerized multimeter system in the 200°-450°C range. Different gas concentrations were obtained by diluting NO_2 with dry air through an MKS147 multi gas controller. The NO_2 concentration in the downstream was measured by a Dräger electrolytic sensor. The gas sensitivity S is defined as the ratio between resistance R_g in gas and R_a in air, i.e. $S = R_g/R_a$.

RESULTS AND DISCUSSION

Microstructure

Our previous research activities [7-8] has established that WO_3 thin-film prepared by high vacuum thermal evaporation of pure WO_3 powder and subsequent annealing for 1 h at different temperatures, i.e. 400, 500 and 600 °C in static air, are well crystallized at temperatures above 400 °C, with diffraction peaks belonging to the triclinic phase of WO_3 (JCPDS card no. 20-1323). The crystallite size as estimated from the half width of the (200) peak yielded 21, 23 and 24 nm for the 6 h, 12 h and 24 h respectively.

Figures 1 shows the AFM picture of a 5 x 5 square micron area of the 6 h (a), 12 h (b) and 24 h (c) annealed films respectively. After 6 h annealing the film surface is continuous with a mean roughness of about 1 nm (fig. 1 (a)). Increasing the annealing time at 12 h the mean roughness becomes about three times bigger due to the formation of bumps (lighter regions in figure 1 (b)) and ridges fixing grain boundaries. After 24 h annealing, the mean roughness turns to about 2 nm due to the formation of a smoother surface with microcracks at the grain boundaries.

In order to investigate the topographical features to a sub-micron scale figure 2 shows the AFM picture of a 1 x 1 square micron area of the 6 h (a), 12 h (b) and 24 h (c) annealed films respectively.

From figure 2 (a) it turns out an equiaxed structure with crystallites having different sizes. Figure 2 (b) highlight the formation of ridges probably caused by compression stresses between adjacent grains due to thermal expansion mismatch of the sapphire and the film. Figure 2 (c) shows the existence of microcracks as a consequence of the microstructural rearrangement of the film surface after long time annealing.

Figure 3 shows the microstructure to a sub-micron magnification of the 24 h annealed film. As can be seen well oriented and homogeneous crystallites are formed having 30 x 60 nm dimension, as highlighted by the section analysis profile reported in the figure.

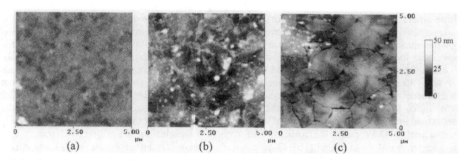

Figure 1. AFM pictures (5 x 5 square micron area) of the 6 h (a), 12 h (b) and 24 h (c) annealed

Figure 2. AFM picture (1 x 1 square micron area) of the 6 h (a), 12 h (b) and 24 h (c) annealed

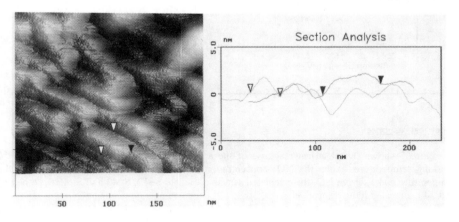

Figure 3. AFM picture (200 x 200 square nm area) of the 24 h annealed and the associated section analysis profile.

Figures 4a and 4b depicts the detailed XPS scan of O 1s and W $4f_{7/2}$ lines of the annealed films and the starting WO_3 powder. The related binding energies of tungsten and oxygen in WO_3 and some others tungsten oxides are also located according to literature data. The line shape of the O 1s spectra (fig. 4a) becomes more symmetric and almost indistinguishable from that of the starting powder during annealing, probably due to the progressive loss with the annealing time of the contribute located at about 532 eV which could be assigned to the presence of water vapor [9]. The binding energy for the W $4f_{7/2}$ of 35.6 eV (fig. 4b) is in good agreement with literature values for W $^{6+}$ (35.8 eV). The 6 h and 12 h annealed samples were found to be slightly oxygen deficient as estimated using the standard XPS quantification method. The 24 h annealed film yields a 2.9 O/W ratio, which, in the limits of the experimental errors, is in good agreement with the theoretical value. However, as reported by other authors, although the XPS analysis gives nearly stoichiometric O/W ratio, the formation of oxygen vacancies on the surface of these WO_3 samples, may not be excluded [10]. Oxygen vacancies may play an important role as adsorption sites for gaseous species and eventually, as reported in literature [11], may greatly enhance the gas sensitivity.

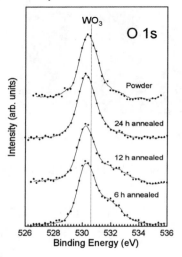

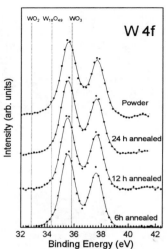

Figure 4a. Detailed XPS scans of the oxygen 1s lines.

Figure 4b. Detailed XPS scans of the tungsten 4 $f_{7/2}$.

Electrical response

Figure 5 shows the electrical response of the annealed films in dry air carrier and 250 °C working temperatures when the NO_2 concentration is changed from 0.7 ppm to 5.4. At this temperature and 1.7 ppm NO_2 the calculated sensitivities yield S=12, S=43 and S=45 for 6 h, 12 h and 24 h annealed films respectively.

Figure 6 shows the electrical responses of the 24 h annealed film when water vapor, at different relative humidity (RH) contents, is the interfering gas. The test has been carried out by exposing the films to: dry air, 0.7 ppm NO_2, dry air, 40% RH air, 0.7 ppm NO_2 and different RH at 40% - 60% - 80% - 40%, 40% RH without NO_2 and finally dry air. From a general point of view, humid air single gas decreases the resistance of the base line. The humidity cross when 0.7 ppm NO_2 are present, is revealed by a decrease of the resistance at saturation as compared to the one measured at 0.7 ppm NO_2 and dry air carrier. The 24 h annealed film shows reduced water vapor cross sensitivity effects as compared with the 6 h and 12 h annealed (not reported for brevity).

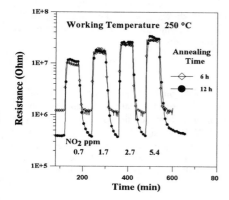

Figure 5. Electrical Response of the films at different NO₂ concentrations.

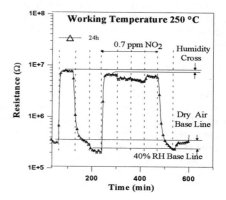

Figure 6. Humidity cross sensitivity of the 24 h annealed film to 0.7 ppm NO₂.

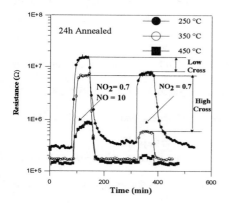

Figure 7. 10 ppm cross sensitivity to 0.7 ppm NO₂ for the 24 h annealed at different temperatures.

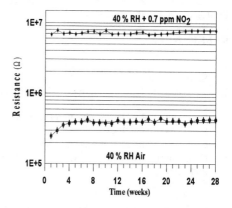

Figure 8. Long term stability of the electrical response over a period of six months.

Figure 7 shows the NO cross sensitivity to 0.7 ppm NO₂ gas for the 24 h annealed film at different working temperatures and dry air carrier. The test has been carried out in order to have the same 0.7 ppm NO₂ concentrations during the conditionings with NO$_x$ and NO₂ mixtures. At 250 °C working temperature the measured sensitivity with NO$_x$ mixture (10 ppm NO and 0.7 ppm NO₂) is S=50. At the same temperature and only 0.7 ppm NO₂ the sensitivity halves to S=26. The same comparison but at 350 °C working temperature yield S=41 (10 ppm NO and 0.7 ppm NO₂) and S=3.6 (0.7 ppm NO₂). As can be seen WO₃ 24 h annealed films are sensitive to NO gas since the measured resistances at 10 ppm NO and 0.7 ppm NO₂, are bigger than the ones measured with only 0.7 ppm NO₂. Moreover the film sensitivity to NO seems strongly affected by the working temperature, being the measure of NO_2 gas at 250 °C temperature less affected by the presence of NO (low cross).

Figure 8 shows the long term stability of the electrical response of the 24 h annealed recorded every week for a period of six months. The 24 h annealed shows a better reproducibility of the electrical response as compared to the 6 h and 12 h. The annealing time seems to have a positive effect in stabilizing the long term stability properties of the sensor.

CONCLUSION

WO_3 thin films prepared by thermal evaporation and subsequent annealing at 6 h, 12 h and 24 h show well developed crystalline phase and microstructural surface features (mean roughness) markedly influenced by the annealing time. After annealing at 24 h the films shows an O/W ratio close to the stoichiometrical one and binding energy of W $4f_{7/2}$ and O 1s close to the one found for the starting WO_3 powder. The 24 h annealed film shows larger sensitivities as respect to the other sensors with S=45 to 1.7 ppm of NO_2 at 250 °C working temperature. Longer annealing decreases the H_2O cross sensitivity to 0.7 ppm NO_2, as well as improves the long term stability properties of the electrical response

ACKNOWLEDGMENTS

This study has been financially supported by the C.N.R. of Italy through the "Progetto Strategico Materiali Innovativi"

REFERENCES

1. M. Akiyama, Z. Zhang, J. Tamaki, T. Harada, N. Miura and N. Yamazoe, Tech. Digest 11th Sensor Symp., Japan, 1992, 181
2. J.Tamaki, T. Fujii, K. Fujimori, N. Miura, N. Yamazoe, Sensors and Actuators **B 24-25**, 396 (1995)
3. L. Armelao, R. Bertoncello, G. Granozzi, G. Depaoli, E. Tondello and G. Battaglin, J. Mater. Chem., **4**, 407 (1994).
4. C. E. Tracy and D.K. Benson, J. Vac. Sci. Technol., **A 4**, 2377 (1986).
5. D. J. Smith, J.F. Vetelino, R.S. Falconer and E.L. Wittman, Sensors and Actuators **B 13-14**, 264 (1993).
6. G. Sberveglieri, L.Depero, S. Groppelli and P.Nelli, Sensors. and Actuators **B 26-27**, 89 (1995).
7. C. Cantalini, H.T. Sun, M. Faccio, M. Pelino, S. Santucci, L. Lozzi, M. Passacantando, Sensors and Actuators **B 31**, 81 (1996).
8. H. T. Sun, C. Cantalini, L. Lozzi, M. Passacantando, S. Santucci and M. Pelino, Microstructural effect on NO2 sensitivity thin film gas sensor, Thin Solid Film
9. Handbook of X-ray Photoelectron Spectroscopy, J. Chastain (ed.) Perkin Elmer corporation, Minnesota (1992).
10. B. Frühberger, M. Grunze, D.J. Dwyer, Sensors and Actuators **B 31**, 167 (1996).
11. N. Yamazoe, Sensors and Actuators **B 5**, 7 (1991).

EFFECT OF NOBLE METALS ON SELECTIVE DETECTION OF LPG BASED ON SnO$_2$

A.RATNA PHANI and M.PELINO
Department of Chemistry and Materials, University of L'Aquila, 67040 Monteluco di Roio
L'Aquila, ITALY. Phani@aquila.infn.it

ABSTRACT

The present investigation deals with the electrical response of noble metal doped SnO$_2$ to improve the selectivity for Liquid Petroleum Gas (LPG) in the presence of CO and CH$_4$. Addition of small amounts of nobel metals (Pd, Pt and Rh) to the base material SnO$_2$ is carried out by co-precipitation method. X-ray diffraction and X-ray photoelectron spectroscopy studies are carried out to find out the crystalline phase and chemical composition of the SnO$_2$. The sensor element has been tested for cross selectivity to reducing gases by measuring sensitivity versus sintering temperatures and sensitivity versus operating temperatures. The sensor elements with the composition of Pd(1.5 wt%) and Pt(1.5 wt%) in the base material SnO$_2$ sintered at 800°C showed high sensitivity towards LPG at an operating temperature of 350°C suggesting the possibility to utilize the sensor for the detection of LPG.

INTRODUCTION

Hydrocarbon gases are being widely used as fuel for domestic and industrial application because of the relatively low cost and the environmentally compatible gaseous combustion emission. However, they are potentially hazardous because of the high possibility of explosion caused by accidental gas leakage. This has resulted in an increased demand of reliable, inexpensive and easy to use sensors able to detect hydrocarbon gases. The term liquid petroleum gas (LPG), is applied to those hydrocarbons the chief component consists of Butane (70 to 80%), Propane (5-10%) and propylene, butylene, ethylene and methane (1-5%). Not much work has been found in literature on LPG sensors. Sieyama [1] reported that the inflammable gases in air could be detected from a change in the electrical resistance of thin film of ZnO, while Taguchi [2] reported that a porous sintered block of SnO$_2$ could also work in the same way. Several investigators have worked on butane (which is the major component of LPG) gas sensors using semiconducting oxides as base material such as SnO$_2$ -TiO$_2$ complex oxides doped with Pt as a catalyst and Sb as a conductive dopant. This sensor is prepared by co-precipitation method and the sensor element is fabricated by screen printing method with enhanced sensitivities to hydrocarbon gases when the loading rate of TiO$_2$ is 3-5 mol%[3]. Sensors based on Fe$_2$O$_3$, with and without Pd doped has been prepared by screen printing method and tested for butane gas and no cross to inferring gases has been mentioned [4]. V$_2$O$_5$ supported on ZrO$_2$ has been tested for hydrocarbon gases and found to be sensitive to the mixture of n-butane and propane [5]. In the present investigation we have conducted the fabrication of selective and sensitive LPG sensors based on SnO$_2$. This oxide reported to be [6] an excellent material for sensor fabrication with the main disadvantage being lack of selectivity, that is it exhibits conductance change on exposure to many reducing gases. A degree of selectivity could be conferred by careful control of sensor operating temperature and use of specific additives. It is well known that addition of small amounts of noble metals such as Pd, Pt, Ir and Rh [7] to the base material can enhance not only of gas sensitivity but also the rate of response. Such promoting effects are undoubtedly related to the catalytic activities of the metals for the oxidation of inflammable gases.

EXPERIMENT

The schematic diagram of the measurement set up shown in the Figure 1. The sensor elements have been prepared by co-precipitation method using the following procedure. 10 ml of $SnCl_4$ are transferred into a 50 ml capacity two necked round bottomed flask connected with a dropping funnel placed in an ice trough. Aqueous solution of ammonia is added drop by drop to $SnCl_4$ till the pH is slightly more than 7.5, so as to obtain a complete precipitation. The flask is kept in ice cold temperature to avoid spillage because the reaction is vigorous. At the same time 0.2235 gm of $PdCl_2$ is dissolved in ethanol and added to the above solution in order to get 1.5 wt% of Pd in SnO_2. The contents are stirred by using a magnetic stirrer. After stirring for about 30 min the contents are filtered. In the present case, Pt (in the form of $H_2PtCl_6, 6H_2O$) and Rh (in the form of $RhCl_3$) are added to SnO_2 in place of Pd to see the effect of gas sensitive properties. The resulting precipitate is filtered and washed thoroughly with hot deionized water to remove Cl^- ions and dried at 120°C for 24h in an electrical oven. The dried calcined powder is ground in an agate mortar and pestle and calcined at different temperatures ranging from 500-1000°C for 5 h. The calcined powder is sieved in 170 mesh to obtain fine particles. In order to improve the mechanical properties, the powder is mixed with few drops of binder Tetraethylorthosilicate (TEOS) and the resulting paste is applied on alumina tube (3mm diameter × 10mm length) attached with two platinum electrodes 8mm apart from each other. The elements are again sintered in the range of 500-1000°C for 5h. For gas sensing experiments the element is kept in a flow apparatus through which air or sample gas is let to flow. The concentration of the gases injected into the system is 1000 ppm. The temperature of the element is controlled with a tungsten heating coil provided inside the alumina tube.

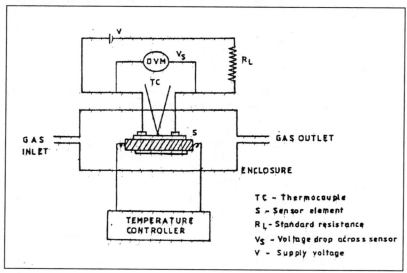

Figure 1 Schematic diagram of the measurement set up of the sensor element

After heat treatment in dry air at 400°C, the element is cooled to a prescribed temperature in the range of 50-400°C at which the electrical resistance of the element in dry air (R_a) is measured by means of conventional circuit in which the element is connected with an external resistor in series. The out put voltage across the external resistor at a circuit voltage of 10V is used to evaluate the electrical resistance of the element.The values of the device resistance are obtained by monitoring the output voltage across the load resistor. A chromel-alumel thermocouple placed on the element indicated the operating temperature.The gas sensitivity (S) is defined as the resistance ratio (R_a-R_g / R_a). Where R_a and R_g are the resistance's of the sensor element before exposed to the gas and after respectively.

RESULTS AND DISCUSSION

Microstructure

X-ray diffractometer (SIEMENS D5000) with copper target, K_α radiation ($\lambda = 1.5406$Å) is used for phase identification. Figure 2 shows the XRD pattern of the SnO_2 calcined at 100°C step in the 500-1000°C temperature range for 5h. The effects of calcination temperature on crystallographic structure and crystallite size for samples calcined at 500°C and 600°C revealed low intensity and broad peaks indicating that the crystallites are not well defined. With further increase in temperature, the samples become highly crystalline.

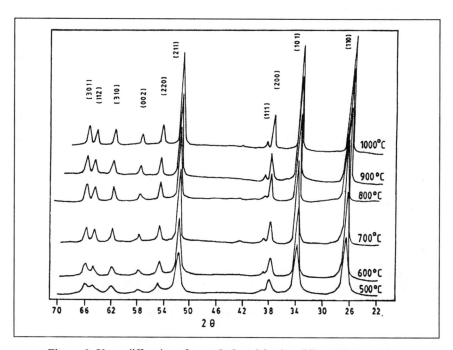

Figure 2 X-ray diffraction of pure SnO_2 calcined at different temperatures

Sharp and intense peaks are observed for the samples calcined above 700°C, indicating a higher degree of crystallinity. All the diffraction lines agree with reported values and match with the rutile structure of JCPDS data confirming the formation of SnO_2. It has been observed that as the calcination temperature increases, the crystallite size °ranges from 126Å at 500°C and 312Å at 1000°C calcination temperature respectively. In order to obtain the most suitable crystallite size and surface area for better sensitivity, SnO_2 calcined at 800°C is chosen as a base material for all the experiments. In case of SnO_2 doped with Pd / Pt / Rh we observed no extra lines other than SnO_2 indicating that there is no formation of any new phase. We do not expect Pd / Pt / Rh lines because their concentrations fall below the level of detection of XRD. X-ray photoelectron spectroscopy study on SnO_2 has been performed using Al K_α radiation source 1253 eV, an x-ray monochromator and a concentric hemispheric33al analyzer.

Figure 3 shows the O1s level of SnO_2 in which we observed a broad peak which is convoluted into two, one at 531 eV and other at 533.07 eV corresponding to O_2 and O_2^{-2}. The observed values for O_2 and O_2^{-2} are in good agreement with the literature values [8]. Apart from the lattice oxygen (O_2^{-2}) we have also observed adsorbed oxygen from the atmosphere. Figure 4 shows the Sn3d core level of SnO_2. The spectrum is resolved into two peaks due to spin orbit splitting i.e., Sn3d$_{5/2}$ ground state and Sn3d$_{3/2}$ excited state, whose values are observed at 486 eV and 495 eV respectively. These values are in good agreement with the literature values [8]. The calculated atomic percentages of Sn and O in the sample are 43% and 57% indicating oxygen deficiency in the sample.

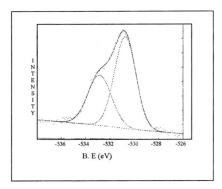
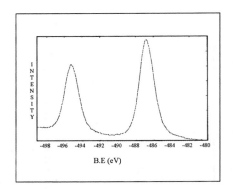

Figure 3 XPS of O1s level in SnO_2 **Figure 4** XPS of Sn3d core level in SnO_2

ELECTRICAL PROPERTIES

We examined sensing characteristics of Pd doped SnO_2 with different loadings of Pd calcined at 800°C for 5 h. It is evident from Figure 5 that 1.5 wt % of Pd in SnO_2 showed maximum sensitivity (0.97) towards LPG at an operating temperature of 350°C and higher Pd concentration drastically decreased the sensitivity towards LPG. Therefore further experiments have been carried out using 1.5 wt% Pd concentration. Figure 6 shows the LPG sensitivity versus sintering temperature of the pure SnO_2 and SnO_2 loaded with Pd, Pt and Rh (1.5 wt%) sensor elements at 350°C operating temperature. In case of pure SnO_2 the sensitivity increases to 0.55 at 800°C sintering temperature then decreases to 0.51 at 900°C. The figure also high lights that the sensitivity for Pd and Pt loaded increases up to 0.97 and decreases to 0.68 at 900°C. The Rh loaded sensing behavior as a function of sintering temperature is similar to pure SnO_2.

It results from the above experiments that as the sintering temperature increases, the sensitivity to LPG also increases up to 800°C and then decreases. The modification of selectivity is also possible by careful choice of noble metal sensitizers because the degree of shift in operating temperature and magnitude of increase in sensitivity to a specific gas is closely related to the kind of nobel metals.

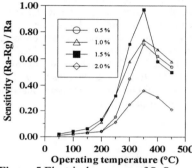

Figure 5 Electrical response of SnO_2 concentrations of Pd for LPG

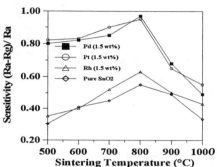

Figure 6 Electrical response of loaded different with pure SnO_2 and loaded with Pd,Pt and Rh operated at 350°C for LPG

Figure 7a shows the sensitivity versus operating temperature of the sensor element SnO_2 / Pd (1.5 wt%) sintered at 800°C for different gases. It is clear from the figure that the sensitivity to LPG gradually increases from 0.1-0.95 and then decreased to 0.57 at operating temperature of 400°C. At the same time the sensitivity to CO , CH_4 gases slightly increases reaching a maximum value of 0.17 at 400°C operating temperature. Figure 7b shows the sensitivity versus operating temperature of the sensor element SnO_2/ Pt (1.5wt%) sintered at 800°C for different gases. This sensor composition is sensitive only to CO gas at an operating temperature of 200°C with sensitivity of 0.43. The sensitivity to LPG reaches its maximum value of 0.97 at 350°C and then decreases to 0.56 at an operating temperature of 400°C.

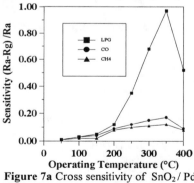

Figure 7a Cross sensitivity of SnO_2/ Pd sintered at 800°C

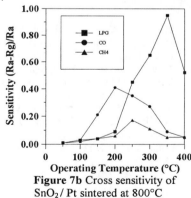

Figure 7b Cross sensitivity of SnO_2/ Pt sintered at 800°C

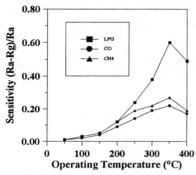

Figure 7c Cross sensitivity of SnO_2/ Rh sintered at 800°C

It is clearly evident that Pt doped SnO_2 can be made selective to a particular gas by changing the operating temperature. In other words by the addition of the catalyst Pt, this composition could be made selective to CO by operating at 200°C and to LPG at 350°C. With Rhodium addition, the sensitivity of LPG increases from 0.1-0.65 but at the same operating temperature 350°C, CO and CH_4 are also sensing with 0.25 and 0.2 values respectively as shown in Figure 7c. Experiments were carried out every month for a period of six months for the sensor elements SnO_2 / Pd (1.5 wt%) and SnO_2 / Pt (1.5 wt%) sintered at 800 °C for 5h at an operating temperature of 350°C. Every time, the sensitivity of the sensor elements were found to be constant indicating the reliability of the sensor with time.

CONCLUSIONS

Addition of nobel metals have been carried out by co-precipitation method in order to obtain high surface area. From the studies of the sensor element with different nobel metal additives, Pd,Pt and Rh on gas sensitivity, it is concluded that SnO_2 loaded with Pd and Pt sintered at 800°C showed reasonably high sensitivity to LPG (0.97) at an operating temperature of 350°C ,whereas Rh provided a sensitivity of 0.65 at the same operating temperature with negligible interference of CO or CH_4 gases.

REFERENCES

1. T.Seiyama, A.Kato, K.Fujiishi and M.Nagatani, Anal.Chem., **34**, 1502 (1962)
2. M.Taguchi, Japan Patent 45-38200 (1962)
3. W.Y.Chang and D.D.Lee, Thin Solid Films. **200**, 329 (1991)
4. D.D.Lee and W.T Chung, Sensors & Actuators. **20**,301 (1989)
5. A.R.Raju and C.N.R.Rao J.Chem.Soc.Chem.Comm., 1290 (1990)
6. N.Yamazoe and T.Seiyama, Proc. of Intel .Conf.on Solid state Sensors and Actuators, Transducers,Boston, 376, (1985)
7. J.C.Kim, H.K.Jun, T.J.Lee and D.D.Lee, Proc.of Second East Asia Conf.on Chemical Sensors, China, p.166 (1995)
8. Hand Book of X-ray photoelctron spectroscopy, C.D.Wagner Ed. (1979)

Synthesis and Spectroscopic Analysis of Smart Photochromic Materials

Yeon-Gon Mo, R. O. Dillon, and P. G. Snyder
Department of Electrical Engineering and Center for Microelectronic and Optical Materials
Research, University of Nebraska, Lincoln, NE, 68588

ABSTRACT

Sol-gel processing was used to dope photochromic materials into metal alkoxide-polymer and pure polymer materials. The films on silicon and quartz substrates were examined, with and without UV irradiation, by UV/VIS spectroscopy, ellipsometry and FTIR.

The UV/VIS spectroscopy showed that the doped matrices were photochromic in the visible and near infrared.

The ellipsometric data were obtained with a variable angle of incidence spectroscopic ellipsometer (VASE). The Cauchy and a combined Cauchy-Lorentz model were used to fit the unirradiated and irradiated films, respectively. The optical constants of the films showed significant changes upon irradiation. This means that the absorption coefficient and hence the emissivity of the films is being modulated with UV irradiation. The VASE-fitted thicknesses of the films were in the range of 1 to 6 microns.

In the FTIR spectra, the spiropyran doped samples have shown IR transmission changes in the two spectral regions (6-7 μm and 7.5-8.5 μm) where changes are expected due to band opening. The transmission ratio for UV irradiated to unirradiated samples decreased by as much as about 24% at a particular IR wavelength.

1. Introduction

In the past few years, photochromic dyes which show reversible photoisomerization have attracted considerable interest from the viewpoint of applications such as novel photometry, optical switching and memory devices[1-3]. Various photochromic dyes have been synthesized and improved reversibility and stability have been reported[4-5].

In this study sol-gel processing was used to develop a smart photochromic material for space applications whose infrared emissivity changes in response to solar radiation. The sol-gel technique can be used to produce both thin and thick films without cracking. This can be attributed to a mechanical relaxation of these materials, caused by various mechanisms such as hydrolysis and polycondensation[6-7].

The mechanical properties and thermal resistance of a pure photochromic dye can be improved by putting it into a matrix. One such matrix is a metal alkoxide-polymer. The metal alkoxide-polymer matrix can be prepared by the incorporation of metal alkoxides into organic polymeric materials via the in-situ polymerization of inorganic alkoxides by the sol-gel processing technique. In several published reports[8-10], various methods have been utilized to promote compatibility between the inorganic network and either an inorganic or organic polymer. These methods include reacting a metal alkoxide directly with organic polymers or oligomers that are end-capped with or that contain functional groups capable of entering into a cross-reaction with the inorganic polymers[11-13]. The other matrix is the polymer poly methyl methacrylate (PMMA). Even though this polymer is not expected to survive well in space, we are using it as a convenient way to test the incorporation of photochromic dyes into polymers. Also, this matrix is potentially useful for applications such as optical recording and switching devices[14-16].

2. Experimental

2.1. Metal alkoxide-polymer matrix

PMMA was used as the binder in which tetraethylorthosilicate (TEOS) was polymerized. The sample was prepared in a glass container by adding the organic polymer dissolved in a co-solvent (tetrafuran 5ml + ethanol 1ml). The glass container was immersed in an ice water bath and placed in a magnetic stirrer for a slow chemical reaction. The TEOS (2ml) was added directly to the solution under continuous agitation. Then a stoichiometric amount of acidic water (0.1ml 0.05N HCl) was added. After the addition of photochromic dye to the solution, the solution was stirred for several hours at ambient temperature. In this process, 1,3-dihydro-1,3,3-trimethyl-6-nitrospiro[2H-1-benzopyran-2,2'-(2H)-indoline] (NBPS) (10 wt %) was used as the photochromic dopant. The mixing order of the TEOS, the catalyst, and the photochromic dopant were considered. The addition of polymer into TEOS, resulted in a cloudy solution or precipitation. Too much acidic water caused phase separation. The final solution was used to make the films on silicon and quartz substrates with a 6000 rpm spinning frequency and a time of 1 minute.

2.2. Polymer based matrix

The coating solutions were prepared in a glass container, and NBPS was again used as the photochromic dopant. An NBPS doped PMMA film was obtained from a chloroform solution of PMMA and 10 wt % NBPS. The NBPS mixed with PMMA in chloroform solution was poured onto spinning cleaned quartz and silicon substrates. The spinning frequency and time were 5000 rpm and 1min. The film was dried in a dark place under ambient conditions for a day.

2.3. Spectroscopy

For the UV-VIS and FTIR spectroscopy the spectra of UV-irradiated films were obtained by placing the samples in Xenon-lamp irradiation for about ten minutes and then immediately putting them in the spectrometers.

The ellipsometric data with UV-irradiation was taken with the Xenon lamp irradiating the sample. Experiments with silicon wafers showed that the signal detection apparatus prevented this irradiation from changing the spectra of non-photochromic materials.

3. Results and discussion

3.1. UV/VIS Spectroscopy

UV/VIS transmission spectra of films on quartz substrate, with and without UV irradiation, were measured. Spiropyran compounds such as NBPS show photochromism which results from changes between the spiropyran form and the merocyanine form with C=C double bond conjugation[5], as shown in Figure 1.

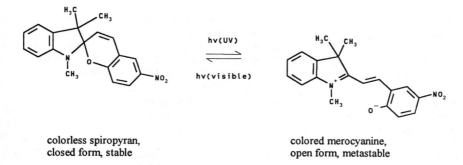

colorless spiropyran,
closed form, stable

colored merocyanine,
open form, metastable

Figure 1. Photochromic transformation of spiropyran (NBPS)

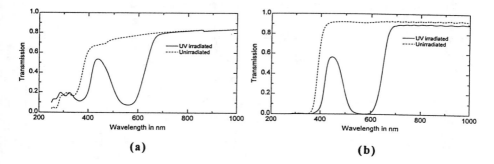

Figure 2. Transmission spectra with and without UV irradiation (a) MP-8 (b) P-9

The UV/VIS spectra of NBPS doped film MP-8, with and without UV irradiation, are shown in Figure 2(a). It does not absorb strongly in the visible region without UV light. In contrast, the irradiated MP-8 shows absorption maxima at 365 and 556 nm. NBPS in polar solvents shows an absorption maximum at 540 nm[5] and evaporated films of NBPS derivatives, which have long alkyl chains, absorb at 380 and 580 nm in the visible region[17]. The absorption shift in the irradiated MP-8 film is explained by the polarity of the NBPS. The polarity of this colored film MP-8 is thought to be higher than that of the NBPS derivatives with long alkyl chains, and lower than those of ethanol or other polar solvents[18].

Spiropyran in the polymer matrix also exhibited photochromism in the UV/VIS spectrum. The spectra of NBPS doped film P-9 are shown in Figure 2(b). The absorption peaks are located in the same positions as in sample MP-8, since the same photochromic dopant is used in both samples. The intensity difference between MP-8 and P-9 was due to the different matrices or different dye concentrations. The reversible photochromism illustrated between the irradiated and unirradiated spectra was repeated several times.

3.2. UV/VIS Ellipsometry

These measurements were made on films that were spun onto silicon substrates. The ellipsometric data of the samples were obtained, with and without UV light irradiation, at an incidence angle of 75 °. The MP-8 ellipsometric data showed a nonuniformity in the film thickness, as evidenced by washing out of interference structure in the spectra[19]. These data were fitted with a model consisting of a single film on a silicon substrate. The unirradiated film optical constants were assumed to follow a Cauchy relation ($n = A_n + B_n/\lambda^2 + C_n/\lambda^4$, $k = A_k \exp[B_k/\lambda]$). The irradiated film required the addition of Lorentz oscillators, representing absorption features. In addition the nonuniformity of the film thickness was modeled. Details of the modeling procedure will be published elsewhere.

The center energies and broadening of the three oscillators were 2.08 (596 nm), 2.34 (530 nm), 3.1 eV (400 nm) and 0.3, 0.3, 0.2 eV, respectively. The fitted thickness of sample MP-8 was 1900 nm, with an 11% thickness nonuniformity. Figure 3 shows the modeled optical constants for sample MP-8.

The P-9 ellipsometric data in Figure 4 show significant changes in Ψ and Δ at 500 through 620 nm and around 400 nm, with UV irradiation. In fitting sample P-9 data, two Lorentz oscillators were added to the Cauchy model to show a good fit with the data.

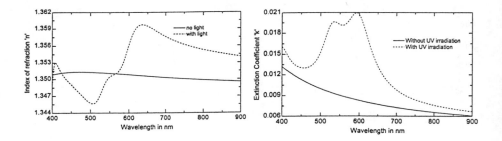

Figure 3. Optical constants of sample MP-8 with and without UV irradiation

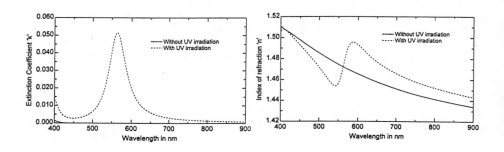

Figure 4. Optical constants of sample P-9 with and without UV irradiation

The center energies and broadening of the oscillators were 2.2 (564 nm), 3.11 eV (400 nm) and 0.21, 0.23 eV, respectively. The modeled optical constants of the film P-9 are presented in Figure 4 and the fitted thickness was 5790 nm, with 5.3% nonuniformity. The P-9 sample showed the biggest changes in optical constants, with irradiation. This is because the film is very thick and hence contains many photochromic molecules.

3.3. Fourier Transform Infrared Spectroscopy

As discussed in section 3.1, spiropyran compounds show photochromism which results from changes between the spiropyran form and the merocyanine form with C=C double conjugation. In this study, both spiropyran doped samples showed absorption changes in FTIR transmission spectra due to UV light. In FTIR transmission spectra for our polymer samples which are corporated with a polymer or a metal alkoxide matrix, we observed major changes at 1200 - 1350 cm^{-1} and 1500 - 1700 cm^{-1} with UV light. The two spectral regions (1700-1500 cm^{-1} and 1350-1200 cm^{-1}) in P-9 transmission spectra are useful regions for spiropyran bond opening. The transmission ratio for the irradiated to unirradiated film P-9 decreased by as much as 24%, as shown in Figure 5, and the average transmission ratio over this wavelength range is decreased. Thus this film could be useful for thermal management since the spacecraft would receive less of the total radiation when in the sun. The MP-8 film show results similar to those for P-9 film in the wavenumber range between 800 and 1700 cm^{-1}.

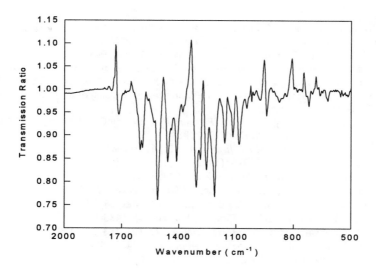

Figure 5. FTIR transmission ratio: unirradiated to irradiated sample P-9

4. Conclusions

The ellipsometric data showed significant changes in Ψ and Δ with UV irradiation. The Cauchy and Cauchy-Lorentz combined model were used to fit the unirradiated and irradiated films, respectively. It was found that UV light irradiation produced absorption bands near 400 and 560 nm. This means that the absorption coefficient and hence the emissivity of films is being modulated with UV light. The calculated thicknesses of the films were in the range of 1 to 6 microns.

The FTIR data showed that a modulated IR transmittance was achieved in the spiropyran based materials. A key factor in this study is the conversion of spiropyran to merocyanine under a UV lamp since it causes changes in the infrared as well as the visible. The objective of modulating IR transmittance and the emissivity of films with spiropyran dopants was achieved. Wide band modulation from 6 to 15 microns was realized with specific wavelengths reaching transmittance changes of about 24 %. For space applications the robustness of the materials needs to be tested, and possibly further work needs to be done to achieve space-compatible materials. The incorporation of organic polymeric materials into inorganic oxides by polymerization process is one way to achieve robust materials.

Acknowledgement

We thank the National Aeronautics and Space Administration (NASA) for supporting this research under grant No. NAGW-4414.

References

1. S. L. Gilat, S. H. Kawai, and J. M. Lehn, *J. Chem. Soc. Chem. Commun.* 1439(1993).
2. F. Ebisawa, M. Hoshino, and K. Sugekawa, *Appl. Phys. Lett.*, **65**(23), 2119(1994).
3. T. Koshido, T. Kawai, and K. Yoshino, *Synthetic Metals*, **73**, 257(1995).
4. *Photochromism, Molecules and Systems*, Edited by H. Durr (Elsevier, New York, 1990).
5. *Photochromism, Techniques of Chemistry*, Edited by G. H. Brown (Wiley, New York, 1971).
6. C. J. Brinker and G. W. Scherer, *Sol-Gel: The Physics and Chemistry of Sol-Gel Processing* (Academic Press, New York, 1990).
7. L. E. Scriven, in *Better Ceramics Through Chemistry II*, Edited by C. J. Brinker, D. E. Clark, and D. R. Ulrich (Mat. Rec. Soc., Pittsburg, Pa., 1988).
8. M. Wenzel and G. K. Atkinson, *J. Am. Chem. Soc.*, **111**, 6123(1989).
9. C. J. T. Landry et al, *Polymer*, 33(7), 1496(1992).
10. V. A. Krogauz and E. S. Goldburt, *Macromolecules*, **14**, 1382(1981).
11. J. J. Fitzgerald, C. J. L. Landry, and J. M. Pochan, *Macromolecules*, **25**, 3715(1992).
12. B. K. Bradley, J. J. Fitzgerald, and V. K. Long, *Macromolecules*, **26**, 3702(1993).
13. C. J. I. Landry, B. K. Coltran, and B. K. Bradley, *Polymer*, **33**, 1486(1992).
14. S. S. Xue, G. Manivannan, and R. A. Lessard, *Thin Solid Films*, **253**, 228(1994).
15. V. Weiss, A. A. Friesem, and V. A. Krogauz, *J. Appl. Phys.*, **74**(6), 4248(1993).
16. Y. Atassi, J. A. Delaire, and K. Nakatani, *J. Phys. Chem.*, **99**, 16320(1995).
17. S. Hayashida, H. Sato, and S. Sugawara, *Poly. J.*, **18**(3), 227(1986).
18. S. Hayashida, H. Sato, and S. Sugawara, *Jpn. J. Appl. Phys.*, **24**(11), 1436(1985).
19. S. Pittal, P. G. Snyder, and N. J. Ianno, *Thin Solid Films*, **233**, 286(1993).

SYNTHESIS AND PROPERTIES OF MAGNETORHEOLOGICAL (MR) FLUIDS FOR ACTIVE VIBRATION CONTROL

PRADEEP P. PHULÉ*, JOHN M. GINDER**, ARUN D. JATKAR*,
* Department of Materials Science and Engineering, University of Pittsburgh, 848 Benedum Hall, Pittsburgh, PA 15261, email: phule@engrng.pitt.edu.
** Ford Research Laboratory, Ford Motor Company, 20,000 Rotunda Drive, SRL MD 3028, Dearborn, Ml 48121-2053, email: jginder@ford.com.

ABSTRACT

Magnetorheological (MR) fluids represent an exciting class of smart materials for use in active vibration control and other applications. This paper discusses some of the fundamental materials science concepts that define the scientific basis for designing MR fluids. Preliminary experimental data and observations concerning the synthesis as well as the rheological behavior of MR fluids based on carbonyl iron and iron oxide particulates are presented and discussed.

INTRODUCTION

Magnetorheological (MR) fluids represent an exciting family of smart materials that have the unique ability to undergo rapid (within a few ms), nearly completely reversible, and significant ($\sim 10^5$ - 10^6 times) changes in their apparent viscosity on application of an external magnetic field[1][2][3]. These fluids typically consist of fine particles of a magnetically soft material (e.g., iron) dispersed in an organic medium such as mineral or silicone oil (Figure 1).

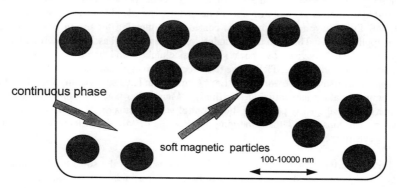

Figure 1. *Schematic of a typical MR fluid. In addition to the magnetic particles and the continuous liquid phase, typical fluids may also contain surfactants and other additives.*

In the *absence* of a magnetic field, MR fluids possess a relatively small apparent viscosity (~ 0.2 - 0.3 Pa-s) and therefore exhibit flow properties similar to those of common dispersions such as paints. However, when an external magnetic field is applied, the originally multi-domain particles, which have little or no net magnetization, are transformed into particles with a net magnetic moment **m**. This introduces an additional interparticle force: the magnetic dipole-dipole interaction (Figure 2). For ferromagnetic and ferrimagnetic particles the magnetic dipole-dipole interaction energy (U) is considerably stronger than other interparticle forces (e.g., Van der Waals, steric, and electrostatic). This interaction energy is a strong function of the interparticle separation R and the angle φ between the magnetic moment **m** and the vector **R** that joins the particle centers.

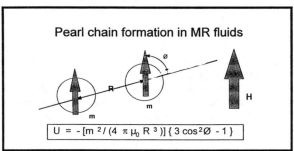

*Figure 2. Energy of the magnetic interaction (U) between a pair of magnetic particles. The magnetic moment of each particle is represented by the vector **m**.*

Since the energy of magnetic dipole-dipole interaction is minimized when the dipole moments of the particles are aligned with the externally applied magnetic field **H**, the particles undergo fibrillation or pearl chain formation. This fibrillated structure leads to a gel-like material which can, depending on the MR fluid composition, possess a substantial yield stress – 10 to 100 kPa – and is therefore capable of dissipating considerable mechanical energy. As a result, MR fluids have recently attracted considerable attention for use in active vibration damping applications. The automotive industry, in particular, has been interested in the development of such devices as shock absorbers, clutches, and engine mounts[4] [5] [6] [7] [8]. Shtarkman and coworkers[5] [7] at TRW have led a recent revival of interest in MR fluids in the United States. Carlson and coworkers of Lord Corporation have successfully introduced one or more commerical devices utilizing MR fluids[9]. Many other potential applications of MR fluids such as optical quality polishing, flow control valves, actuators,[10] and so on have been reviewed by Kordonsky[11]. Other emerging applications of MR fluids include dampers for controlling seismic response[12] [13].

Prior to further discussion, it is useful to emphasize that the so-called *ferrofluids* (or *magnetic fluids*) are very different from MR fluids. Ferrofluids[14] are based on *nanosized* (particle sizes typically less than 10 nm), *superparamagnetic* particles of such materials as iron oxide which do *not* exhibit large changes in their viscosity in an applied magnetic field. In fact, ferrofluids do *not* show any measurable yield stress in the presence of a field. Applications of ferrofluids, which are primarily in the area of sealing devices, are discussed elsewhere[15]. Another noteworthy family of materials that is a close relative of MR fluids is electrorheological or ER fluids[16]. Both ER and MR fluids were first developed in the 1940's, by Winslow[17] and Rabinow[18] respectively.

EXPERIMENTAL PROCEDURE AND DISCUSSION OF RESULTS

Synthesis of MR fluids

Novel MR fluids based on iron and iron oxide particles suspended in polar organic liquids, rather than mineral oil or silicone oil, were prepared[19]. Additives were also incorporated in an effort to enhance the redispersibility and yield stress of these suspensions. Typical volume fractions φ of the solid phase were 0.3 - 0.4, although experiments were also conducted with lower volume fractions (0.075 and 0.15). The process for preparing these MR fluids typical involved introducing the magnetic particles into the base liquid under low shear conditions, followed by ball milling with zirconia (ZrO_2) grinding media for 24 hours. Carbonyl iron powders (ISP Technologies, New Jersey) were used for the synthesis of iron-based MR fluids, while the iron-oxide-based MR fluids used magnetic iron oxide (Fe_3O_4) powder (Steward Magnetics, Chattanooga, TN).

Rheological Characterization of the MR fluids

MR fluids were characterized using a specially designed and built rheometer. The details of the system have been previously reported[20]. This Couette rheometer is based on a concentric-cylinder fixture constructed from low carbon steel; a radial magnetic field in the annular sample is generated using a copper coil driven by an operational amplifier power supply (Kepco 50-8). A search coil in conjunction with a flux meter (Walker MG-3D) was used to estimate the average magnetic induction B in the sample. All MR fluid samples were mixed well before introducing them into the rheometer to minimize the possibility of phase separation due to gravitational settling during storage.

Characterization of magnetic particles used in MR fluids

The carbonyl iron and iron oxide particles used to prepare the MR fluids were characterized using x-ray diffraction (XRD) and scanning electron microscopy (SEM). For XRD analysis a Phillips X-PERT system was used. A vibration sample magnetometer (VSM) was also used to ascertain the ferro- or ferri-magnetic nature of the particles.

Discussion Of Results

The results of XRD analysis (θ-2θ diffraction patterns) revealed that the carbonyl iron and iron oxide powders were single phase. SEM analysis revealed that the Fe particles were spherical in shape and about 5 μm in size. The primary particles of iron oxide were found to be finer, about 2 μm, but aggregated.

Figure 3 shows some of the preliminary rheological data on iron-based fluids. The yield stress data were obtained by using the specially designed and built rheometer to measure the shear stress τ as a function of strain rate γ'. The shear rate was varied up to 200 s⁻¹. Shear stress measurements were repeated with positive and negative shear rates to eliminate effects of drift of the torque sensor signal. Measurements of shear stress as a function of shear rate were conducted at a given coil current chosen to deliver a targeted value of flux density to the sample. The yield stress was obtained by extrapolating the stress to zero strain rate. For iron-based fluids with a particle volume fraction of 0.4, the yield stress achieved is about 120 kPa with a flux density of 1.0 Tesla (Figure 3). Other experiments show that even with relatively small particle volume fractions (7.5 and 15%), the iron-based fluids exhibit relatively high yield stresses (up to 60 kPa, at a magnetic flux of 1 Tesla). At very high flux densities in these dilute fluids some phase separation, potentially induced by the spatial nonuniformity of the magnetic field, can occur; the yield stresses measured at high flux densities thus represent the maximum possible values.

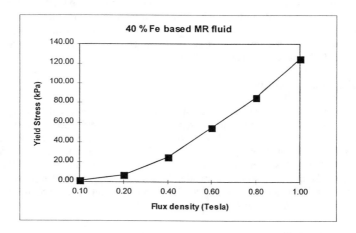

Figure 3. *Preliminary dependence of yield stress on flux density for an MR fluid based on carbonyl iron particles with a volume fraction of 0.4.*

    Preliminary rheological data on iron-oxide-based MR fluids have also been obtained (Figure 4). In these fluids, yield stresses up to ~ 15 kPa have been observed. These yield stress values are higher than those reported by BASF researchers for MR fluids based on ultrafine ferrites (Ref. 10). The BASF MR fluids, however, have an exceptional stability toward sedimentation, which is not observed for the iron oxide fluids discussed here. Qualitatively, the iron-oxide-based fluids do exhibit better stability against sedimentation compared to the iron-based MR fluids. The field-induced yield stresses obtained in the fluids prepared using iron and iron oxide particles are consistent with the models developed by Ginder and coworkers.[20][21] One of the central predictions of these models is that the ultimate yield stress obtainable in an MR fluid scales quadratically with the saturation magnetization of the magnetizable particles. Indeed, we observe that iron-oxide-based MR fluids show relatively lower yield stresses owing to the lower value of saturation magnetization ($\mu_0 M_s \sim$ 0.6 Tesla), while iron-based MR fluids show a considerably higher yields stress ($\sim$ 100 kPa) since the saturation magnetization of iron is significantly higher ($\mu_0 M_s \sim$ 2.1 Tesla).

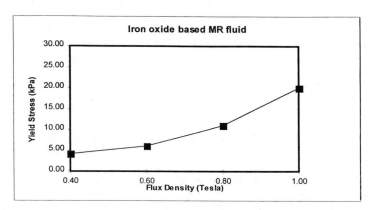

Figure 4. Dependence of yield stress on flux density for an iron-oxide-based MR fluid with a volume fraction of 0.4.

The yield stresses observed for both iron- and iron-oxide-based fluid are significantly higher than the largest yield stresses reported for most practical ER fluids. Obtaining stable and redispersible MR fluids that possess very high yield stresses, however, remains a challenge and is being addressed in our on-going research. We have also built some prototype devices for active damping using both iron- and iron-oxide-based MR fluids. The results concerning vibration damping using these devices as well as the stability and redispersibility of MR fluids will be published elsewhere.

ACKNOWLEDGMENTS

PPP acknowledges the financial support by the Ford Motor Company. Financial support for constructing a rheometer system at the University of Pittsburgh was provided by a special grant from the Office of the Provost, University of Pittsburgh and is acknowledged. PPP would also like to acknowledge the support through William Kepler Whiteford faculty fellowship from the University of Pittsburgh. Contributions from and discussions with Professor W. A. Soffa, Professor R. Marangoni, Professor William Clark, Chi-Chi Odoemene, Dan Kroushl, Amy Nebist, Darrin Schwartz, and Sumukh Patil of the University of Pittsburgh, Dr. Rattya Yalamanchili of IDD Inc., Pittsburgh, Mark Garbowski of Armco Steel, Ohio, and Dr. L. Craig Davis of the Ford Research Laboratory are also acknowledged.

REFERENCES

[1] J.Ginder, "Rheology Controlled by Magnetic Fields," in *Encyclopedia of Applied Physics* (1996) Vol.16, ed. by E. Immergut (VCH, New York, 1996), 487-503.

[2] *Electrorheological Fluids, Magnetorheological Suspensions, and Associated Technology*, Editor W.A. Bullough, (World Scientific, Singapore, 1996): Proc. of the 5th International Conference on ER Fluids and MR Suspensions, Sheffield, UK 10-14 July, 1995.

[3] P.P. Phulé, "An Overview of Synthesis and Applications of Magnetorheological (MR) Fluids," Invited Paper presented at the Annual Meeting of the American Ceramic Society, April 15-18, 1996.

[4] J.D. Carlson et al., U.S. Patent, 5,382,373, issued Jan.17, 1995 to Lord Corporation.

[5] E.M. Shtarkman, U.S. Patent 5,167,850, issued Dec.1, 1992 to TRW Corporation.

[6] K.D. Weiss, T.G. Duclos, J.D. Carlson, M.J. Chzran, A.J. Margida, SAE publication number 932451.

[7] E.M. Shtarkman, U.S. Patent 5,354,488, issued Oct.11, 1994 to TRW Corporation.

[8] M.R. Jolly and J.D. Carlson, in Actuator 96, 5th International Conf. on New Actuators, 26-28 June 1996, Bremen, Germany.

[9] Information on WWW address: http://www.webcom.com/~mrfluid/welcome.html

[10] R. Bölter, H. Janocha, St. Hellbrück, and Cl. Kormann, in Actuator 96, 5th International Conf. on New Actuators, 26-28 June 1996, Bremen, Germany.

[11] W. Kordonsky, J. of Intelligent Materials Systems and Structures, Vol.4, (1993) p. 65-69.

[12] B.F. Spencer, Jr., S.J. Dyke, M.K. Sain and J.D. Carlson, "Phenomenological Model of a Magnetorheological Damper," Journal of Engineering Mechanics, ASCE, in press (1996).

[13] J.D. Carlson, et al., "Seismic Response Reduction using Magnetorheological Dampers," Proc. of the IFAC World Congress, San Francisco, California, June 30 - July 5, 1996.

[14] R.E. Rosensewig, *Ferrohydrodynamics*, (Cambridge University Press, Cambridge, 1985).

[15] R.E. Rosensweig, "Magnetic Fluids: Phenomena and Process Applications," Chemical Engineering Progress, April 1989, p.53-61.

[16] For an overview of the ER technology see, e.g., F.E. Filisko, in *Progress in Electrorheology*, Edited by K.O. Havelka and F.E. Filisko, (Plenum, New York, 1995) p.3; Proceedings of the Symposium on American Chemical Society Symposium on Electrorheological Materials and Fluids.

[17] W.M. Winslow, J. Appl. Phys. (1949) 20, 1137; U. S. Patent 2,417,850 (1947).

[18] J. Rabinow, AIEE Trans. (1948) 67, 1308; U. S. Patent 2,575,360 (1951).

[19] P.P.Phulé, patent application to be filed (1997).

[20] J.M. Ginder, L.C. Davis, and L.D. Elie, "Rheology of Magnetorheological Fluids: Models and Measurements," in *Electrorheological Fluids, Magnetorheological Suspensions, and Associated Technology*, Editor W.A. Bullough, (World Scientific, Singapore, 1996), p. 504-514.

[21] J.M. Ginder and L.C. Davis, Appl. Phys. Lett. (1994) 65, p. 3410-3412.

Part III

Composites

ACTIVE STIFFENING OF COMPOSITE MATERIALS BY EMBEDDED SHAPE-MEMORY-ALLOY FIBRES

J.-E. BIDAUX*, J.-A. E. MÅNSON*, R. GOTTHARDT**
*Laboratoire de Technologie des Composites et Polymères
**Institut de Génie Atomique
Ecole Polytechnique Fédérale de Lausanne, CH-1015 Lausanne, Switzerland

ABSTRACT

The use of shape-memory-alloy (SMA) fibres to actively change the stiffness of a composite beam is investigated on a model system composed of an epoxy matrix with a series of embedded pre-strained NiTi fibres. Stiffness changes are detected through shifts in the natural vibration frequency of the beam. When electrically heated, the pre-strained NiTi fibres undergo a phase transformation. Since the shape recovery associated with the transformation is restrained by the constraints of both the matrix and the clamping device, a force is generated. This force leads to an increase in the natural vibration frequency of the composite beam. Depending on the degree of fibre pre-strain, either ordinary martensite, R-phase or a mixture of the two can be stress-induced. It is found that the R-phase gives rise to the largest change in vibration frequency for a given temperature increase and the most reversible behaviour. Its low transformation strain is also more favourable for fibre-matrix adhesion. The effect of stress relaxation in the polymer matrix on the composite response is discussed.

1. INTRODUCTION

The unique properties of shape memory alloys (SMA), namely shape memory, superelasticity and high damping capacity [1], are all related to a displacive, diffusionless and reversible phase transformation from a high temperature austenitic phase to a low temperature martensitic phase. The austenite-to-martensite transformation starts at the so-called M_s temperature and finishes at the A_s temperature while the retransformation start and finish temperatures are designed A_s and A_f, respectively. These temperatures depend on the alloy composition, on previous thermomechanical treatments as well as on the presence of an applied stress. The martensitic state is characterised by a fine domain structure which can be re-oriented under an applied stress. The domains which produce a strain in the direction of the stress are favoured compared to the other domains which results in a macroscopic deformation of the material. Unlike normal plastic strain, this deformation is totally recovered by heating above the martensite-to-austenite transformation temperature. If the shape recovery is restrained by a bias spring or a clamping device, a stress will appear instead of a deformation and is referred to as the recovery stress. Both shape memory and recovery stress effects can be used to build powerful thermally-activated actuators. Among the advantages of SMA actuators compared to other types of actuators are the very large deformations (>5%) and stresses (>500 MPa) that can be generated.

SMA actuators have recently attracted interest in the field of composite materials. Layers of SMA fibres can be embedded in composites and used to actively control e.g. shape, elastic moduli, internal stress level and natural frequencies of vibration of the composite [2-11]. A large body of experiments has been devoted to these new materials. Active strain energy tuning and active modal modification have been successfully used to modify the vibrational characteristics of composite plates containing SMA fibres [7-14]. SMA-matrix interaction in composites and the effect of the matrix on the phase transformation of the fibre have been investigated [15-16]. Durability issues related to the use of SMA-composite laminates have been studied [4]. Models describing SMA-matrix interaction and predicting thermomechanical properties have been developed [17-20].

In NiTi alloys the martensitic transformation is often preceded by the so-called R-phase transformation [21-24]. The R-phase is a rhombohedral distortion of the original austenitic

Mat. Res. Soc. Symp. Proc. Vol. 459 © 1997 Materials Research Society

structure. It is characterised by a small transformation strain, a very high stress rate dσ/dT and a very small hysteresis. In composite materials, stress rate is of considerable importance, since we are looking for the highest recovery stress as possible for a minimum temperature increase. On the other hand, the transformation strain associated with R-phase transformation is small which should lead to increase durability since the shear stresses at the fibre-matrix interface would be lower. In addition, as shown in Table 1, this transformation occurs with a minimum hysteresis which makes it more suitable for control.

Table 1. R-phase and martensitic phase transformation characteristics in NiTi alloys [22]

transformation	Mart->A	R->A
strain (%)	5.3	0.6
dσ/dT (MPa/°C)	6.4	15.6
hysteresis (°C)	>10	2

In this paper, R-phase transformation is used for the first time to modify the effective stiffness of a composite beam. An inplane tensile stress (recovery stress) is generated by the phase transformation of embedded NiTi fibres. This tensile stress arises because the composite beam is restrained by a clamping device. The result is an increase of the effective stiffness of the beam which reflects in a change of the vibrational characteristics. It is shown that R-phase is very effective in modifying the vibration frequencies of composite materials.

2. EXPERIMENTAL

This work focused on epoxy beams ($150 \times 10 \times 0.7$mm) containing five embedded NiTi fibres placed along the neutral plane of the beam at a distance of 2mm from one another, Fig. 1. The fibres were 300μm diameter NiTi fibres (50.17 at% Ni) supplied by SMA INC., USA. The fibres were annealed for 10min at 475°C in an argon atmosphere and then quenched in water at 20°C. The fibres had the following characteristic transformation temperatures: $M_s \cong -13$°C, $M_f \cong 43$°C, $A_s \cong 30$°C, $A_f \cong 63$, $T_R = 42$°C, according to DSC measurements. The fibres were pre-strained at room temperature and embedded into the epoxy resin. The pre-strain was maintained during curing, restraining the fibres from their normal shape recovery. The resin, supplied by Ciba Geigy, was composed of LY556 bisphenol A epoxy resin, HY917 anhydride hardener and DY070 amine accelerator, mixed in the proportions 100:90:0.5 by mass. The resin was cured 4h at 80°C and then 8h at 140°C, resulting in a glass transition temperature of $Tg \cong 150$°C.

Fig.1. Schematic drawing of the SMA-composite containing five 300μm diameter NiTi fibres embedded in the neutral plane of the beam (scale is in mm)

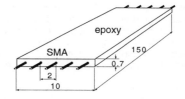

A new experimental set-up was developed to study the change of the vibrational response of SMA-composite beams activated by electrical current heating (Fig. 2). The experimental set-up is composed of a sample holder, an electromagnetic shaker, a Laser Doppler Detector (Brüel & Kjaer) and additional electronic modules to generate, record and treat the measured signals. The U-shaped steel sample holder is fixed on the electromagnetic shaker. Transverse vibrations of the composite beam (displacement perpendicular to the plane of the beam) are excited by a shaker. The fundamental resonance frequency of the beam is extracted by taking the maximum on the Fourier Transform of the measured signal. The sample holder was designed to be sufficiently stiff so as to avoid unwanted resonances in the low frequency range.

The experimental set-up allowed the measurement of the static longitudinal force generated in the plane of the beam by the phase transformation of the SMAs. As will be shown later on, the modifications of the frequency response are mainly due to the existence of this static force. The force transducer was made of four strain gauges glued on the machined, 4mm thick, part of the sample holder. Two strain gauges were mounted on each opposite face and connected in full bridge to compensate for errors due to changes of temperature. The transducer was calibrated using a series of weights. The composite beam could be pretensioned by means of a screw placed on one of the fixtures of the sample holder.

The NiTi fibres were heated by direct electrical current and cooled both by natural convection at the composite surface and by conduction at the composite ends. The current was increased/decreased incrementally. The temperature was allowed to stabilize for a minimum of 30 sec prior to measurements being taken. In the case of the frequency measurements, the stabilization time was longer (2-3 min) since, in that case, the current increments were larger. The fibres were connected in series to ensure a rigorously constant current in all the fibres. The temperature of the composite was monitored using 100 μm diameter thermocouples. The thermocouple was introduced in a small hole drilled in the centre of the composite beam close to the central fibre. The contact between the thermocouple and the sample was ensured by a heat transfer compound composed of silicone grease and metallic oxides. In the case of the plain fibres, a smaller thermocouple, 25μm in diameter, was used to minimise thermal loses through the thermocouple wires.

Additional information on the phase transformation process was obtained by measuring the electrical resistance of the NiTi fibres given by the ratio between the electrical voltage measured between the clamped fibre ends and the heating electrical current. Interestingly, the electrical resistance provides information on the phase transformation process within the fibre without any contribution from the insulating surrounding matrix. DSC thermograms were measured on a Perkin Elmer differential scanning calorimeter (DSC7) at the heating rate of 20°C/min.

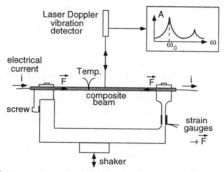

Fig.2. Schematic drawing of the experimental set-up allowing the simultaneous measurement of the longitudinal force, electrical resistance and frequency response of SMA-composite beams

3. RESULTS AND DISCUSSION

3.1 The SMA fibre actuator:

Fig. 3 shows a typical DSC thermogram of a NiTi fibre directly after annealing and without any pre-strain. Two exotherms are observed on cooling, each one corresponding to a particular phase transformation. The first, at $T_R=42°C$, is due to the austenite to R-phase transformation (A->R), the second to the normal martensite transformation (R->M) [22]. On heating, a single endotherm is observed which is due to the martensite to austenite (M->A) reverse transformation. We conclude that directly after the high temperature annealing treatment, the stable phase of the NiTi fibre at room temperature is the R-phase. Its structure is multivariant since no stress (or strain) has been applied to the fibre. If the fibre is elongated as is done when the fibre is pre-strained (Fig. 4), the situation is more complex and depends on the magnitude of the applied strain. Two plateaus are observed on the stress strain curve. The first plateau is related to the re-orientation of the R-phase variants [21], the second to the martensitic transformation. Martensitic transformation does not occur at once but takes place progressively, proportionally to the applied strain. For strains in excess of about 6.5%, the NiTi is fully martensitic and no more R-phase remains in the sample. It can be concluded that the volume fraction of R and M phase in the fibre can be controlled through the degree of pre-strain.

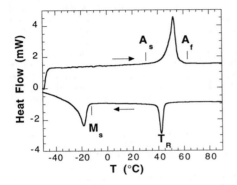 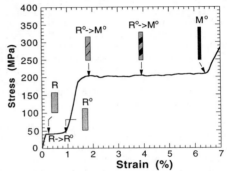

Fig. 3. DSC thermogram of the NiTi fibre measured at the heating/cooling rate of 20°C/min.

T_R indicates R-phase transformation temperature, M_s and M_f martensite start and finish temperatures, A_s and A_f austenite start and finish respectively

Fig. 4. Stress-strain curve of the NiTi fibre measured at room temperature at a constant strain-rate of 0.9%/min. R->R° indicates re-orientation of the R-phase variants and R°->M° the transf. of oriented R-phase to martensite

A typical temperature dependence of the force (recovery force) developed by the SMA fibre actuator is shown in Fig 5 a. The fibre had the two ends clamped and was pre-strained by 2.5% using the screw shown in Fig.2. The fibre was then electrically heated and the recovery force measured without previous unloading. The first thermal cycle exhibits a hysteresis which is not shown here. Fig.5 shows that the subsequent cycles are reversible with almost no hysteresis. The force at room temperature is not zero, for this particular pre-strain value. When the temperature is increased, the force also increases. The major variation of the force occurs between 50 and 65°C. This large force increase is related to the A->R phase transformation. The slope of the force-temperature curve is about 2 N/°C which corresponds to a stress-rate of about 25 MPa/°C. This high stress rate is typical of the R-phase transformation (see Table 1). In the case of martensitic transformation the stress rate would be lower, of the order of 7 MPa/°C (see Table 1). Superimposed to this abrupt force change, an almost linear increase of the force is observed which, for pre-strains equal or below 1.25% is only observed below 50°C but above 2.5% also appears at high temperature (see Fig. 6 a). According to Fig 4, for pre-strains above 1%, some martensite is stress-induced. This small additional increase of the force can therefore be attributed to the martensite to austenite reverse transformation.

The electrical resistance behaviour is shown in Fig 5 b. In agreement with the literature[24], R->A transformation is accompanied by a decrease of the electrical resistance. The increase of the resistance observed after the end of the R-phase transformation (T>65°C) is due to the re-transformation of the stress-induced martensite.

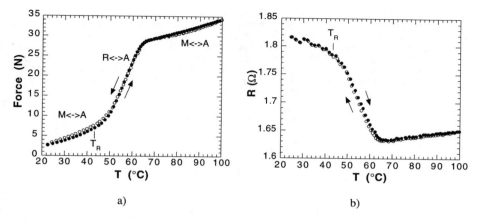

a)   b)

Fig. 5. Recovery force (a) and electrical resistance (b) versus temperature for a NiTi fibre pre-strained 2.5% (second thermal cycle); open circles: heating; closed circles: cooling; T_R indicates R-phase transformation temperature in the plain fibre according to calorimetric measurements

In Fig. 6, the temperature dependence of the recovery force and of the electrical resistance is shown for different pre-strain values. For pre-strains lower than approximately 1%, nearly identical electrical resistance versus temperature curves are observed (Fig. 6b). The electrical resistance decreases by about 15% when the temperature is increased above the R-phase transformation domain, which corresponds to the retransformation of a 100% R-phase sample to the austenite. For larger pre-strains, the decrease is smaller due to the amount of R-phase that has been previously transformed to martensite and does not contribute any longer to the decrease in electrical resistance.

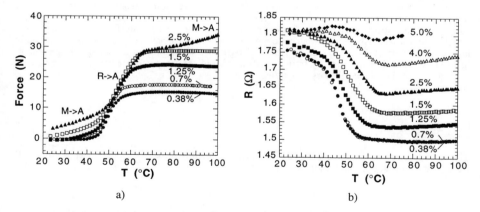

a)   b)

Fig. 6. Recovery force (a) and electrical resistance (b) versus temperature for NiTi fibres pre-strained to various amounts and measured on heating

As shown in Fig. 7, the change of electrical resistance diminishes for pre-strains higher than 1% and tends to zero at about 6.5%. This is consistent with Fig.4 which shows that the plateau stress corresponding to stress-induced martensite is finished near this strain value. This shows that the electrical resistance change is directly proportional to the quantity of R-phase retransformed to austenite. The electrical resistance does not seem to be very sensitive to the degree of orientation of the variants in the R-phase since all the curves are very similar for pre-strains below 1%. In contrast, the recovery force depends on the level of pre-strain, whatever the pre-strain values (Fig. 6 a). Below 1% a single stage increase of the force is observed associated with the R->A retransformation. For larger pre-strains the force changes in the whole temperature domain because of the M->A transformation.

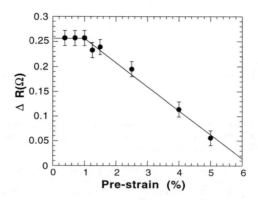

Fig. 7. Change in electrical resistance associated with the R-phase transition versus pre-strain. ΔR is defined as R(70°C)-R(25°C) after subtraction of R(T) slope

These measurements show that R-phase is very appropriate for applications in composites. R-phase transformation gives rise to large stress variations for a minimum temperature change (i.e. minimum electrical power). Moreover, the distortion associated with R-phase transformation is small, below 1%, which is certainly more favourable for fibre-matrix adhesion. Shear stresses which are always present at the fibre matrix interface can be minimised which would lower risk of fibre matrix debonding. On the other hand, R-phase transformation is known to be very resistant to transformation fatigue, which is another advantage. Experiments have shown [1] that millions of transformation cycles can be performed without significant degradation of the shape-memory properties.

3.2 The SMA-composite:

3.2.1. Force

The total force measured at the extremities of a clamped-clamped composite containing NiTi fibres can as a first approximation be written as:

$$F_{tot}(T) = F_0 - F_{th}(T) + F_{rec}(T) \qquad (1)$$

where F_0 is the external applied force (if any), $F_{th}(T)$ is the sum of the longitudinal thermal forces and $F_{rec}(T)$ is the total recovery force produced by the martensite-to-austenite transformation in the embedded NiTi fibres. It can be shown that the forces related to thermal expansion of the fibres are negligible compared to the forces related to matrix thermal expansion. Accordingly, the thermal forces can be written by the following equation:

$$F_{th}(T) = \alpha_m(T - T_0)E_m A_m \qquad (2)$$

where E_m, A_m and α_m are the matrix elastic modulus, cross-sectional area, and thermal expansion coefficient, respectively. T_0 is the room temperature. The recovery force $F_{rec}(T)$ is a complex function which depends mainly on the temperature and on the amount of pre-strain but also on the previous thermomechanical treatment. For plain NiTi fibres, $F_{rec}(T)$ is given by the curves of Fig. 6a. In embedded NiTi fibres, the transformation behaviour can be modified by the surrounding matrix and $F_{rec}(T)$ can differ from the behaviour shown in Fig. 6a.

Fig. 8(a) shows the temperature dependence of the longitudinal force for a composite containing 2.5% pre-strained NiTi fibres. At this pre-strain value, the structure of NiTi is a mixture of R and martensitic phase (see Fig. 4). However, the behaviour is still dominated by R-phase since according to Fig. 7, about 80% of the fibre volume is in the R-phase against about 20% in the martensitic phase. Starting at an initial tension, F_0, of about 20N, the force initially decreases when the temperature increases. This is due to the thermal expansion of the polymer matrix (see Eqs.1 and 2). Because thermal expansion is prevented by the clamping, a longitudinal compressive force exists in the matrix. As has been shown in a previous paper [12], in the absence of the recovery forces of the NiTi fibres, this compressive force eventually leads to a buckling of the composite beam. At a temperature of about 42°C which, according to calorimetric measurements, corresponds approximately to the R-phase transformation temperature, an increase of the force is observed. According to Eq. 1, it means that at that temperature, the tensile forces produced by the R-to-austenite transformation in the SMA fibres exceed the thermal compressive forces generated by the surrounding matrix. For temperatures above 42°C the slope of the force versus temperature curve is positive, the behaviour being dominated by the SMA fibres. Upon subsequent cooling, the curve follows nearly the same path as during heating with a small hysteresis. In Fig 8 (b) we observe that the total electrical resistance change, about 10%, is comparable to the one observed for the plain NiTi fibre pre-strained to the same degree (Fig.5 b). This seems to indicate that the R-phase transformation is not very much modified by the embedding.

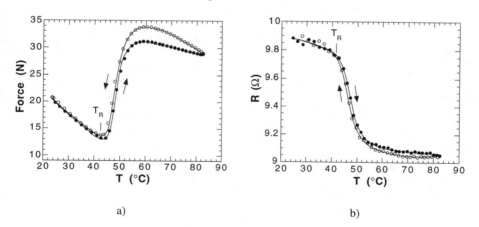

a)   b)

Fig. 8. Force (a) and electrical resistance (b) versus temperature for a composite with NiTi fibres pre-strained 2.5%; closed circles heating; open circles cooling; T_R indicates R-phase transformation temperature in the plain fibre according to calorimetric measurements

3.2.2. Effect of ageing

In order to examine in more detail the effect of the surrounding matrix on the force versus temperature behaviour in Fig. 9, a series of thermal cycles were performed which are schematically represented in Fig. 9 c. For each cycle, the hold time, t_a, at the maximum temperature of the cycle, T_a, was kept constant and equal to 1 hour. From cycle 1 to cycle 14, T_a was increased at each subsequent cycle, whereas from cycle 15 to cycle 22, T_a was decreased at each subsequent cycle. In the case of cycles 4 and 18, where T_a is below 80°C, the behaviour is totally reversible. In contrast, for cycles 11 and 14, where T_a is close to the matrix glass transition temperature, T_g, an increase of the force is observed during ageing which is not recovered during subsequent cooling. As shown in Fig 9 b, where the residual force F_{res} at room temperature is plotted versus T_a, the most important change occurs in the vicinity of T_g. This irreversible increase of the force can be explained by the relaxation of the thermal stresses which build up when the temperature is changed. Let us neglect the internal stresses generated during processing. When the temperature is increased, thermal compressive forces build up during heating mainly because of the constraint of the clamping device but also due to the NiTi fibres. These thermal forces are responsible for the decrease of the force seen in the initial stage of the curve 4 in Fig 9a. Since the polymer is still in the glassy state, these stresses cannot relax. When the temperature is increased in the vicinity of the glass transition temperature, the stresses begin to relax. Since the forces are compressive, the relaxation of the forces will give rise to an increase of the total measured force, which is what we observe in Fig. 9 a. This shows that, if a reversible change of the force or frequency is desired, the temperature should be kept well below the glass transition temperature. If the glass transition temperature is too close to the working temperature, irreversible changes will occur. In our case, the maximum temperature should not exceed 80°C.

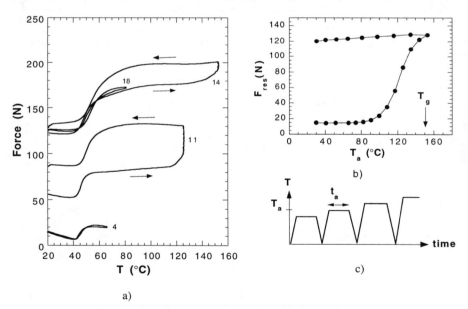

a)

b)

c)

Fig. 9. Effect of ageing on the reversibility of the force for composites with NiTi fibres pre-strained 2.5%. The numbers in (a) are the cycling numbers. Only some selected cycles are given as an example. The ageing time t_a was one hour. F_{res} in (b) is the residual force at room temperature after a thermal cycle at T_a

114

3.2.3. Frequency shift

Fig. 10 shows that, together with the force, the natural vibration frequency of the composite beam changes. The change in the resonance frequency is directly related to the change in the longitudinal force. The main point to notice is that here the force is internally generated by a change of the material state rather than by the application of an external force.

In Fig. 10 a, we observe that about 50% increase of the natural vibration frequency can be obtained at 2.5% pre-strain with a fibre volume fraction of only 5%. The change in frequency needs little heating power since the transformation occurs over a temperature range of less than 20°C. Moreover very good reversibility is observed since the R-phase is used and the temperature does not exceed 80°C. In Fig. 10 b, a similar experiment has been performed but using a pre-strain of 5%. At this pre-strain level, the NiTi is 80% martensitic. The behaviour is not as good as for a pre-strain of 2.5%, as expected when taking into account the characteristics of Table 1.

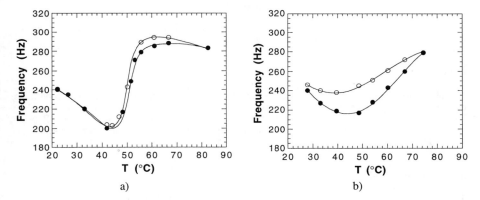

a)   b)

Fig. 10 Effect of the pre-strain on the natural vibration frequency change (fundamental mode) for composites with embedded NiTi fibres (a) 2.5% pre-strain (b) 5% pre-strain; closed circles: heating; open circles: cooling

The relationship between resonance frequency and the axial force of a beam in a clamped-clamped arrangement is given by the solution of the following equation [5]:

$$-2\sqrt{\rho A\lambda} - F_{tot}/\sqrt{EI}\sin(\alpha l)\sinh(\gamma l) + 2\sqrt{\rho A\lambda}\cos(\alpha l)\cosh(\gamma l) = 0$$

with $\quad \alpha = \sqrt{(-F_{tot} + \sqrt{F_{tot}^2 + 4EI\rho A\lambda})/2EI}$ $\qquad\qquad(3)$

and $\quad \gamma = \sqrt{(F_{tot} + \sqrt{F_{tot}^2 + 4EI\rho A\lambda})/2EI}$

where $\lambda = \omega^2$, ρA is the mass per unit length and l is the distance between the clamps [5]. This equation can be solved numerically. The experimental relationship between the fundamental resonance frequency and the longitudinal force is shown in Fig. 11. These experimental data are compared to the theoretical prediction of Eq.3, using the parameters: $l=106$ mm, $E_f=70$ GPa, $E_m=2.9$ GPa, $\rho_f=6.5$ g/cm^3, $\rho_m=1.2$ g/cm^3. The agreement is fairly good taking into account that the above parameters correspond to room temperature values. This shows clearly that, for the geometry used, the frequency change is dominated by the effect of the longitudinal force and that the temperature dependence of the bending stiffness (EI) is negligible.

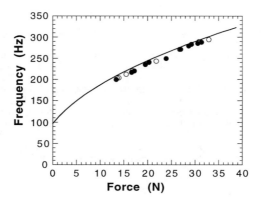

Fig. 11. Vibration frequency versus longitudinal force for the SMA composite with 2.5% pre-strained NiTi fibres (see Fig. 10 a). Full line: theoretical calculation using Eq. 3.

4. CONCLUSIONS

The above results confirm that SMA fibres can be used to control the vibrational behaviour of composites. Reversible increases/decreases of the vibration frequency can be obtained by heating/cooling the fibres using an electrical current. Our results show that the R-phase transformation is particularly effective in modifying stiffness and vibration frequency of composites materials.

5. ACKNOWLEDGEMENTS

The authors thank A. Van den Bossche for his help in performing the experiments presented in this paper. We also acknowledge P. Berguerand, P. Kim and F. Bonjour, from the Laboratoire de Technologie des Composites et Polymères and B. Guisolan from the Institut de Génie Atomique, Ecole Polytechnique Fédérale de Lausanne, Switzerland, for their help in designing and building the sample holder and the composite mould. We also would like to thank Dr A. Karimi and P.-H.Giauque from the Institut de Génie Atomique, for the use of their vibration detector. The financial support of the Swiss Priority Program on Materials Research is acknowledged.

6. REFERENCES

1. T.W. Duerig, K.N. Melton, D. Stöckel, C.M. Wayman, Engineering aspects of shape memory alloys, Butterworth-Heinemann, London, 1990
2. M. V. Gandhi, B. S. Thompson, Smart materials and structures, Chapman and Hall, London, 1992
3. K. Escher, E. Hornbogen, "Aspects of two-way shape memory in NiTi-Silicon composite materials", J. de Physique, 1, C4-pp. 427-432, 1991
4. C.M. Friend, N. Morgan, "The actuation response of model SMA hybrid laminates", J. de Physique IV, 5, C2-pp. 415-420, 1995
5. H. Yoshida, "Creation of environmentally responsive composites with embedded Ti-Ni alloy as effectors", Adv. Composite Mater., 5, pp.1-16, 1995

6. D. A. Hebda, M. E. Whitlock, J. B. Ditman and S. R. White, "Manufacturing of adaptive graphite/epoxy structures with embedded Nitinol wires", J. of Intell. Materials Systems and Structures, 6, pp.220-228, 1995

7. C. A. Rogers, C. Liang, D.K. Barker, "Dynamic control concepts using shape memory alloy-reinforced plates", U.S. Army Research Office Workshop on Smart Mat., Struct., and Math. Issues, Virginia Polytechnic Inst. and State Univ. Blacksburg, Virginia, p. 39, 1988

8. H. G. Mooi, "Active control of structural parameters of a composite strip using embedded shape memory alloy wires", Master Thesis, University of Twente, 1992

9. A. Baz, J. Ro, "Thermodynamic characteristics of Nitinol-reinforced composite beams", Comp. Engineering, 2, Nos 5-7, pp. 527-542, 1992

10. A. Venkatesh, J. Hilborn, J.-E. Bidaux, R. Gotthardt, "Active vibration control of flexible linkage mechanisms using shape memory alloy fibre-reinforced composites", Proc. First European Conf. on Smart Structures and Materials, Glasgow, pp. 185-188, 1992

11. R. Chandra, "Active strain energy tuning of composite beams using shape memory alloy actuators", SPIE vol. 1917, Smart Structures and Intelligent Systems, pp. 267-284, 1993

12. J.-E. Bidaux, N. Bernet, C. Sarwa, J.-A. E. Månson, R. Gotthardt, "Vibration frequency control of a polymer beam using embedded shape-memory-alloy fibres", J. de Physique IV, 5, C8-1177, 1995

13. J.-E. Bidaux, J.-A. Månson, R. Gotthardt, "Active modification of the vibration frequncy of a polymer beam using shape-memory-alloy fibres, Proc. 3rd Int. Conf. on Intell. Materials, Lyon, 1996, pp. 517-522

14. T. A. Weisshaar, M. Sadlowski, "Panel flutter supression with active micro-actuators", Proc. AAAF Int. Forum on aeroelasticity and structural dynamics, Strasbourg, pp. 118-129, 1993

15. J.-E. Bidaux, L. Bataillard, J.-A. Månson, R. Gotthardt, "Phase transformation behaviour of thin shape memory alloy wires embedded in a polymer matrix composite", J. de Physique IV, 3 , C7-p. 561-564,.1993

16. J.-E. Bidaux, J.-A. Månson, R. Gotthardt, "Dynamic mechanical properties and phase transformation in polymer-based shape memory alloy composites", Proc. of the First Int. Conf. on Shape Memory and Superelastic Technologies, Asilomar, Pacific Grove, 1994, A. R. Pelton, D. Hodgson and T. Duerig Eds., pp. 37-42, 1994

17. G. Sun, C. T. Sun, "One dimensional constitutive relation for shape-memory alloy-reinforced composite lamina", J. of Materials Science, 28, pp.6323-6328, 1993

18. L. C. Brinson, R. Lammering, "Finite element analysis of the behavior of shape memory alloys and their applications", Int. J. Solids Structures, 30, pp.3261-3280, 1993

19. J.-E. Bidaux, W.J. Yu, R. Gotthardt, J.-A.E, Månson, "Modelling of the martensitic transformation in shape memory alloy composites", J. de Physique IV, 5, C2-pp. 543-548, 1995

20. R. Stalmans, J. Van Humbeeck, L. Delaey, "Modelling of the thermomechanical behaviour of shape memory wires embedded in matrix materials", Proc. of 3rd Int. Conf. on Intell. Materials, Lyon, 1996, pp. 511-516

21. H. Tobushi, P. H. Lin, K. Tanaka, C. Lexcellent and A. Ikai, "Deformation properties of NiTi shape memory alloy", J. de Physique IV, 5, C2-409, 1995

22. G. Stachowiak, P. G. McCormick, "Shape memory behaviour associated with the R and martensitic transformations in a NiTi alloy", Acta Metall., 36, pp.291-297, 1988

23. H. C. Ling and R. Kaplow, "Stress-induced shape memory changes and shape memory in the R and martensitic transformations in equiatomic NiTi", Metall. Trans. A, 12A, pp.2101-2111, 1981

24. S. Miyazaki, Y. Liu, K. Otsuka, P. G. McCormick, "Electrical resistance change in a NiTi alloy during a thermal cycle under constant load", Proc. of the Int. Conf. on Mart. Transf. (ICOMAT-92), Monterey

MODELLING OF ADAPTIVE COMPOSITE MATERIALS WITH EMBEDDED SHAPE MEMORY ALLOY WIRES

R. STALMANS, L. DELAEY and J. VAN HUMBEECK
Department of Metallurgy and Materials Engineering, KULeuven, De Croylaan 2, B-3001 Belgium, rudy.stalmans@mtm.kuleuven.ac.be

ABSTRACT

Various models and calculation methods for the description of shape memory behaviour have been developed in recent years by different research groups. Some of the models have already been extended towards the thermomechanical and functional behaviour of matrix materials with embedded SMA-wires. The basic concepts of a thermomechanical model which has found widespread use in the literature on smart materials, are critically reviewed.
A recently developed model based on a generalised thermodynamic analysis of the underlying martensitic transformation is discussed more into detail. Experimental verifications indicate that this thermodynamic model can be developed to an effective tool for the materials design of matrix materials with embedded SMA-wires.

1. INTRODUCTION

Shape memory elements ('SMA's') can be used as parts of smart structures in which the actuation function is performed by discrete SMA-elements. SMA-elements, and in particular SMA-wires, can be also easily integrated in matrix materials, yielding smart or adaptive composite materials ('SMA-composites'). In comparison with alternative 'actuating' materials, SMA's offer several important advantages and extra possibilities, such as: reversible strains up to 6%, generation of stresses up to 800 MPa, and storage of large amounts of potential energy. Therefore many experts believe that SMA's offer very promising prospects in this new, rapidly evolving field of materials research. The rapidly increasing interest in these materials is also caused by the fact that thin SMA-wires can be easily embedded into advanced structural materials, such as polymer matrix composites, without losing the structural integrity of the matrix material.
The thermomechanical (and hence the functional) properties of SMA-composites are a direct outcome of the peculiar thermomechanical behaviour of SMA-wires. The essential mechanism on the level of the SMA-wires can be summarised as follows. Prestrained SMA-wires operate during heating against the elastic stiffness of the matrix material. Since the strain recovery of the SMA-wires is biased during heating, high recovery stresses are generated gradually by the SMA-wires (and strain recovery of the SMA-wires is delayed). Vice versa, the recovery stresses decrease during cooling after overcoming a temperature hysteresis (and the SMA-wires become strained).
A first prerequisite to the 'design' of these smart materials and smart structures is that the materials behaviour of the composing elements is known and predictable. An important disadvantage of SMA's in this respect is the complex, non-linear thermomechanical behaviour with hysteresis. Moreover, this complex behaviour is influenced by a large number of parameters. It follows that there are in general no direct and simple relations between the temperature, the position (or strain) and force (or stress). Therefore accurate strain or force control by SMA's is a difficult task and requires the experimental determination of data in combination with advanced control systems. This emphasises the need for comprehensive constitutive models that can accurately describe the thermomechanical behaviour and have moreover a mathematical expression in a form that is amenable to incorporation into other engineering tools. Many models, often of a very elementary level, have been presented in literature.
Especially further developments of Tanaka's modelling [Ta86] have found widespread use in mechanical engineering and in the related liteerature on smart materials. Therefore, this type of modelling is critically reviewed in §3. A recently developed model based on a generalised thermodynamic analysis of the underlying martensitic transformation is discussed more into detail in §4. Experimental verification is an essential part of all types of constitutive modelling. Accurate experimental data are however often lacking. This problem is discussed in §2.

119

2. EXPERIMENTAL DATA ON DISCRETE SMA-WIRES

As explained in the introduction, the functional properties of SMA-composites are directly related to the generation of recovery stresses by prestrained SMA-wires which operate during heating against the elastic stiffness of the matrix material. A remarkable feature of the literature on SMA-composites is that quantitative experimental data on the generation of recovery stresses are very scarce and of dubious quality. Moreover, quantitative data on the generation of recovery stresses by thin SMA-wires, matching the above mentioned biasing and temperature conditions, are completely absent. Many papers refer for quantitative data on generation of recovery stresses to an old NASA-report, published in 1972 [Ja72]. However, the data in this NASA-report refer to preparatory measurements during the first heating cycle for thick Ni-Ti wires in isostrain conditions. These conditions are completely different from the cyclic heating of thin Ni-Ti wires for the complex biasing conditions in SMA-composites.

Experimental data are an essential part of constitutive modelling. Firstly, some parameters in the modelling have to be determined experimentally. Secondly, the accuracy of the modelling has to be verified and optimised by comparison with experimental results. For these purposes, the authors have performed numerous series of experiments. The experiments have been performed on a fully computerised apparatus which offers the possibilities for complex coupled control of stress, strain and temperature and for real-time data acquisition and processing [St92].

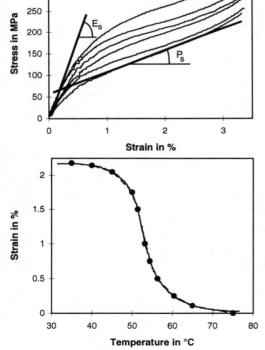

Figure 1: Stress-strain curves of discrete SMA-wires during loading, starting from the complete austenitic condition, for different temperatures. The temperatures (in °C) are respectively 47.2, 49.2, 53.2, 55.2, 57.2 and 61.2. The stress levels increase with temperature. The elastic modulus E_s and pseudoelastic modulus P_s, both used in the thermodynamic modelling, are determined by a linear approximation as indicated in the figure.

Figure 2: Strain-temperature curve of the SMA-wire during free recovery. The sample was (i) heated to 353 K, which is above A_f, (ii) loaded with a stress equal to 235 MPa, (iii) cooled to 303 K, which is below M_f, (iv) unloaded and (v) heated to 363 K. The strain-temperature curve during the latter heating is shown. The dots correspond to data that have been used as input data in the thermodynamic modelling.

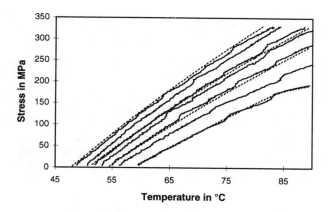

Figure 3: Stress-temperature curves measured on discrete SMA-wires, showing the generation of recovery stresses. In these experiments, steps (i) to (iv) are identical to the experimental procedure in figure 2. In step (v) free recovery was allowed until a programmed contact strain e_c was obtained. During further heating the strain e was controlled as a linear function of the temperature (e = e_c + $\alpha*\Delta T$). Parameter α was in all experiments equal to $100*10^{-6}$/K. Results for the following values of e_c (in %) are shown; 1.90, 1.75, 1.60, 0.92, 0.75, 0.55, and 0.30. Lower starting temperatures correspond with higher e_c. Results calculated with the thermodynamic modelling are given in dashed lines for the following values of e_c (in %); 1.90, 1.60, 0.75 and 0.30.

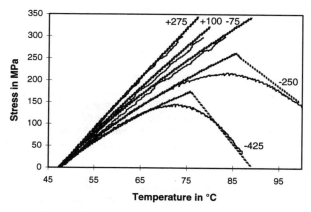

Figure 4: Stress-temperature curves of discrete SMA-wires, measured during generation of recovery stresses. The experimental procedure is identical as in fig.2. The contact strain was in all experiments equal to 1.9%. Results for the following values of α (in 10^{-6}/K) are shown, as indicated above the curves: 275, 100, -75, -250 and -425. Curves at lower stresses correspond with lower values of α. Results calculated with the thermodynamic modelling are given in dashed lines.

Experimental results obtained on discrete SMA-wires (Ni-Ti-6wt%Cu) with a diameter of 1.15 mm are shown in figures 1-4. The samples have been aged for 20' at 823K, and underwent afterwards a training treatment (100 thermomechanical cycles) to obtain a stable shape memory behaviour.

Figure 1 shows the stress-strain curves during uniaxial tensile loading, starting from the austenitic condition, for different temperatures. The elastic and pseudoelastic modulus have been determined as indicated on the figure.

Figure 2 shows a strain-temperature curve during free recovery. The sample was (i) heated to 353 K, which is above A_f, (ii) loaded with a stress equal to 235 MPa, (iii) cooled to 303 K, which is below M_f, (iv) unloaded and (v) heated to 363 K. The strain-temperature curve during the latter heating is shown on the figure.

Figure 3 and figure 4 show the generation of recovery stresses. In these experiments, steps (i) to (iv) are identical to the experimental procedure above. In step (v) free recovery was allowed until a programmed contact strain e_c was obtained. During further heating the strain e was controlled in the following way, with T the temperature and α a parameter:

$$e = e_c + \alpha * \Delta T \qquad |1|$$

Figure 3 shows the results for different contact strains e_c. Figure 4 shows the results for different values of the parameter α. The thermomechanical conditions applied in these experiments on discrete SMA-wires were selected in order to simulate the biasing and temperature conditions in SMA-composites.

3. CONSTITUTIVE MODELLING BASED ON ADDITION OF STRESS COMPONENTS

Based on continuum mechanics, Tanaka developed a unified constitutive modelling of the thermomechanical behaviour of shape memory alloys [Ta86]. This modelling has been used, modified and further extended by many other research groups (see e.g. [Li90, Li90b, Ta92, Ta93, Br93, Br94, Su93, Be94, Sh94, Bo94, Bo95]). The basics of this type of modelling can be summarized as follows. According to this model, the stress σ is determined by the total strain e, the temperature T and the fraction of material ξ that is transformed to martensite. Differentiation of σ gives:

$$\dot{\sigma} = \frac{\partial \sigma}{\partial e} * \dot{e} + \frac{\partial \sigma}{\partial T} * \dot{T} + \frac{\partial \sigma}{\partial \xi} * \dot{\xi} \qquad |2|$$

This is further simplified to:

$$\dot{\sigma} = D * \dot{e} + \Theta * \dot{T} + \Omega * \dot{\xi} \qquad |3|$$

in which D, Θ, Ω are in general assumed to be material constants, which are respectively denoted as Young's modulus, thermoelastic tensor and transformation tensor.

The stress is composed of three simple components: an elastic component given by $\{D*\dot{e}\}$, a thermal dilatation component $\{\Theta*\dot{T}\}$, and a transformation component $\{\Omega*\dot{\xi}\}$.

The thermoelastic tensor is related to the thermal expansion coefficient α_t:

$$\Theta = -D * \alpha_t \qquad |4|$$

In general it is also assumed that the shape memory component of the strain, e_{sr}, is proportional to ξ. From this assumption follows directly a simple relationship between Ω and D:

$$\Omega = -e_1 * D \qquad |5|$$

where e_1 is the maximum residual strain of the SMA, which is also a material constant.

The transformation kinetics are described by simple mathematical functions. In most cases cosine functions are applied. The kinetics of the martensite to austenite transformation are in that case given by:

$$\xi = (\xi_M/2) * \{\cos[a_A*(T-A_s) + b_A*\sigma] + 1\} \qquad |6|$$

and for the austenite to martensite transformation:

$$\xi = (0.5 - \xi_A/2) * \cos[a_M*(T-M_f) + b_M*\sigma] + 0.5 + \xi_A/2 \qquad |7|$$

in which b_A and b_M are material constants.

The constants a_A and a_M follow directly from the transformation temperatures:

$$a_A = \Pi/(A_f - A_s), \qquad a_M = \Pi/(M_s - M_f) \qquad |8|$$

This type of constitutive modelling has several important advantages. Firstly, the derivation of this constitutive modelling is relatively simple. Secondly, the constitutive description has a simple mathematical expression, and only includes a limited number of material constants, which can be easily quantified. Therefore, this constitutive modelling can be easily incorporated in engineering design procedures. Moreover, it has been claimed that this constitutive description can provide a quantitative prediction of the complete shape memory behaviour (see e.g. [Li90]) and can be applied for the description of free recovery, constrained recovery and pseudoelastic cycling.

There are however also several questionable points in this type of modelling.
An essential but also crucial point in this type of modelling is the simplification from |2| to |3|, in which it is assumed that Ω is a material constant. It can be easily deduced from the experimental results in fig.1-4 that Ω is not a material constant; modelling of pseudoelasticity (fig.1), free recovery (fig.2) and of constrained recovery (fig.3-4) require completely different values of Ω. This inconsistency has been discussed more into detail by Tang and Sändström [Ta94, Ta95]. The different slopes of the curves in figure 3 indicate that Ω is not only a function of the type of the transformation process but also of the strain conditions. Hence, the assumption that Ω is a material constant is clearly disproved by experimental results.
A second questionable point is the assumed proportionality between the strain e and the fraction of transformed material ξ. This assumption is only correct for the special case of single crystals, but not for the general case of polycrystalline SMA's (see e.g. [De76]) which makes the equality in |5| invalid.
The kinetics of the martensitic transformation are very complex and are influenced by many parameters: simple relationships as those used in |6| and |7| can only give a very rough approximation of the kinetics.
Based on these inconsistencies, significant inaccuracies can be expected when this modelling is applied for complex constraining conditions and more in particular for the constraining conditions imposed by matrix materials on embedded SMA's.

4. CONSTITUTIVE MODELLING BASED ON A THERMODYNAMIC ANALYSIS

The functional properties of shape memory alloys are closely linked to a temperature or stress induced, crystallographically reversible, thermoelastic martensitic transformation [De91, Ah86]. This thermoelastic transformation has been the subject of many thermodynamic studies. However, most thermodynamic models also fail in computing the thermomechanical behaviour of polycrystalline shape memory alloys. The main reason for this failing is that most of these thermodynamic models have only been developed for very specific purposes such as the analysis of calorimetric measurements [Or88] or for very simple cases such as the behaviour of non-cycled single crystals [Wo79]. Therefore a generalized thermodynamic model of this transformation has been developed in order to predict the thermomechanical behaviour of SMA's. Details of this generalised thermodynamic model can be found elsewhere [St93, St94, St95]. Important points of difference between this and previous thermodynamic models are that the contributions of the stored elastic energy and of the crystal defects are also included. In addition, the mathematical approach and the assumptions in this model have been selected in such a way that the equations yield close approximations of the real system, and that the final mathematical equations are relatively simple. Additional advantages are the clear physical meaning of the parameters and the large area of application. The use of this thermodynamic model has already resulted in an increased fundamental understanding of shape memory behaviour, and especially of the relationships between training, the two way memory effect, and related effects [St93, St94, St95]. As an example of quantitative calculations, it is shown in the following that this thermodynamic modelling can be further developed to an effective tool in the "design" of multifunctional materials consisting of shape memory elements embedded in matrix materials.

4.1. Notions and definitions

Some notions and definitions of the thermodynamic modelling have to be introduced. The variable ξ represents the martensite mass fraction. However, this internal variable only gives incomplete[1] information on the transformation. The volume function $[\xi(r)]$ gives additional information on the strains linked with the transformation; $[\xi(r)]$ represents in any point r the phase and the crystallographic orientation of the phase. Hence, this function also describes the martensite variant arrangement. $\Delta e_{tr}(m_j-a)$ represents the transformation strain linked with transformation from austenite to the martensite variant m_j. $\Delta e_{tr\Sigma A}([\xi])$ represents the average transformation strain on all austenite-martensite interfaces corresponding with a transformation fraction $[\xi]$. The mathematical definition is:

$$\Delta e_{tr\Sigma A}([\xi]) = \frac{\displaystyle\int_{\Sigma A} \Delta e_{tr}*\rho*dydA}{\displaystyle\int_{\Sigma A} \rho*dydA} \qquad |9|$$

where ΣA represents a summation on all martensite-austenite interfaces dA which are in thermodynamic equilibrium; dy represents an infinitesimal displacement perpendicular to dA; and ρ the local mass density.

From the first and second law of thermodynamics, a global and local equilibrium condition can be deduced [St93]. Differentiation of the global equilibrium condition gives a generalised Clausius-Clapeyron equation [St93, St94]:

$$\{d\sigma_s/dT\}_{[\xi]=cst} = -(\rho_0*\Delta s)/\Delta e_{tr\Sigma A}([\xi]) \qquad |10|$$

The material constant Δs is the entropy change during transformation from austenite to martensite. The material constant ρ_0 is the mass density of the SMA's in stress-free conditions. An important remark is that the transformation strain, $\Delta e_{tr\Sigma A}([\xi])$, in the denominator is not a material constant: $\Delta e_{tr\Sigma A}([\xi])$ decreases in general with increasing martensite mass fraction and is also influenced by other factors, including the heat treatment and the thermomechanical history [St93, St94]. From the derivations follows that the Clausius-Clapeyron equation is only valid for conditions of constant $[\xi]$. When $[\xi]$ is constant, the right-hand part of $|10|$, $\{-(\rho_0*\Delta s)/\Delta e_{tr\Sigma A}([\xi])\}$, is also constant. It follows that the stress σ_s increases in direct proportion with the temperature.

4.2. The considered system

The general procedures are illustrated by the example of the hybrid element depicted in figure 5. This beam shaped element with length L is composed of straight SMA-wires uniformly distributed in an isotropic matrix material. The external uniaxial forces acting on this hybrid element are divided into a constant force F_c and a linear bias force with spring constant k^2. A completely rigid, resp. free element can be modelled by taking the spring constant equal to infinity, resp. zero. The recoverable deformation of the SMA-wires at zero stress at the starting temperature T_0, which is below M_f, is given by e_{s0}. The cross section of the matrix is given by Q_m and the total cross section of the SMA-wires is given by Q_s. The task here is to calculate the stresses and strains in the hybrid element and in the composing elements during heating, starting from the temperature T_0.

---

[1] As stated before, in many papers it is incorrectly assumed that the shape memory strain is directly proportional to the martensite mass fraction.

[2] This configuration allows a uniaxial analysis. Derivations for more complex systems can be performed along the same lines. However, the equations would become more complex, with the result that the analysis becomes less clear.

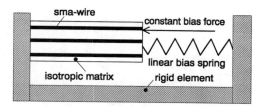

Figure 5: Schematic drawing of the SMA-matrix beam. Straight SMA-wires are uniformly arranged in the isotropic matrix material. The external bias forces are divided into a constant force and a linear spring.

4.3. Derivation of the thermomechanical boundary equation

The total strain of the SMA's, indicated by e_s, can be divided into a recoverable shape memory strain e_{sr}, a thermal dilatation e_{sa} and an elastic strain e_{se}:

$$e_s = e_{sr} + e_{sa} + e_{se} = e_{sr} + a_s*(T-T_0) + \sigma_s/E_s \qquad |11|$$

with T the temperature and σ_s, a_s and E_s resp. stress, thermal dilatation coefficient and elasticity modulus of the SMA-wires.
The mathematics can be simplified by defining the corrugated shape memory strain e_{sc} as:

$$e_{sc} = e_s - e_{sa} = e_{sr} + e_{se} \qquad |12|$$

Similarly, the total strain of the matrix, e_m, can be divided into a thermal dilatation e_{ma} and an elastic strain e_{me}:

$$e_m = e_{ma} + e_{me} = a_m*(T-T_0) + \sigma_m/E_m: \qquad |13|$$

with σ_m, a_m and E_m resp. stress, thermal dilatatation coefficient and elasticity modulus of the matrix material.
The matrix material and external conditions hamper the shape change of the SMA's. This imposes a boundary condition on the thermomechanical action of the SMA's. A mathematical formulation of this boundary condition follows directly from the equations that express stress and strain equilibrium in the hybrid element:

$$e_s = e_{s0} + a_m*(T - T_0) + \{F_c - k*L*a_m*(T-T_0) - \sigma_s*Q_s\}/(E_m*Q_m + k*L) \qquad |14|$$

This boundary condition corresponds to a plane in the thermomechanical $\{e_s\text{-}T\text{-}\sigma_s\}$-space, and can be simplified to:

$$e_s = C_{s1} + C_{s2}*T + C_{s3}*\sigma_s \qquad |15|$$

with C_{s1}, C_{s2} and C_{s3} three system constants respectively given by:

$$C_{s1} = e_{s0} - a_m*T_0 + (F_c + k*L*a_m*T_0)/(E_m*Q_m + k*L) \qquad |16|$$

$$C_{s2} = a_m - k*L*a_m/(E_m*Q_m + k*L) \qquad |17|$$

$$C_{s3} = -Q_s/(E_m*Q_m + k*L) \qquad |18|$$

4.4. Derivation of the constitutive equations

Three temperature regions can be distinguished in the thermomechanical behaviour of the SMA's. The SMA's are in the martensitic state between T_0 and T_{st}, the temperature at which the reverse martensitic transformation starts. The retransformation continues until a temperature T_{se}, the temperature at which the retransformation ends. Above T_{se}, the SMA's are in the parent state

125

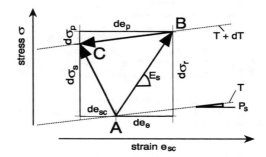

Figure 6: The process of a temperature increase dT at the imposed boundary conditions (A→C) can be replaced by a temperature increase at constant transformation fraction [ξ] (A→B), followed by an unloading at constant temperature (B→C).

('austenite'). The equations in the retransformation region $\{T_{st} \to T_{se}\}$ are derived first.
Since the retransformation proceeds during heating of the hybrid element, the conditions for using the Clausius-Clapeyron equation (|10|) are not fulfilled. A modified Clausius-Clapeyron that is also valid for non-constant [ξ], can be derived as follows. An infinitesimal temperature increase dT of the SMA's induces a stress increase $d\sigma_s$ and a strain change de_{sc}. This infinitesimal step from A (T, σ_s, e_{sc}) to C ($T+dT$, $\sigma_s+d\sigma_s$, $e_{sc}+de_{sc}$) can be also achieved in two intermediate steps, as indicated in figure 6.
The first step is a temperature increase dT *at constant transformation fraction [ξ]*. This temperature increase yields an increase of the stress by $d\sigma_r$, and an increase of the elastic strain by de_e (A→B). The stress increase $d\sigma_r$ follows directly from the Clausius-Clapeyron equation (|10|). The elastic strain de_e follows from:

$$de_e = d\sigma_r/E_s \qquad |19|$$

The second step is an unloading by $d\sigma_p$ *at constant temperature {T+dT}*. Since the constrained heating corresponds to retransformation of martensite to austenite, this unloading results in a pseudoelastic decrease of the strain by de_p (B→C). The pseudoelastic strain de_p can be approximated by:

$$de_p = d\sigma_p/P_s \qquad |20|$$

with P_s the pseudoelastic modulus, given by the slope $d\sigma_s/de_s$ during pseudoelastic unloading (see fig.1).
It follows that the stress increase $d\sigma_s$, resp. strain change de_{sc}, are given by:

$$d\sigma_s = d\sigma_r + d\sigma_p \qquad |21|$$

$$de_{sc} = de_e + de_p \qquad |22|$$

Combination of the boundary condition (|14|), the Clausius-Clapeyron equation (|10|) and equations |19-22| results finally in the differential constitutive equation which is valid *during retransformation* in the temperature region $\{T_{st} \to T_{se}\}$:

$$d\sigma_s/dT = \frac{\{(a_m-a_s) + (P_s-E_s)* \rho_0*\Delta s/(\Delta e_{tr\Sigma A}([\xi])*P_s*E_s)\} - \{k*L*a_m/(E_m*Q_m + k*L)\}}{1/P_s + Q_s/(E_m*Q_m + k*L)} \qquad |23|$$

The shape memory strain e_{sr} is *before retransformation* equal to the initial value e_{s0}. From equations |11-14| can be deduced that σ_s in the temperature region $\{T_0 \rightarrow T_{st})$ is given by:

$$\sigma_s = E_s*\{E_m*Q_m*(a_m-a_s)*(T-T_0) - k*L*a_s*(T-T_0) + F_c\}/(E_s*Q_s + E_m*Q_m + k*L) \qquad |24|$$

By integration of the Clausius-Clapeyron equation (|10|) it can be deduced that the relationship between the temperature T_{st} and the stress σ_s is given by:

$$T_{st} = T_{tr}(e_{s0}) - \sigma_s*\Delta e_{tr\Sigma A}([\xi_0])/(\rho_0*\Delta s) \qquad |25|$$

with $T_{tr}(e_{s0})$ the retransformation temperature corresponding with an initial deformation e_{s0} in stress-free conditions.

For simplification, a system constant K_1 is introduced:

$$K_1 = E_s/(E_s*Q_s + E_m*Q_m + k*L) \qquad |26|$$

Combination of |24|, |25| and |26| results in the following equation, from which T_{st} can be determined:

$$T_{st} = \frac{T_{tr}(e_{s0}) + K_1*[\Delta e_{tr\Sigma A}([\xi])/(\rho_0*\Delta s)]*\{E_m*Q_m*(a_m-a_s)*T_0 - k*L*a_s*T_0 - F_c\}}{(1 + K_1*[\Delta e_{tr\Sigma A}([\xi])/(\rho_0*\Delta s)]*\{E_m*Q_m*(a_m-a_s) - k*L*a_s\}} \qquad |27|$$

The retransformation ends at the temperature T_{se}. At T_{se}, the shape memory strain e_{sr} becomes equal to zero. From equation |11-14| can be deduced that σ_s *after retransformation* at temperatures above T_{se} is given by:

$$\sigma_s = E_s*\{E_m*Q_m*(a_m-a_s)*(T-T_0) - k*L*a_s(T-T_0) +$$
$$(E_m*Q_m+k*L)*e_{s0} + F_c\}/(E_s*Q_s+E_m*Q_m+k*L) \qquad |28|$$

The solution of the above set of equations requires only a limited set of data, which can be simply determined. First, an array of $\{e_{s0}-T_{tr}\}$ data at zero external stress from which $T_{tr}(e_{s0})$ in |27| can be determined. The dotted data in figure 2 are an example of this and are used for the calculated results in figure 3, figure 4 and figure 7. Second, the slopes $d\sigma_s/dT$ for a low and a high value of e_{s0} during heating of the SMA-wires in constrained conditions; the term $\{\rho_0*\Delta s/(\Delta e_{tr\Sigma A}([\xi])\}$ in |23| and |27| can be determined with sufficient accuracy from these slopes by linear interpolation. The two extreme curves in figure 3 have been used for this purpose. Third, the pseudoelastic modulus P_s is calculated from a linear approximation of the slope $d\sigma/de$ during pseudoelastic loading, as indicated in figure 1. Fourth, average values of the thermal dilatation coefficient a_s and the elasticity modulus E_s are also required.

The hybrid element is characterised by the following set of input variables. The input variables of the system are: the cross section of the matrix material Q_m, the total cross section of the SMA-wires Q_s, the bias force F_c, the bias spring constant k, the length of the hybrid element L, the starting temperature T_0, and the initial deformation e_{s0} of the SMA-wires. The input variables for the matrix material are the thermal dilatation coefficient a_m and the elasticity modulus E_m.

For such sets of input variables, the stresses (σ_s and σ_m) and strains (e_s, e_{sc}, e_{sr}, e_{se}, e_{sa}, e_m, e_{me} and e_{ma}) during heating can be computed from the equations above, as follows. In the temperature region $\{T_0 \rightarrow T_{st}\}$, the stress σ_s follows directly from |24|. The transition temperature T_{st} can be calculated from |27|. In the temperature region $\{T_{st} \rightarrow T_{se}\}$, the stress increase $\Delta\sigma_s$ at a small temperature increase ΔT is obtained directly from |23|. The shape memory strain gradually decreases and becomes equal to zero at T_{se}. From that temperature, σ_s can be calculated from |28|. When σ_s is known, the different strain components and stress σ_m can always be calculated from |11-14|. A simple computer program 'SMA-hybrid' has been developed that, by use of these equations, calculates the stresses and strains in the hybrid element and in the composing elements during heating. This computer program can be used to calculate the complex behaviour of SMA-matrix beams during heating for different sets of input variables.

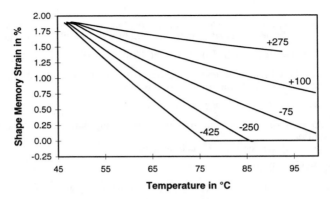

Figure 7: Calculated values of the shape memory strain e_{sr} versus temperature. The thermomechanical conditions are identical as in fig. 4. The values (in 10^{-6}/K) of the parameter α are given above the curves. Curves at lower strains correspond with lower values of α.

4.5. Evaluation of the modelling

Accurate, direct measurements of the stresses and strains developed by SMA-wires in a composite matrix are at the moment not yet available. Important to notice is however that the thermodynamic modelling and the related computer programs can be also used for calculation of the thermomechanical behaviour of discrete SMA-wires in complex constraining conditions.

Quantitative agreement between the calculated results and experimental results has already been shown for different boundary conditions on Cu-base SMA-wires in previous papers [St95].

Calculations have also been performed for NiTi-base SMA-wires. Examples hereof for diverse input variables are shown in figure 3 and figure 4 by dashed lines. A comparison of the computed results with the experimental results shows an accurate *quantitative* agreement.

The experimental results in figure 3 and figure 4 are measured on discrete SMA-wires but the experimental conditions are closely related to the boundary conditions which are imposed by the composite matrix on embedded SMA-wires. The experimental conditions in figure 3-4 simulate the case of a single SMA-wire embedded in a wide matrix without any external loading. The effect of e_c on the generated stress in figure 3 should resemble the effect of the prestrain e_{sO} on the stress σ_s generated by the SMA-wires in a composite matrix. The effect of different values of α in figure 4 should closely resemble the effect of different thermal dilatation coefficients a_m on σ_s.

The excellent quantitative agreement in figure 3 and figure 4 shows that the modelling gives correct results for boundary conditions closely related to the conditions in SMA-composites.

The differential constitutive equation |23| can also be written as:

$$d\sigma_s/dT = P_s*(E_m*Q_m*(a_m-a_s)-k*L*a_s)/(P_s*Q_s + E_m*Q_m + k*L) +$$
$$K_2*\{\rho_0*\Delta s/\Delta e_{tr\Sigma A}([\xi])\} \qquad\qquad |29|$$

in which the system constant K_2 is defined as:

$$K_2 = (E_m*Q_m+k*L)*(P_s-E_s)/\{E_s*(E_m*Q_m + P_s*Q_s + k*L)\} \qquad\qquad |30|$$

It can be easily shown that in general circumstances K_2 is close to $\{-1\}$. The first term in equation |29|, $\{P_s*(E_m*Q_m*(a_m-a_s) - k*L*a_s)/(P_s*Q_s + E_m*Q_m + k*L)\}$, is the result of the external stresses and the difference in thermal dilatation between matrix and SMA-wires. It can be easily shown that the absolute value of this term is in general smaller than 0.2 MPa/K. Since K_2 is close to $\{-1\}$, the second term $\{K_2*(\rho_0*\Delta s/\Delta e_{tr\Sigma A}([\xi]))\}$ closely approximates the right-hand part of the Clausius-Clapeyron equation (|5|). Typical values of this term are 3 to 20 MPa/K. It follows that the first term is in general relatively small compared to the 'Clausius-Clapeyron' term. This indicates that the stress rate $\{d\sigma_s/dT\}$ *during retransformation* is mainly determined by the 'Clausius-Clapeyron' term, as confirmed in fig.4. Figure 4 shows clearly that the slopes $\{d\sigma_s/dT\}$ at the starting temperature (48°C) are hardly influenced by large changes of the parameter α, which is used as a simulation of the thermal dilatation coefficient a_m of the matrix material. The parameter α has however an important influence on the progress of the transformation, expressed by $\{de_{sr}/dT\}$, as shown by the calculated data in figure 7.

The stress rate $\{d\sigma_s/dT\}$ before and after retransformation can be obtained by differentiation of |24| and |28|, which both give the same result:

$$d\sigma_s/dT = E_s*\{E_m*Q_m*(a_m-a_s) - k*L*a_s\}/(E_s*Q_s + E_m*Q_m + k*L) \hspace{2cm} |31|$$

In a first approximation it follows from a comparison of |31| with |23| that changes of the stress rate $d\sigma_s/dT$ as a result of changes of the input variables are proportional to P_s during retransformation and proportional to E_s before and after transformation. Since P_s is much smaller than E_s, it follows that the input variables $(Q_m, Q_s, k, L, a_m$ and $E_m)$ have a much smaller influence on the stress rate $\{d\sigma_s/dT\}$ during retransformation than before or after retransformation, as confirmed in previous papers [St95] and also in figure 4. A numerical comparison of the curves for α equal to -250 and -425 can clarify this. The difference in stress rate at the starting temperature is about 0.7 MPa/K, which is close to $\{P_s*\Delta\alpha\}$. After retransformation, when the stress is decreasing at increasing temperatures, the difference in stress rate is about 6 MPa/K, which is close to $\{E_s*\Delta\alpha\}$.
This important difference between |31| and |23| is however not taken into account in most other modelling such as the constitutive modelling based on addition of stress components.

5. CONCLUSIONS AND FURTHER DEVELOPMENTS

Although the constitutive modelling based on the addition of stress components is widely used, it has been shown that this type of modelling shows several inconsistencies, which might also have implications for the accuracy of the calculations.
Experimental verification has confirmed that the constitutive modelling based on a thermodynamic analysis can give accurate results for discrete SMA-wires in complex thermomechanical boundary conditions. This indicates that this thermodynamic model can be developed to an effective tool for the materials design of matrix materials with embedded SMA-wires.

The main advantages of this modelling are: (i) a high accuracy combined with a very large field of application, (ii) although that the thermodynamic derivation is relatively complex, the final mathematical equations are relatively simple, and (iii) the calculations only require a limited number of simple to determine shape memory data.

The modelling and computer programs are further developed and improved. Examples of improvements in short term are:
(i) the equations have to be slightly modified when the martensite mass fraction becomes close to zero or one; a subroutine will be added to solve this problem,
(ii) pseudoelastic unloading is approximated by a constant P_s in the present treatment; the results can be improved by replacing the constant P_s by an array of $\{P_s-e_s\}$ data.

Also, experiments on SMA-composites will be performed in short term, allowing the further evaluation and optimisation of this type of modelling.

ACKNOWLEDGEMENTS

R. Stalmans and J. Van Humbeeck acknowledge the F.W.O. Vlaanderen for a grant as Postdoctoral Fellow, resp. Research Leader.

REFERENCES

[Ah86] M. Ahlers, Progress in materials science 30, pp. 135-186 (1986).

[Be94] A. Bekker and L. C. Brinson, in 'Mechanics of phase transformations and shape memory alloys', Ed. L. C. Brinson and B. Moran, The American society of mechanical engineers, New York, USA, pp. 195-213 (1994).

[Bo94] Z. Bo and D. Lagoudas, in 'Adaptive structures and composite materials - analysis and application', Ed. E. Garcia, H. Cudney and A. Dasgupta, The American society of mechanical engineers, New York, USA, pp 9-21 (1994).

[Bo95] Z. Bo and D. Lagoudas, in 'Smart structures and materials 1995, Smart materials', Ed. A. P. Jardine, Proc. SPIE 2441, pp. 118-130 (1995).

[Br93] L. C. Brinson, Journal of intelligent material systems and structures 4, pp. 229-242 (1993).

[Br94] W. Brand, C. Boller, M. S. Huang and L. C. Brinson, in 'Mechanics of phase transformations and shape memory alloys', Ed. L. C. Brinson and B. Moran, The American society of mechanical engineers, New York, USA, pp. 179-193 (1994).

[De76] L. Delaey and J. Thienel, in: Shape Memory Effects in Alloys, Ed. J. Perkins, Plenum Press, New York, pp. 341-349 (1975).

[De91] L. Delaey, in: Materials science and technology, Vol.5, Phase transformations in materials, Ed. P. Haasen, VCH Verlagsgesellschaft mbH, Weinheim, Germany, pp. 339-404 (1991).

[Ja72] C. Jackson, H. Wagner and R. Wasilewski, 55-Nitinol - The alloy with a memory, NASA-SP-5110 (1972).

[Li90b] C. Liang and C. A. Rogers, J. of Intell. Mater. Syst. And Struct. 1, pp. 207-234 (1990).

[Li90] C. Liang, The constitutive modelling of shape memory alloys, Ph.D. Dissertation, Dep. of Mech. Engng., Virginia Polytech. Univ., Blacksburg, USA (1990).

[Or88] J. Ortìn and A. Planes, Acta Metall. 36, pp. 1873-1889 (1988).

[Sh94] A. R. Shahin, P. H. Mecki and J. D. Jones, in 'Adaptive structures and composite materials - analysis and application', Ed. E. Garcia, H. Cudney and A. Dasgupta, The American society of mechanical engineers, New York, USA, pp. 227-234 (1994).

[St92] R. Stalmans, J. Van Humbeeck and L. Delaey, Acta metall. mater. 40, pp. 501-511 (1992).

[St93] R. Stalmans, Shape memory behaviour of Cu-base alloys, Doctorate Thesis, Catholic University of Leuven, Department of Metallurgy and Materials Science, Heverlee, Belgium (1993).

[St94] R. Stalmans, J. Van Humbeeck and L. Delaey, Mechanics of Phase Transformations and Shape Memory Alloys, AMD-VOL.189/PVP-VOL292, ASME, Chicago, USA, pp. 39-44 (1994).

[St95] R. Stalmans, J. Van Humbeeck and L. Delaey, Journal de physique IV, Colloque C8, Vol. 5, pp. 203-207 (1995).

[Su93] G. Sun and C. T. Sun, Journal of materials science 28, pp. 6323-6328 (1993).

[Ta86] K. Tanaka, Res Mechanica 18, pp. 251-263 (1986).

[Ta92] K. Tanaka, T. Hayashi and Y. Itoh, Mechanics of materials 13, pp. 207-215 (1992).

[Ta93] K. Tanaka, F. D. Fischer and E. Oberaigner, in 'Proceedings of the international conference on martensitic transformations '92', Ed. C. M. Wayman and J. Perkins, Monterey institute of advanced studies, Carmel, USA, pp. 419-424 (1993).

[Ta94] W. Tang and R. Sandström, in 'Shape memory materials '94', Ed. Chu Yougi and Tu Hailing, International Academic Publishers, Beijing, China, pp. 535-540 (1994).

[Ta95] W. Tang and R. Sandström, Journal de Physique IV, complement to Journal de Physique III, Vol. 5, pp.185-190 (1995).

[Wo79] P. Wollants, M. De Bonte and J. R. Roos, Z. Metallk. 70, pp. 113-117 (1979).

MECHANICAL PROPERTIES OF NiTi-TiC SHAPE-MEMORY COMPOSITES

D.C. DUNAND*, K.L. FUKAMI-USHIRO¶, D. MARI#, J.A. ROBERTS§, M.A. BOURKE§
* Department of Materials Science and Engineering, Massachusetts Institute of Technology, Cambridge, MA 02139, ¶ Raychem Corp., Menlo Park CA 94025, # ACME, 1015 Lausanne, Switzerland, § LANSCE, Los Alamos National Laboratory, Los Alamos, NM 87545.

ABSTRACT

This paper reviews recent work on the mechanical behavior of martensitic NiTi composites reinforced with 10-20 vol.% TiC particulates. The behavior of the composites is compared to that of unreinforced NiTi, so as to elucidate the effect of mismatch due to matrix transformation, thermal expansion, twinning or slip, in the presence of purely elastic particles. The twinning and subsequent thermal recovery of deformed composites, measured both macroscopically (by compressive testing and by dilatometry) and microscopically (by neutron diffraction), are summarized.

INTRODUCTION

The intermetallic NiTi plays a central role in the area of smart materials, because it is capable of large scale recovery after deformation upon heating (i.e. the shape-memory effect) or upon mechanical unloading (i.e. the superelastic effect), depending on stoichiometry and temperature. Both these effects are the result of the allotropic nature of NiTi and its unusual deformation mechanisms which, unlike dislocation motion, are reversible: twinning (leading to shape-memory recovery by thermally-induced phase transformation) and stress-induced transformation (leading to superelastic recovery by a stress-induced back-transformation). Recently, there has been interest in NiTi-based composites with shape-memory capabilities. Most of the research has focused on imbedding NiTi wires or foils within an organic [1-4] or metallic [5, 6] matrix. On the other hand, we have investigated metal matrix composites with a martensitic NiTi matrix and inert ceramic particulates as a second phase, with the aim to examine the effect of mismatch stresses between matrix and reinforcement upon the deformation and subsequent shape-memory recovery of the composites. Since the allotropic transformation responsible for the unique properties of NiTi is thermoelastic, these mismatch stresses can be expected to strongly affect the mechanical response of the matrix.

This review summarizes recent work done at MIT in the area of mechanical properties of TiC-reinforced metal matrix composites with a martensitic NiTi matrix. More details can be found in the original articles describing the thermal transformation behavior [7, 8], the bulk mechanical properties in compression [9], the subsequent shape-memory recovery [10] and the study by neutron diffraction of twinning deformation and shape-memory recovery [11, 12].

EXPERIMENT

While the detailed experimental procedures can be found in the original publications [7-12], a brief summary is given in the following. Billets were fabricated by hot-pressing (followed in some cases by hot-isostatically pressing) of blended powders of prealloyed NiTi (51.4 at % Ti) and TiC particles (between 44 μm and 100 μm in size) with volume fractions 0%, 10% or 20%. Compression samples (respectively labeled in what follows NiTi, NiTi-10TiC and NiTi-20TiC) were cut from the billets by electro-discharge machining and annealed at 930 °C for 1 hour.

Mechanical testing was performed in compression between graphite-lubricated carbide plattens with strain measured by strain gauges, a linear variable displacement transducer or an extensometer. Specimens were thermally recovered in air in a Orton dilatometer outfitted with quartz sample-holder and push-rod. Heating occurred from 20°C to a maximum temperature of at least 275°C at a rate of 1 K/min., followed by cooling to 20°C at the same rate.

Mat. Res. Soc. Symp. Proc. Vol. 459 © 1997 Materials Research Society

Neutron diffraction measurements were performed in time-of-flight mode at the pulsed neutron source of the Los Alamos Neutron Science Center. Two samples (NiTi and NiTi-20TiC) were deformed in compression up to a stress of -280 MPa, unloaded and subsequently recovered above A_f and cooled to room temperature. Strains parallel and perpendicular to the loading axis were calculated from the shift of individual Bragg reflections at a series of stress values, using two detectors forming an angle with respect to the incident beam of -90° (scattering vector Q parallel to the load) and +90° (Q perpendicular to the load). The unstressed, annealed bulk samples were used as a stress-free reference. Changes in the intensities of individual reflections, which correspond to changes in the fraction of NiTi variants in the diffraction condition, were quantified with a Normalized Scale Factor defined as the ratio of a peak intensities under load and prior to loading. The average material response was determined using the Rietveld approach, where the intensities and positions of all Bragg peaks are predicted from an assumed crystal structure.

RESULTS

Compression Behavior

A typical stress-strain curve for NiTi is shown in Fig. 1 with compressive stress and strain given as positive values. Phases are identified as follows: β denotes the austenitic parent phase; M refers to the annealed martensitic phase consisting of the 24 possible variants; and the deformed, twinned martensite is labeled M'. Dominant mechanisms can be separated by the five regimes shown in Fig. 1. On loading in Region A, the martensite M deforms elastically. Region B corresponds to deformation by twinning (M → M'). Deformation in Region C includes contributions of both Region B and Region D (elastic and plastic deformation by slip of oriented martensite M'). Finally, upon unloading in Region E, elastic recovery and some reverse twinning of oriented M' are responsible for strain recovery. Figure 2 shows the stress-strain curves of all samples deformed at room temperature far below M_f (M_f = 57-64 °C).

Shape-Memory Behavior

In dilatometric recovery experiments conducted on specimens deformed at room temperature, the recovery R was defined as:

$$R = \frac{L(T) - L_M(T)}{L_A(T) - L_M(T)} \qquad (1)$$

where L(T) is the measured length of the specimen as the temperature T, $L_A(T)$ is the length of the fully recovered austenitic sample and $L_M(T)$ is the hypothetical length of the deformed, unrecovered martensitic sample. Another recovery parameter directly comparable to the mechanical prestrain e_p is the recovery strain e, defined as:

$$e = \frac{L(T) - L_M(T)}{L_0} \qquad (2)$$

where L_0 is the initial specimen length.

Figure 3 depicts a typical dilatometry curve which can be separated in regions with different slopes. Some strain is recovered during initial heating of the deformed martensite M' in Region 1. The main recovery by the (M' → β) transformation is characterized by a large expansion in Region 2 between A_s and A_f, while limited post-transformation recovery takes place in Region 3, where most of the matrix consists of austenite. Upon cooling, negligible recovery is measured in Region 4, while the austenite β transforms back to martensite in Region 5 between M_s and M_f. The martensitic structure obtained after the first thermal recovery cycle is referred to as M", which is associated with a net contraction or expansion in Region 5, depending on the magnitude of the prestrain.

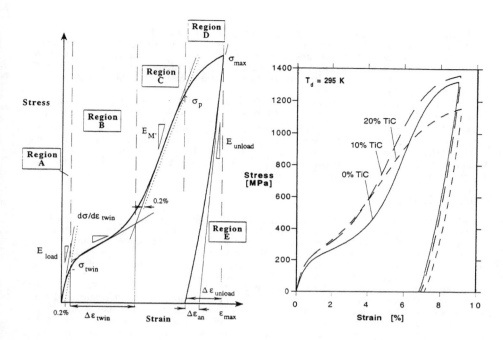

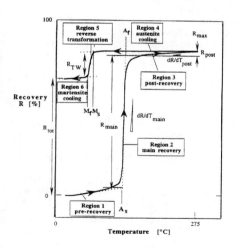

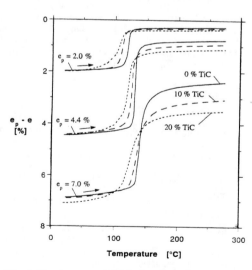

Fig. 1: Schematic compression stress-strain curve for NiTi

Fig. 2: Stress-strain curve of NiTi and NiTi-TiC composites at room temperature.

Fig. 3: Schematic dilatometry curve with its 6 main regions.

Fig. 4: Recovery strain upon heating of specimens deformed to various prestrains.

Figure 4 shows the strain recovered upon heating for samples deformed to three different mechanical prestrains. It is apparent that (i) with increasing prestrain, the transformation temperatures A_s and A_f increase, and recovery strain e increases while recovery extent R decreases; (ii) with increasing TiC volume fraction, recovery strain e and recovery extent R decrease, and transformation temperatures A_s and A_f decrease while the transformation temperature range $A_f - A_s$ increases.

In Fig. 5, the cooling behavior of NiTi-20%TiC is shown as a function of prestrain, with strains emanating from a value of 0% at a temperature of 100 °C for comparison purpose. The ($\beta \rightarrow$ M') recovery is negative for all non-zero prestrains, indicating the Two-Way Shape-Memory Effect (TWSME). A qualitatively similar behavior was found for NiTi-10%TiC and NiTi. However, compared to unreinforced NiTi, the composites exhibit lower transformation temperatures M_s and M_f, wider transformation temperature range $M_s - M_f$ and a smaller total strain recovery (Fig. 6) but larger TWSME strain (Fig. 7).

Neutron Diffraction

Figure 8 shows for NiTi-20TiC the residual strains for the most important planes under no applied stress after mechanical unloading and after shape-memory recovery. While the sign of the residual strains varies between these two states, their magnitude and average are similar. Also, residual strains for all measured planes are similar for NiTi and NiTi-20TiC. Figure 8 also shows the average residual strain after unloading and recovery, calculated as the mean of the residual strains in the three cell directions determined from crystallographic constants.

Figure 9 shows the normalized scale factors of the most intense reflections as a function of the plastic strain during loading. Because the scale factors are proportional to the integrated peak intensities, they are a measure of the volume fraction of variants with planes in the Bragg condition, and thus describe preferred orientation. The systematic increase or decrease of the scale factors indicates that a texture develops upon deformation as a result of twinning. The composite exhibits the same preferred orientation behavior as the bulk matrix, indicating that the TiC particles have only a small effect on the average twinning behavior of the matrix.

Upon unloading however, the scale factors of NiTi and NiTi-20TiC evolve quite differently, as depicted in Figs. 10 (a,b). Moreover, in contrast to loading (Fig. 9), there is upon unloading no linear relationship between scale factors and plastic strain: in many cases, the scale factor /plastic strain curves exhibit a change of slope sign, indicative of a complex twinning behavior. In most cases, however, a trend to reversion towards the undeformed state is observed for the scale factors upon shape memory recovery, but not in the reversible fashion suggested by the dotted line in Figs. 10 (a,b).

DISCUSSION

Compressive Behavior

Assuming that the TiC particles do not induce additional twinning in the matrix and that elastic load transfer from the matrix is operative, the composite stiffness tensor can be predicted by the Eshelby's theory [13]. With an apparent elastic modulus $E_M = 68$ GPa for NiTi (incorporating both elastic and twinning contributions in Region A), theory predicts for the composites the elastic modulus values given in Table I, which also lists the experimental values measured by strain gauge. For an experimental error estimated at about ±3 GPa, the measured elastic modulus for NiTi-10TiC is in agreement with predictions from the Eshelby theory. This shows that stiffening results from load transfer to the TiC particles without creation of mismatch stresses exceeding the critical stress for twinning σ_{twin}, which would induce significant additional matrix twinning and a lower apparent elastic modulus. However, the elastic modulus of NiTi-20TiC, is significantly smaller than predicted by the Eshelby theory, indicating that stiffening by load transfer to the TiC particles is partially canceled by the additional twinning resulting from relaxation of elastic incompatibility between matrix and reinforcement. Only a small twinning strain ε_t is needed to decrease the modulus from a theoretical value $E = 94$ GPa to an apparent macroscopic average

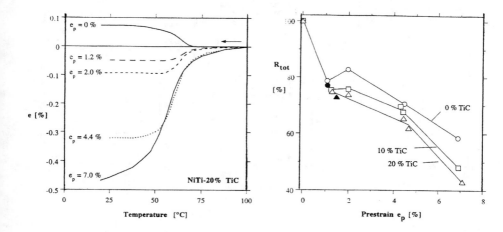

Fig. 5: Strain upon cooling of NiTi-20% TiC deformed to different prestrains.

Fig. 6: Total extent of recovery after a full cycle as a function of prestrain.

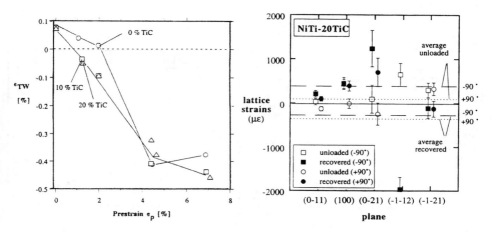

Fig. 7: TWSME recovery strain as a function of prestrain.

Fig. 8: Residual strain measured by neutron diffraction for NiTi-20TiC after mechanical unloading (empty symbols) and shape-memory recovery (filled symbols), with average values determined by Rietveld refinement.

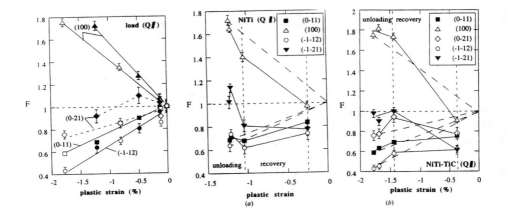

Fig. 9: Normalized scale factor upon mechanical loading as a function of plastic strain for NiTi (filled symbols) and NiTi-20TiC (empty symbols).

Fig. 10: Normalized scale factor upon mechanical unloading as a function of plastic strain
(a) NiTi
(b) NiTi-20TiC.

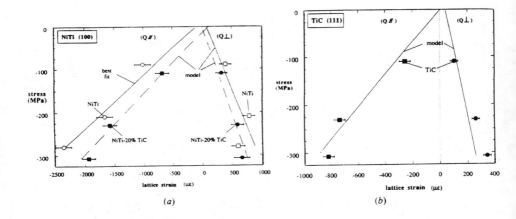

Fig. 11: Measured elastic gradients for (100) planes in both phases and both directions. The linear regression fits are shown as full lines and the predicted behavior from the Eshelby's method as dotted lines.
(a) NiTi lattice strains for NiTi and NiTi-20TiC (b) TiC lattice strains for NiTi-20TiC.

E_{app} = 79 GPa: for an applied stress $\sigma = -50$ MPa, corresponding to the apparent macroscopic yield stress, this strain is less than 0.02%. Early twinning in NiTi-20TiC is similar to the continuous yielding behavior observed in metal matrix composites deforming by slip. Upon application of a small external stress, the yield stress is reached internally as a result of the elastic mismatch between the two phases, resulting in a plastic zone around the particulates. Upon overlap of these plastic zones, the composite yields at an applied stress much lower than the yield stress of the unreinforced matrix, and it may exhibit little or no measurable elastic deformation. In NiTi composites, localized matrix twinning, relaxing mismatch strains developed around the stiff TiC particles will also lead to a low apparent elastic modulus at small strains. At large strains, localized twinning does not preclude load transfer from the compliant matrix to the stiff reinforcement,.

This load transfer can also be calculated by the Eshelby method [13] for the case of a matrix deforming by slip. Assuming that generalized yield in the composite occurs at an applied stress σ_{yc} for which the matrix average stress in the composite is equal to the yield stress of the unreinforced matrix, σ_y, strengthening is described by a parameter P :

$$\sigma_{yc} = \frac{\sigma_y}{P} \qquad (3)$$

P can be calculated from the Tresca yield criterion [13]. For NiTi composites, the yield stress σ_y can be replaced by σ_{twin}, the stress for the onset of twinning. The measured twinning stress values in Table I confirms that additional twinning is taking place in the composites, because of relaxation by twinning of the mismatch stresses between matrix and particles. Indeed, the measured value σ_{twin} is independent of the volume fraction TiC, in disagreement with the prediction of Eq. (A2) based on load transfer without relaxation. This is in contrast to the behavior of the metal matrix composites deforming by slip, whereby mismatch relaxation by dislocation punching strengthen the matrix and results in yield stress values higher than predicted by Eq. (A2).

With increasing fractions of TiC particles, the curves exhibit the following trends: increase of $d\sigma/d\epsilon_{twin}$, and decreases of $\Delta\epsilon_{twin}$, $E_{M'}$ and σ_p. These observations can be explained by the stabilization of martensite M, as TiC particles prevent twinning (M → M') at strains above about 2%. Because twinning is inhibited by dislocations exerting a friction stress on the moving twin boundaries [14], dislocations created to accommodate the plastic mismatch between the matrix and reinforcement are expected to stabilize the martensitic matrix M. Finally, the significant decrease of the slip yield stress σ_p exhibited by NiTi-20TiC can be understood if the yield stress for plastic deformation by slip of M, which exists in larger quantities in NiTi-20TiC, is lower than that of M'. This suggests that twinning of M, which also produces dislocations in the product M', induces subsequent hardening upon plastic deformation by slip of M'.

Table I: Elastic modulus upon loading E_{load} and critical stress for the onset of twinning σ_{twin}, averaged for samples tested at room temperature with strain gauges.

	E_{load} (GPa)		σ_{twin} (MPa)	
	measured	predicted (Eshelby)	measured	predicted (Eq. 3)
NiTi	68	68*	180	180*
NiTi-10TiC	78	80	180	197
NiTi-20TiC	79	94	186	215

* assumed equal to the measured value

Shape-Memory Recovery

Mechanical strain in martensitic NiTi can be separated in three components: (i) plastic strain by twinning which is recoverable, (ii) plastic strain by slip which is unrecoverable and (iii) residual elastic strains, which do not on average contribute to the total macroscopic prestrain, but may be recovered by relaxation during phase transformation. These three types of strains affect the kinetics and thermodynamics of the transformation as follows: (i) twinning reduces the number of martensitic variants, thus facilitating their subsequent transformation to austenite, and may also induce residual elastic stresses, due to residual mismatch strains between twinned variants; (ii) dislocations introduced by slip both stabilize the deformed martensite and induce the TWSME by their elastic stress field; and (iii) elastic stresses stabilize the martensite, as described by the Clausius-Clapeyron equation, and can also induce the TWSME.

Figure 6 shows that the magnitude of the recovery R after a complete thermal cycle is significantly less than 100% for all materials. Incomplete recovery of about 50% was also reported by Johnson et al. [15] for NiTi samples fabricated by powder metallurgy after tensile prestrains of 2 to 10%. The lack of complete recovery after a complete cycle in our samples is not due to the TWSME, because the recovery at the maximum temperature is at most about 85 %. It thus appears that, upon compressive deformation, substantial unrecoverable deformation by slip occurs concurrently with recoverable deformation by twinning, even for prestrains as low as 1.2 %. Plastic deformation by slip represents about 15 % of the total plastic deformation for prestrain values e_p between 1% and 4.4%, i.e. in a regime where twinning is expected to be the dominant deformation mechanism (Fig. 2).

The mismatch between the NiTi matrix and the purely elastic TiC particles cannot be relaxed by the reinforcement, provided it does not fracture in the bulk or at the interface. However, relaxation of the elastic and plastic mismatches between the two phases may occur by twinning or slip of the matrix. In the first case, thermal transformation of mismatch twins formed during mechanical deformation to accommodate the elastic particles is similar to thermal transformation of twins responsible for the overall shape change of the specimen. Thus, the overall shape-memory recovery behavior of the composite is expected to differ little from that of the unreinforced matrix. In the second case, where matrix slip relaxed the mismatch with the reinforcement, dislocations are expected to stabilize the deformed martensite and enhance the TWSME. Furthermore, if relaxation by slip or twinning is incomplete after deformation, the resulting residual elastic stresses are also expected to affect the macroscopic recovery behavior by stabilizing the martensite and by inducing the TWSME. Residual elastic stresses after deformation, however, are small on average as measured by neutron diffraction (Fig. 8).

The effect of TiC particles can now be discussed in light of the mechanisms described above. First, enhanced slip deformation of the matrix to accommodate the mismatching particles can explain the following observations:

(i) With increasing TiC content, the extent of recovery R_{max} decreases for prestrains larger than $e_p = 1.4\%$. This indicates that unrecoverable deformation by slip increases in the composites.

(ii) The strain recovery during the pre- and post-transformation (Regions 1 and 3, Fig. 3) increases with increasing TiC fractions, and the recovery gradient de/dT_{main} during the main recovery in Region 2 decreases. These two phenomena broaden the recovery temperature interval, as expected if the martensitic variants close to the particles contain more mismatch dislocations (and are thus more stabilized) than the variants far from the particles.

(iii) The TWSME strain increases with increasing TiC content (Fig. 7), as expected if the composites contain more dislocations, the elastic stress field of which biases the formation of oriented martensite.

Second, stored elastic stresses can explain the following observations:

(iv) All transformation temperatures A_s, A_f, M_s, and M_f decrease with increasing TiC content, in agreement with the effect of elastic stresses on a thermoelastic transformation [16].

(v) The TWSME strain increases with increasing TiC content (Fig. 7). While this effect can be explained by plastic deformation (see (iii) above), elastic stresses due to incomplete relaxation in the composite can also have the same effect.

We note that plastic deformation in the composites lowers the total extent of recovery after a full cycle R_{tot} (Fig. 6) by decreasing the magnitude of the recoverable plastic strain by slip (effect (i)) and by increasing the magnitude of the TWSME strain (effect (iii) or (v)).

We thus conclude that all the trends observed in the recovery characteristics of the composites can be explained by an increase in plastic strains in the matrix and increased residual elastic stresses due to the mismatch between the matrix and the reinforcement during compressive deformation. In most cases, however, the difference in recovery properties between reinforced and unreinforced NiTi is relatively minor, indicating that most of the mismatch is accommodated by twinning. These conclusions are similar to those drawn from the shape of the stress-strain curves, as discussed earlier.

Neutron Diffraction

As for any metal matrix composite cooled from an elevated processing temperature, NiTi-20TiC is expected to exhibit residual thermal mismatch stresses due to the mismatch in coefficients of thermal expansion between reinforcement and matrix. Furthermore, additional mismatch stresses are expected due to the allotropic transformation of NiTi. These thermal and allotropic mismatch contributions calculated by the Eshelby method are given in Table II at room temperature for both phases. The expansion of the matrix upon allotropic transformation, results in room-temperature residual strains smaller by about 30% than if no transformation had taken place (Table II). We thus conclude that the residual strains are virtually the same in the undeformed NiTi and the undeformed NiTi-20TiC, since the value of the average matrix strain due to phase mismatch in the composite ($\varepsilon = 1.28 \cdot 10^{-4}$, Table II) is on the order of the diffraction measurement accuracy.

Fig. 11 (a) shows for NiTi-20TiC the elastic behavior of the (100) planes, for which the measurement error is small and whose elastic gradient can be assimilated to $1/S_{11}$. For both directions parallel and perpendicular to the applied load, good agreement is found between the measured diffraction data and the elastic behavior of the matrix predicted from load transfer by the Eshelby theory. The matrix lattice strain for the highest applied stress seems to be slightly lower than expected if elastic load transfer were solely taking place. This indicates that additional load transfer may be taking place, as discussed later. Fig. 11 (b) shows the predictions for the TiC phase in the NiTi-20TiC sample, which are also in satisfactory agreement with diffraction measurement. We thus conclude that, while elastic load transfer is taking place in the NiTi-20TiC composite in quantitative agreement with predictions by the Eshelby theory (Figs. 11(a, b)), no macroscopic stiffening is observed in the apparent elastic region of the stress-strain plot, as described earlier, because of the enhanced matrix twinning needed to relax the mismatching particles.

Plastic deformation of the matrix in the presence of elastic particles results in a plastic mismatch which is additive to the elastic mismatch contribution. Thus, load transfer should be enhanced compared to a case where both phases are elastic, with a concomitant increase in the gradient of the applied stress versus the lattice strain curve of the matrix (Fig. 11a). Respectively

Table II: Predicted mean internal stress $\overline{\sigma}$ and mean internal strain $\overline{\varepsilon}$ for NiTi (matrix M) and TiC (inclusion I) at room temperature due to thermal mismatch and transformation mismatch between the two phases of NiTi-20TiC.

	$\overline{\sigma}_M$ (MPa)	$\overline{\sigma}_I$ (MPa)	$\overline{\varepsilon}_M$ ($\cdot 10^6$)	$\overline{\varepsilon}_I$ ($\cdot 10^6$)
thermal mismatch	75	-302	182	-416
transformation mismatch	-22	89	-54	123
total mismatch	53	-213	128	-293

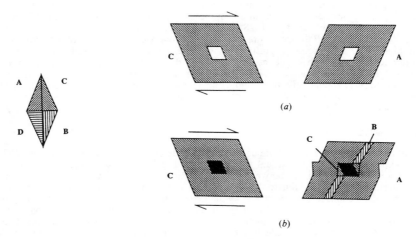

Fig. 12: Two-dimensional illustration of the accommodation of a mismatching rigid particle within a martensitic crystal with 4 possible variants.
(a) crystal with orientation C containing a hole, deformed by twinning to orientation A;
(b) crystal with orientation C containing a rigid particle, deformed by twinning to orientation A with accommodation twins C and B for the undeformed particle.

a decrease is expected for the elastic gradient of the reinforcement. These effects were observed by neutron diffraction in aluminum matrix composites deforming plastically by slip [13]. In the case of NiTi-20TiC deforming by twinning, these effects are much less than predicted by theory (Fig. 11), indicating that, unlike composites deforming by slip, NiTi-20TiC deforming by twinning relaxes very efficiently the plastic mismatch between matrix and reinforcement.

The *in-situ* neutron elastic and orientation measurements (Figs. 9 and 11) all indicate that the addition of 20 vol.% TiC particles has little effect on the plastic behavior of the matrix over the strain range explored here. This is in contrast with the increased strengthening, rate of strain hardening and plastic load transfer between matrix and reinforcement observed in metal matrix composites with a matrix plastically deforming by slip [13]. This insensitivity of the NiTi matrix to the presence of a large volume fraction of stiff, mismatching particles can be interpreted as the result of the greater ease, as compared to slip, for mismatch relaxation by twinning. When accommodation takes place by dislocation punching, strain hardening of the matrix and concomitant increase in strength result [17]. In contrast, complete relaxation of the mismatch between the elastic particle and the plastic matrix is possible by local twinning in martensitic NiTi composites. This is illustrated schematically in Fig. 12, for an hypothetical two-dimensional crystal with 4 possible variants. If the crystal contains a hole and is deformed by twinning from variant C to variant A, a concomitant change in the shape of the hole results. However, if the hole contains a rigid particle and the crystal is again twinned from variant C to variant A, complete accommodation of the resulting mismatch is possible by retaining some of the variant C in contact with the particle and accommodating the A-C boundary by a B twin (Fig. 12(b)). The net result is an overall macroscopic deformation similar to that of the particle-free case (Fig. 12(a)), with a small volume fraction of accommodating variants (B and C in Fig. 12 (b)). Because such accommodating variants are expected to also exist for particle-free, polycrystalline NiTi to relax mismatch between grains of different orientations, the twins necessary to accommodate elastic particles in a polycrystalline composite sample may represent only a small fraction of the twins already existing in particle-free polycrystalline samples. This hypothesis explains the lack of significant difference between NiTi and NiTi-20TiC in terms of macroscopic behavior and average variant orientation by twinning (Fig. 9). Furthermore, unlike relaxation by slip, no strain hardening is expected as a result of localized twinning relaxing the mismatch.

TiC particles do not significantly interfere with the shape-memory effect for prestrains less than about 4%, corresponding to deformation by twinning. First, the fraction of strain recovered by shape-memory heat-treatment is only slightly smaller for NiTi-20TiC than for NiTi, as shown in Fig 6. This result confirms that the mismatching TiC particles do not induce a significant amount of slip in the material during deformation, since the strain resulting from slip is unrecoverable. Second, comparison of Figs. 10 (a,b) indicates that the preferred variant orientation after shape memory recovery is similar for both materials, despite the differences in variant orientation before shape-memory recovery. Third, the residual strains for all planes are, within experimental error, the same for both materials. If martensite deformation occurs purely by twinning of variants, which also accommodates the mismatch of TiC particles as depicted in Fig. 12, thermal transformation to austenite after twinning is indeed expected to be mostly unaffected by the TiC particles. Upon heating to a temperature above A_f, the composite is returned to the same state as before deformation, for which little thermal mismatch strains are expected (Table II). Subsequent cooling to room-temperature induces again small transformation mismatch strains, which can be minimized by self-accommodation through an appropriate combination of variants at the interface, similarly to the case of strain-induced deformation illustrated in Fig. 12(b).

SUMMARY

This paper gives an overview of recent work performed on martensitic NiTi composites containing 0, 10 and 20 vol.% TiC particulates. Mechanical deformation by twinning and subsequent strain recovery by shape-memory heat-treatment are investigated both macroscopically (by mechanical testing and by dilatometry) and microscopically (by neutron diffraction). The mismatch stresses between the elastic TiC particulate and the NiTi matrix due to matrix transformation and thermal expansion (during thermal excursions) and due to matrix twinning or slip (during mechanical deformation) are discussed.

ACKNOWLEDGMENTS

The work summarized in this paper was performed at MIT (with the financial support of the NSF Materials Research Laboratory, grant DMR90-22933, administrated through MIT's Center for Materials Science and Engineering) and at the Los Alamos Neutron Science Center (supported in part by DOE contract W-7405-ENG-36). DCD also gratefully acknowledges financial support by Daimler Benz Research and Technology and helpful discussions with Mr. Heinz Voggenreiter from that company.

REFERENCES

1. C.A. Rogers, C. Liang, and J. Jia, Comput. Struct. **38**, p. 569 (1991).

2. E. Hornbogen, M. Thumann, and B. Velten, in Progress in Shape Memory Alloys, edited by S. Eucken (DGM, Oberursel, Germany, City, 1992), p. 225.

3. J. Ro and A. Baz, Composite Eng. **5**, p. 61 (1995).

4. J.E. Bidaux, J.A. Manson, and R. Gotthardt, in First International Conference on Shape Memory and Superelastic Technologies, edited by A.R. Pelton, D. Hodgson, and T. Duerig (MIAS, Monterey CA, City, 1995), p. 37.

5. Y. Furuya, A. Sasaki, and M. Taya, Mater. Trans. JIM **34**, p. 224 (1993).

6. Y. Yamada, M. Taya, and R. Watanabe, Mater. Trans. JIM **34**, p. 254 (1993).

7. D. Mari and D.C. Dunand, Metall. Mater. Trans. **26A**, p. 2833 (1995).

8. D. Mari, L. Bataillard, D.C. Dunand, and R. Gotthardt, J. Physique IV **5**, p. 659 (1995).

9. K.L. Fukami-Ushiro and D.C. Dunand, Metall. Mater. Trans. **27A**, p. 183 (1996).

10. K.L. Fukami-Ushiro, D. Mari, and D.C. Dunand, Metall. Mater. Trans. **27A**, p. 193 (1996).

11. D.C. Dunand, D. Mari, M.A.M. Bourke, and J.A. Goldstone, J. Physique IV **5**, p. 653 (1995).

12. D.C. Dunand, D. Mari, M.A.M. Bourke, and J.A. Roberts, Metall. Mater. Trans. **27A**, p. 2820 (1996).

13. T.W. Clyne and P.J. Withers, An Introduction to Metal Matrix Composites, Cambridge University Press, Cambridge, 1993.

14. S. Miyazaki and K. Otsuka, Metall. Trans. **17A**, p. 53 (1986).

15. W.A. Johnson, J.A. Domingue, S.H. Reichman, and F.E. Sczerzenie, J. Physique **43**, p. C4 (1982).

16. R.J. Salzbrenner and M. Cohen, Acta Metall. **27**, p. 739 (1979).

17. M.F. Ashby, Phil. Mag. **21**, p. 399 (1970).

MECHANICAL PROPERTIES OF SMART METAL MATRIX COMPOSITE BY SHAPE MEMORY EFFECTS

K. HAMADA *, J. H. LEE **, K. MIZUUCHI ***, M. TAYA *, K. INOUE**
*Dept. of Mech. Engr., Univ. of Washington, Seattle, WA 98195, hamada@u.washington.edu
**Dept. of Matls. Science & Engr., Univ. of Washington, Seattle, WA 98195
***Osaka Municipal Tech. Res. Inst., Osaka 536, JAPAN

ABSTRACT

The thermomechanical behavior of TiNi shape memory alloy fiber reinforced 6061 aluminum matrix smart composite is investigated experimentally and analytically. The yield stress of the composite is observed to increase with prestrain given to the composite. Analytical model is developed by utilizing a shape memory alloy constitutive model of exponential type for the thermomechanical behavior of the composite. The model predicts that the composite yield stress increases with increasing prestrain, and the key parameters in affecting the composite yield stress are prestrain and matrix heat treatment. The model predicts reasonably well the experimental results of the enhanced composite yield stress.

INTRODUCTION

A smart composite, which utilizes the shape memory alloy (SMA) fiber as a reinforcement, has been designed and successfully developed [1-6]. The design concept of smart composite is shown in our previous study [2]. SMA fibers in composite are loaded at room temperature, making the austenitic phase of TiNi SMA fiber changed to martensitic phase (forward transformation). Then they are heated so as to induce the martensite to austenite phase transformation (called reverse transformation). With this reverse transformation, the SMA fibers shrink in the composite, that induces tensile stress in the fibers and compressive stress in the matrix. This compressive stress in the matrix is the main source of the tensile properties enhancement of the smart composite. The internal stress analysis is the key step in the precise evaluation of the smart composite.

Recently we have developed two types of smart composite utilizing TiNi SMA fiber as a reinforcement, and demonstrate successfully the enhancement of the yield stress of TiNi/aluminum composite and fracture toughness of TiNi/epoxy composite at high temperature [1, 3, 5]. In our previous studies [2, 4, 6], pure aluminum was used for the matrix due to easiness of processing into a composite and demonstrating the enhancement of the composite yield stress. For applications of smart composites to structural composite, use of pure aluminum as a matrix metal is not advantageous due to its low flow stress (yield stress). Hence in the present study, use of 6061 aluminum alloy and T6 heat treatment are adopted. For the processing of metal matrix smart composite, we used pressure casting of molten aluminum matrix previously [4], while here, we will use a vacuum hot press method to enhance strong interfacial bonding between TiNi fibers and aluminum alloy matrix through a solid-solid diffusion bonding process.

In our previous model, the martensite phase induced in the prestrain process is assumed to show reverse transformation to austenite above austenite starting temperature, A_s, and the TiNi fiber is fully austenitic at temperatures higher than A_f [2, 6]. In this paper, we develop a model including stress effects, especially, on the transformation during the heating process, because this transformation is strongly effected by temperature and stress [7].

143

The objectives of this paper are: <1> to demonstrate the enhanced mechanical performance of TiNi/6061 aluminum smart composite at high temperature, which is also aided by the efficiency of high performance matrix and of vacuum hot pressing; and <2> to develop one dimensional model for more accurate stress analysis of smart composite under thermomechanical loading.

EXPERIMENTAL PROCEDURE

The shape memory alloy fiber used is Ti - Ni 50.3 at% fiber of diameter 200μm. To obtain the mechanical properties and transformation data, fiber tensile testing is performed at various temperatures between 297 and 373 K. First the TiNi fibers are treated at 773K in air for 30 min. and water quenched for shape memorized treatment, followed by fiber tensile tests in air, at a constant strain rate of 1×10^{-4}/s. The strain is measured by the crosshead displacement. From these test results, the transformation starting and finishing stresses at selected temperatures can be obtained, and differential scanning calorimetry is utilized for the measurement of transformation temperature without stress. The experimental results of the phase transformation stress-temperature diagram are summarized in Fig. 1.

For the processing of the smart composite, we use the vacuum hot press method where the degree of vacuum is up to 10^{-3} torr attained by a mechanical pump. The processing is made at 773K for 30min at a 54MPa press pressure and then specimen is cooled down in the furnace while keeping press pressure constant. The volume fraction of TiNi fibers is 5.3% in the present experiment. Flat bar type tensile specimens are cut off from as-processed composite. To enhance the mechanical properties of the matrix, T6 heat treatment is performed. That is, specimens are solution treated at 803K in air for 30 min. and water quenched, then aging treated at 443K in air for 18 hours followed by another water quenching. As the final step of the processing, three different prestrains are given to composite specimens. Mechanical testing at a strain rate of 1×10^{-4} /s is then conducted on a composite specimen with a given prestrain upon heating to T=373K.

For providing various amount of prestrain, loading and unloading process is applied to tensile specimens at 293K in air. Loading process is performed at a constant strain rate of 1×10^{-4} /s and tensile tests are made at 373K in air and at the same strain rate. Specimen temperature is measured at the surface of each specimen using a thermo-couple and displacement of the specimen is measured by an extenso-meter whose gage length is 10mm.

ANALYTICAL MODELING

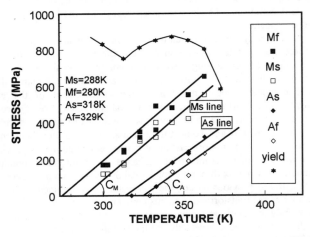

Fig. 1 Temperature-stress diagram of TiNi shape memory fiber

144

<u>One Dimensional Composite Model</u>

Assuming no sliding between fiber and matrix, the isostrain condition requires the following condition,

$$\varepsilon_c = \varepsilon_f = \varepsilon_m \tag{1}$$

where ε_c, ε_f and ε_m are the total strain of composite, fiber and matrix, respectively. The force equilibrium equation of composite along the fiber axis direction is given in equation (2),

$$\sigma_c = V_f \times \sigma_f + (1 - V_f) \times \sigma_m \tag{2}$$

where σ_c is the applied stress of composite, σ_f and σ_m are the stress of fiber and matrix, respectively, and V_f is the volume fraction of fiber.

Because the fiber is expected not to yield in our experiment, the total strain of fiber can be described in the next form,

$$\varepsilon_f = \varepsilon_f^{el} + \varepsilon_f^{CTE} + \varepsilon_f^{trans} \tag{3}$$

where ε_f^{el}, ε_f^{CTE} and ε_f^{trans} are the fiber elastic strain, thermal strain and transformation strain, respectively. The matrix strain consists of elastic (ε_m^{el}), thermal (ε_m^{CTE}) and plastic strain(ε_m^{pl}), and is expressed in the next form,

$$\varepsilon_m = \varepsilon_m^{el} + \varepsilon_m^{CTE} + \varepsilon_m^{pl} \tag{4}$$

The elastic constitutive equations of fiber and matrix are given in the following forms,

$$\sigma_f = E_f \times \varepsilon_f^{el} \tag{5}$$

$$\sigma_m = E_m \times \varepsilon_m^{el} \tag{6}$$

where E_f and E_m are the elastic modulus of fiber and matrix, respectively. The thermal strain can be described as a function of a temperature change, ΔT, which is expressed in the following forms,

$$\varepsilon_f^{CTE} = \alpha_f \times \Delta T \tag{7}$$

$$\varepsilon_m^{CTE} = \alpha_m \times \Delta T \tag{8}$$

where α_f and α_m are the coefficients of thermal expansion of fiber and matrix, respectively.

The work hardening behavior of the matrix can be approximated by a power law of matrix plastic strain and is given in the following form,

$$\sigma_m = \sigma_{my} + K \times (\varepsilon_m^{pl})^n \tag{9}$$

where σ_{my} is the yield stress of the matrix, and K and n are material constants. From equation (9), the incremental matrix strain can be obtained as a function of the incremental matrix stress, which is expressed in the following form,

$$d\varepsilon_m = \frac{d\sigma_m}{E_m} + \frac{d\sigma_m}{nK} \left(\frac{\sigma_m - \sigma_{my}}{K}\right)^{\frac{n-1}{n}} \tag{10}$$

Phase Transformation of Shape Memory Fiber

It is well known that the volume fraction of martensite (ξ) in a SMA mainly depends on its stress and temperature. Different types of equations have been proposed to describe the thermomechanical behavior of SMAs [8, 9], and in this paper, ξ during the austenite to martensite (A to M) transformation is defined by Tanaka [8] and it is expressed in the next form,

$$\xi(T,\sigma_f) = 1 - \exp\left[a^M \times (M_s - T) + b^M \times \sigma_f\right] \tag{11}$$

$$a^M = \frac{\ln(0.01)}{Ms - Mf}, b^M = \frac{a^M}{C_M} \tag{12}$$

where M_s and M_f are the martensite transformation starting and finishing temperatures without stress, respectively, and C_M is the slope of the temperature-stress phase diagram from austenite to martensite. The phase diagram is shown in Fig. 1. Here the A to M transformation end is defined as $\xi=0.99$ (due to the singular behavior in the constitutive equation, $\xi=1$ was avoided). From equation (11), the increment in ξ can be expressed in terms of ΔT and $d\sigma_f$ and is expressed in the next form,

$$d\xi(T,\sigma_f,dT,d\sigma_f) = \\ \exp\left[a^M \times (M_s - T) + b^M \times \sigma_f\right] - \exp\left[a^M \times (M_s - T - dT) + b^M \times (\sigma_f + d\sigma_f)\right] \tag{13}$$

For the martensite to austenite (M to A) transformation, ξ is also given by Tanaka [8] and expressed in a similar manner to equation (11);

$$\xi(T,\sigma_f) = \exp\left[a^A \times (A_s - T) + b^A \times \sigma_f\right] \tag{14}$$

$$a^A = \frac{\ln(0.01)}{As - Af}, b^A = \frac{a^A}{C_A} \tag{15}$$

where A_s and A_f are the austenite transformation starting and finishing temperatures without stress, respectively, and C_A is the slope of the temperature-stress phase diagram from martensite to austenite. Here the end of the M to A transformation is defined as $\xi=0.01$. From equation (14), the increment in ξ can be expressed in terms of ΔT and $d\sigma_f$, which is shown in the next form,

$$d\xi(T,\sigma_f,dT,d\sigma_f) = \\ \exp\left[a^A \times (A_s - T - dT) + b^A \times (\sigma_f + d\sigma_f)\right] - \exp\left[a^A \times (M_s - T) + b^A \times \sigma_f\right] \tag{16}$$

The coefficient of thermal expansion (CTE) and the elastic modulus of shape memory alloys are known to depend on the martensite volume fraction. Assuming a linear dependence of α_f and E_f on ξ, the CTE and elastic modulus can be expressed in the following forms,

$$\alpha_f(\xi) = \xi \times \alpha_f^{MAR} + (1-\xi) \times \alpha_f^{AUS} = \alpha_f^{AUS} + \xi \times (\alpha_f^{MAR} - \alpha_f^{AUS}) \tag{17}$$

$$E_f(\xi) = \xi \times E_f^{MAR} + (1-\xi) \times E_f^{AUS} = E_f^{AUS} + \xi \times (E_f^{MAR} - E_f^{AUS}) \tag{18}$$

where α_f^{MAR} and α_f^{AUS} are the coefficients of thermal expansion of martensite and austenite, respectively, and E_f^{MAR} and E_f^{AUS} are the elastic moduli of martensite and austenite, respectively.

The phase transformation strain of shape memory alloy fiber is assumed to have a linear function of the martensite volume fraction change, which is simply expressed by the next form,

$$\varepsilon_f^{trans} = d\xi \times \varepsilon^{t\,max} \qquad (19)$$

where ε^{tmax} is the maximum transformation strain from 100% of austenite to 100% of martensite and vise versa.

RESULTS AND DISCUSSION

The material properties used for the thermomechanical process are shown in Table 1, where the indication of -F means no heat treatment. The parameters of the matrix yield stress and work hardening are measured from specimens without fiber but processed using the same processing parameters as those for the composite. The tensile testing of the composites is performed at 373K to obtain a complete set of stress-strain curves, from which the composite yield stress (as determined at 0.2% offset strain) are measured and plotted in Fig. 2 as a function of prestrain. The experimental data show that the composite yield stress increases with increasing prestrain, whose dependency is affected by the mechanical properties of the matrix. The composite yield stresses obtained analytically are also shown in Fig. 2, indicating reasonably good agreement with the experiment. Some potential reasons for the dependency between predictions and experimental values are supposed to be as follows.

One reason is the temperature dependence of the material properties. Current material parameters used for theoretical calculation are the data given at room temperature, but the elastic modulus and yield stress should be effected by temperature. Another reason is the difference in properties of fiber between being embedded in the composite and being not embedded, because of some interfacial reactions between fiber and the matrix. In addition, the

Table 1. Materials parameters.

Composite Properties
V_f 0.053

6061 Aluminum Properties
E_m 70GPa
α_m 23.6×10^{-6} /K
σ_y 35MPa (-F), 245MPa (-T6)
K 445MPa (-F), 85MPa (-T6)
n 0.49 (-F), 0.2 (-T6)

TiNi Fiber Properties
M_s: 288K, M_f: 280K
A_s: 318K, A_f: 329K
C_A: 7.1MPa/K, C_M: 7.7MPa/K
E_f^{MAR}: 26.3GPa, E_f^{AUS}: 67GPa
α_f^{MAR}: 6.6×10^{-6}/K, α_f^{AUS}: 11×10^{-6}/K
ε^{tmax} 0.06

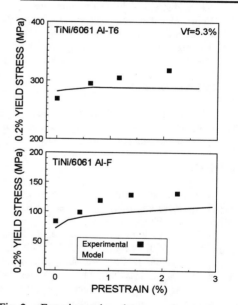

Fig. 2 Experimental data and computed predictions of the 0.2% yield stress dependence on prestrain of TiNi/6061 Al composites

heat treatment history of composite is different from that of the fiber used for evaluation of its properties. Thus, the constitutive properties of the fiber in composite are supposed to be different from the evaluated value. Microstructural change of fiber should also cause significant effects on transformation properties including transformation temperatures and stress

CONCLUSIONS

TiNi SMA fiber reinforced 6061 aluminum alloy matrix smart composite with prestrain has been successfully processed and the dependence of the composite yield stress on prestrain is theoretically evaluated. A one-dimensional model is developed by utilizing an exponential type SMA constitutive model for the thermomechanical behavior of smart composite containing SMA fiber. The main findings of this study are summarized as follows:

1. The shape memory alloy fiber reinforced smart composite provides a matrix compressive residual stress while being heated above A_f temperature. This is mainly because of the fiber shrinkage associated with the martensite to austenite transformation during heating process, and partially because of the misfit coefficient of thermal expansion between fiber and matrix. This compressive residual stress formed in the matrix is the primary source of improvement in the yield stress of the smart composite.
2. Enhancement of the yield stress of the TiNi/6061 Al-F and -T6 composite is observed to increase with increasing prestrain.
3. The present model successfully predicts the observed composite yield stress dependence on prestrain.

ACKNOWLEDGEMENT

The authors would like to acknowledge the support of National Science Foundation, Grant No. CMS-94-9414696. They also thank Dr. J. K. Lee (presently Hyundai Motor Company, Korea) for his fruitful advise to the model.

REFERENCES

1. A. Venkatesh, J. Hilborn J. E. Bidaux and R. Gotthardt, in Proc. 1st European Conf. On Smart Structure and Materials, edited by B. Culshaw, P. T. Gardiner and A. McDonach (IOP, Bristol, UK, 1992) pp. 185-188.
2. M. Taya, Y. Furuya, Y. Yamada, R. Watanabe, S. Shibata and T. Mori, in Proc. Smart Materials, ed. V. K. Varadan, SPIE, 373 (1993).
3. J. G. Boyd and D. C. Lagoudas, J. Intelligent Material Systems and Structures, 5, 333 (1994).
4. Y. Furuya, A. Sasaki and M. Taya, Materials Trans., JIM, 34, 227 (1993).
5. W. D. Armstrong, J. Intelligent Material Systems and Structures, 7, 448 (1996).
6. M. Taya, A. Shimamoto and Y. Furuya, in Proc. 10th International Conf. of Composite Materials, V, 275 (1996)
7. For example, K. Otsuka, C. M. Wayman, K. Nakai, H. Sakamoto and K. Shimizu, Acta Metall., 24, 207 (1976)
8. K. Tanaka, Res Mechanica, 18, 251 (1986)
9. C. Liang and C. A. Rogers, J. of Intelligent Material Systems and Structures, 1, 207 (1990)

ON THE THEORY OF ADAPTIVE COMPOSITES

A.L. ROYTBURD
Department of Materials and Nuclear Engineering
University of Maryland, College Park, MD 20742, U.S.A.

ABSTRACT

Thermodynamics of an adaptive layer composite with a poldomain (polytwin) ferroelectric component is considered. It is shown that there are λ-singularities on the thickness dependencies of compliance and susceptibility at some critical thicknesses of the composite layers.

INTRODUCTION

An adaptive composite contains an active component with a variable microstructure changing under external field. The model of a layer adaptive composite with ferroelectric active component is presented in Fig. 1. The active layer of the composite consists of an alternation of 90^0-domains, or twins. This polydomain, or polytwin, microstructure is a result of a para-ferroelectric transformation in a constrained layer. The formation of domains minimizes the misfit between layers of a composite and decreases the elastic energy of internal stresses. Under external mechanical or electrostatic fields the equilibrium domain structure should change in such a way that the fraction of the domain with preferrent orientation with respect to the field should increase (Fig. 2). The misfit and the elastic energy may increase, but this increment is compensated by the decrease of the potential energy of the composite in the external field.

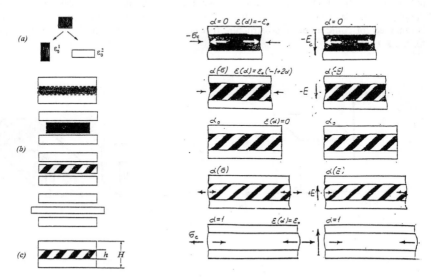

Figure 1.   Formation of an adaptive composite.

Figure 2.   Deformation and polarization of an adaptive composite.

Mat. Res. Soc. Symp. Proc. Vol. 459 © 1997 Materials Research Society

If the mobility of interdomain interface is high, this thermodynamic effect results in additional contributions to the deformation and the polarization, increasing the compliance ("superelasticity") and the susceptibility of the composite [1,2]. We show below that the increase of compliance and susceptibility should have critical λ-type behavior if the thickness of the composite layer approaches to some critical values.

THEORETICAL MODEL

Consider a composite which consists of a periodic alternation of plane-parallel layers of an active ferroelectric phase and a passive phase. An elementary unit of the composite is shown in Fig.1 c. $\beta = h/H$ and $1-\beta = 1-h/H$ are relative thicknesses of the layers. The ferroelectric phase has a polytwin periodic microstructure, α is the fraction of domain 2. For simplicity, a 2-dimensional composite is considered [1]. The self strains of a para-ferroelectric transformation are:

$$\hat{\varepsilon}_0^1 = \begin{pmatrix} -\varepsilon_0 & 0 \\ 0 & \varepsilon_0 \end{pmatrix} \qquad \hat{\varepsilon}_0^2 = \begin{pmatrix} \varepsilon_0 & 0 \\ 0 & -\varepsilon_0 \end{pmatrix} \tag{1}$$

with the twin strain:

$$\hat{\varepsilon}_0 = \hat{\varepsilon}_0^2 - \hat{\varepsilon}_0^1 = \begin{pmatrix} 2\varepsilon_0 & 0 \\ 0 & 2\varepsilon_0 \end{pmatrix} \tag{2}$$

The chosen form of the self-strain allow us to use the 2D model for the description of the 3D polydomain microstructure, formed by the two domains of a tetragonal phase with the c-axis in the plane of the ferroelectric layer with $\varepsilon_0 = \varepsilon_0' - \varepsilon_0''$ (Fig.3) [3]. For both cases the minimum misfit is obtained for equidomain microstructure if it is assumed that the minimum misfit equals zero for the 2D case.

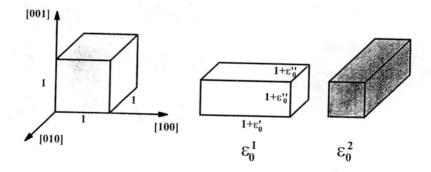

Figure 3.   Two domains of a tetragonal phase formed from a cubic crystal.

ENERGY OF COMPOSITE

The energy of the composite is a sum of the elastic energy of long-range internal stresses, the elastic energy of short-range microstresses near the interfaces between the composite layers and the energy of the interdomain interfaces [3]. The expression for the energy per unit area of the composite and the graphical presentation of each term in that expression is schematically illustrated in Table I.

The first term describes energy of elastic interaction between the layers and has a minimum at $\alpha=1/2$. e_{12} is the energy of indirect interaction between domains [3]. $e_{12}=1/2E_0(2\varepsilon_0)^2$, where E_0 is the Young's modulus, for the 2D model. e_0 is the misfit energy of single domain layer and is equal to $1/2E_0\varepsilon_0^2$ for the 2D model.

The second term describes approximately the energy of the microstresses [4]. The microstresses are localized within a thin layer near the interfaces between the composite phases; its thickness is approximately equal to the period D of the polydomain microstructure. The microstresses and their energy have been calculated in many papers [5-9]. However, the analysis of the results of the calculations shows that the simple expression above gives a fairly good estimation of microstress energy for polydomain microstructures with a period less than the thickness of the polydomain layer and not very small domain fractions.

The third term expresses the energy of the interdomain interfaces with γ as their specific energy.

Table I. Energy of a Polydomain Heterostructure.

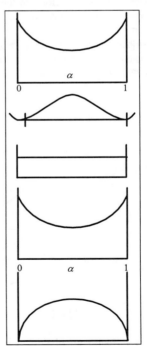

$$e'H = H\beta(1-\beta)\left[e_0 - \alpha(1-\alpha)e_{12}\right]+$$

$$D\xi\alpha^2(1-\alpha)^2 e_{12} +$$

$$\frac{h}{D}2\gamma.$$

$$e''H = H\beta(1-\beta)\left[e_0 - \alpha(1-\alpha)e_{12}\right]+$$

$$\sqrt{lh}\,2\xi\alpha(1-\alpha)e_{12} +$$

The second and the third term together have a minimum at an equilibrium domain period $D_0=[\alpha(1-\alpha)]^{-1}(lh)^{1/2}$, where $l=2\gamma/\xi e_{12}$ is a characteristic length (~10-100 nm), ξ is a coefficient of order 1. The energy of the equilibrium polydomain microstructure is described by the two quadratic functions of α: the energy of macrostress which is proportional to H and the combined energy of microstresses and domain interfaces, which is proportional to $h^{1/2}$ ($e''H$ in Table I). Therefore, there is a critical value of an effective thickness $\mathcal{H}=\beta^2(1-\beta)h=(1-\beta)^2\beta H$:

$$\mathcal{H}_{cr} = 4\xi^2 l \tag{3}$$

At $\mathcal{H}=\mathcal{H}_{cr}$ the domain microstructure is unstable.

Formation of new domain interfaces is necessary to establish the equilibrium domain period. In real cases the formation of domain interfaces is usually a more difficult process than their displacement. Therefore, the microstructure can be in a partial equilibrium when the minimum of the energy is obtained due to the interface movement at a fixed domain period. The following expression gives the energy density e' (see Table I) as a function of deviation of domain fraction α from the equilibrium value $\alpha_0=1/2$:

$$e' = \beta(1-\beta)e_{12}\left\{\left(1-\frac{\xi D}{2\beta(1-\beta)H}\right)(\Delta\alpha)^2 + \frac{\xi D}{\beta(1-\beta)H}(\Delta\alpha)^4\right\} + e_0 \tag{4}$$

The expression (4) has the form of the Landau's expansion for a free energy in the theory of second order transitions with $\Delta\alpha=\alpha-1/2$ as an "order parameter". The energy dependencies on $\Delta\alpha$ at different relative thicknesses $t=2\beta(1-\beta)H/\xi D$ are shown in Fig. 4.

At the critical point $t=1$ a minimum at $\Delta\alpha=0$ (a unstressed state) disappears and the two minima corresponding to stressed states arise: a spontaneous transition from a symmetrical state to non-symmetrical states proceeds.

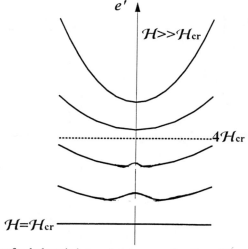

Figure 4. Energy of polydomain heterostructure as a function of domain fraction.

ELASTIC COMPLIANCE AND SUSCEPTIBILITY OF ADAPTIVE COMPOSITES

As soon as the energy as a function of $\Delta\alpha$ is known, a response of the composite to external fields affecting the domain fraction can be determined. The standard thermodynamic procedure leads from the enthalpy Φ of the composite under mechanical stress σ or electrostatic field strength $\vec{E}$, to the following expressions for an average strain ε and polarization as well as the compliance S and the susceptibility χ.

$$\Phi\left(\Delta\alpha, \frac{\sigma}{\vec{E}}\right) = e(\Delta\alpha) - \begin{cases} \sigma\beta\varepsilon_0\Delta\alpha - \dfrac{1}{2}S_0\sigma^2 \\ \vec{E}\beta P_s\Delta\alpha \end{cases} \tag{5}$$

$$\varepsilon = -\frac{\partial\Phi}{\partial\sigma} = S_0\sigma + \beta\varepsilon_0\Delta\alpha_0 \qquad \left.\frac{\partial\Phi}{\partial\Delta\alpha}\right|_{\Delta\alpha_0} = 0 \tag{6}$$

$$P = -\frac{\partial\Phi}{\partial\vec{E}} = \beta P_s\Delta\alpha_0$$

$$S = S_0 + \left(\beta\varepsilon_0\right)^2 \Big/ \frac{\partial^2 e}{\partial(\Delta\alpha)^2}$$

$$\chi = \beta^2 P_s^2 \Big/ \frac{\partial^2 e}{\partial(\Delta\alpha)^2} \tag{7}$$

where S_0 is the compliance of the passive phase or a single domain ferroelectric ($S_0 \cong 1/E_0$). P_s is the saturation polarization. It is assumed that depolarization field is canceled by free charges. The thickness dependencies of the compliance and the susceptibility for the relative thickness $t \gg 1$ are presented graphically in Fig. 5 and analytically as follows:

2D: $\qquad e = \dfrac{1}{2}E_0\left(\varepsilon_0\Delta\alpha\right)^2, \qquad S = S_0 + \dfrac{\beta}{(1-\beta)E_0}, \qquad E = (1-\beta)E_0 \tag{8}$

3D: $\qquad e = \dfrac{1}{2}\dfrac{E_0}{1-\nu}\left(\varepsilon_0\Delta\alpha\right)^2, \qquad S = S_0 + \beta\Big/\dfrac{2(1-\beta)E}{1-\nu}, \qquad \chi = \beta P_s^2\Big/\dfrac{2(1-\beta)E_0\varepsilon_0^2}{1-\nu} \tag{9}$

For t close to 1, a critical behavior should be observed if the mobility of domain interfaces is sufficiently high to establish the partial equilibrium (Fig. 6). The critical behavior near $t=1$ is described by the standard Landau theory. Particularly, the elastic modulus should drop at the critical thicknesses of the active or passive layers and linearly approach to zero in the vicinity of $t=1$. The critical properties will be considered in detail elsewhere [10]. A similar singularity of χ for very thin ferroelectric film has been found in [11] by numerical calculations.

CONCLUSION

The thermodynamic analysis presented predicts the critical softening of an adaptive composite at deformation and polarization. The critical singularities of the compliance (modulus) and the susceptibility should be observed at low and high critical thicknesses of the active polydomain layer.

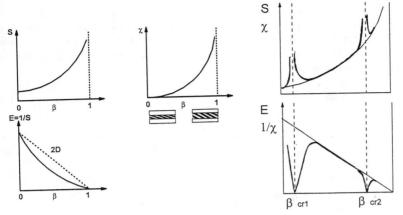

Figure 5. Elastic compliance (S), modulus (E)
susceptibility (χ): t>>1.

Figure 6. Critical behavior of elastic
Compliance (S), modulus (E)
and susceptibility (χ).

ACKNOWLEDGMENTS

This work is supported by Office of Naval Research and NSF.

REFERENCES

1. A.L.Roytburd in Shape Memory, edited by C.T. Lu and M. Wuttig (Mater. Res. Soc. Proc. 246, 1992), p. 91-103.
2. A.L.Roytburd and Y. Yu in Twinning in Advanced Materials, edited M.H. Yoo and M. Wuttig (TMS 1994), p. 217.
3. A.L. Roytburd in Heteroepitaxy of Dissimilar Materials, edited A. Zangwill (Mater. Res. Soc. Proc. 221, 1991), p. 100; A.L. Roitburd, Phys. Stat. Sol. (a) 37, p. 329 (1976).
4. A.L.Roitburd in Solid State Physics Vol:33, edited by H. Ehrenreich, F. Seitz and D. Turnbull, Academic Press, New York, 1978, p. 317; A.L. Roytburd, Phase Transitions 45, p. 1 (1993).
5. A.G. Khachaturyan, Theory of Structural Transformation in Solids, John Wiley and Sons, New York (1983).
6. J.S. Speck, A.C. Daykin, A. Seifert, A.E. Romanov, and W. Pompe, J. Appl. Phys. 78, p. 1696 (1995).
7. A.E. Romanov, W. Pompe, J.S. Speck, J. Appl. Phys. 79, p. 4037 (1996).
8. N.A. Pertsev and A.G. Zembilgotov, J. Appl. Phys. 78, p. 6170 (1995).
9. N. Sridhar, J.M. Rickman and D.J. Srolovitz, Acta Mater. 44, p. 4085, p. 4097 (1996).
10. A.L. Roytburd, Phys. Rev. Lett., submitted.
11. N.A. Pertsev, G. Arlt, and A.G. Zembilgotov, Phys. Rev. Lett. 76, p. 1364 (1996).

FINITE ELEMENT MODELLING OF ADAPTIVE COMPOSITE

Y. WEN*, S.DENIS*, E. GAUTIER* and A. ROYTBURD**
* Laboratoire de Sciences et Génie des Matériaux Métalliques, CNRS URA 159, Ecole des Mines Parc de Saurupt 54042 Nancy France
** Department of Materials Science and Engineering, Maryland University, College Park, USA

ABSTRACT

The deformation of adaptive composite with active polydomain components has been calculated by finite element modelling (FEM). The average deformation, the distribution of stresses and the elastic energy of the composite as a function of different extent of twinning of the composite active component are calculated. Comparing the results of the FEM with the results of the analytical theory demonstrates the effect of the microstresses on the mechanics of the adaptive composite.

INTRODUCTION

The concept of adaptative composite with polydomain active component has been formulated and the theory of this composite has been developed in papers [1-3]. The theory predicts some unusual physical and mechanical properties of the composite, particularly, considerable decrease of the effective elastic modulus of the composite, non linear stress strain relations, critical behaviour at changing relative thickness of an active layer. Due to importancy of the results obtained theoritically for applications of the composite it is very desirable to verify the theoritical conclusions by numerical modelling. The modelling can avoid some assumption of the theoritical analysis. The 2D finite element model developed in [4, 5] is an ideal tool for numerical analysis of the layer adaptive composite considered theoritically. This paper contains the results of the modelling of adaptive composite and the comparison with the results of the analytical theory.

LAYER ADAPTIVE COMPOSITE : ANALYTICAL RESULTS

The analytical theory of adaptive composite [1] is briefly reviewed. The simplest structure of a layer of adaptive composite is presented in Fig.1a. An "elementary unit" of multilayer composite is shown. It consists of an active layer between two passive ones. Under mechanical stress, the active layer changes its structure by twinning. The change of the structure results in a self strain ε_0. An average self strain of the active layer is then proportional to the change of twin fraction (α). We assume that the active layer with equal fractions of both twin component ($\alpha = 1/2$) fits to the passive layers, so there is no internal stress at $\alpha = 1/2$ (Fig. 1b).

Under deformation, the fraction of the black domains grows, leading to misfit between layers and resulting in internal stresses. According to the theory, the mechanical behaviour of the composite under external stress is determined by its internal stress state. When the black domains become thicker, the elongation of the active layer corresponds to the self strain (Fig. 1c)

$$\varepsilon_0(\alpha) = \varepsilon_0(\alpha - 1/2) \tag{1}$$

An average strain of the composite is :

$$\varepsilon = \beta \, \varepsilon_0 \, (\alpha) \tag{2}$$

where β is the relative thickness of the active layer $\beta = h/H$.

Mat. Res. Soc. Symp. Proc. Vol. 459 © 1997 Materials Research Society

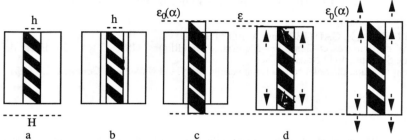

Figure 1 : Schematic changes of the structure and corresponding deformation in the layers.

The density of the elastic energy of internal stress is :

$$e = \beta\,(1-\beta)\frac{1}{2}\,E\,(\varepsilon_0\,(\alpha))^2 = \frac{1-\beta}{\beta}\,\frac{1}{2}\,E\,\varepsilon^2 \qquad (3)$$

where E is the Young's modulus. 2D model with an uniaxial misfit is considered for simplicity.

The internal stress can be determined from the elastic energy. The elastic energy of the composite of thickness H (per unit volume) is

$$e = -\frac{1}{2H}\int_V \varepsilon_0\,(\alpha)\,\sigma\,dV = -\frac{1}{2}\,\varepsilon_0\,(\alpha)\,\overline{\sigma}\,\frac{h}{H} \qquad (4)$$

where the integral is over the volume of the active layer part with the thickness h, $\overline{\sigma}$ is an average stress in the active layer. Then according to (2) :

$$\overline{\sigma} = -\frac{2e}{\varepsilon_0(\alpha)}\,\frac{H}{h} = -(1-\beta)\,E\,\varepsilon_0(\alpha) \qquad (5)$$

Under mechanical stress σ_{ext}, if the composite is in equilibrium there should be no stress in the active layer (otherwise the twin boundaries must move). So,

$$\sigma_{ext} + \overline{\sigma} = 0 \qquad\qquad \sigma_{ext} = -\overline{\sigma} = (1-\beta)E\,\varepsilon_0\,(\alpha) \qquad (6)$$

The strain under external stress is a sum of the elastic strain and the self-strain :

$$\varepsilon_{ext} = \frac{1}{E}\sigma_{ext} + \beta\,\varepsilon_0(\alpha) = (1-\beta)\,\varepsilon_0(\alpha) + \beta\,\varepsilon_0(\alpha) = \varepsilon_0(\alpha) \qquad (7)$$

The stress-strain relation for the composite is :

$$\sigma_{ext} = E_{ef}\,\varepsilon_{ext} \qquad\qquad E_{ef} = (1-\beta)E \qquad (8)$$

The effective Young's modulus of the composite E_{ef} is considerably lower than the modulus of the passive material. After only one component of twin remains, $\alpha = 1$, i.e. after the total deformation $1/2\,\varepsilon_0$, the superelastic behaviour of the composite transfers to the usual elastic one. (Fig.2).

This results corresponds to the 2D model of the adaptive composite. However, the analogous conclusion is obtained for the 3D model too [3].

The theory above is a continuum one. It does not consider discret microstructure of twinned layers. This homogeneous model of twin structure losses the part of elastic energy due to non uniform microstresses at the boundaries of the active layers with the passive ones [3]. These microstresses are localized near the tips of twins. The energy contribution of the nonuniform stress is proportionnal to the ratio D/H, where D is a period of twin structure, and H is the thickness of a "three layers" (Fig.1a). This contribution can become important when the energy of the uniform stresses (eq (3)) is small, e.g. when β is close to 0 or 1 [3].

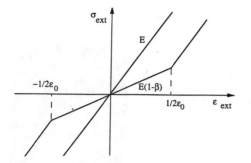

Figure 2 :
Mechanical behaviour of the adaptive composite

In the paper we verify the basic conclusions of the theory of adaptive composites. The internal stress states are modelled and the average stress inside the active layer and the elastic energy of internal stresses are calculated.

FINITE ELEMENT MODEL

A 2-dimensional model (2D) is considered. Fig. 3a is the whole mesh on which the FEM computations are carried out. Two sides named aA and bB are set free. For the two other sides named ab and AB periodic boundaries are considered. Figure 3b is the central layer which represents the adaptive compound. The thickness of the adaptative layer relative to the total thickness of the composite β is equal to 1/3, 1/6 and 1/30. The adaptive layer is composed of different domains with two shear orientations, one sheared in 45°direction and the second one sheared in 225° direction. To each domain is associated a self strain in the X direction equal to - 0.1 for 45° and +0.1 for 225°. Each domain is initially composed of 5 layers. At initial state, the fraction of each domain is the same, i.e. α = 1/2. The calculation of the elastic energy is done in this configuration by giving to each domain its respective self strain. This leads to an elastic strain energy. The strain in the X direction is 0.

In order to increase α, the 45° orientation layers will step by step be transformed in 225° orientation layers corresponding to a change in the self strain in the X direction of 0.2, until one single orientation is obtained. The elastic strain energy as the stress fields are calculated at each step.

The behaviour of each phase is considered as elastic. The Young's modulus is assumed as constant and equal for both phase to 10000 MPa. Plane strain assumption is accepted in this model. The computation is conducted using SYSTUS code which was developed by Framasoft +CSI.

RESULTS

In Fig. 4 are reported the variations of the calculated total deformation in the X direction in function of the self strain in the adaptive layer for the different values of β. A linear relation is obtained between ε and $\varepsilon_0(\alpha)$, and the slope is the one predicted by the continuum theory.

The variations in elastic energy in the total volume are reported Fig. 5a and b in function of the total strain and the self strain respectively. Moreover, in Fig. 5b we report the theoritical calculations of the strain energy considering relation (3). These results show that for α =1/2, the numerical simulation leads to an elastic energy equal to 0.6MJ/m3 ; this value varies very few with the thickness of the layer. As the self strain increases ($\alpha \rightarrow 1$) the elastic strain energy increases. In Fig. 5b we observe that the theoritical and numerical calculated variations of e are almost parallel for β = 1/3 and 1/6 when the self strain increases.

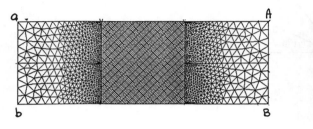
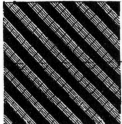

Figure 3 : Mesh used for the calculations by FEM

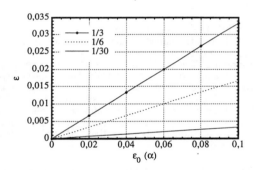

Figure 4 : Variations of calculated total strain versus self strain for different β

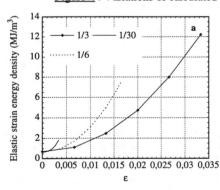
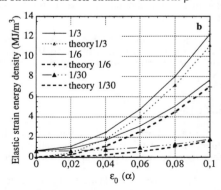

Figure 5 : Elastic strain energy density versus (a) total deformation of the composite and (b) versus self strain.

This can also be shown by comparing the coefficient of the quadratic term of equation (2) with the one obtained from the parabolic law fitted on the calculated results as shown in table 1. A good agreement is obtained for β values of 1/3 and 1/6. However a large discrepancy is evidenced for β = 1/30. The calculated increase in energy is much lower than the theoritical one, and the coefficient of the parabolic curve presents a variation from theory of about 30%.

Table 1 : Comparison of the coefficient of the quadratic term in energy obtained by numerical simulations and theory (values in MPa).

β	analytical theory	numerical simulation
1/3	10000	10365
1/6	25000	25150
1/30	145000	103110

The variations of the internal stresses as a function of the self strain are plotted Fig.6 for both numerical and analytical results. Both results show compressive stresses. Their magnitude increases as the self strain increases. A qualitatively good agreement is obtained between the two results. However the calculated results are always 10 % larger than the theoritical ones.

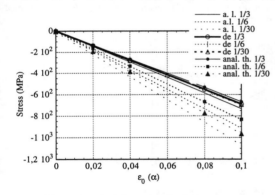

Figure 6 : Variations of the mean stress versus the self strain for different values of β.

a.l. calculated mean stress in the active layer

de results obtained by the derivative of the energy

anal. th. theoritical mean stress in the active layer

To understand these results it is necessary to analyse the distribution of internal stresses in the X direction in the composite. The results obtained by numerical simulation are shown Fig. 7 for two values of β and two values of α. The patterns show that the microstresses are heterogeneously distributed in the layers as in the domains. As the self strain increases, the distribution changes. For a value of β equal to 1/3, the microstresses are confined at the interface between the passive and active layer. As β decreases, the stress distribution in the active layer is highly heterogeneous, even for α = 1. The microstresses will thus largely contribute to the response of the composite by decreasing the elastic strain energy. At last, this difference between the two results for the internal stress in the active layer can be understood if one considers that for the analytical theory it is assumed that the mean microstress in the active layer is 0. For the numerial simulation the mean microstress is 0 in the composite.

This is further evidenced if we calculate the derivative of the elastic strain energy density versus ε, which for the model corresponds to the opposite of the internal stress in the active layer. These values, multiplied per -1, are plotted versus ε_0 (α) and compared Fig. 6 to the values of the internal stress presented previously. We observe that the last values correspond to the analytical ones for β = 1/3 and β = 1/6, and are much lower for β = 1/30.

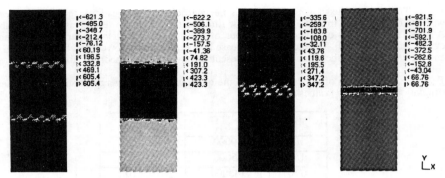

Figure 7 : Stress in the X direction for β = 1/3 (α : 0.5 ; 0.9) and β = 1/30 (α : 0.5 ; 0.9)

The effective Young's modulus which is the slope of the curves presented Fig.6 varies as the analytical one for β = 1/3 and 1/6. For β = 1/30, a deviation from the theoritical curve is obtained due to the microstresses influence as shown Fig. 8. Such a decrease in the effective Young's modulus evidences that a non linearity should be observed in the stress/strain curve of the adaptive composite for the very first strain.

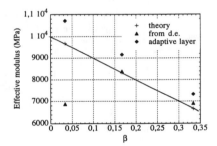

Figure 8 : Variations in the effective Young's modulus versus β.

CONCLUSIONS

The comparison of the numerical results with the theoritical (analytical) ones shows that :

• the elastic strain energy density variations are equivalent for β = 1/3 and β = 1/6 according to equation (3) but considerable discrepancies exist for β = 1/30.

• the calculated mean internal stress in the active layer is 10% larger than the theoritical one. These differences are related to the microstresses which are not taken into account in the analytical theory.

REFERENCES

1. A.L. Roytburd, in <u>Shape Memory</u>, C.T. Lu, M. Wuttig (Eds) Mat.Res. Soc. Symp. Proc. v 246, 1992, pp. 91-103.
2. A.L. Roytburd and Y. Lu, in <u>Twinning in Advanced Materials</u>, M. Yoo, M. Wuttig (Edts), TMS, 1993, pp. 217-224; in <u>Smart Materials</u> Mat. Res. Soc. Symp. Proc. ,V360, 1995
3. A.L. Roytburd this issue
4. Y. Wen, S. Denis, E. Gautier, Journal de Physique IV, Colloque C2, Sup. au Journal de Physique III, vol5, 1995, pp 531-536.
5. Y. Wen, S. Denis, E. Gautier Journal de Physique IV, Colloque C1, Sup. au Journal de Physique III, vol6, 1996, pp 475-483.

EXPERIMENTAL ANALYSIS OF SMART STRUCTURE WITH DAMPING TREATMENT AND SMA

Q. Chen, J. Ma and C. Levy
Mechanical Engineering Department, Florida International University, Miami, FL 33199.

ABSTRACT

The experimental results of a flexible cantilever beam with constrained viscoelastic layer and shape memory alloy layer called smart damping treatment (SDT) are presented. The upper side of the beam is bonded with a viscoelastic layer and then covered with a constraining layer. The lower side is bonded with a shape memory alloy layer, which is used as an actuator. The elastic modulus and loss factor of damping materials are functions of the temperature. The temperature effects on system frequency and loss factor due to heat cycling of SMA layer are evaluated here. It is found that temperature plays an important role on system frequency and loss factor, and thus the temperature effects must be included when discussing such an structure.

INTRODUCTION

Constrained or unconstrained viscoelastic damping treatment has been used for vibration control for many years. Many authors are involved in the study of such structures, [e.g. 1-8]. The problems for such passive application of viscoelastic damping are: 1) damping can only be optimized at specific ranges of temperature and frequency; and 2) the system can not be adjusted after the completion of design.

Composites with SMA have the ability to change material properties, induce large internal forces in the materials, modify the stress and strain state of the structure. These characteristics are very useful for acoustic and vibration control. Their applications in the fields of structural acoustic control and vibration control have attracted more and more attention since the late 1980's [9-16].

In this paper, the experimental results of-a flexible cantilever beam with constrained viscoelastic layer and shape memory alloy layer called smart damping treatment (SDT) is presented. The elastic modulus and loss factor of damping materials are functions of the temperature. The temperature effects on system frequency and loss factor due to heat cycling of SMA layer are evaluated here. It is found that temperature plays an important role on system frequency and loss factor.

EXPERIMENT

Experimental Setup

The experimental setup for the analysis of dynamic characteristics is shown in Figure 1. It is mainly composed of the excitation system, flexible beam structure and data acquisition system. In order to measure the system characteristics (frequency, vibration mode and damping), we use the Bruel & Kjaer impact hammer type 8202 to excite the structure. The response accelerometer with line drive amplifier is mounted on the structure to be tested. The data are measured by a Bruel & Kjaer Dual Channel Signal Analyzer type 2034, and

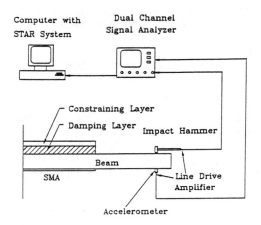

Figure 1 Experiment Scheme

then analyzed.

Beam Structure

The damping material used for the experiments is called DYAD 606 from THE SOUNDCOAT COMPANY, INC. DYAD 606 is Soundcoat's constrained layer damping material for vibration damping at low temperature. DYAD 601 provides maximum damping at about 100^0F (38C), but the material has a useful damping range from 50^0 to 150^0F (10C to 66C). The original uncovered structure is an aluminum beam with dimensions 533.00 mm x 25.4 mm x 3.175 mm. The dimensions for the DYAD 606 damping material are 25.4 mm x 0.6 mm with different lengths of coverage. The constraining layer is an aluminum sheet with dimensions 25.4 mm x 1 mm. The Nitinol SMA material with thickness of 0.833 mm is used in the experiment. The austenite starting temperature A_s=61C and the finishing temperature A_f=72C.

RESULTS AND DISCUSSION

The experiments were carried out for the above structure to obtain the frequency and loss factor of the system. The results for different coverage are shown in Figures 2 - 5.

Coverage Length Effects

The effects of coverage length on the natural frequency of the system are shown in Figure 2. We note from the figure that an increase in coverage length will increase the first and second natural frequencies. This tendency follows the numerical trends found for other types of constrained viscoelastic coverings [6,8].

Figure 3 shows the results of system loss factor versus coverage length ratio. From the figure, we note that an increase in the coverage length ratio will increase the system loss

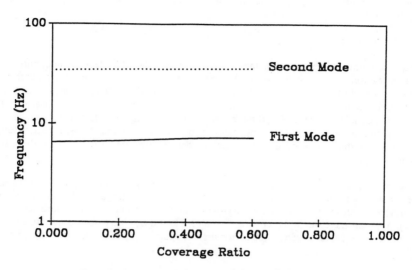

Figure 2 Coverage Ratio vs. Frequency

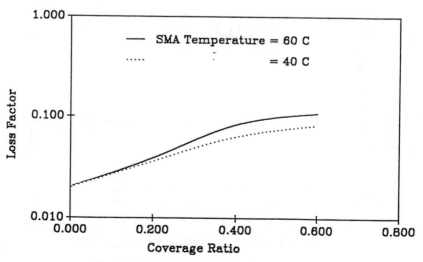

Figure 3 Coverage Ratio vs. Loss Factor

factor because more damping material is used, again in consonance with results previously discussed in [6,8].

Temperature Effects

Figure 4 show the results of system frequency versus the SMA layer temperature. When the temperature is less than the SMA A_s temperature, an increase in temperature will decrease the system frequency. Between the A_s and A_f temperatures, increase of temperature will increase system frequency because of austenite phase transformation effects on Young's modulus. But if the coverage is small (e.g., less than 20%), we can hardly see the increase in system frequency because the SMA austenite transformation effects-- which increase the frequency--are balanced by the temperature effects which decrease the frequency.

Figure 5 shows the results of system loss factor versus coverage ratio. It is noted that there exists a temperature which makes the system loss factor a maximum. This optimal temperature is close to the maximum damping material loss factor temperature as discussed in [15]. Also, an increase in the coverage from 0.2 to 0.6 will increase the system loss factor. This follows the general trends found for the constrained viscoelastic covering [6,8].

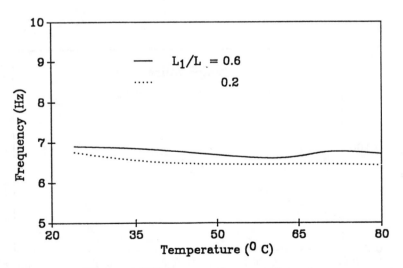

Figure 4 SMA Temperature vs. Frequency

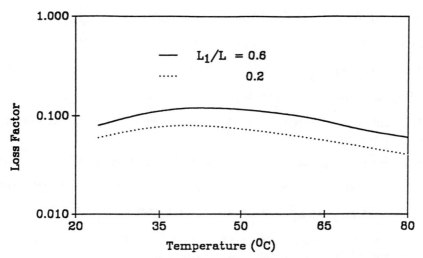

Figure 5 SMA Temperature vs. Loss Factor

CONCLUSION

A flexible cantilever beam with constrained viscoelastic layer and shape memory alloy layer called smart damping treatment (SDT) has been experimentally investigated. The temperature effects on system frequency and loss factor due to heat cycling of SMA layer are evaluated here. It is found that temperature plays an important role in determining the system frequency and loss factor. An analytical model is being developed to investigate the beam that is partially covered with SMA and constrained viscoelastic layer. This model will be reported elsewhere.

ACKNOWLEDGEMENTS

The authors wish to acknowledge the partial support of this work under NASA grant NAG-1-1787.

REFERENCES

1. E. M. Kerwin, J. Acoust. Soc. Amer., **31**, p. 952-962 (1959).

2. D. K. Rao, J. Mech. Eng. Sci., **20(5)**, p. 271-282 (1978).

3. D. J. Mead, J. Sound and Vib.,**83(3)**, p. 363-377 (1982).

4. D. Xi, Q. Chen and G. Cai, ASME J. Vib. Acoust. Stress Rel. Des., **108**, p. 65-68 (1986).

5. M. D. Rao and S. He, AIAA Journal, **31(4)**, p. 736-745 (1993).

6. C. Levy and Q. Chen, J. Sound and Vib., **177(1)**, p. 15-29 (1994).

7. H. Zhou and M. D. Rao, Int. J. Sol. Struct., **30(16)**, p. 2199-2211 (1993).

8. Q. Chen and C. Levy, Int. J. Sol. Struct., **31(17)**, p. 2377-2391 (1994).

9. C. A. Rogers, J. Acoust. Soc. Amer., **88(6)**, p. 2803-2811 (1990).

10. W. R. Saunders, H. H. Robertshaw, and C. A. Rogers, Proc. 31st Structures, Structural Dyn. and Mat. Conf., AIAA p. 90-1090, (1990).

11. C. A. Rogers, C. Liang, and C.R.Fuller, J. Acoust. Soc. Amer., **89(1)**, p. 210-220 (1991).

12. E. J. Graesser and F. A. Cozzarelli, ASCE J. Eng. Mech., **117(11)**, p. 2590-2608 (1991).

13. L. C. Brison, J. Intell. Mat. Sys. Struct., **4**, p. 229-242 (1993).

14. K. H. Wu, Y. Q. Liu, M. Maich, and H. K. Tseng, Proc. SPIE - Int. Soc. Optical Eng., **2189**, p. 306-313 (1994).

15. Q. Chen and C. Levy, Proc. SPIE - Int. Soc. Optical Eng., **2443**, p. 579-587 (1995).

16. Q. Chen and C. Levy, J. Smart Mat. Struct., **5**, p. 400-406 (1996).

Part IV

Poster Session I

SELF-MONITORING OF STRAIN AND DAMAGE BY CARBON FIBER POLYMER-MATRIX COMPOSITE

XIAOJUN WANG, D.D.L. CHUNG
Composite Materials Research Laboratory, State University of New York at Buffalo, Buffalo, NY 14260-4400

ABSTRACT

Self-monitoring of damage and dynamic strain in a continuous crossply (0°/90°) carbon fiber polymer-matrix composite by electrical resistance (R) measurement was achieved. With a static/cyclic tensile stress along the 0° direction, R in this direction and R perpendicular to the fiber layers were measured. Upon tension to failure, R in the 0° direction first decreased (due to increase of degree of 0° fiber alignment) and then increased (due to 0° fiber breakage), while R perpendicular to the fiber layers increased monotonically (due to increase of degree of 0° fiber alignment and delamination). Upon cyclic tension, R (0°) decreased reversibly, while R perpendicular to the fiber layers increased reversibly, though R in both directions changed irreversibly by a small amount after the first cycle. For a 90° unidirectional composite, R (0°) increased reversibly upon tension and decreased reversibly upon compression in the 0° direction, due to piezoresistivity.

INTRODUCTION

Although the placement of strain/damage sensors is common in smart structures, it suffers from poor durability, limited sensing volume, degradation of mechanical properties, and high cost. Self-monitoring refers to the ability of the structural material to monitor itself. By using a self-monitoring material, the disadvantages mentioned above are removed.

A common method of fatigue monitoring is acoustic emission [1,2], which suffers from its inability to monitor dynamic strain. Much less common is the method involving the measurement of the electrical resistance, which increases due to damage and provides a mechanism for self-monitoring. Previous work using electrical resistance to monitor fatigue was carried out on a CaF_2-matrix SiC-whisker composite [3], but dynamic strain monitoring was not performed, probably because this composite's electrical resistivity did not change reversibly with reversible strain. In general, dynamic strain monitoring requires a measurand which changes in value reversibly during reversible straining. In addition, in order for both dynamic strain and damage to be simultaneously monitored with a single method, that method must involve a measurand which changes in value reversibly during reversible straining and changes irreversibly during damage. In this work, we have achieved this by using the electrical resistance as the measurand and continuous carbon fiber polymer-matrix composite as the material.

Continuous carbon fiber polymer-matrix composites are advanced composites which are attractive in that they combine high strength, high modulus and low density. Previous work on a polymer-matrix composite containing a combination of continuous glass and carbon fibers showed that the electrical resistance of this composite increases irreversibly upon damage (due to the fracture of the carbon fibers) [4], but fatigue monitoring and reversible resistivity changes (dynamic strain monitoring) were not explored. Our previous work on continuous unidirectional (0°) carbon fiber polymer-matrix composite [5,6] has shown that both the longitudinal (parallel to fibers) and transverse (perpendicular to fiber layers) electrical resistivities of the composite change reversibly upon dynamic straining (due to the reversible change in the degree of fiber alignment) and change irreversibly upon damage (due to irreversible change in the degree of neatness of the fiber arrangement within a fiber bundle after the first strain cycle). Furthermore, fatigue monitoring simultaneous to dynamic strain monitoring was demonstrated by longitudinal electrical resistance measurement [6]. This paper extends Ref. 5 and 6 in that it addresses unidirectional (90°) and crossply (0°/90°) fiber composites.

EXPERIMENTAL METHODS

Composite samples were constructed from individual layers cut from a 12 in. wide unidirectional carbon fiber prepreg tape manufactured by ICI Fiberite (Tempe, AZ). The product used was Hy-E 1076E, which consisted of a 976 epoxy matrix and 10E carbon fibers. The composite laminates were laid up in a 4 x 7 in. (102 x 178 mm) platen compression mold with laminate configuration $[90]_{32}$ for the unidirectional composite and $[0/90]_{3s}$ for the crossply composite. The individual 4 x 7 in. fiber layers (32 per laminate for the 90° unidirectional composite and 12 per laminate for the crossply composite) were cut from the prepreg tape. The layers were stacked in the mold with a mold release film on the top and bottom of the layup. No liquid mold release was necessary. The density of the laminate was 1.52 ± 0.03 and 1.50 ± 0.03 g/cm^3 for 90° unidirectional and crossply composites respectively. The thickness of the laminate was 4.5 and 1.5 mm for the 90° unidirectional and crossply composites respectively. The volume fraction of carbon fibers in the composite was 57% and 52% for the 90° unidirectional and crossply composites respectively. The laminates were cured using a cycle based on the ICI Fiberite C-5 cure cycle. The curing occurred at 179 ± 6°C (355 ± 10°F) and 0.61 MPa (89 psi) for 120 min. Afterwards, they were cut to pieces of size 153 x 12 mm and 178 x 7 mm for the 90° unidirectional and crossply composites respectively. Glass fiber reinforced epoxy end tabs were applied to both ends on both sides of each piece, such that each tab was 30 mm long and the inner edges of the end tabs on the same side were 100 mm apart. The tensile strength was 62 ± 4.9 and 695 ± 18 MPa for the 90° unidirectional and crossply composites respectively. The tensile ductility was (0.78 ± 0.08)% and (1.06 ± 0.18)% for the 90° unidirectional and crossply composites respectively. The electrical resistivity in the 0° direction was 57.4 and 8.84 x 10^{-2} Ω.cm for the 90° unidirectional and crossply composites respectively; that in the direction perpendicular to the fiber layers was 21.35 and 45.7 Ω.cm for unidirectional and crossply composites respectively.

The electrical resistance R was measured both in the 0° direction and in the direction perpendicular to the fiber layers using the four-probe method while either static or cyclic tension was applied in the 0° direction. Silver electrically conducting paint was used for all electrical contacts. The four probes consisted of two outer current probes and two inner voltage probes. The resistance R refers to the sample resistance between the inner probes. For measuring R in the 0° direction (the stress axis), the four electrical contacts were around the whole perimeter of the sample in four parallel planes that were perpendicular to the stress axis, such that the inner probes were 60 mm apart and the outer probes were 78 mm apart. For measuring R in the direction perpendicular to the fiber layers, the current contacts were centered on the largest opposite faces parallel to the stress axis and in the form of open rectangles of length 75 mm in the stress direction and width 11 mm, while each of the two voltage contacts was in the form of a solid rectangle (of length 20 mm in the stress direction) surrounded by a current contact (open rectangle). Thus, each face had a current contact surrounding a voltage contact. A strain gage was attached to the center of one of the largest opposite faces parallel to the stress axis for samples for measuring R in the 0° direction as well as samples for measuring R in the direction perpendicular to the fiber layers. A Keithley 2001 multimeter was used for DC electrical measurement. The displacement rate was 1.0 and 0.5 mm/min for the crossply composite and the 90° unidirectional composite respectively. A hydraulic mechanical testing system (MTS 810) was used.

RESULTS AND DISCUSSION

Fig. 1 shows the tensile stress, strain and $\Delta R/R_o$ along the stress axis (0° direction) obtained simultaneously for the crossply composite during cyclic tension at a stress amplitude of 35% of the breaking stress. The strain returned to zero at the end of each cycle. Because of the small strains involved, $\Delta R/R_o$ is essentially equal to the fractional increase in resistivity. The $\Delta R/R_o$ decreased upon loading and

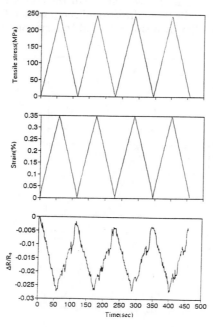

Fig. 1 Variation of $\Delta R/R_o$ (0° direction), tensile stress and tensile strain with cycle number during cyclic tension to a stress amplitude equal to 35% of the fracture stress for crossply composite.

increased upon unloading in every cycle, such that R irreversibly decreased slightly after the first cycle (i.e., $\Delta R/R_0$ did not return to 0 at the end of the first cycle). The behavior is the same as that of the 0° unidirectional composite of Ref. 5 (Fig. 1 of Ref. 5), except that the magnitude of $\Delta R/R_0$ is smaller. That the magnitude of $\Delta R/R_0$ is smaller for the crossply composite than the unidirectional composite at stress amplitudes equal to essentially the same fraction of the corresponding breaking stresses is because the crossply composite has 90° fibers.

A length increase without resistivity change would have caused R to increase upon tension, but R was observed to decrease upon tension. For the crossply composite, the observed magnitude was about 6 times that calculated by assuming that $\Delta R/R_0$ was only due to length increase. Hence, the contribution of $\Delta R/R_0$ from the length increase is negligible compared to that from the resistivity change.

For both 0° unidirectional and crossply composites, the irreversible decrease in $\Delta R/R_0$ in the 0° direction at the end of the first cycle is due to the irreversible decrease in the degree of waviness of the 0° fibers after the first cycle.

Fig. 2 shows the tensile stress, strain (in the 0° direction) and $\Delta R/R_0$ perpendicular to the fiber layers obtained simultaneously for the crossply composite during cyclic tension at a stress amplitude of 35% of the breaking stress. The strain returned to zero at the end of each cycle. The $\Delta R/R_0$ increased upon loading and decreased upon unloading in every cycle, such that R irreversibly increased slightly after the first cycle (i.e., $\Delta R/R_0$ did not return to 0 at the end of the first cycle). The behavior is similar to that of the 0° unidirectional composite of Ref. 5 (Fig. 3 of Ref. 5), except that $\Delta R/R_0$ is smaller by about 85% and the irreversible portion of $\Delta R/R_0$

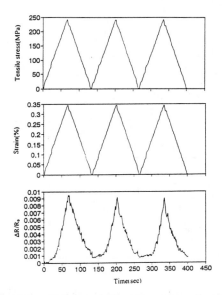

Fig. 2 Variation of $\Delta R/R_0$ (direction perpendicular to the fiber layers), tensile stress and tensile strain with cycle number during cyclic tension to a stress amplitude equal to 35% of the fracture stress for crossply composite.

is positive for the crossply composite but negative for the $0°$ unidirectional composite. The small value of $\Delta R/R_o$ for the crossply composite compared to the unidirectional composite is due to the presence of the $90°$ fibers in the crossply composite and that these fibers make the increase in the degree of $0°$ fiber alignment only slightly affect the chance that adjacent fiber layers touch one another. That the irreversible portion of $\Delta R/R_o$ is negative for the $0°$ unidirectional composite (Fig. 3 of Ref. 5) is due to the irreversible decrease in the degree of neatness of the fiber arrangement within a fiber bundle and the consequent increase in the chance that adjacent fiber layers touch one another. That the irreversible portion of $\Delta R/R_o$ is positive for the crossply composite (Fig. 2) is due to the presence of the $90°$ fibers, which (i) make the irreversible decrease in the degree of neatness of the fiber arrangement have negligible effect on the chance that the adjacent fiber layers touch one another, and (ii) cause the $0°$ fibers to be somewhat wavy (with the $0°$ fibers partly penetrating the gap between the $90°$ fiber bundles of the adjacent $90°$ fiber layer) after composite fabrication by compression molding and this waviness enhances the chance for the $0°$ and $90°$ fiber layers to touch each other. After a cycle of tensile loading in the $0°$ direction, the waviness is irreversibly lessened, thereby decreasing the chance for the adjacent fiber layers to touch each other and thus irreversibly increasing R perpendicular to the fiber layers.

Dynamic tensile loading was conducted on crossply composites. The behavior at different stress amplitudes are similar, though the magnitudes of both reversible and irreversible portions of $\Delta R/R_o$ in both directions increase with increasing stress amplitude (Table 1). Table 1 also shows comparison of $\Delta R/R_o$ between the crossply composite of this work and the unidirectional composite of Ref. 5. For the same stress amplitude (expressed as a fraction of the fracture stress), the magnitudes of both reversible and irreversible portions of $\Delta R/R_o$ in both directions are higher for the unidirectional composite than the crossply composite.

Both dynamic tensile loading and dynamic compressive loading were

Table 1 Effect of stress amplitude and fiber lay-up configuration on the reversible and irreversible parts of $\Delta R/R_o$

Maximum stress ____ Fracture stress	$\Delta R/R_o$ parallel to $0°$ fibers		$\Delta R/R_o$ perpendicular to the fiber layers	
	Reversible	Irreversible	Reversible	Irreversible
*6.1%	-0.0015	0	0.0020	0
*9.2%	-0.0057	-0.00043	0.0030	0.00039
*19.4%	-0.0122	-0.00287	0.0061	0.00054
*35%	-0.0263	-0.00357	0.0090	0.00072
*73.6%	-0.0389	-0.00584	/	/
+35-36%	-0.0680	-0.01200	0.0600	-0.01700

*Crossply composite of this work.
+Unidirectional composite of Ref. 5.

Table 2 Strain sensitivity (or gage factor) in the 0° direction.

Composite type	Maximum stress Fracture stress	Strain sensitivity
0° unidirectional	32%	-35.7*
	36%	-37.6*
Crossply	19.4%	-5.7*
	35%	-7.1*
90° unidirectional	/	2.2*
	/	-2.1+

*Tension   +Compression

conducted on the 90° unidirectional composite.   Stress was applied in the 0°
direction while the electrical resistance in the 0° direction was measured and the
fibers were in the 90° direction.   The resistance increased upon tension and
decreased upon compression. The effect was not totally reversible due to plastic
deformation, which is attributed to the polymer matrix, as the polymer matrix
dominated the mechanical properties perpendicular to the fibers. The reversible
resistance changes under tension and compression are due to piezoresistivity, which
refers to the phenomenon in which the resistivity of a composite with a conducting
filler increases upon increase of the distance between filler units during tension and
decreases upon decrease of this distance during compression.
      The strain sensitivity or gage factor is defined as the reversible part of $\Delta R/R_o$
divided by the strain amplitude. The magnitude of the strain sensitivity is much
higher for the 0° unidirectional composite than the crossply composite and is
smallest for the 90° unidirectional composite (Table 2).   This means that the
contribution of the 90° fibers (i.e., piezoresistivity) to the resistance change observed
in the crossply composite is small. The strain sensitivity for the crossply composite
is not equal to the average of the values for the 0° and 90° unidirectional composites,
due to the interaction between the 0° and 90° fiber layers in the crossply composite.

REFERENCES

1. K.J. Konsztowicz and D. Fontaine, J. Am. Ceramic Soc. **73**, 2809 (1990).

2. E. Santos-Leal and R.J. Lopez, Measurement Sci. Tech. **6**, 188 (1995).

3. A. Ishida, M. Miyayama and H. Yanagida, J. Am. Ceram. Soc. **77**, 1057 (1994).

4. Norio Muto, Hiroaki Yanagida, Masaru Miyayama, Teruyuki Nakatsuji, Minoru
Sugita and Yasushi Ohtsuka, J. Ceramic Soc. Japan **100**, 585 (1992).

5. Xiaojun Wang and D.D.L. Chung, Smart Mater. Struct., in press.

6. Xiaojun Wang and D.D.L. Chung, Smart Mater. Struct., in press.

FABRICATION OF PSEUDO ONE-DIMENSIONAL PHOTONIC BAND SYSTEM BY GRATING PAIR METHOD

K. Todori, and S. Hayase
Materials and Devices Research Laboratories, Toshiba Corporation
1, Komukai Toshiba-cho, Saiwai-ku, Kawasaki 210, Japan, todori@mdl.rdc.toshiba.co.jp

ABSTRACT

We have succeeded in forming a pseudo one-dimensional photonic band by using two gratings facing each other. This pair of gratings had a period of 278 nm. The frequency of the band edge was in the visible region. This method is useful to control the group velocity and the phase velocity of light.

INTRODUCTION

In recent years, the photonic band has been researched by some groups[1]. The conduction bands of semiconductors and metals are formed by the Bragg reflection of the electronic waves at the atoms of these substances. Similarly, photonic bands are formed by the Bragg reflection of electromagnetic waves in a structure which has a periodical change of the refractive index. In the photonic band structure, the group velocity and the phase velocity are expressed by $d\omega/dk$ and ω/k, respectively, where ω is the angular frequency and k is the wave number. When the dispersion relation of the photonic band is not linear, for example the photonic crystal has a band gap, the group- and phase velocities vary with frequency. Therefore, the photonic band is useful to control these velocities.

In this letter, we wish to propose a new photonic band system in order to control the group- and phase velocities of light.

EXPERIMENT

The new system that we propose has a structure in which two gratings face each other (Fig. 1). The grating period and the surface area of the gratings were 278 nm and 25 mm × 25 mm, respectively. The surface was coated with aluminum. The spacing between two gratings was about 10 μm. In this condition, the incident visible light scarcely passes through the spacing between the gratings without being diffracted. We measured the intensity of the light of a halogen lamp transmitted through the grating pair. To obtain the transmittance of the grating pair, we normalized the intensities of transmitted light by that of the light source. The transmittance spectra were measured in the wavelength region from about 450 nm to about 700 nm.

RESULTS AND DISCUSSION

Figure 2 shows the transmittance spectrum of the grating pair. The frequency was normalized by the factor, $\omega d/2\pi c$, where d and c are the grating period and the light velocity

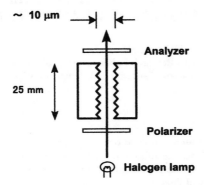

Figure 1. Schematic of experimental setup for a photonic band system.

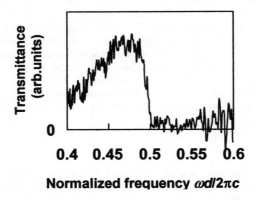

Figure 2. Transmittance spectra of an Al coated grating pair with 278-nm period for H-polarization.

in vacuum, respectively.   When the magnetic vector of the light was parallel to the grating surface (H-polarization), a peculiar structure of transmittance was observed at the normalized frequency of 0.49.   The great change of transmittance near 0.5 is one of the characteristics of the photonic band system.

When the electric vector of the light was parallel to the grating surface (E-polarization), the transmittance spectrum was almost flat.   This dependence of the spectra on the polarization is similar to that of the air-rod photonic band crystal [2].

We compared the observed spectrum with a calculated one in order to judge whether or not the observed spectrum was attributable to the photonic band. We assumed that the grating pair was a one-dimensional (1-D) photonic band crystal, i.e. a layer by layer structure. It has been reported that the dispersion relation of the 1-D photonic band can be simply calculated. Let the refractive index and thickness of the 1-D photonic band crystal be n_1 and a in aluminum layer, and n_2 and b in vacuum layer, respectively. The lattice constant d is $a + b$. Then the wave number $k(\omega)$ is calculated as follows [3]:

$$k(\omega) = \frac{1}{d}\arccos\left[\cos\left(\frac{n_1(\omega)\omega a}{c}\right)\cos\left(\frac{n_2(\omega)\,\omega b}{c}\right) - \frac{n_1 + n_2}{2 n_1 n_2}\sin\left(\frac{n_1\omega a}{c}\right)\sin\left(\frac{n_2\omega b}{c}\right)\right]$$

First, we calculated the dispersion relation, where we substituted the complex indexes of refraction of the aluminum [4] for n_1 in the equation above. We also considered the dispersion of refractive indexes. The lattice constant was fixed at 278 nm. From the dispersion relation, the density of states and the reflections at two surfaces were calculated. Using these two values, we then calculated the transmittance spectra of the 1-D photonic band crystal. By varying a free parameter a/d, we fitted the calculated spectra to the experimental data. An example of the calculated spectra is shown in Fig. 3. In this figure, a/d is 0.05. This spectrum reproduces the experimental one. Despite the simple approximation and the use of only a single free parameter, the calculated spectra agree well with the experimental results. Even when the grating period or coated metal of gratings were changed, the calculated spectra also agreed with the experimental ones using the free parameter value of 0.05 mentioned above. This agreement shows that a photonic band is formed in the grating pair system.

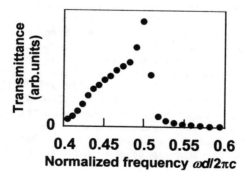

Figure 3. Calculated transmittance spectrum of 1-D photonic band crystals with the lattice constant of 278 nm consisting of aluminum and vacuum. The free parameter a/d is chosen to be 0.05.

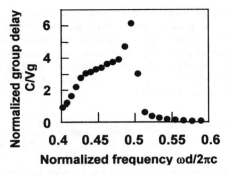

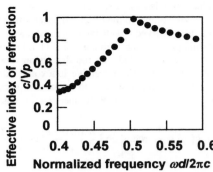

Figure 4. Calculated dispersion of the group delay c/Vg in the 1-D photonic band crystal, where Vg is the group velocity

Figure 5. Calculated dispersion of the effective index of refraction c/Vp in the 1-D photonic band crystal, where Vp is the phase velocity.

Figures 4 and 5 show the calculated group delay c/Vg and the calculated effective index of refraction c/Vp in the 1-D photonic band crystal assumed above, where Vg, Vp are the group velocity and the phase velocity, respectively. The dispersions of c/Vg and the c/Vp change depend on the constituent metal and the lattice constant. This implies that the group- and phase velocities are controllable.

Controlling the group velocity and the phase velocity of the optical pulse can solve several problems. Some organic materials have potential for nonlinear optical devices, but commercial devices have not been fabricated because of several problems. For example, it is difficult to obtain high efficiency for the second harmonic generation (SHG), because crystal growth, cutting, and phase matching are difficult. When the grating pair structure is used, the nonlinear optical polymer can easily be set between two gratings, and further, the phase matching of SHG can be achieved by controlling the phase velocity. For the conventional nonlinear optical devices, the third order nonlinear susceptibilities are too small to fabricate an optical switch. It is considered that the decrease of the group velocity of the optical pulse corresponds to a longer interactive length. This means an increase of the nonlinear optical susceptibilities. Therefore, for example, when the group velocity becomes 1/10, the efficiencies of the phase conjugate wave generation and SHG become 100 times larger. In this case, it is necessary that the size of the photonic band crystal is larger than the optical pulse in spatial size in order to define the group velocity. The grating pair can satisfy this condition.

SUMMARY

In summary, we have succeeded in forming the photonic band by the grating pair method. This method provides useful means for controlling the group- and phase velocities of light.

REFERENCES

1. For example, K. Ohtaka: Phys. Rev. B **19**, 5057 (1979); E. Yablonovitch, Phys. Rev. Lett. **58**, 2059 (1987); S. John: Phys. Rev. Lett. **58**, 2486 (1987).
2. K. Inoue, M. Wada, K. Sakoda, A. Yamanaka, M. Hayashi, and J. W. Haus, Jpn. J. Appl. Phys. **33**, L1463 (1994) .
3. J. P. Dowling and C. M. Bowden, J. Mod. Opt. **41**, 345 (1994).
4. For example, G. Hauss and J. E. Waylonis, J. Opt. Soc. Am.. **51**, 719 (1961); L. G. Schulz and F. R. Tangherlini, J. Opt. Soc. Am. **44**, 362 (1954); L. G. Schulz, J. Opt. Soc. Am. **44**, 357 (1954); G. B. Irani, T. Huen, and F. Wooten, J. Opt. Soc. Am. **61**, 128 (1971).

POROUS SILICON ORGANIC VAPOR AND HUMIDITY SENSOR

M.C. POON *, J.K.O. SIN *, H. WONG **, P.G. HAN *, W.H. KWOK *, Y.C. BOW *
* Department of Electrical & Electronic Engineering, Hong Kong University of Science & Technology, Clear Water Bay, Hong Kong
** Department of Electronic Engineering, City University of Hong Kong, Kowloon Tong, Hong Kong

ABSTRACT

This paper presents new organic vapor sensitive device using anodized porous silicon (PS). The sensor has aluminum (Al)/PS/p-Si/Al Schottky diode structure and sensitivity at room temperature in 2600 ppm acetone, methanol, 2-propanol and ethanol is about 4, 5, 10 and 40 times respectively. The sensitivity in 800-2600 ppm ethanol vapor is 2 to 40 times. The diode sensor can be converted into an Al/PS/Al resistor sensor by switching the electrical contacts, and the sensitivity is about 500 times for a humidity change of 43-75%. All sensors have response time of about 0.5 min. The sensitivity is stable with time and the PS sensor can be integrated into VLSI Si devices to form novel microelectronic systems.

INTRODUCTION

Porous silicon (Si) has many novel and unique properties such as high surface area (>200 m^2/cm^3) and chemical activity, high sensitivity and efficiency in light absorption, and light emitting. It can have numerous applications including opto-electronic devices, micro-electro-mechanical systems (MEMS) and sensors [1-5]. The design and fabrication of porous Si (PS) devices are also compatible with the mature monolithic Si very-large-scale-integrated (VLSI) technologies. However, until now little research has been done on the porous Si devices. In this paper, we present our new results on the systematic study of a novel PS based organic gas and humidity sensor.

EXPERIMENT

Boron doped <100> 15-25 Ω-cm p-Si wafers were chemically cleaned. Aluminum (Al) film of about 1 μm thick was deposited onto the back side of the wafers and annealed at 450°C/30min in dry nitrogen to obtain an uniform ohmic contact. This step is critical in obtaining uniform PS layer by anodization. The Al film was protected with photoresist, and PS layer was formed by anodizing the front Si surface in dilute HF-ethanol solution (HF:C₂H₅OH = 1:1), with platinum electrode and at a current density of 5-15 mA/cm² for 5-10 min in room light and temperature. For better passivation of the PS surface, some PS samples were further boiled for 5-10 min in dilute nitric acid solution (HNO₃:DI=1:9). The photoresist was removed and Al circular contacts with diameter of 1-3mm and thickness of 15-270 nm were then deposited onto the surface of the PS layer to form Al/PS/p-Si/Al "Schottky-like" diodes. To measure the organic vapor sensing characteristics, the diodes were placed in a closed chamber filled with fixed concentration (0-2600 ppm) of vapor mixtures of ethanol, acetone, 2-propanol, and methanol, respectively. Humidity sensitivity was measured in water vapor with 43-75% relative humidity. Sensor responses were studied by measuring the current-voltage (I-V) characteristics of the diodes, and sensitivity of sensor was defined as the ratio of diode current in organic gas and in air for fixed forward applied voltage.

Poly-Si film with thickness of 670 nm was also used to form the PS layer. The poly-Si film was formed by thermal decomposition of silane in a LPCVD reactor at 625°C for 50 min onto p-Si substrates. The film was subsequently doped with 950°C/20min phosphorus diffusion and rendered porous by similar anodization steps.

RESULTS AND DISCUSSION

Fig.1 shows the schematic cross section of the PS sensors with the applied voltages. Thickness of the PS layer is about 3.7 μm which was found by calculation and confirmed by cross-sectional optical microscopy. Fig.2 shows the I-V characteristic curves of the PS diode sensor in air and in presence of 2600 ppm air-organic vapor mixtures of 2-propanol, methanol, acetone and ethanol, respectively. The curves were measured at room temperature and pressure, and measured at 1 minute after exposure to each vapor mixture.

The current responses of the sensor for all gases are rather small (< 3 μA) for reverse applied voltages of 0 to -6V. For forward applied voltages of 3-12V, the sensor has very small forward DC current responses (0.2-2 mA) in air ambient. However, current for exposure to 2600 ppm organic vapor is found to increase very rapidly with increasing forward applied voltages. For ethanol vapor, currents are about 4 to 83 mA and the sensitivity is about 23 to 42 times for voltages of 3 to 12V, respectively. This corresponds to a 180% increase of sensitivity with applied voltages. For methanol, acetone, and 2-propanol, the forward current responses are 0.8-11 mA, 0.7-11 mA, and 1.9-24 mA, and the corresponding sensitivity is about 4, 5 and 10 times respectively. In the results, the sensitivity of sensor due to moist air can be neglected as the current responses are similar to those for the air.

Results in Fig.2 show that the Al/PS/p-Si/Al diode can be a good candidate for organic vapor, specifically ethanol vapor sensor. Ethanol is one of the most popular gases in our daily life and there can be many applications of the sensor such as breath alcohol checkers to monitor the ethanol concentration in the human blood. The results show that the sensor has high sensitivity for exposure to ethanol vapor with thousands ppm of concentration. Nonetheless, preliminary results also show that the sensitivity decreases substantially with ethanol concentration.

Figure 1.   Schematic cross section of (a) PS diode sensor, (b) PS resistor sensor

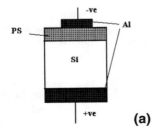

(a)

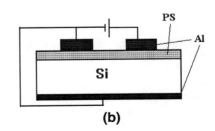

(b)

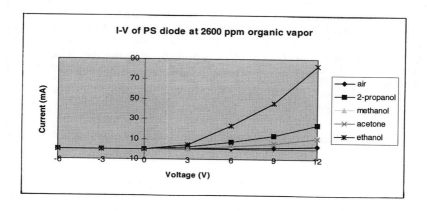

By switching the electrical contacts, the diode sensor (Fig.1a) can be easily converted into an Al/PS/Al resistor sensor (with back Al connected to ground) (Fig.1b). In Fig.3, the PS resistor sensor has been found to have very high sensitivity for humidity. For relative humidity of 43, 50, 55, 70 and 75%, the current responses of sensor are 0.6-2.8 μA, 43-1422 μA, 70-160 μA, 260-1230 μA, and 640-1590 μA, respectively, for forward voltages of 3-12V.

Figure 3. Current responses of PS resistor sensor at different relative humidity

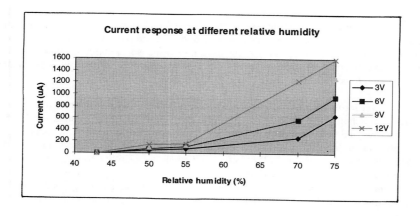

Selectivity of PS resistor sensor at 5200 ppm organic vapors has also been measured and the I-V characteristics are shown in Figure 4. For 2-propanol, methanol, acetone and ethanol, the forward current responses are 3-10 µA, 5-48 µA, 22-110 µA, 62-118 µA and the corresponding sensitivity is about 2, 5, 20, 20 times respectively. Results show that the PS resistor sensor is sensitive with 2-propanol, methanol, acetone and ethanol. The I-V curves are also found to have good linearity. Moreover, the I-V of PS resistor has been found to be unaffected by the substrate terminal, and same results are found for both grounded substrate and floated substrate.

Figure 4   Current responses of PS resistor sensor in air and in 5200 ppm organic vapors

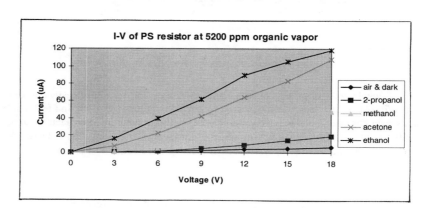

Both the diode and resistance sensors have been found to have very fast response time. The sensor responses are rather fast and abrupt, and the rise and fall response times (time for forward current to rise and drop to 0.7 of the maximum response) are both less than 30 seconds. Moreover, the sensitivity is stable and repeatable with measurements within our experimental period of about 6 months.

PS layer formed from anodized polycrystalline Si film (~700nm) has also been tested, but Al/PS/p-Si/Al Schottky diode shows only low sensitivity of about 150% for all gases (Fig.5). The reasons why PS layer formed from poly-Si film has poor sensitivity are still under investigation in our laboratory, and they might be due to the different porosity and thinner film. We have also studied the effect of other factors to the sensor sensitivity. Results show that the sensitivity increases significantly and almost linearly with decrease in the front Al sensor area. However, the parameters of temperature, light, nitric acid passivation boiling, anodization current density and time, and HF dip before Al deposition, are all found to have little effects to the sensitivity.

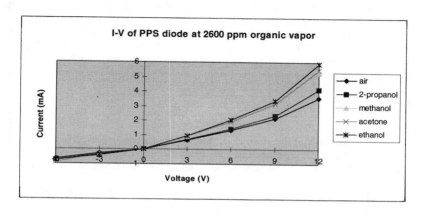

In the literatures, only very few reports have been found on the PS based ethanol and humidity sensors and our results are highly comparable or slightly better than the published works [1-6]. In the work of Watanabe and co-workers [1], the sensitivity is similar but a comb-like contact structure and a more complicated diode structure combining PN junction and PS layer are used. In the work of Anderson and co-workers [4], the sensitivity due to 0-100% humidity change measured by the Schottky barrier PS sensor is only 440%. In the work of O'Halloran et al. [6], sensitivity due to 64% humidity change is 900%. Our sensitivity is still lower than those of the metallic-oxide sensors such as SnO$_2$ [7]. Nevertheless, better sensor performances should be obtainable after further optimization of the design and fabrication of our PS sensors, such as using finger-like structure for larger sensing area. Our new PS sensor can provide more informations on the design of practical PS devices. The sensor can be applied commercially as a very sensitive, low cost and high performance humidity and organic gas sensor (especially as an ethanol/alcohol gas sensor). Moreover, using the Si very large scale integrated (VLSI) technologies, the PS sensor can also be integrated with other novel Si or PS devices/circuits (such as PS light emitting diode) to form numerous revolutionary microelectronic systems.

CONCLUSIONS

In summary, a novel porous Si based gas sensor is presented. It has simple structure, high sensitivity, and fast and stable response for organic vapor and humidity sensing at room environment. The PS sensor can also be integrated with other novel Si or PS devices/circuits to form numerous revolutionary microelectronic systems.

ACKNOWLEDGMENTS

The authors are grateful to Professor Ping Ko, Professor Simon Wong, Dr. Philip Chan, Professor H.S. Kwok, Dr. Jack Lau, Professor K.C. Smith, Dr. Man Wong, and staffs of the Microelectronic Fabrication Center, Material Characterization preparation Center, and Device Characterization Laboratory of the Hong Kong University of Science & Technology, for their valuable comments and help. This work was partially supported by grant number HKUST 691/95E of the RGC of Hong Kong.

REFERENCES

1. Watanabe, T. Okada, I. Choe and Y. Satoh, The 8th International Conference on Solid-State Sensors and Actuators, and Eurosensors IX, Stockholm, Sweden, 1995.
2. Chohrin and A. Kurex, Appl. Phys. Lett., **64**, 481, 1994.
3. Richter, The 7th Int. Conf. Solid-State Sensors and Actuators, Transducers'93, Yokohama, Japan, 1993.
4. Anderson, R.S. Muller and C.W. Tobias, Sensors and Actuators, **21-23**, 835, 1990.
5. Stievenard and D. Deresmes, Appl. Phy. Lett., **67**, 1570, 1995.
6. O'Halloran, P.J. Trimp, P.J. French, H.C. de Graff, 25th European Solid State Research Conference, 1995.
7. Maekawa, J. Tamaki, N. Miura and N. Yamazoe, Sensors and Actuators B, **B9**, 63, 1992.

OPTICAL FIBERS WITH PATTERNED ZnO/ELECTRODE COATINGS FOR FLEXURAL ACTUATORS

S. Trolier-McKinstry,* G. R. Fox, A. Kholkin, C. A. P. Muller, And N. Setter
Laboratory of Ceramics, Swiss Federal Institute of Technology, CH-1015 Lausanne, Switzerland
*Visiting Scientist from the Materials Research Laboratory of the Pennsylvania State University,
University Park, PA, USA, STM1@ALPHA.MRL.PSU.EDU

ABSTRACT

Piezoelectric ZnO coatings were used in this work to develop a flexural actuator for an optical fiber. The basic device geometry was as follows: inner Cr/Au electrodes were evaporated onto a cleaned optical fiber; a thick ZnO coating was then grown by sputtering; finally a set of 2mm ring top electrodes were deposited through a shadow mask. Flexural actuators were made by photolithographically patterning either the inner or outer Cr/Au drive electrodes so that it was split down the length of the fiber. This enables each half of the fiber to be actuated independently. The result is that the optical fiber is forced to flex.

A processing scheme by which 30 μm gaps could be patterned into the electrodes was developed using standard clean room techniques. Such flexural actuators are attractive for scanning near field optical microscopes and in fiber alignment devices.

INTRODUCTION

Piezoelectric thin film coatings on optical fibers are attractive for a wide variety of applications in telecommunications devices, including fast signal modulators. Towards this goal, several designs for optical phase modulators with modulation frequencies up to several hundred MHz [1-3] and tunable Bragg gratings [4] have been reported. In most of these applications, the piezoelectric fiber is used to generate a stress-induced optical path length change in the fiber, so that signals traveling along the fiber can be controlled. ZnO [1-3], piezoelectric polymers [5], and ferroelectric fiber coatings [6] have been investigated for this purpose.

In addition to devices in which the piezoelectric coating is used to modulate the optical signal, it is also interesting to consider the piezoelectric coating as a means to physically displace the fiber (i.e. to have it act as an actuator). Potential applications of such an actuator would be in (i) near field optical microscopes, where the transducer for either scanning or vibrating the tip could be integrated directly on the fiber, and (ii) active fiber alignment devices for integrated optical networks, where fibers should be aligned within microns or tenths of microns [7]. As described elsewhere in this volume [4], ZnO coatings can be used to move the fiber. In this work, the possibility of developing a flexural actuator based on a standard optical fiber was investigated by photolithographically patterning the electrodes for a ZnO piezoelectric coating.

DEVICE DESIGN

Figure 1 shows schematics for the fiber actuators considered here. For both cases, the device consists of three film coatings on the fiber surface: the piezoelectric coating and inner and outer electrodes. If either set of electrodes is patterned along the length with two gaps, then the two halves of the electrode can be driven independently, causing the fiber to flex. In addition, if the unpatterned electrode is driven at a voltage intermediate between the halves of the split electrode, then the strains due to each half of the fiber diameter should couple, producing larger displacements at the unclamped end of the fiber. The two designs shown are for actuators in which the inner and the outer electrode are patterned in this manner.

EXPERIMENTAL PROCEDURE

Standard 125 μm diameter optical fibers were first cleaned in dichloromethane and isopropanol to remove the organic coating. Samples were first mounted on a holder which allows each fiber to be rotated continuously at 2 rpm during deposition so that uniform thickness films could be deposited. Inner electrodes of approximately 13 nm of Cr (to improve adhesion) and

Mat. Res. Soc. Symp. Proc. Vol. 459 © 1997 Materials Research Society

130nm of Au were thermally evaporated. A 6 μm thick, [0001] radially-oriented ZnO film was then deposited by reactive magnetron sputtering in an Ar/O_2 atmosphere with a target-fiber distance of 9 cm. Additional details on the ZnO deposition are given elsewhere [8]. To facilitate subsequent electrical measurements, a portion of the bottom electrode was masked during the deposition of the piezoelectric. The Cr/Au top electrodes were made thick (25 and 400nm thick, respectively) to insure good electrical contact on the rougher fiber surface following the ZnO deposition. All top electrodes were evaporated through another shadow mask to prepare 2mm electrode sections separated by 2mm gaps perpendicular to the fiber axis (See Fig. 1).

To create gaps in the inner electrode, patterning was done prior to deposition of the ZnO, while gaps in the outer electrode were made following all of the coating depositions. Patterning the electrodes was accomplished by dip-coating the fiber in Shipley Microposit 1813 photoresist, using a withdrawal rate of 1mm/sec. After soft-baking for 30min at 90 °C, ~30μm gaps in the electrode were exposed using a contact aligner and a special fixture in which the fiber could be held in contact with a Cr mask. The patterned region is 2 cm long. The fiber was then rotated 180° about its axis and the exposure was repeated. The pattern was developed using the recommended developer solution. Fibers were subsequently postbaked at 120°C for 30 min to harden the resist. For the bottom electrode, the pattern was then transferred into the electrode by etching for 45 seconds in an iodine-based Au etch, and 30 seconds in a $Ce(NH_4)_2(NO_3)_4$ / CH_3COOH / H_2O bath to remove the Cr. Deionized water was used as the etch stop. The protective resist layer was then removed in acetone and the fiber was rinsed in isopropanol. The procedure was similar for the top electrode, with the exception that Au etch was extended to 2 - 4 minutes, and the Cr etch to 1 - 2 minutes to completely clear the gap. Scanning electron microscopy was used to inspect the pattern quality.

For electrical and optical measurements, the fiber was then placed into a slotted Si substrate, and glued into place at one end. Electrical connections were made with silver paint. Impedance measurements were made using an HP4194 Impedance Analyzer. The displacement of the fiber was then measured as a function of the driving frequency using a single beam Mach-Zehnder interferometer.

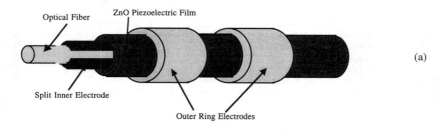

(a)

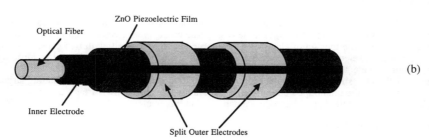

(b)

Figure 1: Schematics for the fiber-based flexural actuators. (a) Geometry I: Split inner electrode, (b) Geometry II: Split outer electrode. In both cases, the electrodes are split on the opposite side of the fiber as well.

RESULTS AND DISCUSSION

Figure 2 shows an example of the photolithographic patterning of an inner electrode. Within the 30 μm gap, the electrode was removed completely. The sidewalls are well defined and reasonably straight, with ~1 μm irregularities along the edge.

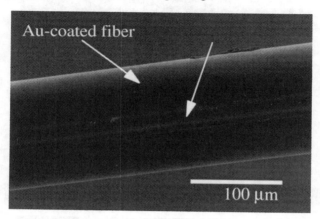

Fig. 2: SEM image of a patterned inner electrode showing the gap with the exposed fiber surface

In patterning the outer electrode, it was found that the Cr etch also dissolved the ZnO, so that gaps were introduced in the piezoelectric coating as well as the electrode. An additional Au/Cr etch cycle was then used to dissolve the exposed inner electrode, producing a more completely decoupled structure. During this process, the ZnO was undercut somewhat, so that the resulting gaps were wider than the exposure mask. Figure 3 shows an SEM micrograph of the result. In Fig. 4, the etch sidewall is shown. The columnar microstructure of the piezoelectric film is evident in the picture. It was also found that while the ZnO in the gap was almost completely removed, there were a few residual ridges of incompletely etched material. In subsequent work, it would be worthwhile to develop more selective etches for the Cr and the ZnO so that the patterning could be optimized.

Following fabrication, the electrical and piezoelectric properties of the patterned fibers were probed. Since wire bonding to the electroded fiber was unsuccessful, air dry silver paint was used for electrical connections. Using impedance spectroscopy, broad radial resonances for the ZnO coating were seen between 21 and 25 MHz, demonstrating that the coatings were, in fact, piezoelectric. To determine whether the coating could be used to position the fiber end, a 2mm section of an 18mm long fiber segment with a patterned inner electrode was excited with an alternating field applied between one section of the inner electrode and the annular outer electrode. The displacement was measured at a 15 mm distance from the clamped end by single beam interferometry as a function of the drive frequency. Fig. 5 shows the resulting response. The first resonance (below 100 Hz) was associated with building vibrations, the remainder are due to the ZnO coating. The resonances appeared at the same frequencies for both 0.5 and 1 V ac drive levels, with the amplitude of the response increasing with field, as expected. Fig. 6 shows the response as a function of the driving voltage for frequencies on and off resonance. For the highest amplitudes, the response is beginning to move out of the linear response of the interferometer. There is also some question about whether the measured data here correspond to the maximum amplitude for the actuated fiber, since it was difficult to insure that the fiber was oriented with the motion parallel to the interferometer axis. It is expected that the induced motion could also be increased considerably if both halves of the fiber were actuated.

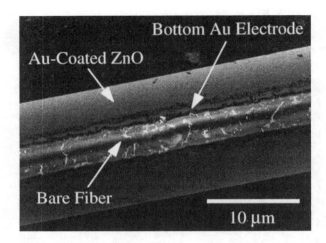

Fig. 3: SEM micrograph of a patterned outer electrode in which the ZnO in the gap has also been patterned. The bright irregular areas in the gap correspond to incompletely etched ZnO bridges.

Fig. 4: SEM micrograph of the sidewall of a film patterned through all three coating layers. The columnar microstructure of the piezoelectric film is clearly visible. The loose material at the base of the sidewall is exposed bottom electrode. This electrode was removed completely through the majority of the gap.

The displacements measured here are large enough to be useful for some scanning near field optical microscopes applications. If four gaps in the electrodes were patterned at 90° intervals around the diameter of the fiber, it should be possible to control motion of the fiber tip along two axes. For fiber alignment systems which require larger static motions than were observed here, either larger segments could be actuated, or a stronger piezoelectric, such as lead zirconate titanate, could be employed as the piezoelectric coating.

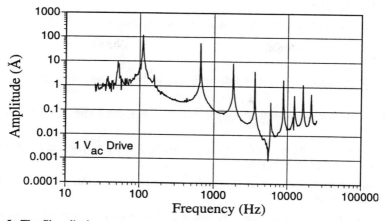

Fig. 5: The fiber displacement measured as a function of excitation frequency, for a film with a split inner electrode with one half of the fiber excited.

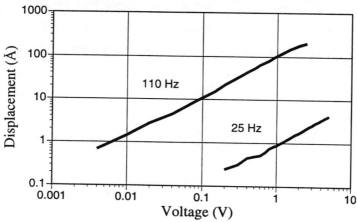

Fig. 6: Displacement as a function of excitation voltage for the same fiber as in Fig. 5 measured on (110 Hz) and off (25 Hz) resonance

CONCLUSIONS

A processing procedure was developed to pattern the electrodes for piezoelectric fiber coatings along the fiber length. The resulting flexural actuators should be capable of producing motion along one axis. Displacements of several hundred Ångstroms was obtained when 2 volts were applied a 2mm segment of a 6 μm thick ZnO coating on a standard optical fiber. By preparing additional splits in the electrode, it should be possible to prepare 2 axis integrated fiber actuators.

ACKNOWLEDGMENTS

The assistance of Markus Kohli is gratefully acknowledged.

REFERENCES

1. N. H. Ky, H. G. Limberger, R. P. Salathe, and G. R. Fox, J. Lightwave Tech. **14,** 23-26 (1996).

2. N. H. Ky, H. G. Limberger, R. P. Salathe, and G. R. Fox, IEEE Phot. Tech. Lett. **8,** 629 (1996).

3. D. S. Czaplak, J. F. Weller, L. Goldberg, F. S. Hickernell, H. D. Knuth and S. R. Young, IEEE Ultrason. Symp. **US-1,** 491 (1987).

4. G. R. Fox, C. A. P. Muller, C. R. Wuthrich, A. Kholkine, D. M. Costantini, and H. G. Limberger, this volume.

5. M. Imai, H. Tanizawa, Y. Ohtsuka, Y. Takase, and A. Odajima, J. Appl. Phys. **60,** 1916 (1986).

6. D. A. Barrow, O. Lisboa, C. K. Jen, and M. Sayer, J. Appl. Phys. **79,** 3323 (1996).

7. R. Jebens, W. Trimmer, and J. Walker, Sensors and Actuators **20,** 65 (1989).

8. G. R. Fox, N. Setter, and H. G. Limberger, J. Mater. Res. **11,** 2051 (1996).

A STUDY OF Ta_2O_5 FOR USE AS HIGH TEMPERATURE PIEZOELECTRIC SENSORS

B.R. JOOSTE, H.J. VILJOEN
Department of Chemical Engineering, University of Nebraska-Lincoln, Lincoln, NE 68588-0126, Tel: (402) 472-9318, Fax: (402) 472-6989

ABSTRACT

Acoustic sensors which can function at high temperatures have important potential uses. In this work we report on the deposition, characterization and qualitative assessment of piezoelectric behaviour of orthorhombic Ta_2O_5. It is shown that orthorhombic Ta_2O_5 belongs to the class 2mm. XRD analysis of films annealed for 1 min., 10 min. and 1 hr at $800°C$ and $900°C$ reveal the formation of (0 0 1), (1 10 0) and (1 11 0) orientations at $800°C$, but the (1 10 0) increases at the expense of the other two as the annealing period is extended. At $900°C$ the dominant orientations are (1 10 0) and (2 9 0). The piezoelectric effect is significantly stronger after annealing and the stronger piezoelectric effect does not correlate with the presence of (1 10 0) and (2 9 0) so much as with the absence of (0 0 1) and (1 11 0).

INTRODUCTION

Amorphous Ta_2O_5 exhibits excellent dielectric properties and has found applications as high-density dynamic memory devices (DRAM)[1]. However, some of the other properties of Ta_2O_5, especially crystalline Ta_2O_5 have been largely overlooked. In 1985 Nakagawa and Gomi[2] first reported the piezoelectric properties of monoclinic β-Ta_2O_5. Surface acoustic wave (SAW) measurements with inter-digital transducers (IDT's) on fused quartz were used to calculate the electromechanical coupling coefficient K_m^2. A value of 0.5% was measured which compares well with the electromechanical coupling of ZnO. In 1990 Nakagawa and Okada[3] published the stiffness, piezoelectric stress and dielectric constants for monoclinic β-Ta_2O_5. Based on the published data the maximum value of K_m^2 for this form of Ta_2O_5 obtained was 2.29%. If one also considers the chemical and thermal stability of Ta_2O_5, it is clear that Ta_2O_5 has the potential to be used as an acoustic sensing device in high temperature, oxidizing environments. These are aggressive conditions and there are currently no acoustic devices available which remain functional. Vibratory or flexural measurements in combustors and turbines, NDE of hot surfaces like nuclear reactors and the monitoring of sintering and densification patterns in green bodies are a few examples of possible applications.

Jehn and Olzi[4] constructed a phase diagram for Ta_2O_5 in the temperature range $1100 - 2500°C$. A low temperature form orthorhombic β-Ta_2O_5 transforms to monoclinic α-phase at $1360°C$. The transformation is sluggish and exists metastably until the melting point of the β-phase at $1785°C$. The low temperature phase is therefore stable over a large temperature range (once it is formed) indicating versatility of an orthorhombic β-Ta_2O_5 sensor in terms of operating temparature range. Despite the efforts of several researchers, the structure of the low- temperature form of Ta_2O_5 still remains obscure. An interesting model was proposed by Harburn et al.[5] . It consists of an intergrowth of two structural types that lends a 'superstructure' to the subcell of orthorhombic symmetry. These two structural types (labeled slabs A and B) consist of pentagons and lozenges. Folding chain-like strings of pentagons, with lozenges between chains form the slab B which are interrupted by slab A structures which consist only of pentagons.

Jooste[6] and Jooste and Viljoen[7] first reported the piezoelectric behaviour of polycrystalline orthorhombic β-Ta_2O_5. Jooste deposited films on a variety of substrates including Si (001), Al, W and stainless steel 304. Since the objective of the work is the manufacture of a high temperature resistant sensor, this study is focused on the deposition of Ta_2O_5 on Inconel. Ta_2O_5 films formed by reactive magnetron sputtering with no intentional substrate heating are generally amorphous in the as-deposited state[7]. Annealing of amorphous Ta_2O_5 at temperatures above 600 °C is needed for crystallization to occur[8] and the crystallization is thermally activated; for example it takes an annealing time of 14 h at 600 °C for Ta_2O_5 to be fully crystalized but only 5 minutes at 900 °C[8].

In the present investigation an XRD investigation of as-deposited Ta_2O_5 films, annealed at different temperatures above 600°C for different periods of time, shows the formation and disappearance of certain crystal orientations. Sensor evaluation tests were done in parallel with the annealing studies to compare the crystal orientation of specimens with their sensing performance and it is shown that orthorhombic $\beta - Ta_2O_5$ shows promising piezoelectrical properties.

EXPERIMENT

Deposition, Annealing and Sensor Evaluation

Ta_2O_5 films were sputter deposited in a DC magnetron sputtering system. Details of the sputtering conditions are reported by Jooste and Viljoen[7]. Polished and chemically cleaned Inconel coupons (0.05x10x30 mm) were used as substrates. Substrates were not heated by an external source during film deposition and as-sputtered films were amorphous. It is interesting to note that despite the lack of order as measured by X-ray diffraction (XRD), the films exhibit (albeit weak) piezoelectric behaviour. This observation is consistent with studies by Shimizu et al.[9] who used microdiffraction to obtain insight into the structure of Ta_2O_5 films considered 'amorphous' by conventional XRD techniques. Microcrystalline regions of β-Ta_2O_5 were found, only a few nanometer in size.

Rapid thermal annealing (RTA) similar to the annealing done by Pignolet et al.[10] was done on the reactively sputtered, amorphous Ta_2O_5 films. The Ta_2O_5 films were placed in a small pre-heated pulsed current furnace, for periods of one minute and removed to cool down under ambient conditions. Longer annealing studies were done at annealing periods of ten minutes and one hour. Two different temperatures were investigated; 800°C and 900°C and three different annealing periods were used for each temperature, 1 min., 10 min. and 60 min. The crystallinity of the annealed films were investigated using standard Rigaku XRD apparatus with Cu-K_α radiation.

In order to evaluate the strength of the piezoelectric effect in the fabricated Ta_2O_5 film sensor, a simple ball-drop test was used. In this test the sensor was electroded and packaged between two small stainless steel plates (1x25x65 mm) which served as the other electrode. An oscilloscope was used to measure the voltage signal from the sensor as a small ball bearing (3 mm diameter) was dropped onto the sensor side of the assembly through a tube to ensure reasonable repeatablility in impact force.

RESULTS

To simplify the identification of a specimen, we will introduce the notation $S(xyz)$ where x denotes the first digit of the annealing temperature and yz denotes the annealing period in minutes. For example $S(901)$ refers to the specimen that was annealed for one minute at 900 $°C$. The XRD spectra for all six specimens are shown in Fig.1. $S(801)$ appears to be amorphous and only an Fe peak of the Inconel is present. $S(810)$ exhibited the following orthorhombic Ta_2O_5 peaks:(0 0 1), (1 10 0), (1 11 0) and two weaker peaks at (2 9 0) and (1 11 1). The intensities of the (0 0 1) and (1 10 0) peaks are dominant. The results change when the annealing period is extended. $S(860)$ shows a strong increase in the (1 10 0) peak intensity, the (2 9 0) peak also became more prominent while the (0 0 1) and (1 11 0) peaks have diminished. There seems to be competition between different peaks and their rates of formation differ. The increase in the intensity of (1 10 0) at the expense of other peaks like (0 0 1) and (1 11 0) indicates that (1 10 0) posseses a lower energy state, but the rate of formation is also slower. This could be ascribed to higher activation energy. The rate of transformation to a specific orientation seems to exhibit an Arrhenius type temperature dependence. $S(901)$ illustrates the temperature dependence well since at 900 $°C$ the (1 10 0) formation rate was faster and higher energy forms like (0 0 1) did not form. At longer periods the (1 10 0) peak became more prominent, compare $S(910)$ with $S(960)$. It is also evident from this comparison that the (2 9 0) peak is growing in intensity. A third peak at (2θ=29.8 $°$) starts to form at longer periods,cf. $S(960)$. The temperature dependance of the rate of crystallization as observed by Oehrlein et al. [9] can be seen clearly by noting that the XRD spectra of S(910) and S(860) look very similar.

The results of the sensor evaluation ball-drop tests (BDT) are shown in Fig. 2. Although $S(801)$ does not have distinguishable orientation, the piezoelectric effect already exists and the BDT shows a weak trace. The maximum peak to peak deviation is approximately 25 mV. Since the scales of Fig.2 were fixed for the sake of comparison, the signal output appears to be very small. The output for $S(810)$ does not differ markedly from $S(801)$, although there is a marked change in the XRD diagrams. The signal output becomes notably stronger for $S(860)$, this coincides with the increase of the (1 10 0) and (2 9 0) peaks and a decrease in the intensity of (0 0 1) and (1 11 0). Although the signal strength has increased with longer annealing periods, it is evident that the order of magnitude remains of all the signals is the same. One should be cautioned in interpreting the results too quantitatively, since there is a variance in the output when the experiments are repeated for the same conditions. However, the signal output for all the specimens which were annealed at 900 $°C$ shows a definite jump from the 800 $°C$ series. Typical peak-to-peak values of 200 mV are measured which constitute an order of magnitude increase in the signal strength.

CONCLUSION

One cannot attribute the increase in the piezoelectric effect to the development of a specific orientation, but it is correct to conclude that films which have been annealed at 900 $°C$, exhibit much stronger piezoelectric behaviour. It is interesting to note that the stronger piezoelectric effect does not correlate with the presence of (1 10 0) and (2 9 0) so much as with the absence of (0 0 1) and (1 11 0).

All the XRD spectra are consistent with the powder diffraction data for orthorhombic Ta_2O_5. The film is polycrystalline orthorhombic and isotropic in the substrate plane. In a

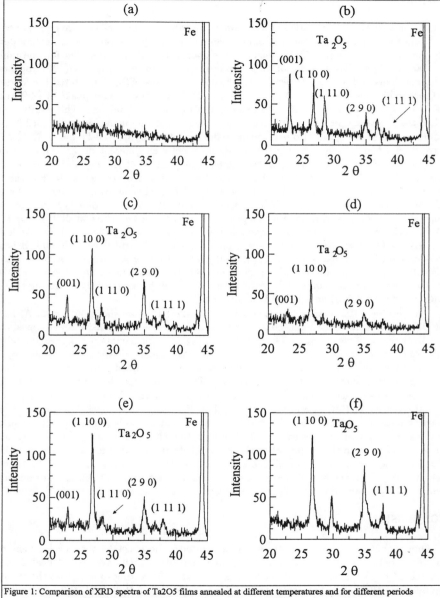

Figure 1: Comparison of XRD spectra of Ta2O5 films annealed at different temperatures and for different periods
(a) Annealed at 800 C for 1 minute, S(801). (b) Annealed at 800 C for 10 minutes, S(810)
(c) Annealed at 800 C for 1 hour, S(860). (d) Annealed at 900 C for 1 minute, S(901)
(e) Annealed at 900 C for 10 minutes, S(910). (f) Annealed at 900 C for 1 hour, S(960)

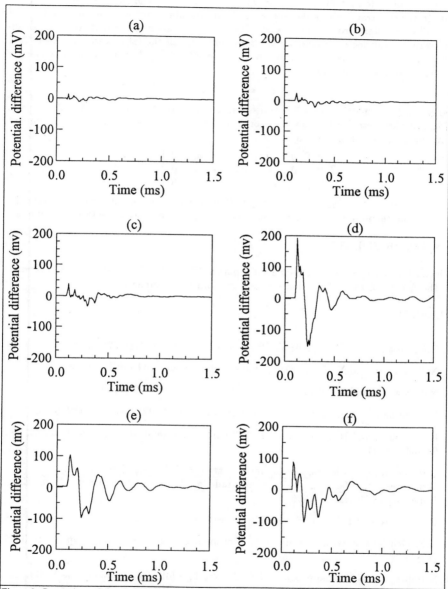

Figure 2: Comparison of signals from Ta2O5 films annealed at different temperatures and for different periods
(a) Annealed at 800 C for 1 minute, S(801). (b) Annealed at 800 C for 10 minutes, S(810)
(c) Annealed at 800 C for 1 hour, S(860). (d) Annealed at 900 C for 1 minute, S(901)
(e) Annealed at 900 C for 10 minutes, S(910). (f) Annealed at 900 C for 1 hour, S(960)

Cartesian coordinate system $X \times Y \times Z$ where $X \times Y$ lies in the substrate plane, with Z normal to the substrate, materials properties should remain intact by arbitrary rotations of the coordinate system around the Z-axis. The only two classes in the orthorhombic group which exhibit piezoelectric properties are 2mm and 222. In the latter case (i.e. class 222) it can be show that the transformation to a system $X' \times Y' \times Z$ changes the piezoelectric strain tensor $\mathbf{d}$ into a null matrix;

$$\mathbf{d}' = \mathbf{adN^T},$$

where N is a Bond matrix and $\mathbf{a}$ is a transformation matrix. In the case of the 2mm class, the form of the piezoelectric strain tensor is preserved, but the number of independent parameters is reduced from five to three:

$$\vec{d} = \begin{bmatrix} 0 & 0 & 0 & 0 & d_{x5} & 0 \\ 0 & 0 & 0 & d_{x5} & 0 & 0 \\ d_{z1} & d_{z1} & d_{z3} & 0 & 0 & 0 \end{bmatrix}$$

The point group of the orthorhombic $\beta\text{-}Ta_2O_5$ films that were investigated must therefore be 2mm. Impedance analyses and SAW experiments are planned to measure these parameters.

ACKNOWLEDGEMENTS

The authors acknowledge the financial support of Globe Metallurgical Inc. BRJ acknowledges the South African Foundation for Research and Development.

REFERENCES

1. G.Q. Lo and D.L. Kwong, Appl. Phys. Lett. **60**(26), pp. 3286-3288 (1992).

2. Nakagawa, Y and Y. Gomi., Appl. Phys. Lett. **46**, pp. 139-140 (1985).

3. Nakagawa, Y and T. Okada., J. Appl. Phys. **68**,pp. 556-559 (1990).

4. H. Jehn and E.Olzi, Journal of the less common metals **27**, pp. 297-309 (1972).

5. G. Harburn, R.J.D Tilley and R.P. Williams, Philosophical Magazine A **68**(4), pp. 633-640 (1993).

6. B.R. Jooste, Piezoelectric Sensors for Vibration and Damage Detection in Structures: Manufacture, Analysis and Modeling, Masters Thesis, Dept. Chem. Eng., University of Nebraska (1996).

7. B.R. Jooste and H.J. Viljoen, J. Mater. Res. (to be published).

8. G.S. Oehrlein, F.M. d'Heurle and A. Reisman, J. Appl. Phys. **55**, pp. 3715-3725 (1984).

9. Shimizu, K., G.E. Thompson and G.C. Wood., Philosophical Magazine B **63**, pp. 891-899 (1991).

10. Pignolet, A., G.M. Rao and S.B. Krupanidhi., Thin Solid Films **285**,pp. 230-235 (1995).

PHOTO-INDUCED POLARIZATION RECOVERY IN PZT THIN FILM CAPACITORS

S.A. MANSOUR, A.V. RAO, A.L. BEMENT. JR., AND G. LIEDL
School of Materials Engineering, Purdue University, West Lafayette, IN 47907-1289

ABSTRACT

Lead Zirconate Titanate (PZT) ferroelectric thin film capacitors were fabricated with metallic platinum and conducting Indium Tin Oxide (ITO) contacts. PZT thin films were fabricated using metallorganic decomposition (MOD) while a combination of MOD and RF-sputtering was used in fabricating the ITO-PZT-ITO capacitors. Photo-induced changes, manifested by an increase in switchable polarization, were studied before and after 10^8 switching cycles fatigue using white and monochromatic light. An increase in photo-induced changes was observed at 3.65eV light energies using monochromatic light using both capacitors. The increase was attributed to the excitation of electrons from PZT valence band into the conduction band causing an increase in film conductivity. However, polarization increase in Pt-PZT-Pt capacitor was more pronounced than ITO-PZT-ITO when white light was used. Some of the response in fatigued Pt-PZT-Pt capacitors was attributed to the excitation of electrons from the platinum Fermi level to oxygen vacancy sites trapped at the Pt-PZT interface by absorption of infrared radiation of white light. The latter observation implied a relationship between PZT fatigue and photo-induced effects.
KEY WORDS: PZT, Fatigue, Photo-Induced Effect, Polarization.

INTRODUCTION

Ferroelectric materials in a thin film capacitor structure are targeted for use in random access memories (RAMs) for non-volatile backup storage and in dynamic random access memories (DRAMs). Fatigue, which is defined as a reduction in the macroscopic polarization under repeated read/write cycling[1] occurs when metallic films are used as contacts to the ferroelectric films. The use of conducting oxide electrodes has extended the life time of these capacitors to up to 10^{12} switching cycles from 10^6 cycles in case platinum electrodes are used[2].

Some lost polarization is reported to be recovered when capacitors are illuminated with light[3,4]. In white light illumination experiments of fatigued Lead Zirconate Titanate, $PbZr_xTi_{1-x}O_3$ (PZT) capacitors[4], it was proposed that the illumination resulted in polarization recovery due to the injection of free electrons from the blocking Platinum (Pt) electrodes into the PZT oxygen vacancies built up at the Pt-PZT interface. The electrons neutralized the accumulated space charge at the interface and resulted in an increase of observed remnant polarization. However, the light energies involved where not defined. Moreover, an increase in polarization from monochromatic (3.4 eV) illumination of PZT capacitors was also observed[5,6].

The interaction of light with PZT capacitors fabricated with conducting metal oxide and metallic contacts, will provide information on the phenomenon of photo-induced polarization changes in PZT. More importantly, comparison of photo-induced recovery of polarization in a capacitor with blocking metal electrodes; one that fatigues rapidly, and in a capacitor with non-blocking oxide electrodes; one that is fatigue resistant, should give evidence towards mechanisms involved in the fatigue phenomenon.

EXPERIMENTAL

Two ferroelectric PZT thin film capacitors were fabricated in the form of Pt-PZT-Pt and Pt/ITO-PZT-ITO/Pt sandwich structures on Pt-coated silicon wafers. Metallorganic decomposition (MOD) was used in PZT synthesis, RF-sputtering in depositing Indium Tin Oxide, $In_{1.8}Sn_{0.2}O_3$ (ITO) bottom and uppper contacts, and DC sputtering in depositing upper Pt-contacts. The capacitors fabrication procedure was published somewhere else[7]. The layout of the capacitors is shown in Fig. 1.

Mat. Res. Soc. Symp. Proc. Vol. 459 © 1997 Materials Research Society

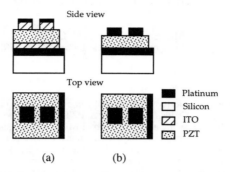

Figure 1: Layout of (a) ITO-PZT-ITO (b) Pt-PZT-Pt capacitors

A Radiant Technology RT66A ferroelectric tester was used to collect hysteresis data and perform standard fatigue on capacitors. A Sawyer Tower circuit in conjunction with a Hewlett Packard digitizing oscilloscope, a desktop computer using Keithley "Viewdac" data acquisition software and a Hewlett Packard LCR meter was used in studying photo-induced changes with light. Continuous monitoring of polarization was done using an input voltage signal in the form of a 5 V sine wave signal operating at 50 Hz. This was done for unfatigued capacitors as well as capacitors which had been cycled for 10^8 cycles.

A 150 W tungsten filament white light source and a fiber-optic guide was used to transmit white light to the film. A 75W xenon light source in conjunction with a monochromator (with a wavelength resolution 1 nm) by Acton Research Corporation and a fiber-optic was used as a monochromatic light source. The output of the light source varied from approximately 190 nm to 750 nm (6.5 eV to 1.65 eV). Long band pass filters were used to eliminate second- and third-order harmonics. A continuous voltage signal fed to the capacitor and the Sawyer Tower circuit recorded the polarization changes.

ITO and Pt films of thickness approximately the same as the electrodes were coated on glass slides. A Magna-IR Spectrometer from Nicolet Analytical Instruments was then used to determine film transmittance between the wavelengths 2.5 μm to 5.0 μm.

RESULTS AND DISCUSSION

Results of the normalized photo-induced recovery of capacitors to white light illumination are shown in Fig. 2. Figure 2 (a) shows an increase of 1.5 μC/cm^2 and 2.9 μC/cm^2 before and after 10^8 switching cycles, respectively using the Pt-PZT-Pt capacitors. The corresponding increases were 2.05 μC/cm^2 and 2.35 μC/cm^2 for ITO-PZT-ITO capacitors as shown in Fig. 2 (b).

Results of monochromatic light tests presented in Fig. 3, however, showed a maximum increase in polarization in the range 3.4 to 3.6 eV. The increases before and after 10^8 switching reversals were 1.6 μC/cm^2 and 1.9 μC/cm^2 in case of the Pt-PZT-Pt (shown in Fig. 3 (a)) and 1.84 μC/cm^2 and 1.88 μC/cm^2 in case of ITO-PZT-ITO capacitors (shown in Fig. 3 (b)), respectively.

The white light source emits light from the ultraviolet region through the visible into the far infrared region. The intensity is low at either end and attains a maximum at a wavelength of about 1.3 μm[7]. Attenuation and reflection of the incident white light intensity through electrodes at various surfaces was made from values of physical and optical parameters such as extinction coefficients and reflectance of the Pt and the ITO films[8-10]. A similar calculation was carried out to estimate the intensity of the light as a function of wavelength in the case of the monochromatic light source. It was found that a much higher (~100 times) intensity of light in the ultra-violet and visible region reaches the ferroelectric-electrode interface when ITO electrodes are used. The calculations also showed that Pt and ITO films transmitted 28% and 15% of the incident infra-red light, respectively.

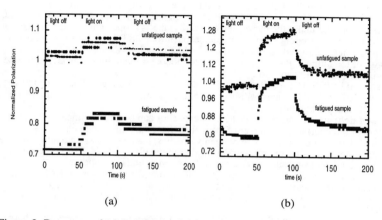

(a)   (b)

Figure 2: Response of (a) Pt-PZT-Pt and (b) ITO-PZT-ITO capacitors to white light

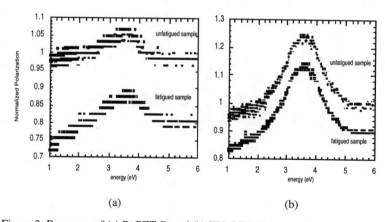

(a)   (b)

Figure 3: Response of (a) Pt-PZT-Pt and (b) ITO-PZT-ITO capacitors to monochromatic
light.

Absorption of energy close to 3.6 eV as seen in Figure 3 was attributed to the jumping of charge carriers across the energy band gap of PZT. This will result in an increase in the conductivity of the PZT film and is measured by the current integrator as an increased charge stored in the capacitor. The intensity of the 3.6 eV light reaching the electrode-ferroelectric interface was calculated to be higher when ITO contacts are used. The changes induced will, hence, also be intensified leading to a more prominent response from the ITO-PZT-ITO capacitors as compared to the Pt-PZT-Pt capacitors.

The response of the ITO-PZT-ITO capacitors to white light is independent of the number of switching cycles as seen in Figure 2. This response is attributed almost entirely to increased conductivity of the PZT. The response from the Pt-PZT-Pt, on the other hand, nearly doubles after 10^8 switching cycles. The area under the curves in Figures 2 and 3 is proportional to the amount of light energy absorbed by the capacitors. Hence fatigued capacitors with Pt contacts absorb greater amounts of light energy in the fatigued state as compared to the unfatigued state.

The response of the fatigued Pt-PZT-Pt capacitor to monochromatic light is nearly the same as that of the unfatigued Pt-PZT-Pt capacitor. It is hence evident that the exact energy being absorbed by the fatigued Pt-PZT-Pt capacitor is not in the 1.65 to 6.5 eV range. This leads to the conclusion that a fatigued Pt-PZT-Pt capacitor absorbs light in the infra-red region (less than 1.65 eV) and recovers some lost polarization.

An analysis of light absorption by the Pt-PZT-Pt capacitor can be done by means of an energy band diagram of the interface. A number of intrinsic defects exist in ferroelectric ceramics after fabrication. In PZT thin films these can be oxygen vacancies, lead vacancies, and other mobile ions . The defects can migrate at different rates under an alternating electric field. The defects move until they are trapped at a defect sink such as domain walls, grain boundaries, and/or electrode interfaces. The majority of defects, however, are trapped at the interface between the PZT and the electrode producing an internal field which opposes the externally applied field[11] or creating an oxygen depleted volume which renders some PZT cells unswitchable due to the distortion in the oxygen octahedron[4].

Figure 4 shows the energy band diagram of Pt-PZT interface with some defects and charge traps shown. There are two transitions that can potentially absorb infrared radiation and cause recovery of polarization in a fatigued capacitor. The first involves the defect levels at 0.8 eV above the valance band which was reported by others [11]. The nature of the defect is unknown and it is possible that it is formed from an interaction between oxygen vacancies and interface defects.

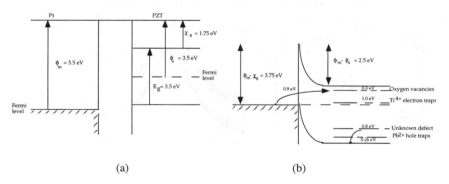

Figure 4: Band diagrams of (a) Pt and PZT and (b) The interface between Pt and fatigued PZT.

Absorption of 0.8 eV will cause an electron to be excited from the valence band into this defect level and reduce the charge on this defect. This will lead to a recovery of some of the lost polarization.

We propose a second and more probable transition occurring when an electron jumps from the Fermi level of Pt into the oxygen vacancy defects just below the conduction band in the PZT. If tunneling occurs this transition will require less than 1.45 eV. The exact energy required depends on oxygen vacancy concentration (n-character) of the fatigued PZT film. In Figure 4(b) the n-character of the PZT is assumed to raise the Fermi level by a nominal 0.5 eV. The energy required to excite the electron will now be about 0.95 eV. This shift of Fermi level would require an oxygen vacancy concentration at the interface of at least 2×10^{15} cm^{-3}. Vacancy concentrations 4 orders of magnitude higher have been reported in fatigued films[12].

Capacitors with ITO contacts do not exhibit any increased polarization recovery after switching. This is attributed to the absence of oxygen vacancy pile-up at the ferroelectric-electrode interface. It is proposed that oxygen vacancies are not trapped at the PZT-ITO interface and that they move into the ITO film and are neutralized.

CONCLUSIONS

1. Photo-induced polarization recovery with various extents was observed in PZT ferroelectric capacitors.
2. An increase in photo-induced changes was observed at 3.65 eV light energies using monochromatic light using both capacitors. The increase was attributed to the excitation of electrons from PZT valence band into the conduction band causing a drop in film resistance.
3. Polarization increase in Pt-PZT-Pt capacitor was more pronounced than ITO-PZT-ITO when white light was used. Some of the response in fatigued Pt-PZT-Pt capacitors was attributed to the excitation of electrons from the platinum Fermi level to oxygen vacancy sites trapped at the Pt-PZT interface by absorption of infrared radiation of white light.
4. A relationship between PZT fatigue and photo-induced effects is implied.

REFERENCES

1. J. Carrano, J. Lee, C. Sudhama, V. Chikarmane, A. Tasch, W. Shepherd, and N. Abt, IEEE transactions on Ultrasonics, Ferroelectrics and frequency control" **38**, 690 (1991).

2. R. Ramesh, A. Inam, W.K. Chan, F. Tillerot, B. Wilkens, C.C. Chang, T. Sands, J.M.Tarascon, V.G. Keramidas, Appl. Phys. Lett., **59**, 3542 (1991).

3. D. Dimos, W.L. Warren, B.A.Tuttle, Mat. Res. Soc. Symp. Proc., **310**, 87 (1993).

4. C.R. Peterson, S.A. Mansour, A.Bement Jr., Integ. Ferro. **7**, 139 (1995).

5. D. Dimos, R.W. Schwartz, Mat. Res. Soc. Symp. Proc., **243**, 73 (1992).

6. C.R. Peterson, S.A. Mansour, A. Bement Jr., Integ. Ferro. **10**, 295 (1995).

7. A. Rao, S.A. Mansour, A.L.Bement Jr., accepted, Materials Letters (1996).

8. R. Resnick, D. Halliday, Physics for students of Science and Engineering, 1960.

9. W. Driscoll, W. Vaughnan, Handbook of Optics, Optical Society of America, McGrawhill Book Co., New York (1978).

10. E. Palik, Handbook of optical constants of solids, Academic Press, Orlando (1985).

11. I.K. Yoo, S.B. Desu, J. of Intelligent Mater Sys and Struc, **4**, 490 (1993).

12. J.F. Scott, C.A. Araujo, B.M.Melnick, L.D.Mcmillan, R.Zuleeg, J. Appl. Phys. , **70** (1), 382 (1991).

Hard and Soft Composition Lead Zirconate Titanate Thin Films Deposited by Pulsed Laser Deposition

Johanna L. Lacey* and Susan Trolier-McKinstry
Intercollege Materials Research Laboratory, The Pennsylvania State University, University Park, PA 16803
*jll122@email.psu.edu

ABSTRACT

Lead zirconate titanate thin films offer considerably larger piezoelectric coefficients than do ZnO, and so are attractive for microelectromechanical sensors and actuators. To date, much of the research in this field has concentrated on undoped PZT. In this work, PZT films grown from both hard and soft PZT targets have been deposited on platinum coated silicon wafers by pulsed laser deposition so that the effect of doping on the properties can be determined. Dielectric constants of 1000-1500 are regularly achieved in both types of films, with loss values varying from 0.01 for soft films to 0.03 for hard films. Remanent polarizations are typically 30 $\mu C/cm^2$ for both types of films with no observable difference in coercive fields. When subjected to ~140 MPa of biaxial tension and compression, only small (~5%) reversible changes were observed, indicating a lack of substantial domain reorientation in the films.

INTRODUCTION

Incorporation of lead zirconate titanate (PZT) thin films into microelectromechanical devices holds great promise due to their piezoelectric coefficients which are much larger than those of commonly used materials like ZnO. Of the family of PZT compositions, soft PZT's have the highest piezoelectric coefficients and are, therefore, of particular interest. Soft PZT's are donor doped with materials like Nb. This gives them pinned defect dipoles resulting in mobile domain walls. This is in comparison to hard PZT which is acceptor doped, often with Fe 3+, causing mobile defect dipoles which stabilize the domain configuration causing pinned domian walls. The enhanced properties of soft bulk PZT are typically attributed to motion of the mobile domain walls [1]. In a thin film, several factors can influence domain wall mobility including residual stress [2], dopant distribution [3], high defect concentrations, and small grain sizes [4].

It is hoped that by preparing films under identical conditions from differently doped targets, their electrical properties will be an indication of whether or not the properties of soft bulk PZT can, in fact, be achieved in a thin film.

EXPERIMENTAL PROCEDURE

PZT films were grown by pulsed laser deposition on Pt coated Si wafers. A 248nm wavelength 250 mJ Lambda Physik EMG-150 krypton-fluoride laser was focused down using a 50 cm focal length plano-convex lens. Commercial composition targets of hard and soft PZT (Piezokinetics 8 and 5A, respectively) were ablated in 40-60 mTorr of 10% O_3 in O_2. Substrates were held at either room temperature or 400°C to produce amorphous films. Substrates were heated by a resistive element block heater with its temperature measured by a type K thermocouple embedded in the block. The substrate was affixed to the block with silver paint for good thermal contact. After deposition, the films were cooled to room temperature in 140 Torr of O_2. Films were crystallized in a rapid thermal processor (RTP) (AG Associates 210T-03) at 650°C for 60 seconds.

Pt electrodes were DC diode sputtered onto the films using a shadow mask to produce 0.3mm diameter electrodes. After electrodes were deposited, the films were reprocessed in the RTP for 60s at 650°C to improve contact at the film/electrode interface.

Electrical properties were measured using a Radiant Technologies RT66A ferroelectric tester for hysteresis behavior and an HP 4275A LCR bridge for capacitance measurements. The thickness of the films was measured by a Tencor Alpha-Step 500 surface profiler.

Mat. Res. Soc. Symp. Proc. Vol. 459 © 1997 Materials Research Society

Biaxial tensile and compressive stresses were applied to the films using the apparatus in Figure 1. A PZT film on a 3" Pt coated Si substrate was sandwiched between two viton O-rings and clamped on top of a bored out aluminum round. The cavity behind the wafer could either be evacuated by a 1μm roughing pump or pressurized by a gas bottle/regulator combination to apply compressive or tensile biaxial stress to the wafer. At a pressure of 15psi, the maximum stress at the center of the wafer was calculated to be 140MPa using large deflection plate theory [5].

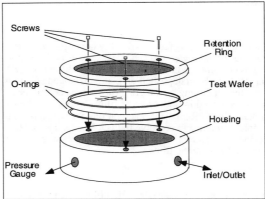

Figure 1. Illustration of stress apparatus

RESULTS

Film Facts

Figure 2 shows a glancing angle x-ray diffraction patterns of hard and soft composition films grown at room temperature and post deposition annealed at 650°C for 60 seconds. Films grown at room temperature typically exhibited partial (100) preferred orientation, but films grown at 400°C exhibited random orientation. Typical film thickness was ~4000 Å, and a typical surface grain size was ~0.1μm.

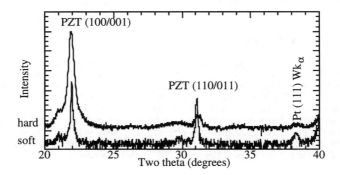

Figure 2. Glancing angle x-ray diffraction patterns of hard and soft films grown at room temperature and post deposition crystallized

Electrical Properties

Electrical properties for both hard and soft films were found to be similar. Typical hysteresis loops for both types of films are shown in figure 3. Both compositions show similar coercive field and remanent polarization values. Dielectric constants of films in this study ranged between 1100 and 1500 with no trend of higher or lower values based on composition. The loss tangent, however, was found generally to be lower for soft films than for hard films. For the two films in figure 3, the loss tangents were 0.01 and 0.03 for soft and hard compositions, respectively.

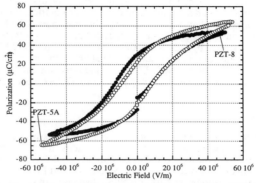

Figure 3. Hysteresis loops for hard and soft composition PZT thin films

The observance of similar coercive field is notable because it is a prime indicator of domain wall mobility. Similar coercive field leads to the conclusion that the domain walls have comparable mobility in the hard and soft composition films. The increased loss tangent in hard composition films also supports this conclusion. In bulk hard and soft PZT, the loss is higher in soft PZT due to the contribution of domain wall motion, and lower in hard PZT for the lack of domain wall motion [1]. The trend seen in these films is the opposite. If the contribution of domain walls were similar in both cases, other contributions must be considered to explain the differences. An important contribution to consider is conduction. Since conduction in undoped PZT is generally p type, hard (acceptor) doped materials would be expected to exhibit increased conduction due to an increase in hole concentration. Soft PZT on the other hand, which is donor doped, should exhibit a reduced amount of conduction due to suppression of carriers from electron-hole recombination [1].

Factors Influencing Domain Wall Mobility

It is believed that any or all of three main factors influence the domain wall mobility in these films. The first is grain size. In a TEM study of grain size vs. domain size, Cao and Randall [4] observed that in ceramics of less than 1μm grain size, fewer domain variants were present than in larger grained soft PZT ceramics. Similarly, Kim [6] observed in a systematic study of properties of PZT vs. grain size that below 1μm a reduction in remanent polarization and increase in coercive field were observed. Part of this was attributed to a decrease in domain wall mobility associated with the changes in domain structure with size. Since the surface grain size in our films is on the order of 0.1 μm, the films are well into the range in which this domain variant reduction has been observed.

The second factor that can affect the coercive field in the films is incorporation of dopants. If the dopants were for some reason not incorporated into the grains, (for instance, if they had segregated to the grain boundaries) the films would effectively be undoped. In such a situation, they would behave intrinsically hard. It has also been shown that in Nb doping of sol-gel films [3] that the Nb tends to inhibit the transformation from pyrochlore to perovskite, resulting in microstructural anomalies like surface pyrochlore layers. It is not certain that no such anomalies exist in our films. Compositional analysis by TEM to explore grains and boundaries is necessary to determine the location of dopants in the films.

The next factor believed to influence domain wall mobility is residual stresses in the film. In the most general of considerations, any local stress present in the film can pin walls. So for any applied stress, whether it be electrical or mechanical, the effect of this local stress must be overcome before the domain walls can become mobile. We know that PZT films grown on silicon substrates are under tensile stress due to the thermal expansion coefficient mismatch of the Si and the PZT. It is well documented that stress will change the properties of bulk hard and soft PZT [1,7,8]. It has also been found that the effect is somewhat greater on soft PZT than on hard PZT depending on the type of stress applied. In biaxial compression up to 392 MPa, hard PZT was shown to undergo a 50% reduction in capacitance with almost a full recovery to its original properties upon removal of the stress. Soft PZT under the same amount of compressive stress is shown to undergo a 60% reduction and will only recover to ~70% of its original value when the stress is removed [7].

Effects of Applied Stress

In an effort to sort out what mechanisms are responsible for the similar electrical behavior in our films, stress experiments were carried out. In looking at the results of these experiments, two things are worth noticing from the stress studies on bulk materials. First, the magnitude and direction of the change in dielectric constant, and second, that soft PZT was more affected by stress than hard PZT. In the studies on bulk ceramics, changes of this magnitude were attributed in large part to permanent domain reorientation. It follows that if domain reorientation is the dominant mechanism, soft PZT should be more affected than hard PZT since domains are more reorientable in soft PZT. However, much like the observation of loss tangent, the response to stress in thin films is found to be somewhat different than that observed in bulk ceramics.

Figure 4 shows the results of biaxially applied compressive (4a) and tensile (4b) stress on both hard and soft composition PZT thin films. In all cases, the changes ranged between 3-7% of the total unstressed properties and are essentially reversible. The direction of change is also opposite that observed in bulk ceramics. This leads to the conclusion that domain reorientation is not the mechanism causing the changes in the dielectric constant.

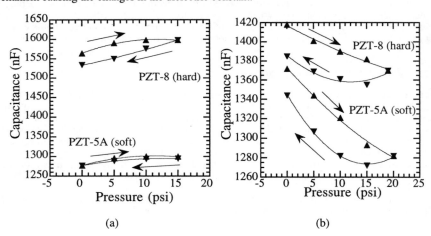

(a)   (b)

Figure 4. Capacitance vs. applied stress for biaxial compression (a) and biaxial tension (b).

Similar behavior has been observed in undoped sol-gel films [5], which should act intrisically hard. We believe that this is a strong indication that the domain walls in films of both composition are pinned, and therefore are not able to contribute to the dielectric properties.

CONCLUSIONS

PZT films were grown under identical conditions by pulsed laser deposition from hard and soft composition targets to explore whether or not properties of soft PZT could be achieved in a thin film. Both compositions have shown similar coercive field and remanent polarization values. In addition, both types of films show small reversible changes to applied tensile and compressive biaxial stress. The dielectric constants were found to be equal within experimental scatter, but the loss tangents were found to be lower for soft compositions. These observations are contrary to what is observed in bulk materials and are taken to indicate that the domain walls in these films are not mobile. It is concluded these films do not show the properties of soft PZT.

FUTURE WORK

We would like to be able to pinpoint the mechanism reponsible for the lack of domain wall mobility in our films so that possibly a growth technique can be designed which will allow it to be circumvented. This work includes TEM composition analysis to determine dopant location and also microstructure, further stress analysis including high field experiments to try and better understand the mechanism causing the observed changes, and also capacitance vs. temperature measurements to explore any changes in ferroelectric to paraelectric transition temperature that might be an indication of size effects.

REFERENCES

[1] B. Jaffe, W. Cook, and H. Jaffe, *Piezoelectric Ceramics*, Academic Press, 1971.

[2] B. Tuttle, T.J. Garino, J. Voigt, T. Headley, D. Dimos,and M.O. Eatough,*Science and Technology of Electroceramic Thin Films*, O.Auciello and R. Waser, eds. Kluwer Academic Publishers (1995).

[3] R.D Klissurska, K. Brooks, I. M. Reaney, C. Pawlaczyk, M. Kosec, and N. Setter, *Journal of the American Ceramic Society* , 78 **6** pp.1513-1520, (1995).

[4] W. Cao and C. Randall, *J. Phys. Chem. Solids*, 57 **10** pp 1499-1505 (1996).

[5] J. Shepard and S. Trolier-McKinstry, Materials Research Society Meeting, Fall 1996.

[6] N. Kim, Ph.D. Thesis, The Pennsylvania State University, (1994).

[7] R.F.Brown, *Can. J. of Phys.* **39** (1961).

[8] Q.M.Zhang, J. Zhao, K. Uchino, J. Zheng, *J. Mat. Res.*, submitted April 1996.

CHARACTERIZATION OF SrBi$_2$Nb$_2$O$_9$ THIN FILMS PREPARED BY SOL GEL TECHNIQUE

E. CHING-PRADO[*], W. PÉREZ[**], A. REYNÉS-FIGUEROA[**], R.S. KATIYAR[**], D. RAVICHANDRAN[+], AND A.S. BHALLA[+].
[*]Department of Applied Physics, Technological University of Panamá, Tocumen-Panamá, and Department of Physics, University of Panamá, Box 10761 University Post Office, Panamá.
[**]Department of Physics, University of Puerto Rico, P.O Box 23343, San Juan, P.R. 00931.
[+]Materials Research Laboratory, The Pennsylvania State University, University Park, P.A. 16802.

ABSTRACT

Thin films of SrBi$_2$Nb$_2$O$_9$ (SBN) with thicknesses of 0.1, 0.2, and 0.4 μ were grown by Sol-gel technique on silicon, and annealed at 650°C. The SBN films were investigated by Raman scatering for the first time. Raman spectra in some of the samples present bands around 60, 167, 196, 222, 302, 451, 560, 771, 837, and 863 cm^{-1}, which correspond to the SBN formation. The study indicates that the films are inhomogeneous, and only in samples with thicknesses 0.4 μ the SBN material was found in some places. The prominent Raman band around 870 cm^{-1}, which is the A$_{1g}$ mode of the orthorhombic symmetry, is assigned to the symmetric stretching of the NbO$_6$ octahedrals. The frequency of this band is found to shift in different places in the same sample, as well as from sample to sample. The frequency shifts and the width of the Raman bands are discussed in term of ions in non-equilibrium positions. FT-IR spectra reveal a sharp peak at 1260 cm^{-1}, and two broad bands around 995 and 772 cm^{-1}. The bandwidths of the latter two bands are believed to be associated with the presence of a high degree of defects in the films. The experimental results of the SBN films are compared with those obtained in SBT (T=Ta) films. X-ray diffraction and SEM techniques are also used for the structural characterization.

INTRODUCTION

The SrBi$_2$Nb$_2$O$_9$ material belongs to the ABi$_2$B$_2$O$_9$ perovskite-like layer, which consists of two layers of octahedrons. Compounds with a layered structure have, apparently, a face-centered orthorhombic unit cell, which in the first approximation can be considered as a body-centered tetrahedral unit cell. Because perovskite-like layers, consisting of NbO$_6$ octahedrons, are present in the lattice of this material, spontaneous polarization can take place in the plane of these layers. Thus, SrBi$_2$Nb$_2$O$_9$ corresponds to a ferroelectric material with high temperature phase transition, Tc, found around 420°C [1].

In recent years, there has been a great interest in combining oxides based perovskite-like structures in thin film form with the semiconductor materials, which result in a wide variety of applications. In particular, the layered bismuth compounds are currently one of the most technologically interesting materials because of its applications to fatigue free ferroelectric memories [2,3]. Numerous data exist on macroscopic properties (mechanical, electrical, thermal, and optical) of some of these materials, but only a few deal with the spectroscopie measurements [4,5]. The microscopic mechanism of these systems is not completely understood, in particular the lattice dynamics of a wide group of Bi$_2$A$_{m-1}$B$_m$O$_{3m+3}$, where A= Na, K, Ba, Sr; B= Fe, Ti, W,

Nb, and m and m-1 are the numbers of oxygen tetrahedron and perovskite-like units in a perovskite-like layer, respectively [1]. In fact, the low-frequency modes have not yet been studied, and not all crystals of such compounds have been grown yet. Moreover, the mechanism of the phase transition has not been clarified. In this paper we report the study of $SrBi_2Nb_2O_9$ material in thin film form on silicon substrate. This, we believe, is the first study of SBN thin film, prepared by sol-gel method, using micro-Raman scattering. Other experimental techniques were also employed in order to characterize the films.

VIBRATIONAL MODES

As a first approximation, it is helpful to assign tetragonal symmetry to the $SrBi_2Nb_2O_9$ material above the Curie temperature (420°C). It corresponds to a $I_{4/mmm}$ space group. In this symmetry the following Raman and IR modes should be active [5].

$$4A_{1g} (R) + 2B_{1g}(R) + 6E_g(R) + 7A_{2u}(IR) + B_{2u}(IR) + 8E_u (IR).$$

Below the Curie temperature (T<Tc) an orthorhombic symmetry is expected in SBN, so that each Eg mode is split into $B_{2g} + B_{3g}$. A splitting, at room temperature, in two components must be consistent with the orthorhombic distortion, as have been found in familiar compounds, such as: $SrBi_2Ta_2O_9$, Bi_3TiNbO_9, and $Bi_2La_2Ti_3O_{12}$ [4,5].

EXPERIMENTAL

SBN films were prepared using the metal-organic precursors, such as Bi (2,5 ethyl hexonate), Nb (ethoxide) and metalic Sr granules (Aldrich chemicals 99.99 % purity). Initially stoichiometric amounts of Bi (2,5 ethyl hexonate) was refluxed with 2-methoxyethanol (2-MOE) in a three neck flask at 125°C for 6 hours in an argon atmosphere.

Bi (2,5 ethyl hexonate) + 2 methoxyethanol ® Bi (methoxy ethoxide)$_3$ + ethanol

In a similar manner Nb (ethoxide) was refluxed with 2-MOE at 125°C for 6 hours in argon gas. The contents of the flask were cooled to room temperature and made to react with Bi-sol to form the double alkoxide [6].

$Nb (OC_2H_5)_5 + 2 CH_3OCH_2CH_2OH ® Nb (OCH_2CH_2OC_2H_5)_5 + 5 C_2H_5OH$

Finally metallic Sr was refluxed with 2-MOE by flowing argon at 125°C for 6 hours. The contents of the flask were cooled, then Sr $(OCH_2CH_2OC_2H_5)_2$ was reacted with the double alkoxide of Bi-Nb sol for 12 hours in flowing argon gas.

$Sr_2 + CH_3OCH_2CH_2OH ® Sr (OCH_2CH_2OC_2H_5)_2 + 2 C_2H_5OH$

To fabricate thin-films, a portion of the solution was concentrated to 0.3 M and hydrolyzed with H_2O/2-MOE in the ratio 1:2. To obtain crack free films 4% (by volume) formamide was added and stirred well. Prior to spin coating Si substrates was cleaned using ultrasound in isopropanol for 5-10 minutes and then dried. Films were spun at 4000 rpm for 15 seconds using Integrated

Technology P-6204 spin coater. Thin-films of SBN deposited on Si substrates were pyrolysed at 300-400°C to remove the volatile organic matter and annealed at 650°C for 24 hours. Thus, films with thicknesses of 0.1μ (SBN1), 0.2μ (SBN2), and 0.4μ (SBN4) were obtained. Also a thicker film (SBNt), much larger than SNB4, was produced. Raman scattering measurements were performed using a Raman microprobe system, from Instrument S.A. The excitation source was 514.5nm radiation from an argon ion laser manufactured by Coherent Inc. The spectra were measured in the back scattering geometry with 100x microscope objective. In addition, scanning electron microscopy (SEM), X-ray diffraction (XRD), and Fourier transform infrared (FT-IR) techniques were used for the structural characterization of the SBN thin films.

RESULTS AND DISCUSSION

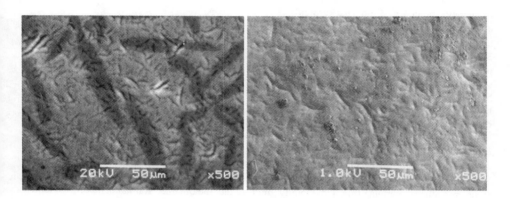

Figure 1: SEM micrograph of SBN (a) with thickness of 0.1 μ and (b) with thickness of 0.4 μ.

SEM micrographs of SBN1 and SBN4 are shown in Fig. 1. Significant difference can not be observed in the morphology of the SBN samples with thicknesses of 0.1, 0.2, and 0.4 μ, respectively. A relative rough surface can be observed in all of these films, where no well defined grains can be identified. Figure 2 shows the SEM picture of SBNt, in which a distribution of grains (1-5 μ) is observed. This SBN film presents the co-existence of other phases, which can correspond to chunck particles of Pt or to mobile clusters of Bi metal which has been found in SBT films. The SEM study reveals that the morphology of the films changes with the inclusion of

Figure 2: SEM micrograph of SBN/Si with Pt coating with thickness more than .4μ.

215

platinium coating. XRD study of SBN2 and SBN4 show a very broad peak around $2\theta = 15°$, which seems to be characteristic of an amorphous material. This broad band seems to be related with the (040) plane of the well crystallized SBN material [7]. X-ray analysis of SBNt film also shows the broad band around $2\theta = 15°$, however it is well defined respect to that found in the rest of the films. Figures 3, 4, 5, and 6 show the Raman spectra of the SBN samples. Using micro-Raman spectroscopy measurements in different places a topographic study was possible in each sample. In the SBN1 sample (Figure 3) the measurements in different points do not show substantial differences in the Raman spectra. They are similar to that obtained in pure Si, except that in low frequency range a high background is obtained. In addition, in some places a small band around 844 cm^{-1} is observed. In the SBN2 sample (Figure 4) the Raman spectra present significant differences between spatial points in the film. In SBN4 (Figure 5) Raman spectra show well defined bands, however, differences can also be observed in different places in the same film. The results

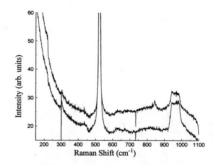

Figure 3: Raman spectra of SBN/Si film with a thickness of .1μ in different points in the film.

obtained in the SBNt sample (Figure 6) are similar to SBN4 sample, but sharper bands were found in SBNt. Note that due to the platinum coating on Si substrate in the SBNt film, the silicon Raman line (521.5 cm^{-1}) is not found in their spectra in comparison to those in the SBN/Si thin films. Comparison among Raman spectra of SBNt film SBT (T=Ta) ceramic and films, and other materials of the same family, indicate that the obtained Raman bands correspond to formation of SBN structure [4,5]. Raman study indicates that the films are polycrystalline and inhomogeneous. The films with thicknesses of 0.1 and 0.2 μ are practically amorphous in nature, while the film with thickness around 0.4 μ have places with SBN structure, although others places without Raman bands associated to the SBN structure were also found. The SBNt, which corresponds to the film with Pt/Si substrate, is the film with the formation of the SBN material in most of the places. The results suggest that the formation of SBN compound increases with increasing in thickness and with the presence of platinum coating on Si substrate. In term of the Raman analysis, it can be observed that the band around 450 cm^{-1} is more sensitive to the SBN crystallization degree. The frequency of this band change from 433 cm^{-1}

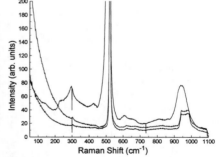

Figure 4: Raman spectra of SBN/Si film with a thickness of .2μ in different points in the film.

to 457 cm^{-1}, the highest frequency value seems to be associated with the well crystallized SBN material. This band is assigned to the Nb-O$_3$ torsional mode [8,9]. The band around 163 cm^{-1} also appears to the sensitive to the SBN crystallization, changing from 170 cm^{-1} to 163 cm^{-1}.

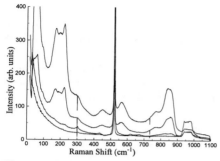

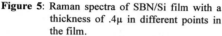

Figure 5: Raman spectra of SBN/Si film with a thickness of .4μ in different points in the film.	Figure 6: Raman spectra of SBN/Pt/Si film, with a thickness of more than .4μ, in different points in the film.

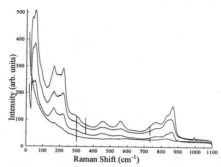

This band is also found in SBT sample and has been assigned to A_{1g} mode [5]. The frequency-variations of these two Raman bands (457 and 163 cm^{-1}) are associated with the ions in the non-equilibrium positions, resulting in changes in the force constants of the modes. The prominent feature at 850 cm^{-1} corresponds to A_{1g} mode and is associated with the symmetric stretching of the NbO$_6$ octahedral [5]. However, our results indicate that this strong feature, at room temperature, consists of two bands: around 838 cm^{-1}, and 870 cm^{-1}. The splitting of this band must be related with the change of tetragonal to orthorhombic symmetry. The sharp and intense Raman band at 60 cm^{-1} corresponds to the rigid-layer mode also found in familiar systems [4]. The bands at 224 and 290 cm^{-1} correspond to the O-Nb-O bending modes, which are likely to be of $B_{2g} + B_{3g}$ origin due to the lifting degeneracy of Eg species and Nb shares its oxygen atoms with another Nb atom, phenomenon similar to those found in perovskite niobates [10].

Figure 7 shows the FT-IR external reflection of the SBN samples in comparison with that of a SBT film prepared at 700°C. The spectra present basically three bands around 770, 960, and 1260 cm^{-1} , respectively. Note that the halfwidth of the bands at 770 and 960 cm^{-1} in the samples SBN1, SBN2, and SBN4 are very broad in comparison to SBNt and SBT samples. Thus, the halfwidth of these

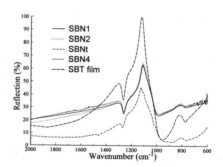

Figure 7: FT-IR reflectivity spectra of SBN and SBT films.

two bands seem to be sensible to the SBN crystallization degree. The sharp band at 1260 cm^{-1} can be associated to the presence of other phase co-existing with SBN material. A possible explanation is the formation of SiO$_2$ compound, where IR-band at 1200-1260 cm^{-1} range has been found in this material [11]. A good crystallization degree is obtained in SBN films annealed at 800°C. A thickness dependence study of the Raman spectra is in progress and will be reported in near future.

CONCLUSION

Thin films of $SrBi_2Nb_2O_9$ (SBN) were grown at $650^{\circ}C$. The films present high amorphous contents, and the SBN formation increases with increasing thickness and with the presence of platinum coating on the Si substrate. Raman studies indicate that the films are polycrystalline and inhomogeneous. Raman band changing from 433 to $457cm^{-1}$ is interpreted as ions in non-equilibrium positions associated with the crystallization degree. A similar explanation is given to the results obtained from the infrared study, where the broad bands around 770 and 960 cm^{-1} compared to those in SBN/Pt/Si film and those in SBT (T=Ta) film annealed at $770^{\circ}C$, suggest a high concentration of defects.

ACKNOWLEDGEMENTS

The authors wish to thank J. Garcia-Orozco for technical support. This research is supported by NSF-EPSCoR Grant, and DOE GRANT DE-FG02-94ER75764.

REFERENCE

1- G.A. Smolenskii, V.A. Isupov, and A.I. Agranovskaya, Soviet Physics-Solid State, 3, 651 (1961).

2- K. Amanuma, T. Hase, and Y. Miyasaka, Mat. Res. Soc. Symp. Proc., 361, 21 (1995).

3- D.P. Vijay, S.B. Desu, M. Nakata, X. Zhang, and T.C. Chen, Mat. Res. Soc. Symp. Proc., 361, 3 (1995).

4- S. Kojima, R. Imaizumi, S. Hamazaki, and M. Takashige, Jpn. J. Appl. Phys., 33, 5559 (1994).

5- P.R. Graves, G. Hua, S. Myhra, and J.G. Thompson, J. Solid State Chem., 114, 112 (1995).

6- D.C. Bradley, R.C. Mehrotra, and D.D. Gaur, Metal Alkoxide, p.308, Academic Press, London (1978).

7- S.Y. Chen, X.F. Du, and I.W. Chen, Mat. Res. Soc. Symp. Proc., 361, 15 (1995).

8- M.T. Vandenborre, E. Husson, and M. Chubb, Spectrochim. Acta, 40, 361(1984).

9- P.R. Graves, S. Myhra, K. Hawkins, and T.J. White, Physica C, 181, 265(1991).

10- E. Husson, L. Abello, and A. Morell, Mat. Res. Bull., 25,539(1990).

11- E.I. Kamitsos, Phys. Rev. Abs., 27, 20(1996).

SURFACE RELAXATION IN FERROELECTRIC PEROVSKITES: AN ATOMISTIC STUDY

SIMON DORFMAN*, EUGENE HEIFETS#, DAVID FUKS**, EUGENE KOTOMIN$, and JOSHUA FELSTEINER*

*Dept. of Physics, Technion, 32000 Haifa, ISRAEL
Institute of Chem. Phys., University of Latvia, Riga, LV-1586, LATVIA
** Mater. Engin. Dept., Ben Gurion University, Beer Sheva, ISRAEL
$ Institute of Solid State Physics, University of Latvia, Riga, LV-1063, LATVIA

ABSTRACT

The effect of the [001] surface relaxation on the polarization of the paraelectric $BaTiO_3$ is simulated in the framework of the shell model. Our atomistic simulations show a large polarization of ions in the first several layers nearby the surface and confirm the possibility of co-existence of Ti-and Ba-terminated [001] $BaTiO_3$ surfaces which have very close surface energies.

INTRODUCTION

Although there exist several theoretical studies of defects in technologically important perovskites [1, 2, 3], they were devoted mainly to the investigation of point defects. However, any crystalline surface, even perfect, is nothing but 2D defect which may lead to unusual behavior of perovskite films and nanocrystals, including the changes of the thermodynamic and the kinetic properties of first-order phase transitions. The relaxation of the surface atoms may turn out to be large enough for affecting the thermodynamic parameters characterizing the phase transitions, as compared with their bulk values. In particular, the atomic relaxation nearby the surface of paraelectric $SrTiO_3$ may result in its *ferroelectric* reconstruction at the finite temperatures, as suggested in ref. [4].

The aim of this communication is to study in details the surface relaxation of a large (up to ten planes) near-surface region and, on its basis, to demonstrate creation of a considerable *polarization* induced by a surface in perovskites. We use cubic $BaTiO_3$ as a prototype. Although *ab initio* calculations are known to be efficient in the study of the oxide properties [5], their use is essentially restricted by a relatively small number of surface layers which could be realistically handled. This is why in this paper a simpler, the so-called *shell-model* technique is used [6]. This approach was previously successfully applied to the investigation of defects in numerous ionic crystals including perovskites [1, 2, 3]. Its advantage is that the shell model is very-well suited for the treatment of the polarization effects which are a central issue of our study.

SIMULATION TECHNIQUE

In the present simulations we have studied a periodic two-dimenional slab of $BaTiO_3$.

Table 1: Short-range potential parameters for BaTiO₃ used in our simulations.

Interaction	A(eV)	ρ(Å)	C(eV/Å^6)
Ba^{2+} - O^{2-}	1214.4	0.35220	8.0
Ti^{4+} - O^{2-}	877.2	0.38096	9.0
O^{2-} - O^{2-}	22764.0	0.1490	43.0

Ion	Y(e)	k(eV/Å^2)
Ba^{2+}	1.848	29.1
Ti^{4+}	-35.863	65974.0
O^{2-}	-2.389	18.41

To study the surface relaxation, we have optimized the atomic positions in several (varied from one to ten) [001] surface planes placed into the electrostatic field of the remainder of the crystal (simulated by six additional planes whose atoms were fixed in their perfect lattice sites). The number of these additional planes was chosen to reach a convergency of the crystalline field in the surface planes.

The interatomic interactions are described, as usual, by the core-core, core-shell and shell-shell pair potentials, representing the shell-model (Table 1). In this approach each ion has a charged core and electronic shell. The sum of the core and shell charges is equal to the charge of the corresponding ideal ion. The spring constant k connects the core and the shell of the same ion. The value of this spring coefficient and the shell charge Y are chosen to describe correctly the polarizability α of the ion in the crystal:

$$\alpha = Y^2/k. \tag{1}$$

The interactions between the cores and between cores and shells of different ions are of entirely Coulombic nature. At the same time, the interactions between the shells of different ions besides the Coulombic part contain also the short-range potentials accounting for the effects of the exchange repulsion as well as van-der-Waals attraction between them. The short-range potential is

$$W_{sh} = A \cdot exp(-r/\rho) - C/r^6. \tag{2}$$

Parameters A, ρ, and C depend on the nature of the interacting ions and on the crystalline environment. The use of integer ionic charges does not imply restrictions to ionic materials: the short-range potential in Eq.(2) takes into account the covalency and charge-transfer effects. The detailed description of the shell model can be found in [6]. The relevant shell model parameters were earlier carefully fitted to lattice constant of the cubic bulk BaTiO₃, its elastic properties (C_{11}, C_{12}, C_{14}), and low/high frequency dielectric constants [1]. Our calculations are done by means of the MARVINS computer/code [7]. This code effectively realizes the shell-model technique for simulations of the surface structures. In our slab calculations we simulated both Ti- and Ba-terminated [001] surfaces.

CALCULATIONS AND MAIN RESULTS

Our calculations show that Ti^{4+}, Ba^{2+} and O^{2-} ions are displaced from their crystal sites quite differently. This leads to creation of *a dipole moment at the surface*. The induced dipole moment in both possible cases (Ti- or Ba-terminated surface) is directed perpendicularly to the surface. This is because all ions move only along the z axis perpendicular to the surface thus retaining the unrelaxed surface symmetry.

Simulations show that the values of the surface dipole moment *oscillate* as a function of the number of layers allowed to relax (Fig. 1). These oscillations practically vanish when the number of relaxed layers reached six. Note that the same six layers is necessary to reach convergence in the crystalline field in the surface region. This clearly shows that the surface affects 5-6 near-surface planes of a $BaTiO_3$ crystal. The value of the surface dipole moment converges to 0.271 $e \cdot Å$ for the Ba termination and to -0.755 $e \cdot Å$ for the Ti-termination.

Calculations also demonstrate the dependence of the very *sign* of the dipole moment on the terminating atom type: for the Ba ions the induced dipole moment is positive whereas for the Ti-terminated surface it is negative. This is true irrespective of the number of the relaxed surface layers.

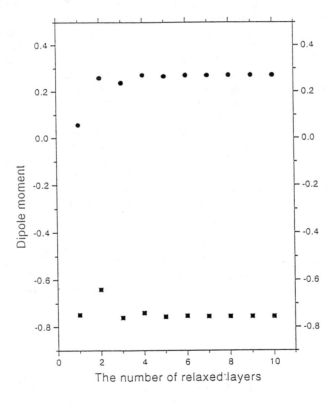

Fig. 1. Calculated surface dipole moments in the cubic $BaTiO_3$ as a function of a number of planes allowed to relax on the Ti- terminated (filled squares) and Ba terminated surface (filled circles). The dipole moments are in units of $e \cdot Å$.

In Table 2 we listed the displacements of ion cores and shells for 6 top layers for Ba-terminated surface. One can see that displacements of ions in the top layer are directed inwards whereas those in the second layer are pointed outwards. The same results were obtained for Ba- and Ti-terminated surfaces.

For the Ba-terminated case the surface Ba ions move inwards, approximately by 3.5% of the lattice constant (a_0), in the third layer their displacements are reduced to 0.5 %. Ti ions move outwards the crystal by $\sim$1.3% in the 2nd layer and 5 times less in the 4th layer. O^{2-} ions move inwards on Ba-terminated surface but outwards on the Ti-terminated one. Magnitudes of O^{2-} displacements fall down in an order from the first to the forth layers. These displacements in the third and the forth layers are very small.

The total positive sign of the surface dipole moment appears because of large displacements of negative O^{2-} ions in the top layer inwards, and positive Ti^{4+} ions in the second layer outwards the crystal. Ba^{2+} ions have twice smaller charges and thus their movement inwards the crystal cannot compensate the dipole moment created by the movement of both Ti^{4+} and the surface O^{2-} ions. The very similar trends in ion displacements takes place for the Ti-terminated surface. The displacement of the surface Ti ions is $\sim$2.7% inwards the crystal, and the displacement of Ba ions in the second layer is $\sim$2.1% outwards. O^{2-} ions move inwards in the surface layer but slightly outwards in the second one. Displacements of ions in the deeper layers fall down approximately by an order of magnitude and practically vanish in the 5th-6th layers. The negative sign of the surface dipole moment is determined by inward displacement of the strongly charged Ti ions on the surface.

In both cases of surface termination we observe a large *electronic* polarization of ions in the first two layers of the surface. It exhibits itself in large core-shell relative displacement of ions. This reaches, for example, $\sim$3% for O^{2-} ions in the surface layer at the Ba-terminated surface. Although the differences between core and shell displacements of Ti ions look sometimes very small, one has to remember that the charge of Ti shell is quite large in the employed shell-model parameters (Table 1). This leads to the large dipole moment of Ti ions, even with a relatively small relative shift of its shell from the core position. A large polarization of ions in a $BaTiO_3$ crystal is an evidence of a large electric field created in suface layers.

The surface energy for the Ba-terminated surface (16.588 J/m^2) is only slightly larger than that for the Ti-terminated one (16.538 J/m^2). The total energy of relaxed surface *per surface unit cell* with the same number of relaxed layers is only by 0.05 eV lower for the Ti-terminated surface. This difference appears only due to different relaxation energies of the surfaces in both cases.

CONCLUSIONS

Our simulations have clearly demonstrated that polarization of the surface in the $BaTiO_3$ crystal is determined by three main factors:

- Large electric field appears at the perovskite surface because of destroyed force balance for the surface atoms (short-range vs long-range, Coulomb forces). It pushes ions of the surface layer inwards and the second layer ions outwards the surface.

- Quite different ionic radii lead to their different displacements.

Table 2: Relaxation of 6 layers for the Ba-terminated surface. Totally 10 layers were relaxed. Coordinates and displacements are given in the lattice parameter units of the unrelaxed lattice.

No. of layer	Ion	Type	Coordinates of site			Displacement along z axis
1	Ba^{2+}	core	0.5	0.5	-0.5	-0.0372
		shell	0.5	0.5	-0.5	-0.0343
	O^{2-}	core	0.0	0.0	-0.5	0.0099
		shell	0.0	0.0	-0.5	-0.0276
2	Ti^{4+}	core	0.0	0.0	-1.0	0.0125
		shell	0.0	0.0	-1.0	0.0123
	O^{2-}	core	0.0	0.5	-1.0	0.0076
		shell	0.0	0.5	-1.0	0.0103
	O^{2-}	core	0.5	0.0	-1.0	0.0076
		shell	0.5	0.0	-1.0	0.0103
3	Ba^{2+}	core	0.5	0.5	-1.5	-0.0051
		shell	0.5	0.5	-1.5	-0.0048
	O^{2-}	core	0.0	0.0	-1.5	0.0016
		shell	0.0	0.0	-1.5	-0.0026
4	Ti^{4+}	core	0.0	0.0	-2.0	0.0020
		shell	0.0	0.0	-2.0	0.0019
	O^{2-}	core	0.0	0.5	-2.0	0.0011
		shell	0.0	0.5	-2.0	0.0015
	O^{2-}	core	0.5	0.0	-2.0	0.0011
		shell	0.5	0.0	-2.0	0.0015
5	Ba^{2+}	core	0.5	0.5	-2.5	-0.0003
		shell	0.5	0.5	-2.5	-0.0002
	O^{2-}	core	0.0	0.0	-2.5	0.0007
		shell	0.0	0.0	-2.5	0.0002
6	Ti^{4+}	core	0.0	0.0	-3.0	0.0006
		shell	0.0	0.0	-3.0	0.0006
	O^{2-}	core	0.0	0.5	-3.0	0.0005
		shell	0.0	0.5	-3.0	0.0005
	O^{2-}	core	0.5	0.0	-3.0	0.0005
		shell	0.5	0.0	-3.0	0.0005

- The large difference in cation charges (Ti^{4+} and Ba^{2+}) does not allow to compensate a dipole moment created by Ti^{4+} displacements. There is no complete cancellation possible also because of the opposite displacements of ions of both types.

We would like to stress that the Ti-terminated surface is only slightly more stable than the Ba-terminated one. However, this energy difference is very small and herefore both types of surfaces should co-exist in the reality. Note that macroscopic depolarization fields were not accounted for in our calculations focusing on microscopic effects for a finite number of near-surface planes.

ACKNOWLEDGEMENTS

Authors thank David Gay and Andrew Rohl from the Royal Institution of Great Britain for supplying us with the MARVINS computer code. Financial support through Latvian State Program on "New materials for microelectronics" is appreciated by E.Kotomin.

References

[1] G.V. Lewis, and C.R.A. Catlow, J. Phys. C **18**, 1149 (1985); G.V. Lewis, and C.R.A. Catlow, J. Phys. Chem. Solids, **47**, 89 (1986).

[2] M.Cherry, M.S.Islam,J.D.Gale and C.R.A. Catlow, J. Phys. Chem., **99**, 14614 (1995); M.Cherry, M.S.Islam and C.R.A. Catlow, J. of Solid State Chem. **118**, 125 (1995).

[3] H.Donnenberg and M.Exner, Phys. Rev. B **49**, 3746 (1994).

[4] N. Bickel, G. Schmidt, K. Heinz, and K. Müller, Phys. Rev. Lett., **62**, 2009 (1989).

[5] D.J.Singh and L.L.Boyer, Ferroelectrics, **136**, 95 (1992); D.J.Singh, Ferroelectrics, **164**, 143 (1995).

[6] *Computer Simulation of Solids*, edited by C.R.A. Catlow and W.C.Mackrodt, Lecture Notes in Physics, **166**, (Springer-Verlag, Berlin, 1982).

[7] D.H.Gay and A.L.Rohl, J. Chem. Soc. Faraday Transactions **91**, 925 (1995).

A TECHNIQUE FOR THE MEASUREMENT OF d_{31} COEFFICIENT OF PIEZOELECTRIC THIN FILMS

JOSEPH F. SHEPARD JR.* , PAUL J. MOSES, AND SUSAN TROLIER-McKINSTRY
Intercollege Materials Research Laboratory, The Pennsylvania State University, University Park, Pennsylvania 16802

ABSTRACT

This paper describes a new technique by which the d_{31} coefficient of piezoelectric thin films can be characterized. Silicon substrates coated with lead-zirconate titanate (PZT) are flexed while clamped in a uniform load rig. When stressed, the PZT film produces an electric charge which is monitored together with the change in applied load. The mechanical stress and thus the transverse piezoelectric coefficient can then be calculated. Experiments were conducted as a function of poling field strength and poling time. Results are dependent upon the value of applied stress, which itself is dependent upon the mechanical properties of the silicon substrate. Because the substrate is anisotropic, limiting d_{31} values were calculated. In general, d_{31} was found to be ~20 pC/N for field strengths above 130 kV/cm and poling times of less than 1 minute. d_{31} was increased more than a factor of three, to ~77 pC/N, when poled at 200 kV/cm for ~21 hours.

INTRODUCTION

There has, in recent years, been a great deal of interest paid to the design and construction of microelectromechanical systems (MEMS). Intended applications for novel MEMS designs are often focused on implementation as sensors and/or actuators. Of the possible actuation mechanisms, piezoelectricity is particularly attractive due to the high energy densities which can be achieved. Numerical descriptions of the degree of piezoelectricity a material displays are provided via characterization of a material's longitudinal (d_{33}) and transverse (d_{31} or d_{32}) piezoelectric coefficients. The current state of the art in d_{31} measurement of thin film materials relies upon the converse piezoelectric effect to excite a millimeter size cantilever beam [1] or free standing membrane [2], the deflection of which is then measured using a laser beam interferometer (or related technique) and correlated to the specific conditions of the experiment. In contrast to optical techniques, the procedure and instrumentation presented here utilize the direct piezoelectric effect to generate an electrical signal proportional to the amount of mechanical stress applied to a piezoelectric film i.e.

$$Q_3 = d_{31}(\sigma_1 + \sigma_2) \qquad (1)$$

where Q_3 is the charge produced in the direction parallel to the poling direction, σ_1 and σ_2 represent biaxial stresses applied perpendicular to the poling direction, and d_{31} is the transverse piezoelectric coefficient.

EXPERIMENTAL

PZT Film Deposition

Sol-gel PZT films were synthesized by spin-coating 0.5 molar sol onto pre-annealed platinized silicon wafers (3" diameter, {100} configuration) at 3000 rpm. Following pyrolysis at 300°C, a lead oxide (PbO) overlayer was deposited to aid in the minimization of lead volatilization during the subsequent crystallization step [3]. Completed films were rapid thermal annealed between 650 and 750°C, yielding films of final thicknesses on the order of 0.3 μm. Platinum top electrodes were sputter coated through a 1.5 mm diameter shadow mask and post annealed in the RTA unit at 700°C for 60 seconds.

---

* jfs12@email.psu.edu

Mat. Res. Soc. Symp. Proc. Vol. 459 © 1997 Materials Research Society

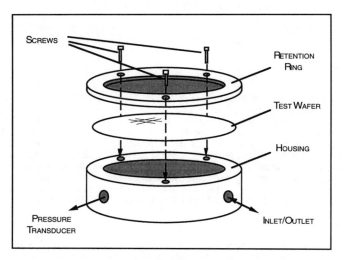

Figure 1. Uniform Pressure Rig

Piezoelectric Characterization

Following deposition, coated wafers were placed in a uniform pressure rig, a schematic of which is given in Figure 1, designed to apply a constant gas pressure over the surface of a 3" silicon substrate [4]. The pressure rig is constructed from a bored out aluminum round with an internal support diameter (i.e. wafer support) of 2.5". Test wafers are placed on the open end of the chuck and clamped in position with an aluminum ring of identical internal dimensions. Evacuation or pressurization of the cavity behind the wafer then flexes the sample in either a concave or convex fashion, thus placing the PZT on the surface of the wafer in a controlled state of biaxial tension or compression. The magnitude of the stresses applied to the film are calculated using small deflection plate theory for a clamped circular plate. Note that the substrate is anisotropic and results presented here were calculated under the assumption that the wafer behaves as a homogenous isotropic material. The piezoelectric coefficients were therefore calculated for elastic moduli and Poisson's ratios which correspond to the two limiting cases for a {100} wafer. Calculations performed thus represent the possible maxima and minima of applied stress and d_{31} coefficient.

The models which describe the bending stress of a clamped circular plate subjected to a uniform pressure p_o are given as [5]:

$$\sigma_r = \frac{3 p_o z}{4 t^3}\left[(1+v)a^2 - (3+v)r^2\right] \qquad (2)$$

$$\sigma_t = \frac{3 p_o z}{4 t^3}\left[(1+v)a^2 - (1+3v)r^2\right] \qquad (3)$$

where σ_r and σ_t are the radial and tangential stresses on the plate, z is the distance from the neutral axis, t is the plate thickness, v is Poison's ratio, a is the support radius, and r is the distance from the center of the plate.[1]

Mechanical stress is applied to the wafer, and thus the PZT, via a variation of gas pressure within the cavity behind the coated sample. Gas pressure is changed with a change of rig volume using a 60 cc plastic syringe attached to an access port on the base of the aluminum housing. Pressure is determined from the change in output voltage of a variable resistance pressure transducer, excited and monitored with an EG&G 7260 lock-in amplifier.

The manual variation of internal pressure changes the magnitude of stress on the PZT film which produces a proportional change of polarization as described by equation (1). The change in PZT polarization is monitored using a charge integrator set to measure the variation of voltage on a capacitor of known size placed in series with the stressed sample. Voltage output from the PZT film is monitored in real time, together with the change in bridge voltage from the pressure transducer using a Hewlett Packard 54600A oscilloscope. Figure 2 is an illustration of the complete experimental setup.

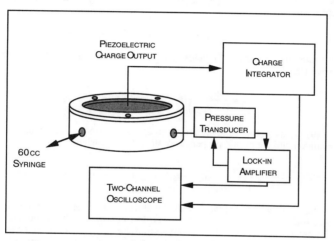

Figure 2. d_{31} Measurement Setup

The stresses calculated from equations (2) and (3) must be corrected for the difference in mechanical properties between the PZT film and the silicon substrate. The assumption was made that deformation of the composite plate was not affected by the presence of the metallization layer or the PZT, i.e. deformation is governed by the mechanical properties of the silicon. Generalized Hooke's law for the silicon substrate is then written as;

$$\varepsilon_1^{Si} = \frac{\sigma_1^{Si}}{E_{Si}} - v_{Si}\frac{\sigma_2^{Si}}{E_{Si}} \quad (4)$$

$$\varepsilon_2^{Si} = \frac{\sigma_2^{Si}}{E_{Si}} - v_{Si}\frac{\sigma_1^{Si}}{E_{Si}} \quad (5)$$

---

[1] Equations (2) and (3) are descriptive of a clamped plate which deflects a maximum of no more than 20% of its thickness. Deflection beyond that point produces membrane stresses which can not be neglected and the small deflection models quoted, if used, would over approximate the actual stresses applied to the plate.

where σ^{Si}_1 is applied stress in the 1 direction (1 and 2 are the principle directions corresponding to the tangential or radial orientation), ε_1^{Si} is the strain in the 1 direction, ν_{Si} is the Poisson's ratio of the silicon, and E_{Si} is the modulus of elasticity of the silicon.

If the assumption is made that all strain applied to the substrate is transferred to the PZT film then Generalized Hooke's law for the film is written in a similar manner and;

$$\varepsilon_1^{PZT} = \frac{\sigma_1^{PZT}}{E_{PZT}} - \nu_{PZT}\frac{\sigma_2^{PZT}}{E_{PZT}} \qquad (6)$$

$$\varepsilon_2^{PZT} = \frac{\sigma_2^{PZT}}{E_{PZT}} - \nu_{PZT}\frac{\sigma_1^{PZT}}{E_{PZT}}. \qquad (7)$$

From small deflection plate theory and knowledge of the mechanical properties of silicon equations (4) and (5) may be solved. Equating the film and substrate strains (eqs. 4-7) and then solving (6) for σ_1^{PZT} gives;

$$\sigma_1^{PZT} = \varepsilon_1^{Si}E_{PZT} + \nu_{PZT}\sigma_2^{PZT} \qquad (8)$$

which when substituted into equation (7) yields an equation for σ_2 in terms of the strain applied to the silicon substrate i.e.

$$\sigma_2^{PZT} = \frac{E_{PZT}}{(1-\nu_{PZT}^2)}\left(\varepsilon_2^{Si} + \nu_{PZT}\varepsilon_1^{Si}\right). \qquad (9)$$

The expansion of (9) yields a more fundamental form of the equation written in terms of the elastic properties of the silicon and the elastic properties of the PZT film where;

$$\sigma_2^{PZT} = \frac{E_{PZT}}{(1-\nu_{PZT}^2)}\left[\frac{\sigma_1^{Si}}{E_{Si}}\left(\nu_{PZT} - \nu_{Si}\right) + \frac{\sigma_2^{Si}}{E_{Si}}\left(1-\nu_{PZT}\nu_{Si}\right)\right]. \qquad (10)$$

From equation (10) it is clear that the calculated film stresses are dependent upon the elastic properties of both the film and substrate materials. Inaccuracies in those quantities will carry through to the d_{31} calculation via the stress analysis and the corresponding terms in the right side of eq. (1). Mechanical constants for thin film PZT are not widely characterized, and the few published reports vary considerably. This could account for much of the discrepancy among d_{31} values reported in the literature. The values used in this study are given in Table I.

TABLE I.
ELASTIC PROPERTIES OF PZT THIN FILMS AND {100} SILICON

	Thin Film PZT	Silicon <100> [8]	Silicon <110> [8]
Young's Modulus (GPa)	101 [6]	130	169.5
Poison's Ratio	0.3 [7]	0.28	0.064

RESULTS

Strength of Poling Field

Table II shows the variation of d_{31} with the magnitude of the applied poling field. Results were obtained within 5 minutes after poling for less than 1 minute and suggest the existence of some threshold field (between 65 and 135 kV/cm for a ~0.3 μm film thickness) above which d_{31} is independent of field strength. Since the calculated d_{31} coefficients are dependent upon the values of elastic modulus and Poisson's ratio used in eqs. (2)-(10), the extreme values are given in Table II. The difference between the calculated limits is a factor of 1.2 however numbers reported are still in good agreement with published values, often quoted at about 30 pC/N [e.g. ref. 6 and 9].

TABLE II.
d_{31} COEFFICIENTS AS A FUNCTION OF POLING FIELD

Poling Field (kV/cm)	<100> Elastic Properties		<110> Elastic Properties	
	d_{31} (pC/N) Sample 1	d_{31} (pC/N) Sample 2	d_{31} (pC/N) Sample 1	d_{31} (pC/N) Sample 2
67	8	5	10	6
133	18	15	21	18
200	19	17	23	20
267	18	18	22	21
333		16		19

<u>Poling Time</u>

Results of experiments on the influence of poling time (at 200 kV/cm) are presented in Figure 3. The data indicate a rapid increase of the piezoelectric coefficient for increased poling times from 1 to 20 minutes. For exposure times greater than 20 minutes however, the rate of change of d_{31} slows. The maximum value achieved was between 70 and 84 pC/N for a poling time of ~21 hours.

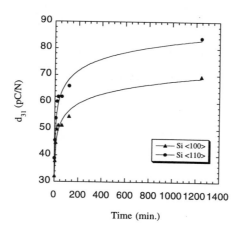

Figure 3. d_{31} vs. Poling Time at 200 kV/cm

CONCLUSIONS

The uniform pressure technique is well suited to rapid characterization of the transverse piezoelectric coefficient and results obtained are in reasonable agreement with those previously published. Calculated d_{31} coefficients are on the order of 20 pC/N in the just poled state (i.e. less than 1 min.) to a maximum of about 77 pC/N (average of two extremes) after poling for ~21 hours at 200 kV/cm. The accuracy of the technique relies upon the validity of the elastic properties of the PZT film used in the stress calculations. Comparison of this with other more elaborate techniques (i.e. laser techniques) shows a similar dependence of d_{31} on the mechanical properties of the experimental film. The scatter of reported coefficients within the literature is thought to result, at least in part, from discrepancies among the values of the elastic modulus of PZT used in the respective calculations and that observation should form a starting point for future d_{31} experiments.

ACKNOWLEDGMENTS

The authors would like to acknowledge Tao Su for his assistance in depositing the PZT films used in this study. This work is funded by DARPA under contract DABT63-95-C-0053.

REFERENCES

1. S. Watanabe, T. Fujiu, and T. Fujii, Appl. Phys. Lett. **66**, p. 1481-1483 (1995).

2. P. Muralt, A. Kholkin, M. Kohli, T. Maeder, Sensors and Actuators **A53**, p. 398-404 (1996).

3. T. Tani and D. A. Payne, J. Am. Ceram. Soc. **77**, p. 1242-1248 (1994).

4. J. Shepard, S. Trolier-McKinstry, M. Hendrickson, R. Zeto, to be published in, Proceedings of 10th International Symposium on Applications of Ferroelectrics, New Brunswick, New Jersey, 1996.

5. A. Ugural, Stresses in Plates and Shells, McGraw-Hill, New York, 1981.

6. T. Tuchiya, T. Itoh, G. Sasaki, T. Suga, J. Ceram. Soc. Jap. **104**, p. 153-163 (1996).

7. G. Spierings, J. Appl. Phys. **78**, p. 1926-1933 (1995).

8. W. Brantley, J. Appl. Phys. **44**, p. 534-535 (1973).

9. M. Toyama, R. Kubo, E. Takata, K. Tanaka, K. Ohwada, Sensors and Actuators **A45**, p. 125-129 (1994).

COMPLEX PIEZOELECTRIC COEFFICIENTS OF PZT CERAMICS: METHOD FOR DIRECT MEASUREMENT OF d_{33}

T. L. JORDAN*, Z. OUNAIES** and T. L. TURNER***

*Data Systems and Instrument Support Branch, NASA Langley Research Center, Hampton, VA 23681
**National Research Council, NASA Langley Research Center, Hampton, VA 23681
***Structural Acoustics Branch, NASA Langley Research Center, Hampton, VA 23681

ABSTRACT

A fiber-optic device is used to determine the magnitude and phase of the strain in the poling direction of lead zirconate titanate (PZT) ceramic wafers. This information yields the real and imaginary components of the piezoelectric strain coefficient d_{33}. The measurement hardware and software are described and results from the measurements of d_{33} for PZT 4 and 5 H wafers are presented. This method has the advantages of being direct, inexpensive and relatively simple to use. Verification of the results is provided through the use of the resonance method.

INTRODUCTION

Active control of structural vibrations and acoustic fields has received a great deal of attention in the last several years. Significant advancements have been made in various supporting areas, but current sensor and actuator technology limits further progress [1]. Practical limitations such as acceptable excitation voltages, mechanical durability, and control system complexity and stability are driving research for sensor and actuator improvement. Development of performance measurement techniques for piezoelectric devices is a key component of this research as recent capabilities have been relatively inaccurate and incomplete. Impedance techniques are effective at high frequencies (resonance of the piezoelectric device), but measurement of performance characteristics at low frequencies (1Hz - 3 kHz) is needed for noise and vibration control applications. Furthermore, effects of dispersion at the low frequencies as well as nonlinearity at the high fields make it necessary to develop a method to measure the strain as a function of electric fields for a broad frequency and electric field range.

Direct measurements of piezoelectric coefficients have been widely reported using strain gages, interference methods and capacitance techniques [2,3]. These methods are generally used to yield the magnitude of the response only, and no information on phase is reported. Pan et al.[4] used laser interferometry to calculate the real and imaginary parts of the piezoelectric coefficients from the measured strain amplitude and phase delay of the strain with respect to the applied electric field for PZT ceramics. To ensure stability and precision in the phase detection, evaporating or affixing mirrors to the samples was necessary. All of the above techniques tend to be complex and require tedious sample preparation and considerable effort to maintain accuracy. We saw a need in our laboratories to develop a simple, inexpensive and more direct method to characterize PZT wafers at low frequencies and both low and high fields. This paper presents a direct method for the measurement of the real and imaginary piezoelectric strain coefficient d_{33} using the converse effect.

Mat. Res. Soc. Symp. Proc. Vol. 459 © 1997 Materials Research Society

METHOD

The measurement system consists of a fiber-optic displacement sensor, a computer (Toshiba T4700CT) running LabVIEW, a function generator, an amplifier and signal conditioning hardware as shown in figure 1. The PC is used to specify the amplitude and frequency of a driving voltage for the device under test (in this case, the PZT wafer). The displacement sensor then measures the change in thickness of the PZT wafer and produces an analog voltage proportional to the displacement. The PC records the output of the displacement sensor along with the driving voltage. The raw data is then displayed on the screen and the phase angle and d_{33} value are calculated and displayed.

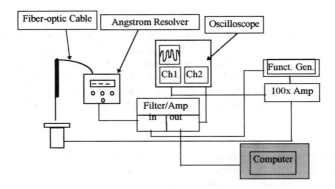

Figure 1. Block diagram of measurement system.

Hardware Description

The computer controls a function generator through an IEEE interface to supply a voltage of known frequency, amplitude, and offset. The voltage is then routed to an amplifier with a fixed gain of 100. The amplified voltage is next applied across the PZT wafer which is held in a test fixture.

The fiber-optic displacement measuring device measures the change in thickness of the wafer. This instrument was chosen over other displacement measuring devices, such as a laser interferometer and capacitance based devices, because of its low cost (~ $2500), high resolution and accuracy, ease of set-up and adaptability to different wafer sizes, shapes, and surface finishes. The device, a two channel Angstrom Resolver, has an analog output which is a function of the distance of the tip of the probe to the target. As the tip-to-target distance is increased from zero, the output increases, reaches a maximum and then decreases. Each channel therefore has two linear ranges over which it may be used, corresponding to the front and back slope of the output voltage. One channel of the Angstrom Resolver was chosen to maximize the range of the instrument (600 μm) and the other channel was chosen to maximize the resolution (2 pm at 1 Hz). The reflectivity of the sensing surface can be calibrated out by normalizing the output of the probe. No special preparations of the surface of the PZT wafers are necessary. The analog output of the instrument makes it simple to integrate into common data acquisition systems.

The output of the Angstrom Resolver is amplified and filtered through a low-pass filter (Stanford research amp/filter model SR640) to block any high frequency noise which may be present. The cutoff frequency of the filter is set to one and a half times the driving frequency. To compensate for the phase delay which is introduced by the filter (θ_F), the drive signal from the function generator is also routed to the filter and compared to the unfiltered signal. The phase delay due to the filter can then be subtracted from the total phase (θ_T).

An oscilloscope is used for visual verification of the drive signal and the filtered output of the Angstrom Resolver. These signals are also routed to a data acquisition board. This board (a National Instruments DAQ-1200) has four channels of analog input with programmable range, 12 bit resolution, and is capable of data acquisition rates of up to 50 kHz.

Software Description

The software for this measurement system was developed in LabVIEW and serves as instrument control, data acquisition and reduction. The user can control the frequency, amplitude, and offset of the driving voltage which is applied to the wafer. The controller is also connected to the oscilloscope and controls the horizontal (time) and vertical (voltage) scales of the scope allowing the user to visually check the driving and displacement voltage signals. Also controlled through an IEEE interface is the filter/amplifier. The user sets the gain of the amplifier from the front panel of the software. The cutoff frequency of the filter is automatically set by the software at 1.5 times the driving frequency.

Once the driving voltage has been applied to the wafer, the software acquires several cycles of the driving voltage and Angstrom Resolver output. Algorithms in LabVIEW then analyze the data to determine the AC component of the displacement signal, the peak-to-peak value of the driving voltage and the phase difference between the two. Once the amplitude of the displacement signal has been determined it is then multiplied by the appropriate constants to measure the peak-to-peak displacement of the PZT wafer. The value of d_{33} is then calculated from the driving voltage (V) and the displacement of the wafer (Δt) by the following relationship:

$$d_{33} = S/E = (\Delta t/t) / (V/t) \qquad (1)$$

where S is the induced strain and E is the electric field. The displacement of the wafer is displayed graphically and the d_{33} value displayed numerically on the screen.

The phase angle associated with the wafer (θ_P) is then calculated by subtracting the phase due to the filtering process (θ_F) from the overall phase (θ_T) as follows:

$$\theta_P = \theta_T - \theta_F \qquad (2)$$

The phase angle associated with the PZT wafer is then displayed numerically on the screen.

Wafer Measurement

Two PZT compositions are measured to test the apparatus (PZT4, 5H). Figure 2 shows the sample geometry and mounting. Proper sample mounting is crucial to the measurement technique. The wafer must be fixed in such a way that there is little or no restriction in the movement of the wafer, especially in the direction of the desired displacement. Deformations in modes other than the one in question, however, must be isolated from the measurement system. A holder was

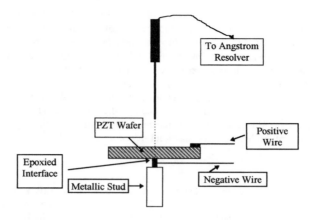

Figure 2. Sample Mounting.

designed in which the wafer is mounted with conductive epoxy to a small (~3mm diameter) horizontal flat surface. The wafer is centered in the horizontal plane above the mounting surface so that it is held at a node. The Angstrom Resolver probe is then positioned above the surface of the wafer directly above the mounting surface. By measuring the deformation at this point, displacements due to deformations other than changes in thickness (bending or twisting) are minimized. The mounting stud acts as the bottom electrode for the wafer and a thin wire is epoxied to the top surface to act as the upper electrode. The probe is held in an optical mount which has both coarse and fine adjustments for ease in positioning the probe. The entire sample holder sits atop a one inch pad of dense wool and is supported by a pneumatic vibration isolation table to isolate the holder from external disturbances.

The measurement is independent of sample dimension because the measurement principle is based on point displacement. This is another flexibility of the present apparatus in that it allows the strain measurement as a function of position with very little adjustment to the Angstrom Resolver.

If a sinusoidal field is applied to the sample, the complex piezoelectric coefficient can be expressed as:

$$d = d' + j\, d'' = (S/E)\, e^{j\theta_P} \qquad (3)$$

Separating the real and imaginary parts:

$$d' = (S/E)\cos\theta_P \ , \ d'' = (S/E)\sin\theta_P \qquad (4)$$

RESULTS

Using the fiber-optic measurement method, strain measurements and corresponding phase values at a field strength of 2 kV/cm are collected. Equations 3 and 4 are used to yield the real and imaginary part of the piezoelectric strain coefficient d_{33}. Figure 3 presents this data for both d'_{33} and d''_{33} as a function of frequency for PZT 5H and PZT 4 wafers.

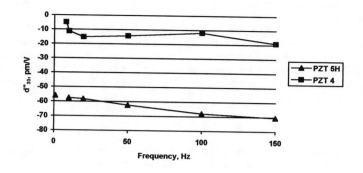

(a)

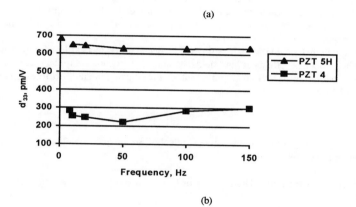

(b)

Figure 3. d_{33} coefficient as a function of frequency, (a) imaginary part; (b) real part.

At each frequency (i.e., 1 Hz to 150 Hz), repeatability is within 1% for the value of magnitude of d_{33} and 8% for the value of the phase. Inspection of Table I shows a good agreement between the range of values for d'_{33} and d''_{33} in the low frequency regimes and those at resonances for both compositions (PZT4 and 5H). The data reduction for resonance method measurements was accomplished using the Piezoelectric Resonance Analysis Program [5]. Resonance frequency for the PZT4 wafers is 61.5 kHz and for the PZT 5H is 56.4 kHz. These results successfully validate the proposed fiber-optic method as a direct measurement tool of the converse piezoelectric coefficient.

Table I. d'_{33} and d''_{33} from direct method and resonance method.

	PZT 4	PZT 5H
d'_{33}(direct method)	220 - 300 pm/V	630 - 680 pm/V
$|d''_{33}|$ (direct method)	5 - 18 pm/V	55 - 70 pm/V
d'_{33}(resonance method)	205 pm/V	678.5 pm/V
$|d''_{33}|$ (resonance method)	14 pm/V	46.3 pm/V

Measurements at higher frequencies were unstable and not repeatable, mostly due to the presence of significant response at frequencies other than the driving frequency in the displacement signal from the Angstrom Resolver. This noise in the signal may be due to external disturbance or unsteadiness/resonance in the sample holder. Future efforts will concentrate on extending the method to higher frequencies and improving repeatability. For example, the response signal will be further analyzed to investigate applicability of other filtering techniques, such as notch filtering or lock-in amplification. Additionally, nonlinear effects at high electric field strength will be investigated.

CONCLUSION

A versatile tool for the determination of the complex piezoelectric strain in the poling direction for low frequencies and a wide range of field strengths has been developed. Sample measurements of the d_{33} coupling coefficients for PZT 4 and 5H wafers are presented. Repeatable magnitudes and phases of the piezoelectric coefficient are obtained for both compositions. Resonance measurements are obtained for both wafer types which provided validation of the direct method technique.

Future work will include a systematic study of the dispersion effects at low frequencies, as well as understanding nonlinear effects at high fields. Mounting concepts and replacement of low-pass filter with band-pass filter or lock-in amplifier will be investigated in an effort to extend the frequency range to 1 kHz.

REFERENCES

1. R. A. Burdisso, ed., Recent Advances in Active Control of Sound and Vibration, Technomic Publishing Company, Lancaster, 1993.

2. IEEE Standard on Piezoelectricity, The Institute of Electrical and Electronics Engineers, Inc., New York, 1988.

3. Q. M. Zhang, W. Y. Pan and L. E. Cross, J. Appl. Phys. **63**, p. 2429 (1988).

4. W. Y. Pan, H. Wang and L. E. Cross, Jpn. J. Appl. Phys. **29**, p.1570 (1990).

5. Piezoelectric Resonance Analysis Program, TASI Technical Software (1996).

LASER IRRADIATION EFFECT ON SAW PROPERTIES OF LAYERED STRUCTURE OF OXIDE/PIEZOELECTRIC SUBSTRATE

Yo ICHIKAWA, Masatoshi KITAGAWA, Kentaro SETSUNE & Syun-ichiro KAWASHIMA
Central Research Laboratories, Matsushita Electric Ind., Co., Ltd.
3-4 Hikaridai, Seika, Soraku, Kyoto 619-02, Japan

ABSTRACT

Using surface acoustic wave (SAW) propagating in the layered structure of oxide/piezoelectric substrate, a responsibility of the oxide thin films for the laser irradiation has been investigated. Amorphous Ti-O, Si-O and Si-X-O, where X is other metal elements, films were formed on the surface of the SAW device composed of quartz or LiTaO$_3$ substrate and several hundred Al electrode fingers for oscillating and detecting the SAW. A KrF excimer laser with 248nm in wavelength was used for the irradiation. The center frequency of the SAW devices was immediately decreased by the irradiation of the laser pulses. Although the response to the irradiation was reversible for lower laser energy, the change of the center frequency was irreversible for the laser energy density higher than 20mJ/cm^2. It is considered that the response appeared in the frequency shift is generated by a change an elastic stiffness of the films lowered by an absorption of the laser energy.

INTRODUCTION

SAW devices are widely applied to components of high frequency communication systems. Especially, for SAW filters, a control of operating frequency (center frequency) is a serious problem because the frequency is defined by the accuracy of the device dimension. The center frequency of SAW filter is strongly dependent on the thickness of the electrodes. In general, the frequency linearly decreases as the thickness increases. Therefore, development of an in-situ frequency controlling method is desirable for improving the efficiency of device processing. On the other hand, studies on the interaction between ceramic materials and UV light for improving the materials are interesting because most materials have an optical band gap in UV region. Irradiation with UV light can enhance the adhesion strength of metal films to ceramic substrates [1, 2]. In the study of irradiation using a XeCl excimer laser with 308nm wavelength on Al-O substrates, the generation of point defects during laser irradiation was suggested experimentally [3]. Laser irradiation can be expected to be used for a selective improvement of oxide thin films for electronic devices. As is well known, SAW devices have a high sensitivity to surface conditions of a piezoelectric substrate where an acoustic wave propagates. In this study, for preliminary experiments, dielectric thin films of amorphous Ti-O and Si-O induced other metal elements were deposited on SAW filters, and effects of the irradiation on the operational properties of the SAW filters have been investigated using UV laser

Mat. Res. Soc. Symp. Proc. Vol. 459 © 1997 Materials Research Society

pulses of KrF excimer laser with 248nm wavelength in order to develop an frequency controlling method.

EXPERIMENTAL TECHNIQUES

Oxide thin films were prepared by using ion beam sputtering [4] on surfaces of SAW filters, which are very sensitive to the conditions of the substrate surface [5-7]. Sputtering conditions are shown in Table I. An ion source was the Kaufman-type (Commonwealth Co.) with beam diameter of 30mm at outlet of the source. The oxide films were formed reactively by supplying oxygen gas to the substrate from a nozzle. Si-X-O films were prepared by sputtering the Pyrex glass (Corning code no.;7740) and by co-sputtering Si and metal targets with two ion sources. The metal elements X, Al and Ti were selected in order to obtain glass (Si-X-O) films which can absorb UV light. A deposition rate was monitored by quartz crystal oscillators. The substrate temperature increase due to bombardment by sputtered particles was negligible.

TABLE I. Ion beam sputtering conditions.

Acceleration voltage	500V-1kV
Beam current	20mA (Ar gas; 2-3SCCM)
Sputtering area of target surface	15cm^2
Target-substrate distance	15cm
Target	Ti-, Si-, Al-metal, Pyrex glass (75mm in diameter)
Oxygen partial pressure	1×10^{-4}torr
Substrate temperature	room temperature
Deposition rate	0.5-1nm/min

In the beginning, optical property and crystal structure were investigated for the oxide films formed on fused glass and other single crystal substrates (sapphire and quartz). Crystal structure of the films was identified by X-ray diffraction (XRD) analysis. The microstructural changes induced by the irradiation were investigated using transmission electron microscopy (TEM). Compositional ratio of metal elements in the Si-X-O films was identified by electron probe micro-analysis (EPMA) calibrated by inductively coupled plasma optical emission spectroscopy (ICP).

The UV light irradiation was conducted with a KrF excimer laser with a wavelength of 248nm. The laser light was pulsed to 20nsec (FWHM), and irradiated on the surface of the oxide films at 10Hz typically. The laser energy density was 20-60mJ/cm^2 at the film surface.

In this study, the oxide films were deposited on 315MHz- and 872MHz-SAW filters. The deposited thin films induce a frequency shift depending on thickness, Young modulus, Poisson's ratio and residual stress in the films [8, 9]. The SAW filter is composed of a piezoelectric substrate and several hundred Al electrode fingers for oscillating and detecting the surface acoustic waves. Figure 1 shows a cross-sectional sketch of the SAW filter and electrode-fingers. The cross sectional width and thickness of an electrode is several μm.   For the 315MHz- and 872MHz-SAW filters, the piezoelectric substrate was quartz and LiTaO₃, respectively.   An oscillator circuit was constructed with the SAW filter and an amplifier. The frequency of the microwave resonating in the circuit was measured by a frequency counter and a spectrum analyzer.   The frequency was defined as the "center frequency" of the SAW filter.   The measurement was carried as the laser pulses were irradiated on the film deposited on the SAW filter.

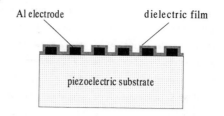

FIG.1. Cross-section of SAW filter with dielectric film.

EXPERIMENTAL RESULTS AND DISCUSSION

The XRD analysis revealed that the crystal structure of the as-deposited films is in amorphous state. Figure 2 shows transmission spectra of as grown Ti-O ((a)) and Si-X-O ((b) and (c)) films on fused glass substrates.   The thickness of the films was 20nm. In Fig.2, the spectrum (b) was obtained from the film by sputtering the Pyrex glass target ("Pyrex film"), and the spectrum (c) was obtained from the film by co-sputtering the Si and the Ti targets.   For the Si-Ti-O film, the compositional ratio was Si:Ti=9:1.   It was observed that these films have an optical band gap around 300nm in wavelength, though the gap is not well defined.   UV-laser pulses were irradiated on the as grown films.   However, there was little change in the optical transmission shown in Fig. 2 and in the results obtained by XRD analyses before and after the irradiation.

Figure 3 shows the change in center frequency of the 315MHz-SAW filter overcoated by the Ti-O film versus the number of laser pulses.   At that time, the laser energy density was 35mJ/cm²(10Hz).   Before the

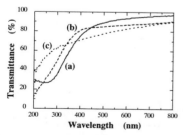

FIG.2. Transmittance of as-grown films on fused glass:(a)Ti-O, (b)Pyrex and (c)Si-Ti-O.

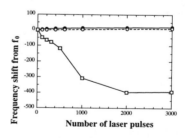

FIG.3. Shift of center frequency of 315MHz-SAW filter by irradiation of KrF excimer laser pulses. Film thickness:(●)0nm, (○)1nm and (□)2nm.

irradiation of the laser pulses, the center frequency decreased at a rate of about 5kHz/nm with increasing film thickness.   This decrease in the frequency is thought to be due to the "mass effect"[8].   Although the change at the film thickness of 1nm was as small as that without coating, the frequency was abruptly changed at the film thickness of 2nm as shown in Fig. 3. It is considered that the irradiation causes little damage to the Al electrodes, the surface of the quartz substrate, or the bonding of the leading wires for measurements because the change of the frequency is very small for the case of the non-coated specimen. The change seems to be saturated around 2500 laser shots. Same results for the irradiation of the laser pulses were obtained when the Si-X-O films were coated and the 872MHz-SAW filter used. For the 872MHz-SAW filters coated with Ti-O and Pyrex films, the frequency versus the weight of the films is shown in Fig.4. The weight was defined by (thickness)$\times$(density ρ). In this case, 4.5 and 2.3g/cc were used as density ρ of Ti-O and Pyrex, respectively. As shown in Fig.4, both results show same tendency before and after the irradiation of the laser pulses (35mJ/cm^2, 100 shots by 10Hz). These results reveal that the decrease in the frequency is dependent on the "mass effect".

The relation between the frequency shift and the laser energy density was investigated. Figure 5 shows the change in the frequency versus the energy density of the laser pulse when pulses of 100 shots at 10Hz were irradiated on the Ti-O films of 10nm in thickness. It was clear that the frequency is changed at the threshold laser energy density of about 25mJ/cm^2. On the other hand, the frequency change seems to saturate for laser energy densities higher than 50mJ/cm^2. For the laser energy density over 70mJ/cm^2, damage was clearly observed on the surface of the Al electrodes, and the SAW filter did not operate as a filter.

The center frequency of a SAW device is proportional to the sound velocity of the surface acoustic wave. When a film is coated on the surface of the SAW substrate, the frequency is dependent on elastic properties of the film and residual stress in boundary between the film and the substrate. As is well known, it is difficult to simulate an electric property of the SAW device with a layered structure[8-11]. It is expected that the SAW velocity decreases monotonically with increasing film thickness when the film formed on a

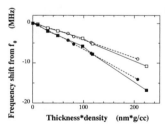

FIG.4. Frequency shift of 872MHz-SAW filter versus thickness*density of oxide films:
($\square$) and ($\blacksquare$) are Ti-O, ($\bigcirc$) and ($\bullet$) are Pyrex.
($\square$) and ($\bigcirc$):before laser irradiation.
($\blacksquare$) and ($\bullet$):after laser irradiation.

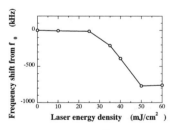

FIG.5. Frequency shift of 315MHz-SAW filter versus energy density of irradiated laser pulse.

surface of the SAW substrate is smaller than the periodicity of the Al electrode fingers[8].   In this study, the thickness of the oxide films is small enough for the SAW velocity to decrease.   It is considered that the crystalline structure of the films was improved, elastic properties of the films were changed uniformly and the SAW velocity was decreased.   In this case, the change in the frequency generated by the irradiation of the UV-laser pulses is dependent mainly on the change of the velocity of the SAW propagating into the oxide film due to the elastic property and the stress of the film against the quartz substrate, because the mass of the film is not changed.   In Figs. 3 and 5, the frequency decreased monotonically until the laser flux reached to the saturation region.   It is considered that the improvement of the elastic properties of the film was completed in the this region.

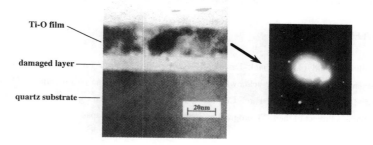

FIG.6. Cross sectional TEM image of laser irradiated Ti-O film/quartz substrate(left) and electron diffraction pattern of the Ti-O film(right).

In order to investigate the microstructure of the films, TEM analysis was carried out   for the specimens composed of the Ti-O film on ST-cut quartz substrates.   Figures 6 shows the cross sectional image of the TEM photographs (left) and the electron diffraction patterns (right) of the Ti-O film of the irradiated specimen.   The specimen shown in Fig.6 was irradiated by five hundred pulses.   In the TEM photograph, a damaged layer of the quartz substrate is observed under the Ti-O layer.   The damaged layers are thought to be generated from a boundary between the Ti-O layer and the substrate by an irradiation of the electron beam used in analysis.   However, an electron probe micro-analysis (EPMA) carried out for the damaged area showed no emission from Ti particles.   In addition, emission from Si atom was not observed in the Ti-O film.   Therefore, it not appear that an interdiffusion of the Si or Ti atoms through the boundary between the Ti-O layer and the substrate was generated by the laser irradiation or the electron beam irradiation of the analysis. The generation of crystalline granules are observed in the TEM image as shown in Fig. 6.   The diameter of these granules is several nm.   The electron diffraction image shown in the right of Fig. 6 indicates the existence of polycrystalline phases in the Ti-O film.   The diffraction spots identified the structure as TiO_2 of the rutile type.   The electron diffraction pattern was a halo indicating an amorphous state of the Ti-O crystal for the film before the laser irradiation.

CONCLUSIONS

In this study, the effect of pulsed KrF excimer laser irradiation on the frequency of SAW filters coated by amorphous oxide films was investigated. Oxide films, as amorphous Ti-O, Si-O and Si-X-O, where X is other metal elements, were formed on the surface of the SAW filter. The center frequency of the SAW devices was immediately decreased by an increase in the number of laser shots. By the TEM analysis, it was observed that crystalline granules, identified as rutile type TiO_2 structure, are produced in the Ti-O film by the irradiation of the laser pulses. It is considered that the frequency shift is generated by a lowering of the elastic stiffness of the films by absorption of the laser energy. It is expected that this technology may be used for in-situ frequency control of electronic devices using piezoelectric materials of SAW.

ACKNOWLEDGMENTS

The authors would like to thank Drs. K. Kugimiya and T. Nitta for their support and constant encouragement. We also gratefully acknowledge the TEM and EXAFS analyses of Mrs. T. Okano and Y. Umetani of Matsushita Techno. Research Co., Ltd. This research has been supported by NEDO, under the Synergy Ceramics Project of the ISTF program promoted by AIST, MITI, Japan.

REFERENCES

1. H. Esrom, Chemical Perspectives of Microelectronic Materials II, edited by L. V. Interrante, K. F. Jensen, L. H. Dubois, and M. E. Gross (Mater. Res. Soc. Symp. Proc. 204, Pittsburgh, PA, 1991), 457.

2. M. J. Desilva, A. J. Pedraza, and D. H. Lowndes, J. Mater. Res. 9, 1019 (1994).

3. A. J. Pedraza, J. W. Park, H. M. Meyer III, and D. N. Braski, J. Mater. Res. 9, 2251 (1994).

4. Y. Ichikawa, H. Adachi, K. Setsune, and K. Wasa, Appl. Surf. Sci. 60/61, 749 (1992).

5. K. Setsune, O. Yamazaki, and K. Wasa, Elect. Lett. 20, 433 (1984).

6. I. P. Raevskii, A. N. Rybyanets, M. A. Malitskaya, V. G. Poltavtsev, and A. V. Turik, Sov. Phys. Tech. Phys. 37, 475 (1992).

7. K. Komine, N. Araki, and K. Hohkawa, Proc. IEEE Ultrasonics Symp. 253 (1993).

8. B. K. Sinha, and S. Locke, Proc. IEEE Trans. Ultrasonics, Ferroelectrics, and Freq. Cont. UFFC-34, 29 (1987).

9. K. Yamanouchi, H. Satoh, T. Meguro, and Y. Wagatsuma, IEEE Trans. Ultrason., Ferroelectrics, and Freq. Cont. UFFC-42, 392 (1995).

10. G. S. Kino, and R. S. Wagers, J. Appl. Phys. 44, 1480 (1973).

11. Z. Wang, J. D. N. Cheeke, and C. K. Jen, IEEE Trans. Ultrason., Ferroelectrics, and Freq. Cont. UFFC-43, 844 (1996).

COMBINATION OF DIFFERENT METHODS TO CHARACTERIZE MICROMECHANICAL SENSORS

R. BUCHHOLD, U. BÜTTNER, A. NAKLADAL, K. SAGER *
E. HACK, R. BRÖNNIMANN, U. SENNHAUSER **
A. SCHROTH ***
* Dresden University of Technology, Inst. of Solid-State Electronics, 01062 Dresden, Germany
** EMPA, Department of Metrology, Überlandstrasse 129, CH-8600 Dübendorf, Switzerland
*** Mechanical Engineering Laboratory, AIST, MITI, Namiki 1-2, Tsukuba, Ibaraki, 305 Japan

ABSTRACT

In the last years many micromachined sensors for measuring quite different quantities have been developed (e.g. [1][2]). The successful design of silicon micromachined sensors and their conversion into commercial products is still limited by the lack of full understanding of complex fault mechanisms. In order to detect those mechanisms it seems to be advantageous to investigate interrelated phenomena by different analysis methods. By combining results of several means the true nature of disturbing effects can be determined much more easily.

This paper describes a comprehensive approach for investigating failure mechanisms in piezoresistive pressure sensors. Sophisticated methods of signal analysis were combined with special semiconductor techniques, determination of thermo-mechanical properties, Finite-Element-(FE-) simulation and Michelson interferometry. By that it was possible to separate disturbing mechanical as well as electrical effects and their relation to sensor output and sensor accuracy, respectively. In particular, we discovered complex influences of passivation layer systems (differing in geometrical and technological parameters) on sensor accuracy.

INTRODUCTION

Micromachined piezoresistive sensors were developed in the early 1960s [3]. Restricted to static measurements of pressure in the first time (Fig. 1a), several sensor principles (based on the piezoresistive effect) for detection of different quantities have been developed over the last years. In 1994 Sager introduced a piezoresistive sensor even for chemical quantities that means for water content in humid air [4].

The investigation has been focused on piezoresistive pressure sensors. The operation principle can be simply explained by separating the whole sensor system into subsystems (Fig. 1c):
- the mechanical subsystem represented by the diaphragm of a micromachined pressure sensor,
- mechano-electrical subsystem represented by the (mostly) implanted piezoresistors and
- electrical subsystem represented by a Wheatstone bridge.

Due to different disturbing effects real sensor transfer functions deviate from the ideal curves (Fig 1b). The determinate deviations (e. g. nonlinearity and temperature dependence of piezo-resistive coefficients) have been successfully explored by several groups (e.g. [5][6]). In contrast to that, stochastic effects in the subsystems (noise) have not been understood completely yet. The random disturbances are obviously strongly influenced by the passivation system chosen [6]. Because stochastic effects cannot be compensated, they restrict the sensor accuracy attainable [7]. For further optimization of piezoresistive sensors it is absolutely essential to investigate the true nature of stochastic effects in passivation layers.

Mat. Res. Soc. Symp. Proc. Vol. 459 © 1997 Materials Research Society

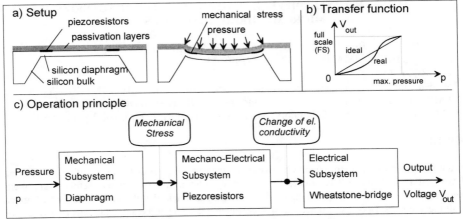

Fig. 1: Setup, transfer function and operation principle of piezoresistive pressure sensors

APPROACH

The performance of piezoresistive pressure sensors is limited by several physical effects [1], e. g.:
mechanical effects:
- geometric nonlinearities of the mechanical subsystem,
- thermo-mechanical stress,
- relaxation and recrystallization of passivation or contact layers
electrical effects:
- space charge induced resistance modulation (SCIRM),
- formation of channels by positive (negative) surface charges due to depletion (enhancement),
- parasitic currents due to adsorbed and absorbed charges or due to insufficient isolation,
- "aging" due to carrier-induced electrochemical reactions.
Due to the complexity of the sensor setup and manufacturing it is hardly possible to assign instabilities to the related physical effects as described above.

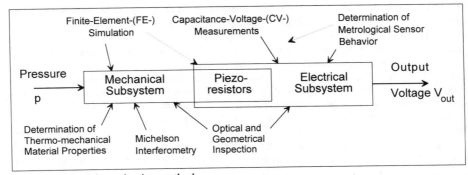

Fig. 2: Survey on investigation methods

Measurement of metrological behavior (output voltage) phenomenologically records the totality of relevant instability mechanisms. Considering the actual boundary conditions (e. g. stimulation regime, material properties or manufacturing data) this totality can be separated by signal analysis. This separation allows the determination of dependences of sensor performance from design and technological parameters.

The combination of this knowledge with results from other methods (Fig. 2) represents the basis for the correct interpretation of ongoing physical mechanisms.

As mentioned above the major part of the investigation is the determination and analysis of metrological sensor behavior consisting of the following steps:
- stimulation with multidimensional regimes (pressure, temperature, relative humidity),
- regression of the multidimensional transfer function,
- heuristic modeling of separated transfer channels,
- evaluation of determinate dynamic (L) and static (V_s) transfer coefficients,
- evaluation of channel noise V_{st} (stimulation-induced) and V_{in} (intrinsic noise) as deviation of measured data from transfer function.

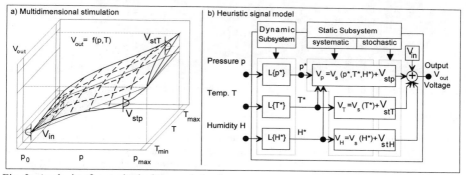

Fig. 3: Analysis of metrological sensor behavior

Determinate environmental influences and nonlinearities of transfer functions can in generally be compensated by appropriate means (microcontrollers, ASIC's).

The overall noise may be in the range up to 1% full scale (FS) of sensor output and cannot be compensated. Therefore the accuracy of piezoresistive pressure sensors is limited by induced (V_{stp}, V_{stH}, V_{stT}) and intrinsic (V_{in}) noise.

Interpretation of data obtained with the described approach will allow performance improvement via development of suitable passivation concepts and optimization of processing technology.

MECHANICAL SUBSYSTEM

Determinate Effects

Several mechanisms in the mechanical subsystem cause deviations from ideal transfer function:
- deviations of geometric parameters due to manufacturing technology,
- thermo-mechanical stress in passivation and packaging components,
- mechanical-induced stresses in sensor chips (clamping on packaging carriers).

Last years those mechanisms have been investigated extensively. As a result very sophisticated models were developed and published [3][5][7]. For instance, in [7] an approximation formula was introduced, that expresses sensor offset and temperature coefficient as functions of passivation layers properties obtained by measurement (Fig. 2).

However, in the low pressure range the accuracy of those models is limited by deviations from ideal design caused by technology. Fig. 4 shows the inhomogeneous deflection of a sensor diaphragm that causes a determinate but not predictable offset.

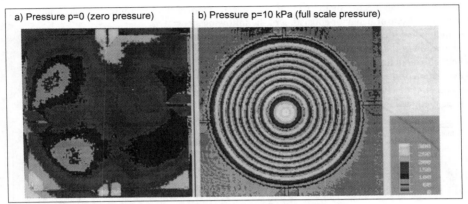

Fig. 4: Measurement of diaphragm deflection via Michelson interferometry

Stochastic Effects

Besides determinate mechanical effects also changes of material properties (e. g. relaxation or recrystallization) influence sensor behavior. For instance they cause hysteresis, nonreversibility and dependencies on load history. Methods to measure those effects on-chip are not available.

For practical application (commercial sensors) it is convenient to express the sum of those phenomena as load-dependent noise V_{stp} (Fig. 3b). The value of V_{stp} represents a unique and worthful characteristic of quality of the mechanical subsystems.

ELECTRICAL SUBSYSTEM

Determinate effects

The transfer behavior of the electrical subsystem also depends on processing and environmental conditions. Determinate effects are:
- dependence of resistance of the piezoresistors from implantation depth and dose,
- temperature coefficient and nonlinearity of piezoresistive coefficients.
Similar to the mechanical subsystem it is possible to describe the mechanisms with existing models after having measured the material properties [3][7].

Stochastic effects

The measured data deviate significantly from the multidimensional transfer function (Fig. 3a). This cannot be explained with determinate effects and is therefore regarded as noise.

With means of data processing the noise of different channels was separated and correlated to single design and process parameters (Fig 5). By that it could be proved that the noise of the sensors is chiefly influenced by passivation concepts.

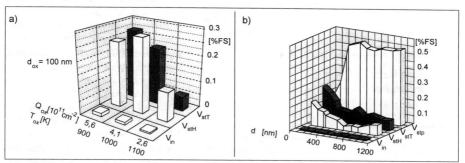

Fig. 5: Influence of passivation layer properties on noise of piezoresistive pressure sensors
    a) Influence of oxidation temperature of the primary oxide layer
    b) Influence of overall passivation layer thickness

The following information was extracted:
- Intrinsic noise V_{in} is negligibly influenced by passivation concept.
- Higher deposition temperatures for oxide passivation layers cause decrease of humidity V_{stH} and temperature induced noise V_{stT}.
- Increased overall passivation system thickness causes decrease of humidity V_{stH} and temperature induced noise V_{stT}, but higher pressure induced noise V_{stp} in the mechanical subsystem.
However, without a theoretical based interpretation of the obtained data sensor optimization will be half-measure. Therefore, an analysis of possible mechanisms was carried out. Fig. 6 shows several mechanisms influencing V_{in}, V_{stH} and V_{stT}.

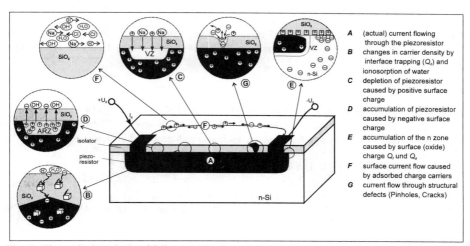

Fig. 6: Theoretical analysis of failure mechanisms

Stochastic effects in the electrical subsystem are chiefly caused by electrical charges (Fig. 6). Therefore the charge densities of the oxide layers deposited at different temperatures (Fig. 5a) were determined with Capacitance-Voltage-(CV-) measurements (Fig 7b).

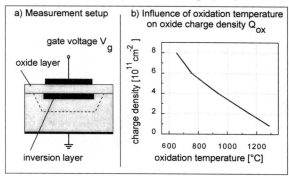

Fig. 7: Determination of charge densities in SiO_2 layers

It could be proved that the charge densities of silicon oxide layers (thermally grown) are significantly reduced at higher deposition temperatures. By that less electrical charges interact with the electrical subsystem, and the noise V_{stH} and V_{stT} is reduced (Fig. 5a).

Rise of pressure induced noise with passivation system thickness (Fig. 5b) can be explained with an increased impact of recrystallization and relaxation effects in the layers.

CONCLUSIONS

Nowadays accuracy of piezoresistive pressure sensors can be further improved by the reduction of sensor noise. This noise is made up of the single noise of every different channel and can be obtained by analyzing multidimensional transfer functions. It is possible to correlate channel noise with processing or design parameters. By that new design rules can be derived.

It was found out that stochastic effects are strongly influenced by passivation systems. Oxidation temperatures of primary oxide layers should be as high as possible. A compromise between stability, V_{stH} and V_{stT} on the one side and V_p on the other is achieved with passivation layer thicknesses in the range of 400 nm.

ACKNOWLEDGMENTS

The research was sponsored by the German Bundesministerium für Bildung, Wissenschaft, Forschung und Technologie BMBF (grant no. 16SV248-0) and by the Deutsche Forschungsgemeinschaft DFG (grant no. Ge 779 2/2).

REFERENCES

[1]   F.-P. Steiner, et al., Proc. of Transducer '95 in Stockholm (Sweden), 814-817 (1995).
[2]   H. Takao, et al., Proc. of Transducer '95 in Stockholm (Sweden), 683-686 (1995).
[3]   G. Gerlach, K. Sager and A. Nakladal, Technisches Messen. 63 (11), 403-412 (1996).
[4]   K. Sager, G. Gerlach and A. Schroth, Sens. & Act. 18 (1-3), 85-88 (1994).
[5]   G. Gerlach, PhD thesis, Dresden University of Technology, 1990.
[6]   J. Bryzek, Proceedings of SENSOR '95 in Nuremberg (Germany), 45-50 (1995).
[7]   A. Schroth, PhD thesis, Dresden University of Technology, 1996.
[8]   G. Barbottin and A. Vapaille : Instabilities in Silicon Devices. (Elsevier Science Publishers, Amsterdam 1989).

APPLICATION OF POROUS SILICON TO BULK SILICON MICROMACHINING

G. KALTSAS, A. G. NASSIOPOULOS

Institute of Microelectronics, NCSR Demokritos,
P.O. Box 60228, 15310 Aghia Paraskevi Attikis,
Athens, GREECE

ABSTRACT

A fully C-MOS compatible process for bulk silicon micromachining using porous silicon technology and front-side lithography is developed. The process is based on the use of porous silicon as a sacrificial layer for the fabrication of deep cavities into monocrystalline silicon, so as to avoid back side lithography. Cavities as deep as several hundreds of micrometers are produced with very smooth surface and sidewalls. The process is used to produce : a) suspended monocrystalline silicon membranes, b) free standing polysilicon membranes in the form of bridges or cantilevers with lateral dimensions from a few µms to several hundreds of µms. Important applications to silicon integrated devices as sensors, actuators, detectors etc., are foreseen.

I.   INTRODUCTION

Porous silicon is a material which has been intensively studied since 1990 [1] after the discovery that it can emit light very efficiently at room temperature. Foreseen applications were in silicon integrated optoelectronics. However, due to its easy fabrication by electrochemistry in a C-MOS compatible way, other interesting applications are foreseen. In this work we use porous silicon as a sacrificial layer for bulk silicon micromachining with potential applications in sensor fabrication. C-MOS compatible chemicals were used in all processing steps concerning also the dissolution of the porous silicon sacrificial layer. KOH, which was used by other authors [2,3], has been avoided. Two different processes will be described in this paper. The first one is used for the formation of polycrystalline silicon membranes in the form of bridges or cantilevers suspended on deep cavities into silicon and the second one is used for the fabrication of free standing monocrystalline silicon membranes. The process is described in detail below.

II.   PROCESSING FOR MEMBRANE AND CANTILEVER FABRICATION

II.1   Polycrystalline silicon membranes and cantilevers

The process for producing free standing polycrystalline silicon membranes and cantilevers has been described in detail elsewhere [4,5]. Porous silicon is produced by electrochemical dissolution of monocrystalline silicon on selected areas of the silicon wafer. As masking material for the local electrochemical reaction we use a thin layer of undoped polycrystalline silicon on top of a thin silicon dioxide layer. This bilayer was found to be a perfect mask for anodization of silicon in an ethanoic HF solution. The use of the silicon dioxide layer under polysilicon revealed to be necessary in order to avoid initiation of porous silicon formation under the mask, due to a small current passing through polysilicon [4]. The HF

Mat. Res. Soc. Symp. Proc. Vol. 459 © 1997 Materials Research Society

Top View   Cross Section

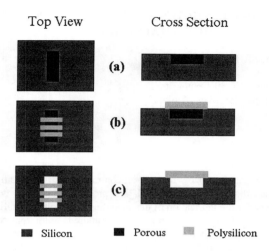

■ Silicon   ■ Porous   ■ Polysilicon

Figure 1 : Process flow for polysilicon membrane formation (a) Porous formation in selected areas (b) Polysilicon deposition and patterning (c) Porous removal.

concentration in the solution was 60%. Under the above conditions, very thick porous silicon layers were grown isotropically into silicon at thicknesses exceeding one hundred microns. After dissolution of porous silicon, cavities with very smooth bottom surface and sidewalls were

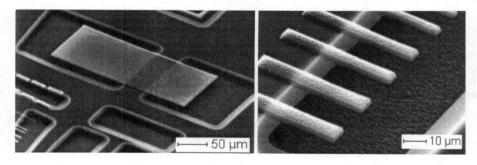

Figure 2 : Polycrystalline silicon micromechanical structures produced by porous silicon technology.

obtained.

     Polycrystalline silicon membranes were produced by depositing a polycrystalline film by Low Pressure Chemical Vapour Deposition on top of the wafer. This film was patterned in order

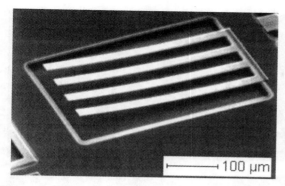

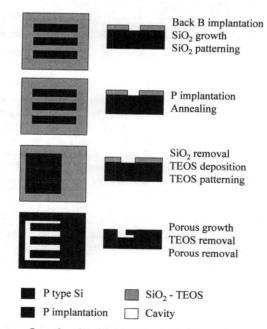

to define the membrane area. Porous silicon was then dissolved and free standing membranes were obtained.

Figure 1 shows schematically the process flow for polysilicon membrane formation and figure 2 shows examples of micromechanical structures produced by this process. The membrane area is quite flat. This is achieved by minimizing strain within polysilicon [6]. The effect of strain on the membranes is to cause bending as illustrated in figure 3.

|⊢————————⊣ 100 µm

Figure 3 : Effect of polysilicon strain on the membranes

II.2 Monocrystalline silicon bridges and cantilevers

In some applications it is interesting to use silicon membranes composed of monocrystalline silicon. A slightly different process was developed in this respect as follows : It is well known that n-type silicon is not dissolved during electrochemical dissolution of p-type

Back B implantation
SiO_2 growth
SiO_2 patterning

P implantation
Annealing

SiO_2 removal
TEOS deposition
TEOS patterning

Porous growth
TEOS removal
Porous removal

■ P type Si   ▨ SiO_2 - TEOS
■ P implantation   ☐ Cavity

Figure 4 : Process flow for the fabrication of free standing n-type monocrystalline membranes.

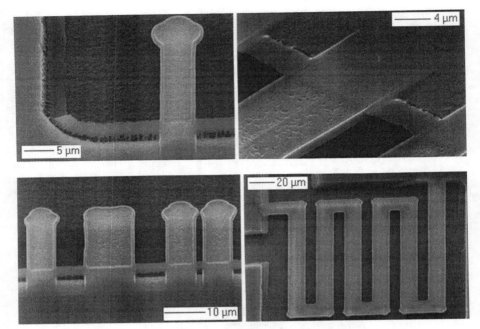

Figure 5 : N-type monocrystalline silicon membranes, produced by porous silicon microma-chining

silicon in the dark. So it is possible to produce porous silicon locally on selected p-type areas without affecting neighboring n-type areas. This property has been used to fabricate free standing n-type monocrystalline silicon membranes and cantilevers. Phosphorous is implanted on patterned areas, through an SiO_2 mask so as to produce locally n-type silicon. A TEOS silicon dioxide is then deposited on top and patterned appropriately in order to be used as mask for porous silicon formation. Due to the isotropic behaviour of the anodization process, porous silicon is formed also under the n-type layer, thus leaving this layer free standing after porous silicon dissolution. The whole process is shown schematically in figure 4. The porous thickness was of the order of a few microns and so deep were in this case the cavities. Much deeper cavities may be produced by using the bilayer polysilicon/SiO_2 mask.

Examples of n-type monocrystalline silicon membranes, produced as described above, are shown schematically in figure 5. These are the first attempts to produce such structures and further improvements are possible.

IV. CONCLUSION

Porous silicon technology has been successfully used for bulk silicon micromachining by using only front side lithography. Two different processes were developed for producing a) polycrystalline and b) n-type monocrystalline silicon membranes and cantilevers suspended on deep cavities into silicon. Cavities were as deep as ~120 μm, with very smooth bottom silicon surface and sidewalls if care was taken to avoid strain within polysilicon and as large as 230 μm x 550 μm, with a thickness of 0.5 - 3 μm. Monocrystalline silicon membranes were fabricated by exploiting the fact that n-type and p-type silicon have different dissolution rates for anodization in the dark, for porous silicon formation. The developed micromachining technology is fully C-MOS compatible and opens new important possibilities to integrated sensor fabrication.

REFERENCES

[1] T. Canham, Appl. Phys. Lett., Vol. 57, (1990), pp.1046-1048.
[2] Bischoff, G. Móller and F. Koch, Eurosensors X, Leuven, Belgium, 1996, Proceed., pp.211-214.
[3] Navarro, J. M. Löpez-Villegas, J. Samitier and J. R. Morante, Eurosensors X, Leuven, Belgium, 1996, Proceed., pp. 235-238.
[4] G. Kaltsas, A. G. Nassiopoulos, Sensors and Actuators A (submitted).
[5] G. Kaltsas, A.G. Nassiopoulos, Microelec. Eng. (accepted).
[6] Guckel, T. Randazzo and D. W. Burns, J. Appl. Phys., 57 (5), (1995), pp. 1671-1673.

FLEXIBLE GRAPHITE AS A STRAIN/STRESS SENSOR

XIANGCHENG LUO, D.D.L. CHUNG
Composite Materials Research Laboratory, Furnas Hall, State University of New York at Buffalo, Buffalo, NY 14260-4400

ABSTRACT

Flexible graphite sandwiched by copper, after stabilization by two cycles of compressive stress, is an effective piezoresistive compressive strain/stress sensor for stresses up to 4 MPa and strains up to 25%. The stress sensitivity (fractional change in resistance per unit stress) is up to 5.4 MPa^{-1} and strain sensitivity (fractional change in resistance per unit strain) is up to 6.2 in the direction perpendicular to the sheet. The electrical resistance decreases reversibly upon compression, due mainly to reversible decrease in the contact resistivity between graphite and copper. Stabilization removes most of the irreversible effects. The strain/stress sensitivities decrease with increasing strain/stress.

INTRODUCTION

Strain/stress sensors that give an electrical output that relates to the strain/stress are important due to the advent of smart structures, which require sensing for the sake of either structural health monitoring or control. The electrical output can be a change in electrical resistance, a change in dipole moment per unit volume, a change in capacitance, etc. A change in electrical resistance is experimentally simpler to detect than the other changes.

Piezoresistivity refers to the phenomenon in which the electrical resistivity changes due to strain/stress. The most common type of piezoresistive material is a composite material with an electrically non-conducting matrix (usually a polymer) and a conducting filler (particles or fibers) [1-4]. Upon tension, the distance between adjacent filler units increases, so the resistivity increases; upon compression, the distance between adjacent filler unit decreases, so the resistivity decreases. Another type of piezoresistive material is a semiconductor, the energy band gap of which changes with strain/stress [5]. This paper provides a new type of piezoresistive material, namely flexible graphite (a flexible sheet made by compressing a collection of exfoliated graphite flakes without a binder [6-10]) sandwiched by a metal. The piezoresistivity of the sandwich stems from the change in contact resistivity in the sandwich. Flexible graphite is mostly used industrially as a gasket for fluids. Compared to other gasket materials, it is attractive for its high temperature resistance, excellent chemical resistance and environmental safety. Resilience in the direction perpendicular to the sheet is an important property that makes flexible graphite a reusable gasket material. This property is exploited in this work in using flexible graphite as a reusable sensor of strain/stress in the direction perpendicular to the sheet. Compared to the other types of piezoresistive strain/stress sensors, flexible graphite is attractive in the large stress sensitivity, i.e., that a small stress produces a large fractional change in resistance, in addition to the high temperature resistance and excellent chemical resistance.

Mat. Res. Soc. Symp. Proc. Vol. 459 © 1997 Materials Research Society

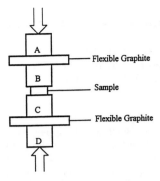

Fig. 1 Set-up for strain/stress sensitivity testing.

EXPERIMENTAL

Flexible graphite sheets (Grade GTB) of thicknesses 1.6 and 3.1 mm were provided by EGC Enterprises, Inc. (Mentor, Ohio). According to the manufacturer, the ash content is < 5.0%; the density is 1.1 g/cm^3; the tensile strength in the plane of the sheet is 5.2 MPa; the compressive strength (10% reduction) perpendicular to the sheet is 3.9 MPa; the thermal conductivity at 1093°C is 42 W/m.K in the plane of the sheet and 3 W/m.K perpendicular to the sheet; the coefficient of thermal expansion (CTE) (21-1093°C) is -0.4 x 10^{-6} °C^{-1} in the plane of the sheet.

The electrical resistivity is 7.5 x 10^{-4} Ω.cm in the plane of the sheet, as measured in this work by the four-probe method, using silver paint for the electrical contacts. This value is essentially the same as the value of 8 x 10^{-4} Ω.cm given by the manufacturer. The electrical resistivity perpendicular to the sheet is 0.4 Ω.cm according to the manufacturer. However our measurement using the four-probe method, with silver paint for the electrical contacts and with the two current probes in the form of loops on the opposite faces in the plane of the sheet and the two voltage probes in the form of dots at the centers of the loops, gave values of 0.011 ± 0.001 Ω.cm for the sheet of thickness 1.6 mm (based on four samples cut from the sheet) and 0.037 ± 0.011 Ω.cm for the sheet of thickness 3.1 mm (based on four samples cut from the sheet). The values obtained by our measurement rather than those given by the manufacturer were used in this work.

The strain/stress sensing ability of flexible graphite was evaluated using the four-probe (A,B,C,D) set-up illustrated in Fig. 1. A rectangular piece of flexible graphite was sandwiched between two copper cylinders (diameter = 12.8 mm, height = 9.9 mm) labeled B and C in Fig. 1. Two larger pieces of flexible graphite cut from the original sheet (thus same thickness) as the abovementioned smaller piece were sandwiched between copper cylinders A and B and between copper cylinders C and D (Fig. 1). These larger pieces extended beyond the circumference of the copper cylinders, whereas the smaller piece was within the circumference. Silver paint was applied to the interface between each piece of flexible graphite and its adjacent copper cylinder. A copper wire was soldered to each of the four copper cylinders. The contacts to cylinders A and D were for current (DC) to pass through, whereas those to cylinders B and C were for measuring the voltage. A Keithley 2001 multimeter was used. In this way, the resistance between cylinders B and C was measured. This resistance consisted of the volume resistance of the flexible graphite between cylinders B and C in the direction of the cylinder axis and the contact resistance at each of the two interfaces between cylinder (B or C) and the flexible graphite between cylinders B and C, as the volume resistance of each cylinder is negligible. This resistance measurement was conducted while cyclic compressive loading (under load control) was applied to the stack in Fig. 1 in the direction of the cylinder axis, using a screw-type mechanical testing system (Sintech 2/D), which gave the stress and strain during the test. The strain of the flexible graphite between cylinders B and C in the direction of the cylinder axis was obtained by dividing the crosshead movement of the mechanical testing machine by 1.83, as the stack contained 3 pieces of flexible graphite and the area of the flexible graphite between cylinders B and C was about half of that of the cylinder. The crosshead speed was 0.5 mm/min.

RESULTS AND DISCUSSION

Fig. 2 shows the fractional increase in resistance ($\Delta R/R_o$, negative for decrease), strain (negative for shrinkage) and stress (negative for compression) simultaneously obtained during the first 10 loading cycles at a constant stress amplitude of -0.42 MPa for a flexible graphite sample (7.6 x 7.1 mm) of thickness 1.60 mm. The resistance (initially 0.27 Ω) decreased during loading and increased during subsequent unloading in every cycle, such that the resistance change was mostly reversible, but had an irreversible portion at the end of the first cycle. Furthermore, the maximum magnitude of $\Delta R/R_o$ was larger in the first cycle than in subsequent cycles. The irreversible portion at the end of a cycle slightly increased after the subsequent cycle, but did not further change after that. Thus, it took 2 cycles for the cyclic resistance change to stabilize. When stabilized, the magnitude of the reversible portion of $\Delta R/R_o$ was 41%, while the magnitude of the irreversible portion of $\Delta R/R_o$ was 12%. The magnitude of strain increased with the magnitude of stress during each loading, such that the strain magnitude increase was partly reversible and partly irreversible at the end of the first cycle. Moreover, the maximum magnitude of strain was larger in the first cycle than in subsequent cycles, probably due to strain hardening. The irreversible portion of strain did not further increase in subsequent cycles. Thus, it took only 1 cycle for the cyclic strain change to stabilize. When stabilized, the magnitude of the reversible portion of strain was 7%, while the magnitude of the irreversible portion of strain was 8%. The $\Delta R/R_o$ variation closely corresponded to the strain variation, such that (i) a larger magnitude of $\Delta R/R_o$ was associated with a larger magnitude of strain, (ii) the irreversible portion of $\Delta R/R_o$ was associated with the irreversible portion of strain, and (iii) the reversible portion of $\Delta R/R_o$ was associated with the reversible portion of strain. After the initial two loading cycles, the flexible graphite provided reproducible strain/stress sensing.

Similar results were obtained with the same sample as Fig. 2, but at stress amplitudes of -1.25, -2.06 and -2.48 MPa. The corresponding reversible $\Delta R/R_o$ was -68%, -70% and -72% respectively; the corresponding reversible strain was -13%,

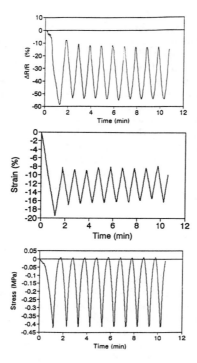

Fig. 2 Fractional resistance increase ($\Delta R/R_o$), strain and stress simultaneously obtained during the first 10 compressive loading cycles (under load control) of flexible graphite of thickness 1.6 mm and initial resistance 0.27 Ω. The stress amplitude was -0.42 MPa.

-17% and -19% respectively; the corresponding stress sensitivity was 0.54, 0.34 and 0.29 MPa^{-1} respectively; the corresponding strain sensitivity was 5.4, 4.2 and 3.8 respectively. Thus both stress sensitivity and strain sensitivity decreased with increasing stress amplitude.

When the stress amplitude reached -4.12 MPa, the irreversible portion of the $\Delta R/R_o$ magnitude as well as that of the strain magnitude increased after every cycle and the maximum strain magnitude of a cycle increased in every cycle. This is due to irreversible shrinkage of the flexible graphite in every cycle at the large stress amplitude. Similar behavior was observed at a stress amplitude of -5.78 MPa. Flexible graphite could not serve as a reliable strain/stress sensor when the stress magnitude exceeded 4 MPa or when the strain magnitude exceeded 25%.

The stress and strain sensitivities for various combinations of sample thickness, initial resistance and stress amplitude are listed in Table 1. In most cases, the sensitivities decreased with increasing initial resistance. When the initial resistance was too high, the $\Delta R/R_o$ and strain changes had significant irreversible portions, even after considerable prior cycling. Therefore, a small initial resistance

Table 1 Stress and strain sensitivities for various combinations of flexible graphite thickness, initial resistance and stress amplitude.

Thickness (mm)	Initial resistance (Ω)	Fraction of initial resistance due to volume resistance	Stress amplitude (MPa)	Sensitivity	
				Stress (MPa^{-1})	Strain
1.6	0.042	0.079[b]	-0.12	2.0	5.4
1.6	0.051	0.065[b]	-0.172	1.3	3.4
1.6	0.27	0.012[b]	-0.42	1.0	6.2
1.6	0.36	0.009[b]	-0.11	a	a
3.1	0.034	0.58[c]	-0.076	5.4	3.4
3.1	0.084	0.23[c]	-0.166	0.72	1.5
3.1	0.106	0.18[c]	-0.24	a	a

a   Significant irreversible portions for both $\Delta R/R_0$ and strain.
b   Volume resistance = 0.0033 Ω.
c   Volume resistance = 0.020 Ω.

is preferred. This dependence on the initial resistance is attributed to the fact that the contact resistance between sample and the copper cylinder is an important part of the measured resistance. As shown by calculation of the volume resistance of the sample based on the separately measured volume resistivity perpendicular to the sheet, the volume resistance constitutes only a small part of the measured resistance for most of the samples (Table 1). A low initial resistance reflects mainly a low contact resistance. The contact resistivity varies from one sample to another, while the volume resistivity variation is much smaller. The observed reversible resistance decrease in most of the samples is mainly due to a decrease in the contact resistivity, associated with reversible conforming of the flexible graphite surface to the surface topography of the copper cylinder upon compression. The reversibility is possible due to the resilience of flexible graphite. The contributions to the observed resistance decrease by a possible reversible resistivity decrease (microstructural change) in the flexible graphite upon compression and by the thickness decrease are small in most of the samples. A high initial resistance corresponds to a high contact resistivity (i.e., a poor interface), which in turn corresponds to a large irreversible portion of $\Delta R/R_0$ (associated with some irreversible tendency to conform to the surface topography of the copper cylinder). The irreversible $\Delta R/R_0$ corresponds to irreversible strain, which is 5.3% in Fig. 2. This irreversible strain is associated partly with the irreversible tendency to conform to the surface topography of the copper cylinder and partly with irreversible thickness decrease. A larger thickness appears to yield higher sensitivities, as suggested by the outstandingly high stress sensitivity for the sample with both a large thickness (3.1 mm) and a low initial resistance (0.034 Ω). Thus, the best conditions for high stress and strain sensitivities are low stress amplitude, low initial resistance, and probably large thickness as well.

The stress sensitivity of flexible graphite is exceptionally high compared to other piezoresistive stress sensors, while the strain sensitivity is lower than those of some piezoresistive strain sensors. For example, a short carbon fiber epoxy-matrix composite (a state-of-the-art piezoresistive strain/stress sensor) has a stress sensitivity of 0.02 MPa^{-1} and a strain sensitivity of 29-31 (stress amplitude ranging from -15 to -40 MPa) [4]. The high stress sensitivity of flexible graphite is due to the low stress required for a substantial resistance change -- a consequence of the softness of flexible graphite. However, flexible graphite can sense compressive stress/strain, but not tensile stress/strain, due to the difficulty of applying tension in the direction perpendicular to the sheet. In contrast, conventional piezoresistive sensors work both in compression and in tension.

CONCLUSION

Flexible graphite (sandwiched by copper) which has been stabilized by two cycles of compressive stress is an effective piezoresistive compressive strain/stress sensor that exhibits stress sensitivity up to 5.4 MPa^{-1} (outstandingly high compared to other piezoresistive sensors) and strain sensitivity up to 6.2 in the direction perpendicular to the sheet. The resistance perpendicular to the sheet decreases reversibly upon compressive loading perpendicular to the sheet, due mainly to reversible decrease of the contact resistivity between flexible graphite and copper. The stabilization serves to largely eliminate the irreversible resistance and strain changes. The strain/stress sensitivities decrease with increasing strain/stress. A low stress amplitude (< 4 MPa) and a low strain amplitude (< 25%) is necessary in order to minimise irreversible deformation of the flexible graphite. The strain/stress sensitivities are enhanced when the contact resistivity is smaller, as indicated by a smaller measured resistance.

REFERENCES

1. J. Kost, M. Narkis and A. Foux, J. Appl. Polymer Science **29**, 3937-3946 (1984).

2. S. Radhakrishnan, Sanjay Chakne and P.N. Shelke, Mater. Lett. **19**, 358-362 (1994).

3. P.K. Pramanik, D. Khastgir, S.K. De and T.N. Saha, J. Mater. Sci. **25**, 3848-3853 (1990).

4. X. Wang and D.D.L. Chung, Smart Mater. Struct. **4**, 363-367 (1995).

5. V.A. Gridchin, V.M. Lubimsky and M.P. Sarina, Sensors Actuators A **49**, 67-72 (1995).

6. J.H. Shane, R.J. Russell and R.A. Bochman, US Patent 3 404 061 (1968).

7. Z. Huang, Runhua Yu Mifeng **6**, 28-32 (1981).

8. L. Shi and Y. Fan, Runhua Yu Mifeng **27**, 17-24, 70 (1981).

9. R.K. Flitney, Tribology Int. **19**, 181 (1986).

10. D.D.L. Chung, J. Mater. Sci. **22**, 4190 (1987).

HIGH SENSITIVE STRAIN MICROSENSOR BASED ON DIELECTRIC MATRIX WITH METAL NANOPARTICLES

I.A. KONOVALOV*, R.D. FEDOROVICH**, S.A. NEPIJKO**, L.V. VIDUTA**
*Toronto, Canada, cl591@freenet.toronto.on.ca
**Institute of Physics, 46 Pr. Nauki, Kiev, Ukraine

ABSTRACT

A dielectric matrix, containing metal nanoparticles with interparticle spacings of 1-2 nm, is a system with tunnel mechanism of electrical conductivity. Its electrical resistance is very sensitive to deforming of matrix because it leads to changes in spaces between particles and as a result the potential barrier transperancy is varied.

Different metals (Mo, Cr, Ta, Au, Pt, Bi, Al) and their films morphology structure were studied in order to get high sensitive strain sensors. Metal nanoparticles were deposited on elastic dielectric substrates. Strain coefficients were measured for a wide range of strains and temperatures. Variation of matrix structure gives possibilities to produce strain sensors with high electrical resistance and weak temperature dependence. The matrix with Au nanoparticles was found to have maximum strain coefficient (>100). These sensors can be manufactured in the miniature scale (sensitive area around 1 micron or less).

INTRODUCTION

Discontinuous metal films represent a system of nanoscale islands on dielectric substrate. They are formed at early stages of vacuum deposition of metals on nonwettable substrates. Conductivity in such films is provided by electron tunneling. It is usually modeled as a system of potential wells and barriers. Thus, the conductivity is very sensitive to the distances between metal islands [1]. When an island metal film is deposited on an elastic substrate , then bending can effectively change the distances between islands (e.g. tunnel barrier width). This causes essential changes in film resistance. Resistance in such system is determined as [2]:

$$R = Aa^2 \exp(4\pi a/h)(2m^* H)^{1/2} \exp \Delta E/kT, \qquad (1)$$

where A - constant, a - distance between islands, h - Planck's constant, m* - effective mass of the tunneling electron, H - tunnel barrier height, ΔE - activation energy, k - Boltzmann's constant, T - absolute temperature

When current passes through an island film, the phenomena of electron and light emission are observed in island metal films under certain conditions [3]. Conducted experiments showed high sensitivity of the electron emission current to substrate deformation.

EXPERIMENT

Island metal films were deposited by vacuum evaporation (residual gas pressure P during vaporizing was P=2 10^{-4} Pa . Steel foils covered by dielectric layer were used as elastic dielectric substrates. A pair of contacts in the form of thin (thickness of 0.2 μm) and narrow strips (width of 3 mm) were deposited on the substrate before island film deposition. The distance between contacts was equal 30 μm (Fig.1)

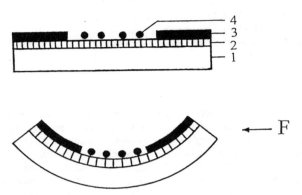

Fig.1. Sketch of the sample with island film and it's possible deformation: 1-elastic substrate, 2-dielectric layer, 3-electrical contact, 4-metal island.

The thin film structure was studied by transmission electron microscopy. The particle size and the interparticle spacing were measured from the microscope photographs. The film structure was varied by using different conditions of evaporation: rate, time and substrate temperature. To protect the as-prepared film from mechanical damage and the influence of atmospheric conditions it was covered by special thin dielectric layer. Electrical conductivity of such system (total thickness was around 50 μm) under different deformations was measured. We determined the strain sensitivity coefficient from:

$$\Delta = \Delta R/R : \Delta L/L = \Delta R/R : \varepsilon, \qquad (2)$$

where $\Delta R/R$ - relative change of resistance, ε - relative deformation.
Electron emission measurements were conducted under vacuum conditions at $P = 2 \cdot 10^{-6}$ Pa.

RESULTS

Strain sensitivity of metal films.

Metals with different melt temperatures were investigated. Metals with high melt temperature (Mo, Ta, Pt) were chosen because of their high stability to studied phenomena. These films had structures with island sizes around 1-2 nm. They had resistance about 10^6-10^8 ohms and coefficient γ was about 20-30 for Pt and 5-10 for Mo and Ta. Al and Bi films showed values of the same order, $\gamma = 5$-10.
More detailed studies were conducted for Au films. Typical current - voltage curves for such films are shown in Fig.2. Below an applied voltage of 4-5 V ohmic behavior takes place, but then non-ohmic one occurs. In this ohmic region γ value does not depend on the

applied voltage. The same thing was observed for the non-ohmic region but the current values were unstable. That's why, a voltage of about 1-2 V was usually used.

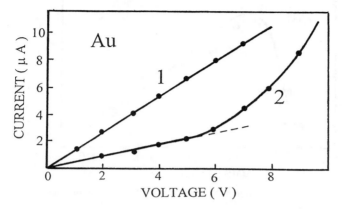

Fig.2. Current -voltage curves of conduction current for Au films with different mass thicknesses: 1 - d=3 nm, 2 - d=6 nm.

Strain sensitivity of Au film strongly depends upon it's mass thickness (Fig. 3), because the island sizes and the distances between them are varying. The strain sensitivity coefficient as a function of average distance between islands is shown in Fig. 4. Both the compressional and the extensional tests gave equivalent results.

The temperature coefficient of resistance for Au films was negative and decreased with increasing film weight thickness. Our experiments showed that the coefficient γ practically does not depend on temperature in the range of 4.2 - 290 K.

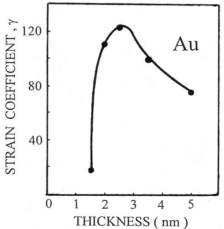

Fig.3. Strain sensitivity coefficient as a function of a mass thickness.

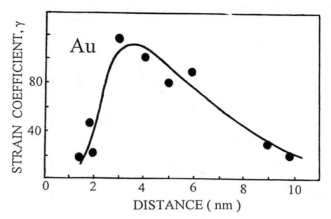

Fig. 4. Strain sensitivity coefficient as a function of the average distance
between islands.

Strain sensitivity of emission current.

As it was mentioned above, electron emission is observed in experiments conducted in vacuum conditions. Measurements, conducted in a special vacuum apparatus, showed high strain sensitivity of the electron emission current. Values γ of 250-300 were obtained for Au films. Emission current as a function of relative deformation is shown in Fig.5.

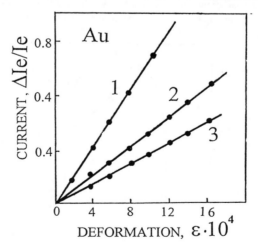

Fig.5. Emission current I_e as a function of relative deformation under different
applied voltages: 1 - U=30 V, 2 - U=25 V, 3 - U=24 V.

CONCLUSIONS

Among the investigated metals, Au films had the highest values of strain sensitivity coefficient (γ=120-130). The most sensitive Au films had structures with average island sizes of 20-30 nm and distances between them about 3-7 nm.

It is possible to protect as-prepared films by covering it with polymeric layers. Up to date results showed 5-10% increasing of resistance during one year storage, and the strain sensitivity coefficient was constant. This type of strain sensors is eligible for use in wide range of deformations and temperatures.

ACKNOWLEDGMENTS

This work was supported by National Academy of Sciences and State Committee for Science and Technologies of Ukraine.

REFERENCES

1. M. Nishiura, S. Yoshida, A. Kinbana, Thin Solid Films, **15**, 133 (1973).
2. J. Morris, Thin Solid Films, **11,** 259 (1972).
3. R. Fedorovich, A. Naumovets, P. Tomchuk, Progr. Surf. Sci., **42**, 189, (1993).

Part V

Shape-Memory Alloys I

EFFECTS OF MAGNETIC FIELD AND HYDROSTATIC PRESSURE ON MARTENSITIC TRANSFORMATIONS IN SOME SHAPE MEMORY ALLOYS

T. Kakeshita*, T. Saburi*, and K. Shimizu**
*Department of Materials Science and Engineering, Faculty of Engineering,
Osaka University, 2–1, Yamada–oka, Suita 565, Japan.
**Department of Materials Science and Engineering, Faculty of Engineering, Kanazawa Institute of Technology, 7–1 Ohgiga–oka, Nonoichi, Ishikawa 921, Japan.

ABSTRACT

The recent works carried by the author's group on the effects of magnetic field and hydrostatic pressure on martensitic transformation are reviewed, which mainly concerned with some shape memory alloys, such as Fe–Pt, Fe–Co–Ni–Ti, Ti–Ni and Cu–Al–Ni alloys. The works clarify the effects of magnetic field and hydrostatic pressure on martensitic transformation temperature, magnetoelastic martensitic transformation and morphology and arrangement of martensites and transformation process of athermal transformation. That is, transformation start temperatures in Fe–Pt and Fe–Ni alloys examined increase with increasing magnetic field, but are not affected in Ti–Ni and Cu–Al–Ni alloys. On the other hand, transformation start temperature decreases with increasing hydrostatic pressure in the Fe–Ni–Co–Ti alloy, but increases in Cu–Al–Ni alloys. The magnetic field and hydrostatic pressure dependences of the martensitic start temperature are in good agreement with those calculated by the equations proposed by our group. In the work on the ausaged Fe–Ni–Co–Ti alloy, the appearance of magnetoelastic martensitic transformation is newly found. In addition, several martensite plates grow nearly parallel to the direction of applied magnetic field in the specimen of an Fe–Ni alloy single crystal. Moreover, we found that in the Cu–Al–Ni alloys exhibiting an athermal martensitic transformation, isothermal holding at a temperature above M_s makes martensitic transformation to start and the incubation time increases with increasing $\Delta T = T - M_s$ (T represents holding temperature). The above results show that the magnetic field and hydrostatic pressure effectively control not only the transformation temperature but also the morphology and distribution of martensites induced, as in the case of uniaxtial stress and compression.

INTRODUCTION

It is well known that martensitic transformation, which occurs in many Fe–, Cu– and Ti–based alloys and ceramics, is one of the most typical examples of the first order structural phase transformations without atom diffusion, and has been widely studied in order to know its characteristics from physical, metallographical and crystallographical points of view. In addition, martensitic transformation has also been studied from technological point of view, partly because fine martensites formed in quenched ferrous alloys and steels result in increase in hardness of the alloys and steels used as structural materials and partly because the shape memory effect and the pseudoelastisity effect have been found to appear in relation to the thermoelastic martensitic transformation and the shape memory alloys with these effects are now supplied to practical uses as functional materials. Thus, the martensitic transformation has recently been more actively studied. According to the studies[1],[2], martensitic transformations are extensively influenced by external fields, such as temperature and uniaxial stress, in transformation temperatures, crystallography and amount and morphology of the product martensites. Therefore, to clarify the effect of external fields on martensitic transformations is very important to understand the essential problems of the transformation, such as thermodynamics, kinetics and the origin of the transformation and is also important to obtain technical information in developing functional and smart materials. Magnetic field and hydrostatic pressure are those of such external fields because there exists some differences in magnetic moment and atomic volume between the parent and martensitic states. Actually, the effects of magnetic field and hydrostatic pressure on martensitic transformations have been studied by many workers, especially in Sadovsky's group in Russia[3], Patel and Cohen[1] and recently in our group[4]–[7]. As a result, we have found many interesting phenomena on them.

In the present paper, some new findings on the effects of magnetic field and hydrostatic pressure on martensitic transformations, especially in some shape memory alloys, such as Fe–Pt, Fe–Co–Ni–Ti, Ti–Ni and Cu–Al–Ni alloys, will be described: (i) the effect of magnetic field on

269

Mat. Res. Soc. Symp. Proc. Vol. 459 ® 1997 Materials Research Society martensitic transformation start temperature, M_s, and the validity of a newly proposed equation by our group to evaluate the relation between M_s and critical magnetic field, H_c, for inducing martensitic transformation. (ii) the effect of magnetic field on magnetoelastic martensitic transformation in an ausaged Fe–Ni–Co–Ti shape memory alloy, which occurs only while a magnetic field is applied and disappears when the magnetic field is removed. (iii) the effect of magnetic field on morphology and arrangement of martensites in Fe–Ni alloy single crystals. (iv) the effect of hydrostatic pressure on martensitic transformation start temperature and the validity of a newly proposed equation by our group to evaluate the relation between M_s and hydrostatic pressure. (v) the morphology of martensite induced by a hydrostatic pressure. (vi) the effects of magnetic field and hydrostatic pressure on the martensitic transformation process.

RESULTS AND DISCUSSION

Effect of Magnetic Field on Martensitic Transformation Temperature

The specimens used were three invar Fe–Ni alloys[4], disordered and ordered Fe–Pt invar alloys[5], non–invar Fe–Ni–C[6] and Fe–Mn–C alloys[7]. Their structural change associated with martensitic transformation are basically those from fcc to bcc. High field magnetization measurements were performed at Research Center for Materials Science at Extreme Conditions, Osaka University, the magnetic field being a pulsed one with its maximum strength of 31MA/m. Details of the ultra high magnetic field instrument have been reported elsewhere[8]. Here we show the typical result of an Fe–31.7at%Ni alloy exhibiting a non–thermoelastic martensitic transformation (M_s is 164K) and that of an ordered Fe–24.0at%Pt alloy exhibiting a thermoelastic martensitic transformation (M_s, M_f, A_s, and A_f are 153, 123, 139 and 177K, respectively, and the degree of order, S, is about 0.8). Figure 1 shows typical magnetization curve ($M(t)$–$H(t)$) for the invar Fe–31.7at%Ni alloy, where ΔT represents the temperature difference between setting temperature, T, and M_s ($\Delta T = T - M_s$). In the figure, an abrupt increase in magnetization is recognized at a certain strength of magnetic field (indicating with an arrow). The strength of magnetic field at the abrupt increase in magnetization corresponds to the critical one, H_c, for inducing the martensitic transformation at T, inversely meaning that the setting temperature, T, corresponds to the martensitic transformation start temperature under the strength of magnetic field of H_c, M_s'. The relation thus obtained between the critical magnetic field and the shift of M_s, ΔM_s, ($=M_s' - M_s$) is shown in Figure 2 (a) with solid squares for the Fe–31.7at%Ni alloy, and is shown in Figure 2 (b) for the Fe–24.0at%Pt alloy with $S \approx 0.8$. It is known from the figures that the shift of M_s increases with increasing magnetic field for both the alloys irrespective of non–thermoelastic and thermoelastic martensitic transformation.

Recently we have proposed the following equation[9] to estimate the relation between the critical magnetic field and the transformation start temperature:

$$\Delta G(M_s) - \Delta G(M_s') = - \Delta M(M_s') \cdot H_c - (1/2) \cdot \chi_h^p \cdot H_c^2 + \varepsilon_0 \cdot (\partial \omega / \partial H) \cdot H_c \cdot B, \qquad (1)$$

where $\Delta G(M_s)$ and $\Delta G(M_s')$ represent the difference in Gibbs chemical free energy between the parent and martensite phases at M_s and M_s' temperatures, respectively, $\Delta M(M_s')$ the difference in spontaneous magnetization between the parent and martensitic states at M_s', χ_h^p the high magnetic field susceptibility in the parent phase, ε_0 the volume change associated with martensitic transformation, ω the forced volume magnetostriction and B the parent bulk modulus. The first, second and third terms on the right-hand side of eq.(1) represent the energies due to the magnetostatic, high field susceptibility and forced volume magnetostriction effects, respectively. Based on the equation, H_c vs. M_s relations have been thermodynamically calculated for the present alloys. In the calculation, the Gibbs chemical free energies for Fe–Ni and Fe–Pt alloys have been obtained by following the equations derived by Kaufman[10] and Tong and Wayman[11], respectively, and spontaneous magnetization in the martensitic states and B for the alloys have been obtained by referring to the previous studies[12],[13]. Other unknown physical quantities involved in the equation were set to be the ones measured in our studies[4]–[7], [14]. The calculated results are shown in Figure 2, where the dotted lines indicated with M.S.E., H.F.E., F.M.E. and (M.S.E.+ H.F.E.+ F.M.E.) mean the H_c vs. M_s' relations calculated for the magnetostatic, high field susceptibility, forced volume magnetostriction and their total effects, respectively. As known from the figure, the calculated relations (M.S.E.+ H.F.E.+ F.M.E.) are in good agreement with the experimental ones for both the alloys. It should be noted that the shift of M_s temperature due to the forced magnetostriction effect is nearly the same order as that due to the magnetostatic effect for both the invar alloys and the shift of M_s due to this effect is a decrease in the ordered Fe–Pt alloy, but an increase in the Fe–Ni alloy. This difference is due to the fact that the volume change associated with martensitic transformation in the ordered Fe–Pt alloy is negative value, but positive in the Fe–Ni alloy.

It can thus be concluded from good agreement between calculated and measured relations that the propriety of the newly derived equation is quantitatively verified. We also applied pulsed high magnetic fields to the Ti–Ni and Cu–Al–Ni shape memory alloys. However, magnetic field–induced martensitic transformations were not recognized in those alloys. The reason for this phenomenon can be explained by eq.(1), that is, the difference in magnetic moment between parent and martensite phases in Ti–Ni and Cu–Al–Ni alloys is so small for inducing martensitic transformation by applying magnetic field (31MA/m) used in the experiment.

Magnetoelastic Martensitic Transformation

In the alloy exhibiting a thermoelastic martensitic transformation, it is known that a martensite crystal grows or shrinks with temperature cycling, that is, it responds to temperature change in a balance between thermal and elastic energies. If a uniaxial stress is applied to such an alloy at temperatures above A_f and released, the alloy exhibits pseudoelastic behavior due to the stress-induced martensitic and its reverse transformation upon loading cycle. Considering this behavior, it can be expected that if a magnetic field is applied to the alloy exhibiting a thermoelastic martensitic transformation above A_f and removed, martensites may be induced only while a magnetic is applied and revert to the parent phase when the magnetic field is removed. We define this type of martensitic transformation as a magnetoelastic martensitic transformation, and actually have found it in an ausaged Fe–31.9Ni–9.8Co–4.1Ti(at%) shape memory alloy[15], as will be described below. Figure 3 shows the spontaneous magnetizations of parent and martensitic states as a function of temperature, which were obtained by magnetization measurements with a low magnetic fields. As known from the figure, the difference in spontaneous magnetization between the two phases is about $0.3\mu_B$/atom at M_s, which is the same order of that in a previous Fe–32.5at%Ni alloy[4]. It is also noted in the figure that M_s and A_f shown with arrows are determined to be about 127 and 159K, respectively and A_s is known to be 60K.

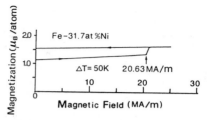

Fig. 1 Magnetization curve of an invar Fe–31.7at%Ni.

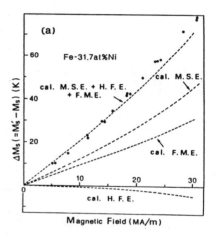

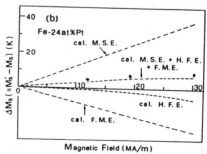

Fig. 2 Calculated and measured shifts of M_s as a function of magnetic field for invar Fe–31.7at%Ni, (a), and invar ordered Fe–24.0at%Pt alloys, (b), where M.S.E., H.F.E. and F.M.E. mean the effects of magnetostatic energy, high field susceptibility and forced volume magnetostriction, respectively.

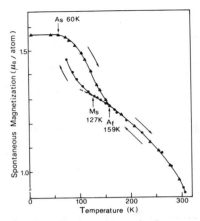

Fig.3 Spontaneous magnetization as a function of temperature in an ausaged Fe–Ni–Co–Ti alloy.

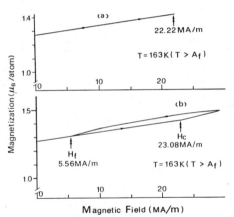

Fig.4 $M(t)$–$H(t)$ curves for an ausaged Fe–Ni–Co–Ti alloy at 163K ($T > A_f$), (a) and (b).

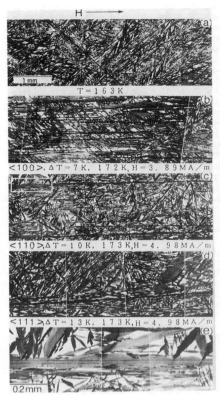

Fig.5 Optical micrographs of thermally–induced, (a), and magnetic field–induced martensites in Fe–31.6at%Ni single crystal specimens with <100>, <110> and <111> orientations, (b), (c) and (d), respectively. (e) is an enlargement of the framed area of (c).

A pulsed high magnetic field was applied to the specimen at a temperature above A_f, 163K (ΔT ($=T-M_s$) = 36K, $T > A_f$) and typical $M(t)$–$H(t)$ curves obtained are shown in Figures 4 (a) and (b). It is noted in (a) that there is no hysteresis of magnetization when a pulsed magnetic field whose maximum strength is 22.22MA/m has been applied and removed. This means that the maximum strength is lower than a critical magnetic field, H_c, to induce martensitic transformation, and therefore that no martensitic transformation occurs under the magnetic field of 22.22MA/m. Then, a higher magnetic field was applied, and the obtained $M(t)$–$H(t)$ curve is shown in (b), which reveals a hysteresis of magnetization. That is, when a magnetic field is applied, the rate of increase of magnetization against magnetic field changes at H_c=23.08MA/m, as indicated with an arrow, and when the magnetic field is removed, the increased magnetization returns to the initial value at about H_f = 5MA/m indicated with another arrow. This means that martensitic transformation is induced at H_c and its reverse transformation is completed at H_f. These observations show that the magnetoelastic martensitic transformation is certainly realized in the ausaged Fe–Ni–Co–Ti alloy, and such behavior is always realized at temperatures above A_f.

In this way, the ausaged Fe–Ni–Co–Ti alloy exhibits a magnetoelastic martensitic transformation as well as shape memory effect, and therefore the alloy may be utilized as a magnetically sensitive device in addition to a thermally sensitive one.

The Effect of Magnetic Field on Morphology and Arrangement of Martensites

The morphology of magnetic field–induced martensites was the same as that of thermally-induced one irrespective of formation temperature and the strength of magnetic field for Fe–Ni[4], Fe–Ni–C[6] and Fe–Mn–C[7] alloys examined. Figure 5 shows the optical micrographs exhibiting the whole view of thermally–induced martensites in Fe–31.6at%Ni alloy[16] single crystals, (a), and magnetic field–induced ones, (b) to (d). Crystal orientation, formation temperature, strength of applied magnetic field and its direction for magnetic field–induced martensites are inscribed in each photograph. Figure (b) and (c) reveal that several martensite plates grow nearly parallel to the direction of applied magnetic field, and some of them run through from one end to the other of the single crystals. However, such a directional growth of martensites plates is not observed in (a). Therefore, the directional growth seems to be characteristic of magnetic field–induced martensites. Figure (e) is an enlargement of the framed area of (c), from which it is clearly known that one plate grows lengthwise along the direction of magnetic field, and that other plates terminate at the directionally grown plates. This means that the directionally grown plates formed first and then the other plates followed. The reason for such a formation of lengthwise grown plates under magnetic field is not clear yet, but a shape magnetic anisotropy effect seems to play an important role.

In this way, by using magnetic field, we can control not only the martensitic transformation temperatures but also the distribution of martensites.

The Effect of Hydrostatic Pressure on Martensitic Transformation Temperature

The specimens used were three invar Fe–Ni alloys[17], an invar Fe–Ni–Co–Ti alloy [18], disordered and ordered Fe–Pt invar alloys[19] and non–invar Fe–Ni–C[20], Cu–Al–Ni[21] and Ti–Ni[18] alloys. Electrical resistivity measurements under a hydrostatic pressure were made in order to examine the hydrostatic pressure dependences of martensitic transformation temperatures in the above alloys. The hydrostatic pressures used in the measurements were generated by a piston cylinder type instrument, in which kerosene and transformer oil in a teflon capsule were

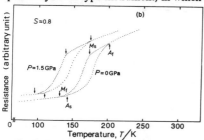

Fig.6 Electorical resistivity vs. temperature relation measured for an Fe–24.0at%Pt alloy with $S \approx 0.8$ under the pressure of 1.5GPa, where that under no pressure is also shown.

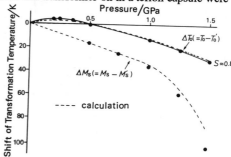

Fig.7 Shifts of equilibrium temperature, ΔT_o of an Fe–24.0at%Pt alloy with $S \approx 0.8$ and of ΔM_s for an Fe –29.9at%Ni alloy, plotted as a function of hydrostatic pressure, and the dotted lines are calculated ones.

used as liquid pressure medium. More details of the pressure instrument have been described el–sewhere[22]. Here, we show the typical results of Fe–Pt, Cu–Al–Ni, Ti–Ni and Fe–Ni–Co–Ti shape memory alloys exhibiting a thermoelastic martensitic transformation and an Fe–Ni alloy exhibiting a non–thermoelastic martensitic transformation for comparison. Figure 6 shows typi–cal electrical resistivity vs. temperature curve of the Fe–24.0at%Pt alloy with $S \sim 0.8$ under the pressure of 1.5GPa, in which another curve under no pressure is also shown for comparison, transformation temperatures being indicated with arrows on each curve. As known from the figure, all the transformation temperatures under pressure shift from those under no pressure. The amount of the shift has been measured by varying the applied pressure. The measured shifts of equilibrium temperature, ΔT_0 ($\Delta T_0 = T_0 - T_0'$, where T_0 is defined as ($M_s + A_f$)/2), for the alloy with $S \sim 0.8$ is shown in Figure 7, as a function of hydrostatic pressure, in which the shift of M_s temperature, ΔM_s ($=M_s - M_s'$), for the previous Fe–29.9at%Ni invar alloy is also shown for comparison[17]. It is noted from the figure that the shift of T_0 for the Fe–24.0Pt alloy with $S \sim 0.8$ increases under the pressures lower than 0.25GPa, but decreases under the pressures higher than 0.25GPa, although the M_s temperature of Fe–Ni alloy only decreases with increasing pres–sure. This means that the volume change associated with the martensitic transformation, ΔV(which is $V^m - V^p$, where V^m and V^p are the volumes of martensite and parent phases, respectively,), for the Fe–Pt alloy with $S \sim 0.8$ changes in sign from negative to positive as pressure increases. This behavior in volume change may be attributed to the invar nature of the alloy, which will be described later. The same measurements were done for Cu–28.8Al–3.8Ni (at%) single crystal alloy ($\beta_1 \rightleftarrows \beta_1'$, M_s, M_f, A_s and A_f under no pressure are 277, 270, 279 and 286K, respectively.), Ti–51.0at%Ni poly crystal alloy (aged at 723K for 3.6ks, B2 $\rightleftarrows$ R $\rightleftarrows$ R19') and the ausaged Fe–31.9Ni–9.8Co–4.1Ti(at%)alloy ($\gamma \rightleftarrows \alpha'$). The results are shown in Figures 8 (a) and (b). It is to be noted that $\beta_1 \rightleftarrows \beta_1'$ equilibrium temperature of the Cu–Al–Ni alloy and R$\rightleftarrows$ B19' transformation temperature of the Ti–Ni alloy increase with increasing pressure although $\gamma \rightleftarrows \alpha'$ transformation temperature decreases. Moreover, B2 $\rightleftarrows$ R transformation temperature of the Ti–Ni alloy is not affected by pressure. These behavior is qualitatively explained by the difference in the sign of ΔV; It is negative for $\beta_1 \rightleftarrows \beta_1'$ transformation of the Cu–Al–Ni and R$\rightleftarrows$ B19' transformation of Ti–Ni alloys and positive for the Fe–Ni–Co–Ti alloy and nil for B2 $\rightleftarrows$ R transformation of the Ti–Ni alloy. These results are the same as those recently studied by Chernenko[23]. A characteristic feature is that the measured hydrostatic pressure dependences of transformation temperature in the non–invar Cu–Al–Ni and Ti–Ni alloys are in good agreement with the relations calculated by using a Patel–Cohen's equation[1], which are shown with dotted lines in Figure 8 (a) and (b),

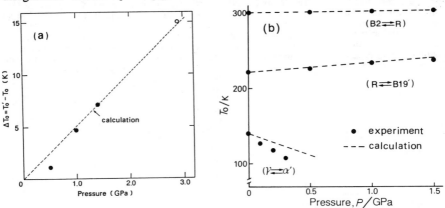

Fig.8 (a) Shifts of equiliblium temperature, $\Delta T_0 (=T_0' - T_0)$, plotted as a function of hydrostatic pressure for a Cu–Al–Ni alloy single crystal, (b) Equiliblium tempera–ture, T_0, of the B2 $\rightleftarrows$ R and the R$\rightleftarrows$ B19' transformations for a short aged Ti–Ni alloy, plotted as a function of hydrostatic pressure, and that of the $\gamma \rightleftarrows \alpha'$ transformations for an ausaged Fe–Ni–Co–Ti alloy. The dashed lines are the calculated ones based on Patel and Cohen's equation.

but those in invar Fe–Ni, Fe–Pt and Fe–Ni–Co–Ti alloys are not (e.g. for the Fe–Ni–Co–Ti alloy, the dotted line in Figure 8 represents the calculated relation, not being in good agreement with the experimental one). These results clearly indicated that the invar effect played an important role in the martensitic transformation of invar alloys under hydrostatic pressures. Then, we speculated[17] that the disagreement for invar alloys is due to the existence of spontaneous volume magnetostriction, ω_s, which brings invar effect and directly influences the volume change associated with martensitic transformations. Whereupon, we derived the following equation[17]including a term due to ω_s to estimate the hydrostatic pressure dependence of transformation start temperatures, by referring to Patel–Cohen's equation[1],

$$\Delta G(M_s, 0) - \Delta G(T, 0) = \int_0^p \{ \frac{1+\omega_s(T, P')}{1+\omega_s(T, 0)} \cdot V_f^p(T, 0) \cdot (1-P'/B^p) - V^m(T, 0) \cdot (1-P'/B^m) \}\, dP', \qquad (2)$$

where $\Delta G(M_s, 0)$ and $\Delta G(M_s', 0)$ represent the differences in Gibbs chemical free energy between the austenite and martensite phases under no pressure at M_s (transformation start temperature under no pressure) and M_s' (transformation start temperature under a pressure, P), respectively, $\omega_s(M_s', P')$ the spontaneous volume magnetostriction at M_s' under P', $V^p(M_s', 0)$ and $V^m(M_s', 0)$ the atomic volumes of austenite and martensite phases at M_s' under no pressure, respectively, and B^p and B^m the bulk moduli of the austenite and martensite phases, respectively. It should be noted from the equation that the atomic volume of austenitic phase is decreased by the factor of $(1+\omega_s(M_s', P'))/(1+\omega_s(M_s', 0))$, because of a decrease in ω_s due to the applied hydrostatic pressure. Based on the equation, hydrostatic pressure dependences of M_s temperatures for the invar Fe–Ni and Fe–Pt alloys have been calculated. In the calculation, temperature and hydrostatic pressure dependences of ω_s is obtained as follow. That is, according to a previous study [24], ω_s of invar alloys is known to be proportional to the square of spontaneous magnetization, $M_0(T, P)$ as $\omega_s(T, P) = A \cdot M_0^2(T, P)$, where A is a constant. Since M_0 can be obtained accurately by magnetization measurement. Thus, the temperature and hydrostatic pressure dependences of ω_s can be calculated from the above equation. The Gibbs chemical free energies for the alloys have been obtained by following the equations derived by Kaufman[10] and Tong and Wayman[11] and B for the alloys have been obtained by referring to the previous studies[12][13]. Other unknown physical quantities involved in the equation were set to be the measured ones in our studies. The calculated relations are shown in Figure 7 with the dotted line, being in good agreement with the experimental ones for all the alloys. It can thus be concluded that the propriety of the newly derived equation is quantitatively verified.

In–situ Observation of Martensitic Transformation under Hydrostatic Pressure

In–situ optical microscopy observations of the transformation behavior under hydrostatic pressure have been done by using a diamond anvil cell type of pressure generator. Figure 9 shows the series of in–situ optical micrographs of martensites formed in the Cu–28.8Al–3.8Ni(at%) alloy single crystal. The applied hydrostatic pressures are indicated in each micrograph and the formation temperature is 292K (the difference beteween M_s and formation temperature is 15K). It is seen in the figure that thin plate like martensites (18R) are induced by applying hydrostatic pressure and disappears when the pressure decreases.

Kinetics of Martensitic Transformation

Martensitic transformations are well known to be classified into two groups, with respect to kinetics, athermal and isothermal ones. The former transformation has a definite transformation start temperature, M_s, and occurs instantaneously at M_s, while the latter one does not have a definite M_s temperature but occurs after some finite incubation time during isothermal holding. Materials exhibiting an isothermal transformation are very few in number, and Fe–Ni–Mn alloys are known to be the typical of such materials. However, there is a view [25] that the isothermal transformation is rather general and the athermal one is its special case, considering that the incubation time needed for the athermal transformation might be undetectable short. Unfortunately, the view has not been clarified yet. Therefore, further investigation is needed for the verification of the above view. We investigated the effects of magnetic field and hydrostatic pressure on the athermal and isothermal martensitic transformations in Fe–Ni–Mn alloys[26][27]. Thus, we found that the isothermal martensitic transformation changes to the

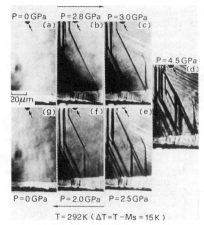

P=0 GPa (a)   P=2.8 GPa (b)   P=3.0 GPa (c)

P=4.5 GPa (d)

20μm

(g)   (f)   (e)

P=0 GPa   P=2.0 GPa   P=2.5 GPa

T = 292 K (ΔT = T − Ms = 15 K)

Fig.9 In–situ optical micrographs showing the appearance and disappearance of hydrostatic pressure–induced martensites,(a)–(g). Formation temperature and pressure are shown for each micrograph.

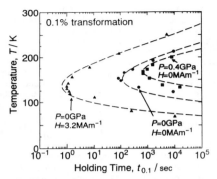

0.1% transformation

P=0.4GPa
H=0MAm^{-1}

P=0GPa
H=3.2MAm^{-1}

P=0GPa
H=0MAm^{-1}

Fig.10 $T\,T\,T$ diagrams of the isothermal martensitic transformation in an Fe–24.9Ni–3.9Mn(mass%) alloy under magnetic field and hydrostatic pressure, and the dotted lines represent the calculated ones with the present theory.

athermal one under a high magnetic field, and that, inversely, the athermal martensitic transformation changes to the isothermal one under a hydrostatic pressure. These results suggest that the two transformation processes are closely related to each other and their difference is not intrinsic but may be explained by one basic rule. Then, we have constructed a phenomenological theory [27], which may give a unified explanation for the two transformation processes, making the following three assumptions: 1) Particles (atom, electron) must acquire a certain critical energy(potential barrier), Δ, before they can change the state from austenite to martensite. 2) The transition probability (Pe) from the austenitic state to martensitic state is proportional to the Boltzmann factor and is expressed as $Pe = P_0 \cdot \exp(-\Delta/k_B T)$, where P_0 and k_B are constants. 3) In case of $\Delta \neq 0$, martensitic transformation does not start even if one particle is excited, but it does when some critical number of particles, n^*, among excited particles, m, make a cluster in the austenite. Based on the assumptions, the probability (P) of the occurrence of martensitic transformation has been derived as,

$$P = \sum_{m(\geq n \geq n^*)}^{N} \sum_{n(\geq n^*)}^{m} f(N, m, n, n^*)(Pe)^m \cdot (1 - Pe)^{N-m},$$ (3)

where N and n^* represent the total number of particles and a minimum number of particles in the cluster which is required to start a martensitic transformation, respectively, and m and n the number of excited particles and $f(N, m, n, n^*)$ the possible number of clusters consisting of n particles within m excited particles. If we assume that the well–known ergodic hypothesis holds in the present analysis, the incubation time at which a martensitic transformation start can be evaluated by the inverse of P, P^{-1}. More details of the theory has been reported elsewhere [27]. Based on the theory, we previously made following some predictions concerned with behavior of athermal and isothermal martensitic transformations [27]: (i) A static magnetic field lowers the nose temperature and increases the incubation time. (ii) A hydrostatic pressure raises the nose temperature and reduces the incubation time. (iii) Also in the materials classified as those which exhibit an athermal transformation, transformation occurs isothermally by holding at a temperature between T_0 and M_s. In fact, we confirmed the above predictions to be appropriate experimentally[28]–[30]. That is, isothermal holding measurements of an Fe–24.9Ni–3.9Mn (mass %)

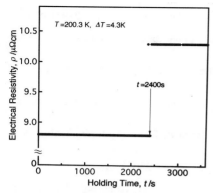

Fig.11 Electrical resistivity as a function of isothermal holding time at $\Delta T(=T-M_s)$ for a single crystal Cu–29.9Al–3.6Ni alloy.

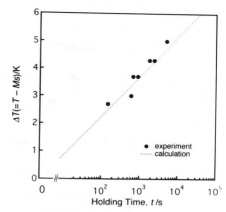

Fig.12 The relation between holding temperature and incubation time required for transformation to start for a single of Cu–29.9Al–3.6Ni alloy. The dotted line is the calculated relation.

alloy exhibiting an isothermal martensitic transformation have been made in order to produce $T\,T\,T$ diagrams under static magnetic fields and hydrostatic pressures[28]. The $T\,T\,T$ diagrams obtained are shown in Figure 10, where the dotted lines represent the calculated $T\,T\,T$ diagrams based on the eq. (3). As known from the figure, the behavior of isothermal martensitic transformation under those external fields is similar to that predicted by the theory mentioned above ((i) and (ii)), suggesting that the theory is confirmed to be appropriated. For the confirmation (iii), we have made isothermal holding measurements above the M_s of Cu–Al–Ni[29] and Fe–Ni[30] alloys exhibiting an athermal martensitic transformation and found that this prediction was realized in those alloys. That is, the electrical resistivity of the alloys has been measured during isothermal holding at a temperature, T, higher than its M_s temperature by ΔT (= $T-M_s$). The typical result of Cu–29.1Al–3.6Ni(at%) alloy single crystal at ΔT = 4.3 K is shown in Figure 11, where an abrupt increase in electrical resistivity due to martensitic transformation with $\beta_1 \overset{\rightarrow}{\leftarrow} \gamma_1{}'$ is recognized after long time holding of 2400sec. We, then, made measurements of the incubation time by varying holding temperature. The obtained results are shown in Figure 12 as a function of ΔT, where the dotted line represents the calculated relation (based on eq. (3))between holding temperature and incubation time required for the martensitic transformation to start. As known from the figure, the incubation time increases with increasing ΔT and is in good agreement with the calculated one. The same behavior was found in Fe–Ni alloys[30]. This means that the prediction of (iii) mentioned above is certainly realized. In addition, the theory can predict that the reverse transformation occurs after some incubation time during isothermal holding at a temperature below A_s but above T_0. We also confirmed this prediction to be appropriate by using the same Cu–29.1Al–3.6Ni alloy[29]. From these results, we conclude that the propriety of the proposed theory is further verified.

REFERENCES

1. J. R. Patel, and M. Cohen, Acta Metall., **1**, p.531–536 (1953).

2. K. Otsuka, H. Sakamoto, and K. Shimizu, Acta Metall.,**24**, p.585–601 (1976).

3. V. D. Sadovsky, L. V. Smirnov, Ye. Fokina, P. A. Malinen, and I. P. Soroskin, Fiz. Met.

Metalloved.,27, p.918–939 (1967).

4. T. Kakeshita, K. Shimizu, S. Funada, and M. Date, Acta Metall., 33, p.1,381–1,389 (1985).

5. T. Kakeshita, K. Shimizu, S. Funada, and M. Date, Trans. Jpn. Inst. Met., 25, p.837–844 (1984).

6. T. Kakeshita, K. Shimizu, S. Kijima, T. Yu, and M.Date, Trans. Jpn. Inst. Met., 26, p.630–637 (1985).

7. T. Kakeshita, H. Shirai, K. Shimizu, K. Sugiyama, K. Hazumi, and M. Date, Trans. Jpn. Inst. Met., 28, p.891–897 (1987).

8. S. Chikazumi, and N. Miura, ed., Physics in High Magnetic Fields (Springer, Berlin, 1981), 44.

9. T. Kakeshita, and K. Shimizu, Proc. ICOMAT-86, Nara, Japan, 230–235.

10. L. Kaufman: referred to Doctor Thesis by M. K. Korenko, MIT, Cambridge, USA,(1973), p.72.

11. H. C. Tong and C. M. Wayman, Acta Met., 22, p.887–893 (1974).

12. J. Grangle, and G. C. Hallame, Proc. Roy. Soc., A272, p.119–132 (1963).

13. G. Oomi, and M. Mori, Physica 119B, p.149–153 (1983).

14. T. Tadaki, K. Katsuki and K. Shimizu, Trans. JIM, 17, p.187–192 (1976).

15. T. Kakeshita, K. Shimizu, T. Maki, I.Tamura, S. Kijima, and M. Date, Scripta Metall., 19, p.973 (1985).

16. T. Kakeshita, S. Furikado, K. Shimizu, K. Kijima and M. Date, Trans. Jpn. Inst. Met., 27, p.477–483 (1986).

17. T. Kakeshita, T. Yamamoto, K. Shimizu, S. Nakamichi, S. Endo, and F. Ono, Materials Transactions, JIM, 36, p.483–489 (1995).

18. T. Kakeshita, K. Shimizu, S. Nakamichi, R. Tanaka, S. Endo and F. Ono, Trans. Jpn. Inst. Met., 32, p.1–6 (1992).

19. T. Kakeshita, K. Shimizu, R. Tanaka, S. Nakamichi, S. Endo, and F. Ono, Trans. Jpn. Inst. Met., 32, p.1,115–1,119 (1991).

20. T. Kakeshita, K. Shimizu, S. Endo, Y. Akahama, and F. E. Fujita, Trans. Jpn. Inst. Met., 30, p.157–164 (1989).

21. T. Kakeshita, Y. Yoshimura, K. Shimizu, S. Endo, Y. Akahama, and F. E. Fujita, Trans. Jpn. Inst. Met., 29, p.781–789 (1988).

22. F. Ono, M. Asano, R. Tanaka, and S. Endo, J.Magnetet.Magn.Mater., 90,91, p.737–739 (1990).

23. V. A. Chernenko, Journal de Physique, 5, C2–p.77 (1995).

24. H. Saito, ed., Physics and Applications of Invar Alloys (Tokyo: Maruzen, 1978), 372.

25. G. V. Kurdjumov, and O. P. Maksimova, Doklady Akademii Nauk SSSR, 61, p.83–93 (1948).

26. T. Kakeshita, K. Kuroiwa, K. Shimizu, T. Ikeda, A. Yamagishi and M. Date, Materials Transactions, JIM,**34**, p.415–422 (1993).

27. T. Kakeshita, K. Kuroiwa, K. Shimizu, T. Ikeda, A. Yamagishi, and M. Date, Materials Transactions, JIM, **34**, p.423–428 (1993).

28. T. Kakeshita, T. Yamamoto, K. Shimizu, K. Sugiyama and S. Endo, Materials Transactions, JIM, **36**, p.1,018–1,022 (1995).

29. T. Kakeshita, T. Takeguchi, T. Fukuda and T. Saburi, Materials Transactions, JIM, **37**, p.299–303 (1996).

30. T. Kakeshita, T. Fukuda and T. Saburi, Scripta Materialia **34**, p.147–150 (1996).

THE INFLUENCE OF HAFNIUM CONTENT, COLD WORK, AND HEAT TREATMENT ON THE R-PHASE TRANSFORMATION OF NITI BASED SHAPE MEMORY ALLOYS

Chen Zhang*, Paul E. Thoma**, Ralph H. Zee*
*ME Dept., MTL Program, Auburn University, AL 36849
**Johnson Controls, Inc., Central Research, Milwaukee, WI 53201-0591

ABSTRACT

The **R**-phase transformation of a Ti-rich NiTi shape memory alloy (SMA) and two ternary SMAs having the compositions $Ni_{49}Ti_{51-x}Hf_X$ with 1at% and 3at% Hf, has been investigated. The influence of cold work (CW) and heat treatment (HT) on the **R**-phase transformation is analyzed thermally using Differential Scanning Calorimetry (DSC). Results show that the **R**-phase transformation depends on the SMA composition as well as the CW and HT conditions in a complex manner. For example, the formation of **R**-phase upon cooling from austenite (**A**) is increasingly suppressed with the substitution of Hf for Ti. For the ternary SMA with 3at% Hf, the **A**→**R** and **R**→**A** transformations are observed only at relatively large amounts of CW (above approximately 40%) and at a high HT temperature (450°C). DSC results also show that for the Ti-rich NiTi and the ternary SMA containing 1at% Hf, the **A**→**R** and **R**→**A** transformation temperatures (TTs) are insensitive to cold work at a HT temperature of 450°C. However, at a lower HT temperature of 350°C, the TTs are found to decrease with increasing CW. For a given CW, the **A**→**R** and **R**→**A** transformations decrease with decreasing HT temperature and the effect is greatest at high CW (>50%) conditions. An effort is made to identify the factors responsible for the observed behavior in the **R**-phase transformation.

INTRODUCTION

The fundamental principle responsible for the shape memory effect in NiTi based alloys is a displasive phase transformation. In binary NiTi SMAs, austenite (**A**) transforms to martensite (**M**) upon cooling with or without a discernible intermediate transformation to R-phase (**R**). There are a number of factors responsible for the presence of a well defined **A**→**R** transformation in NiTi based alloys. Otsuka [1] lists three ways to suppress the **M** transformation temperature (TT) thereby revealing the **A**→**R** transformation on cooling. Cold work (CW) followed by heat treatment (HT) at a temperature between 300°C and 550°C can produce a well defined **A**→**R** transformation [2,3,4,5,6]. **R** phase is also observed in Ni-rich (50.5at% Ni) alloys containing precipitates that are obtained by solution HT followed by aging of the alloys [3]. Miyazaki and Otsuka [7] show that the addition of a third element, such as Fe, suppresses the **M** transformation. The $Ti_{50}Ni_{47}Fe_3$ alloy undergoes a well defined **A**→**R** transformation due to suppression of the **M** transformation. When the **R** phase is heated, a transformation from **R** to **A** occurs. Thoma et al. [8] show that the hysteresis between the **A**→**R** and **R**→**A** transformations for a near equiatomic NiTi SMA is small and relatively constant for different amounts of CW and HT temperatures.

In a previous paper, Zhang et al. [9] discuss the effects of thermal-mechanical processing on the **M** transformation of a Ti-rich binary NiTi SMA and two ternary NiTi based alloys having 1at% and 3at% Hf substituted for Ti. In this investigation, the effects of CW and HT temperature on the **A**→**R** and **R**→**A** transformations of the same Ti-rich binary and ternary alloys are presented and discussed.

EXPERIMENTAL PROCEDURES

A Ti-rich NiTi alloy having an approximate composition of $Ni_{49}Ti_{51}$, and two ternary SMAs having approximate compositions of $Ni_{49}Ti_{50}Hf_1$ and $Ni_{49}Ti_{48}Hf_3$ were investigated. The binary NiTi alloy, with an annealed (700°C HT) austenite finish temperature (A_f) of 116°C, was obtained as a wire from Furukawa Electric Co., Ltd. The Hf containing ternary alloys were made by vacuum arc melting into a cigar shape ingot followed by homogenization at 1000°C for 94.5 hours in vacuum. The ingots were centerless ground into a rod having a uniform diameter. The rods were encased in steel, hot rolled at 900°C and then cold drawn into wires. The cold drawn wires were repeatedly vacuum annealed at 700°C to permit continued cold drawing to obtain different amounts of internal stress (CW), which are listed in Table I. The annealed (700°C HT) Af is 121°C for the $Ni_{49}Ti_{50}Hf_1$ alloy and 132°C for the $Ni_{49}Ti_{48}Hf_3$ alloy. A low speed diamond blade saw was used to prepare differential scanning calorimetry (DSC) specimens from the cold worked wires. The weight of the DSC specimens was 10±1mg. The DSC specimens were heat treated at selected temperatures (see Table I) for 1 hour in an argon atmosphere and furnace cooled in argon to room temperature. TA Instruments Model 2200 DSC test equipment was used to determine the $A \rightarrow R$ and $R \rightarrow A$ TTs of the DSC specimens. The cooling and heating rate during the DSC run was 10°C/min.

Table I. SMA composition, CW and HT temperature used in this investigation

Composition	Af	CW	HT Temperature
Ni49Ti51	116°C	19.6%, 22.7%, 29.5%, 37.6%, 43.6%, 50.4%, 60.3%, 66.8%, 71.5%, 74.1%	300°C 350°C 400°C
Ni49Ti50Hf1	121°C	16.0%, 24.9%, 28.2%, 36.1%, 46.7%, 51.1%, 60.0%	450°C 500°C
Ni49Ti48Hf3	132°C	16.5%, 25.4%, 28.2%, 36.5%, 47.0%, 51.4%	550°C 600°C

RESULTS

Typical DSC curves are shown in figure 1 for the $Ni_{49}Ti_{50}Hf_1$ alloy having 46.7% CW and 450°C HT. The complete cycle ($A \rightarrow R \rightarrow M \rightarrow A$) and partial cycle ($A \rightarrow R \rightarrow A$) with peak TTs are shown. In the complete cycle, A transforms to R on cooling and then to M on further cooling. On heating, M transforms to A. For the partial cycle, the cooling of the SMA is stopped before R transforms to M. Upon heating, R transforms back to A, however, at a temperature lower than the temperature that M transforms to A.

In figures 2 and 3, the influence of CW and HT temperature on the $A \rightarrow R$ and $R \rightarrow A$ transformations for NiTi based alloys with 0, 1 and 3at% Hf is shown. For the same thermal-mechanical processing conditions (CW and HT), the $A \rightarrow R$ and $R \rightarrow A$ TTs decrease with increasing Hf content. For example, for the $Ni_{49}Ti_{51}$ alloy with approximately 51% CW and 450°C HT, the $A \rightarrow R$ and $R \rightarrow A$ TTs are 60°C and 65°C respectively. For the $Ni_{49}Ti_{50}Hf_1$ alloy with the same CW and HT, the $A \rightarrow R$ and $R \rightarrow A$ TTs are 44°C and 50°C respectively. For the $Ni_{49}Ti_{48}Hf_3$ alloy with the same CW and HT, the $A \rightarrow R$ and $R \rightarrow A$ TTs are 14°C and 20°C respectively. Figure 4 shows the relationship between Hf content and the TT of the $A \rightarrow R$ and $R \rightarrow A$ transformations for the SMAs with approximately 51% CW and 450°C HT.

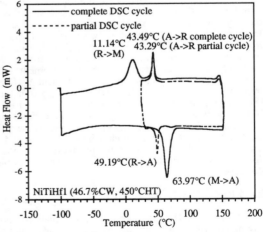

Figure 1. DSC scans for complete and partial cycles of $Ni_{49}Ti_{50}Hf_1$ SMA.

The results shown in figures 2 and 3 also show that for the three alloys investigated, the A→R and R→A transformations are not observed in the SMAs with HT temperatures above 450°C. Also, as the Hf content increases, the A→R and R→A transformations are not observed at lower amounts of CW and lower HT temperatures. For example, the binary NiTi alloy with 0% Hf has an A→R and R→A transformation for all CWs at HT temperatures of 350°C, 400°C, and 450°C. When the HT temperature is 300°C, the A→R and R→A transformations of the above alloy can only be observed for CWs of about 37% and greater. For the $Ni_{49}Ti_{50}Hf_1$ alloy, the A→R and R→A transformations are not observed at a HT temperature below 350°C. For HT temperatures of 450°C, 400°C, and 350°C, the lowest amounts of CW that the A→R and R→A transformations are observed are about 25%, 28%, and 47% CW respectively. For the $Ni_{49}Ti_{48}Hf_3$ alloy, the A→R and R→A transformations can only be observed at a HT temperature of 450°C, and are not observed at CWs below 47%. For the three SMAs, when the A→R and R→A transformations are observed at a specific HT temperature, the R-phase transformation occurs at the maximum amount of CW that can be put into the SMA.

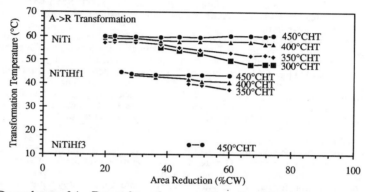

Figure 2. Dependence of A→R transformations on cold work of NiTi based SMAs at different heat treatment temperatures.

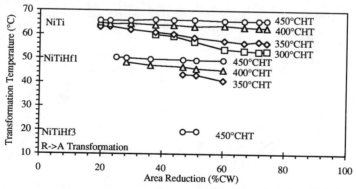

Figure 3. Dependence of R→A transformations on cold work of NiTi based SMAs at different heat treatment temperatures.

Another trend shown in figures 2 and 3 is the influence of HT temperature on the TT at a specific CW. As the HT temperature decreases, the A→R and R→A TTs decrease for a specific CW. At low amounts of CW, HT temperature has only a small influence on the TT. However, at high amounts of CW, HT temperature has progressively larger effects on the TT. This trend was also observed by Thoma et al. [6] for a Ti-rich binary NiTi SMA.

DISCUSSION

AbuJudom et al. [5] and Kao et al. [10] show that NiTi SMAs have transformations (including the A→R transformation) that are influenced by CW and HT temperature. This sensitivity to CW and HT temperature is attributed to the level of internal stress and defect (dislocation) density. These previous investigations [5, 10] show that the A→R TT increases with increasing CW at lower HT temperatures. In a later study by Thoma et al. [6], these results are confirmed for a NiTi SMA having a near equiatomic composition with an annealed Af of 65°C. However, Thoma et al. [6] show that, as the NiTi SMA becomes richer in Ti, the above trend (A→R TT increases with increasing CW) decreases, and when the Ti content reaches the maximum possible level while maintaining single phase NiTi (annealed Af of 116°C), the A→R TT decreases with increasing CW at lower HT temperatures. This investigation confirms that binary NiTi SMA with the maximum possible Ti content has an A→R TT that decreases with increasing CW at low HT temperatures (300°C and 350°C) (shown in figure 2).

Cold work introduces dislocations and causes dislocation entanglement. Heat treatment of a cold worked structure rearranges and removes dislocations (decrease in internal stress). Dislocations and their reorganization are observed in NiTi based SMAs by Miyazaki [11] and Pons et al. [12]. Miyazaki [11] and AbuJudom et al. [5] show that increasing internal stress has the effect of suppressing the transformation to M. Zhang et al. [9] and Zhang [13] show that the transformation to M in binary NiTi and NiTiHf SMAs with 1 and 3at% Hf is conditionally suppressed. For a given CW, the M TT decreases as the HT temperature decreases until a minimum M TT is reached, and then the M TT increases with decreasing HT temperature. For a specific CW, the HT temperature at which the transformation to M begins to increase with decreasing HT temperature becomes higher with increasing Hf content [9, 13]. At lower HT temperatures, for a specific CW, the A→R TT decreases (figure 2) and the M TT increases [9, 13] with decreasing HT temperature. This trend is most significant for the SMA with 3at% Hf. As a result, this opposite movement of the R and M transformations masks the A→R transformation and makes the A→R transformation less detectable. Therefore, for the alloy with 3at% Hf, DSC is able to detect an A→R transformation only when the M TT is at or near a minimum point (shown in figure 5). This trend is also observed in the SMA with 1at% Hf. R is observed at HT temperatures of 350°C, 400°C and 450°C. Zhang et al. [9] and Zhang [13] show that the M TT for the SMA with 1at% Hf and heat treated at 300°C is considerably higher than that heat treated at 350°C. Again, the consequence of the increased M TT for the SMA heat treated at 300°C is a masking of the A→R transformation.

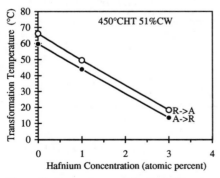

Figure 4. Dependence of A→R and R→A transformations on Hf content of specimens with 51% cold work heat treated at 450°C.

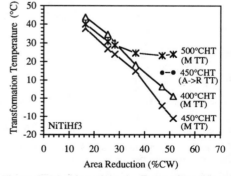

Figure 5. Masking of the A→R transformation by the martensite transformation of $Ni_{49}Ti_{48}Hf_3$.

Another factor contributing to the masking of the **A→R** transformation in the ternary SMAs with Hf is the decrease in the **A→R** TT for a specific CW and HT with increasing Hf content (see figure 4). The decrease of the **A→R** TT with increasing Hf content prevents its detection by DSC until the **M** TT is sufficiently separated from the **A→R** transformation. In the SMA with 3at% Hf, sufficient separation occurs only at a HT temperature of 450°C and high CW (see figures 2 and 5).

The authors speculate that the size difference between the Hf and Ti atoms causes a lattice distortion in the SMAs containing Hf. The interaction of dislocations and other defects in the distorted lattice is probably different from that in the NiTi lattice. This might modify the development of internal stress and the **A→R** TT, which causes the observed results discussed above. It is also conceivable that electronic interactions in ternary NiTiHf alloys between Hf-Ti and Hf-Ni may partially influence the **A→R** transformation characteristics.

At HT temperatures of 500°C and higher, the **A→R** transformation in the binary NiTi SMA and NiTi based SMAs with 1 and 3at% Hf is not observed. The dislocation density and internal stress become lower with increasing HT temperature (onset of recrystallization) [5, 9]. The consequence is an increasing **M** TT which masks the **A→R** transformation.

Thoma et al. [8] show that the hysteresis between the **A→R** and **R→A** peaks of a near equiatomic NiTi SMA is relatively constant (about 6 to 8°C) for different CW and HT temperatures. The results of this investigation show that the hysteresis between the **A→R** and **R→A** DSC peaks for the binary NiTi SMA and the ternary NiTiHf alloys with 1 and 3at% Hf is about 6°C (see figure 4). These results indicate that CW, HT temperature, and Hf content have very little influence on the hysteresis between the **A→R** and **R→A** transformations.

SUMMARY

Results of this investigation show that the **A→R** transformation in NiTi based SMAs is influenced by Hf content, CW and HT temperature. These results can be summarized as follows:

* With increasing Hf content, the **A→R** TT decreases at a specific CW and HT temperature.

* For the Ti-rich binary NiTi SMA and the NiTiHf SMA with 1at% Hf of this investigation, the **A→R** TT decreases with increasing CW at lower HT temperatures, and is not sensitive to CW at higher HT temperatures.

* For the Ti-rich binary NiTi SMA and the ternary NiTiHf SMA with 1at% Hf, at a specific CW, the **A→R** TT decreases with decreasing HT temperature, and this trend is greatest at high CW.

* With increasing Hf content, the **A→R** transformation is increasingly masked by the **M** transformation due to the combined effect from the decreasing **A→R** TT with increasing Hf content, the decreasing **R** TT with decreasing HT temperature at a specific CW, and the increasing **M** TT with decreasing HT temperature at low HT temperature and a specific CW.

* At higher HT temperatures (above 450°C), for all amounts of CW, the **A→R** transformation is not observed due to the merging of the **A→R** and **R→M** transformations.

The results of this investigation also show that the hysteresis between the DSC peaks for the **A→R** and **R→A** transformations of the SMAs studied is about 6°C and is not significantly influenced by Hf content, CW and HT temperature.

ACKNOWLEDGMENTS

The authors are grateful for the technical assistance given by Steven A. Linstead of Johnson Controls, Inc. in preparing the NiTiHf alloys, and for the financial assistance provided by Johnson Controls, Inc. and the National Science Foundation/Alabama EPSCoR Program.

REFERENCES

1. K. Otsuka in <u>Engineering Aspects of Shape Memory Alloys</u>, edited by T.W. Duerig et al., published by Butterworth-Heinemann Ltd., 1990, pp.36-45.

2. V.N. Khachin, V.E. Gjunter, V.P. Sivokha, and A.S. Savvinov in <u>International Conference on Martensitic Transformations (ICOMAT-79)</u> Cambridge, MA, USA, June 24-29, 1979, pp.474-479.

3. S. Miyazaki, Y. Ohmi, K. Otsuka, and Y. Suzuki in <u>International Conference on Martensitic Transformations (ICOMAT-82)</u>, Leuven, Belgium, August 8-12, 1982, L. Delaey and M. Chandrasekaran Eds. (Journal de Physique, Les Ulis Cedex, France, Colloque C4, Supplement au n° 12, Tome 43, Decembre 1982), pp.C4-255-260.

4. S. Miyazaki and K. Otsuka, Metallurgical Transactions A, 17A (January 1986) pp.53-63.

5. D.N. AbuJudom, P.E. Thoma, and S. Fariabi in <u>6th International Conference on Martensitic Transformations (ICOMAT-89)</u>, Sydney, Australia, July 3-7, 1989, B.C. Muddle Ed. (Materials Science Forum, Volumes 56-58, 1990, Part II - Trans Tech Publications, Ltd., Zurich, Switzerland), pp.565-570.

6. P.E. Thoma, D.R. Angst, and K.D. Schachner in <u>International Conference on Martensitic Transformations (ICOMAT-95)</u>, Lausanne, Switzerland, August 20-25, 1995, R. Gotthardt and J. Van Humbeeck Eds. (Journal de Physique IV, Les Ulis Cedex A, France, Colloque C8, Supplement au Journal de Physique III, N°12, Volume 5, Decembre 1995), pp.C8-557-562.

7. S. Miyazaki and K. Otsuka, Philosophical Magazine A, 50, No.3 (1984), pp.393-408.

8. P.E. Thoma, M. Kao, S. Fariabi, and D.N. AbuJudom in <u>International Conference on Martensitic Transformations (ICOMAT-92)</u>, Monterey, CA, USA, July 20-24, 1992, C.M. Wayman and J. Perkins Eds. (Monterey Institute of Advanced Studies, Carmel, CA, USA, 1993) pp.917-922.

9. C. Zhang, R. Zee, and P.E. Thoma in <u>International Conference on Displasive Phase Transformations and their Applications in Materials Engineering</u>, Urbana, IL, USA, May 8-9, 1996. To be published in the proceedings of this conference.

10. M. Kao, S. Fariabi, P.E. Thoma, H. Ozkan, and L. Cartz in <u>Shape-Memory Materials and Phenomena - Fundamental Aspects and Applications</u>, Materials Research Society Meeting, Boston, MA, USA, December 3-5, 1991, C.T. Liu, H. Kunsmann, K. Otsuka, and M. Wuttig Eds. (MRS Vol. 246, 1992), pp.225-233.

11. S. Miyazaki in <u>Engineering Aspects of Shape Memory Alloys</u>, edited by T.W. Duerig et al., published by Butterworth-Heinemann Ltd., 1990, pp.394-413.

12. J. Pons, L. Jordan, J.P. Morniroli, and R. Portier in <u>IIIrd European Symposium on Martensitic Transformations (ESOMAT'94)</u>, Barcelona, Spain, September 14-16, 1994, A. Planes, J. Ortin, and L. Manosa Eds. (Journal de Physique IV, Les Ulis Cedex A, France, Colloque C2, Supplement au Journal de Physique III n° 2, Volume 5, Fevrier 1995), pp.C2-293-298.

13. C. Zhang, <u>Influence of Thermal-Mechanical Processing on NiTi Based Shape Memory Alloys</u>, Master of Science Thesis, Auburn University, Auburn, AL, USA, June 10, 1996.

CHANGE OF M_s TEMPERATURES AND ITS CORRELATION TO ATOMIC CONFIGURATIONS OF OFFSTOICHIOMETRIC NiTi-Cr AND NiTi-Co ALLOYS

Hideki Hosoda*, Toshihiko Fukui**, Kanryu Inoue***, Yoshinao Mishima and Tomoo Suzuki****

Precision and Intelligence Laboratory, Tokyo Institute of Technology, 4249 Nagatsuta, Midori-ku, Yokohama 226, Japan,
* Now with Institute for Materials, Tohoku University, 2-1-1 Katahira, Aoba-ku, Sendai 980-77, Japan.
** Now with NKK Cooperation, Minamiwatarida, Kawasaki-ku, Kawasaki 210, Japan,
*** Department of Materials Science and Engineering, University of Washington, Seattle, WA 98195-2120, USA,
**** Professor Emeritus, Tokyo Institute of Technology, O-okayama, Meguro-ku, Tokyo 152, Japan.

ABSTRACT

Effects of compositional deviation from the stoichiometry and Cr and Co additions on martensitic-transformation-start and austenite-start temperatures (M_s and A_s) of offstoichiometric NiTi alloys are investigated. M_s and A_s are determined using conventional differential thermal analysis (DTA), where the temperature range investigated is between 77K and 423K. Alloys are widely chosen with both Ni- and Ti-rich compositions, to which ternary elements, Cr and Co, are added. It is clearly shown that M_s and A_s in single phase regions are reduced with increasing amount of constituent element, Ni, and ternary elements, Cr and Co. On the other hand, M_s and A_s do not depend on Ti concentration when Ti concentration is more than 52mol.%. NiTi alloys are in two phase region in the case. M_s changes by Co at offstoichiometry are evaluated to be -15K / mol.% in Ni poor side and -30K / mol.% in Ni rich side. These values correspond to -22K / mol.% for the stoichiometric NiTi alloys. Also, effects of Cr on M_s are evaluated to be -65K / mol.% in Ni poor side and -45K / mol.% in Ni rich side. The former is similar to the M_s change in stoichiometric alloys, and the latter is close to our prediction of -30K / mol.% in comparison with the reported value of -120K / mol.% for stoichiometric alloys. It is concluded for offstoichiometric NiTi alloys that effects of ternary additions on M_s can be explained using electronic structures of ternary elements by taking atomic configurations into account, as well as the stoichiometric NiTi alloys. Effect of degree of order is also discussed.

INTRODUCTION

For designing smart materials composed of shape memory alloys (SMA), it is important to control martensitic transformation temperatures and temperature ranges for superelasticity of SMA. Figure 1 shows a schematic deformation map for SMA describing phase stability related to stress and temperature [1]. This map is a modification of the map reported by Otsuka and Wayman [2]. Here, M_s and M_f are martensitic transformation start and finish temperatures, A_s and A_f are austenite transformation start and finish temperatures, and M_d is a maximum temperature below which martensitic transformation can occur by either stress or temperature. Using the map it is easily understood for designing SMA that both transformation temperatures (M_s, M_f, A_s, A_f and M_d) and critical stress for plastic deformation must be controlled.

Recently we have conducted alloy design of intermetallics through controlling atomic configurations, i. e. defect structures at offstoichiometry and site occupation of ternary elements [3-5]. The defect structures and substitution of ternary elements strongly influence mechanical properties as is well known in B2 type NiAl [6]. Also, we have reported a method of controlling M_s using a theoretical approach: M_s of ternary NiTi alloys depends both on occupation sites and electronic structures of ternary elements [1, 7]. The relationships between M_s and electron hole number (N_v) are shown in Figure 2, where the relationships are obtained using experimental data of NiTi alloys containing either 50mol%Ni or 50mol.%Ti. In Figure 2, Ni- and Ti-site substitutions means that ternary elements are expected to occupy Ni and Ti sites, respectively. Such relationships are useful for designing phase transformation temperatures. Control of transformation temperatures (M_s, M_f, A_s, A_f and M_d) and critical stress for plastic deformation can be certainly achieved through alloy designing methods using both offstoichiometry and ternary additions effectively. However, few works have been done about influence of ternary elements on M_s at offstoichiometry. Also, a question remains that why M_s change by Cr in Ni poor side (-120K / mol.%) is quite different from an expected value (about -40K / mol.%) being judged from the other ternary NiTi-X alloys, as seen in Figure 2.

The purpose of the study is to investigate M_s and A_s at offstoichiometric NiTi alloys containing ternary elements, with emphasis on the effect of Cr on M_s. Alloys were widely chosen with both Ni- and Ti-rich offstoichiometry, to which ternary elements, Cr and Co, were added. The data obtained will be discussed in

Mat. Res. Soc. Symp. Proc. Vol. 459 © 1997 Materials Research Society

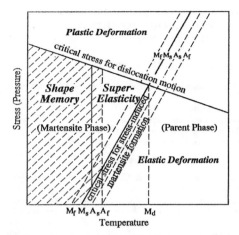

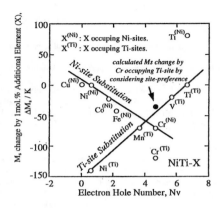

Figure 1 A schematic deformation map for a shape memory alloy (SMA) showing phase stability, transformation temperatures and critical stress for plastic deformation. M_s: martensite start temperature, M_f: martensite finish temperature, A_s: austenite start temperature, A_f: austenite finish temperature, and M_d: maximum temperature below which martensite transformation is induced by stress.

Figure 2 Plots of M_s changes by 1 mol.% additional elements with respect to electron hole number in 3d-orbital (Nv) of ternary elements (X).

connection to the relationships in Figure 2.

EXPERIMENTAL PROCEDURE

Hereafter, compositional regions in ternary NiTi alloys are classified into three as follows: (1) Ni rich side when NiTi alloys contain less than 50 mol.%Ti, (2) stoichiometry when NiTi alloys contain either 50mol.% Ni or Ti, and (3) Ni poor side when NiTi alloys contain more than 50mol.%Ti. NiTi alloys were prepared by arc-melting in an Ar atmosphere using 99.9%Ni, 99.7% Ti, 99.9%Co and 99.99%Cr. The alloy compositions were based on Ni-49mol.%Ti (Ni rich side), Ni-50.5mol.%Ti (near stoichiometry), Ni-52mol.%Ti and Ni-53.5mol.%Ti (Ni poor side). The compositions investigated are indicated by circles in Table 1. Nominal compositions are accepted because of small weight change (less than 0.1%) after melting. The ingots were homogenized at 1223K for 1.8 Ks, followed by hot rolling at 1223K where final reductions were approximately 50% in thickness. Solution treatment was carried out at 1223K for 7.2 ks in a vacuum environment and the ingots were quenched into water. Rectangular specimens of $3x3x5$ mm^3 were cut from the ingots. Differential thermal analysis (DTA) were carried out between 77K and room temperature (R.T.) in a vacuum environment and between R.T. and 423K in air. Heating and cooling rates were chosen to be $3.3x10^{-2}$ K/s below R.T. and $8.3x10^{-2}$ K/s above R.T.

Table 1 Chemical compositions of NiTi alloys, where alloys investigated are shown by circles.

mol.%	binary	0.5Cr	1.0Cr	1.5Cr	0.5Co	1.0Co	1.5Co	3.0Co	4.5Co	6.0Co
49.0Ti	○	○	—	-	○	○	○	—	—	—
50.0Ti	○	—.	—	—	—	—	—	—	—	—
50.5Ti	○	○	○	○	○	○	○	○	○	○
52.0Ti	○	○	○	○	○	○	○	○	○	○
53.5Ti	○	○	○	○	○	○	○	○	○	○

RESULTS AND DISCUSSION

Figure 3 shows some examples of DTA cooling curves of NiTi-Co and NiTi-Cr alloys containing 50.5mol.%Ti (near stoichiometry). Transformation temperatures are defined as intersecting points of two tangent lines as shown in the figure. Some DTA cooling curves are composed of dominant peaks and additional peaks when NiTi alloys contain relatively large amount of ternary elements. As well as NiTi-Fe alloys, the phenomena intend multiphase transformations from parent phases to intermediate phases (incommensurate and/or R phase) and from intermediate phases to martensite phases [8, 9]. However, it was difficult to analyze such DTA curves which separate into two peaks. Here, M_s and A_s are denoted to be the first transformation temperatures during cooling and heating, respectively. It should be mentioned that some alloys did not show so clear DTA peaks during heating that A_s could not be measured. For evaluating the effect of ternary additions on M_s, we use M_s change by additions of 1mol.% of ternary elements (unit: K/mol.%) in this study.

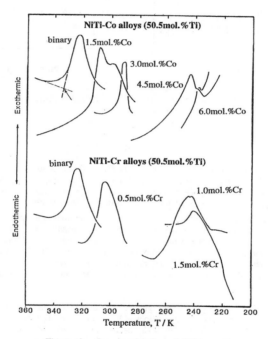

Figure 3   Some examples of DTA cooling curves of NiTi-Co and NiTi-Cr alloys containing 50.5 mol.% Ti (near stoichiometry).

NiTi-Cr alloys

Effect of deviation from stoichiometry   Figures 4 (a) and (b) show Cr concentration dependence of M_s and A_s for NiTi-Cr alloys. It is clearly seen that both M_s and A_s are reduced linearly with increasing Cr concentration, and the composition dependence of A_s and M_s are quite similar each other. It is also seen that (1) M_s and A_s are reduced with increasing Ni content in ternary NiTi alloys as well as binary alloys, and (2) M_s does not depend on Ti concentration when Ti concentration is more than 52 mol.%. Ti concentration of NiTi phase becomes constant even though Ti concentration is more than 51mol.%, because of limited solid solubility [10]. Then, M_s also becomes constant in two phase region regardless of Ti concentration in the case.

289

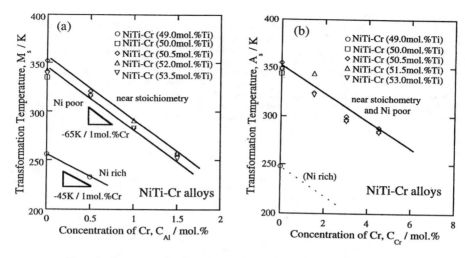

Figure 4   Cr concentration dependence of martensite and austenite transformation temperature for NiTi-Co alloys: (a) M_s and (b) A_s.

M_s change by Cr additions in Ni rich side   As seen in Figure 4, composition dependence of M_s shows different manner in between Ni rich side, where Cr is expected to occupy Ti sites, and Ni poor side including near stoichiometry, where Cr is expected to occupy Ni sites. The M_s change by Cr additions is -65K / mol.% in Ni poor side and -45K / mol.% in Ni rich side in offstoichiometric alloys. Similar composition dependence of M_s is found for stoichiometric NiTi-Cr alloys.   Herein M_s change by Cr additions is experimentally evaluated to be -70K / mol.% in Ti poor side and -120K / mol.% in Ni poor side, respectively [7, 11]. The former (-70K / mol.%) is in good agreement with the present value (-65K / mol.%) in Ni rich side.   Nakata and co-workers have reported that Cr atoms occupy Ni sites in NiTi-Cr alloys containing 50mol.%Ti [12]. Then, all Cr atoms are believed to occupy Ni sites in Ni rich side as well as in the stoichiometric NiTi-Cr alloys, because M_s changes are quite similar in both cases. The substitution behavior of Cr in offstoichiometric NiTi alloys is also in good agreement with our prediction [13].

M_s change by Cr additions in Ni poor side   In Figure 2, M_s change by Cr additions in Ni poor side (-120K / mol.%) is quite different from an expected value (about -40K / mol.%) being judged from the other ternary NiTi-X alloys.   We have explained such anomalous Cr effect in this case by taking atomic configurations into account as follows: (1) Cr atoms equality occupy both Ni and Ti sites in stoichiometric NiTi alloys [13], hence antistructure Ni atoms at Ti site should be induced, (2) the antistructure Ni atoms at Ti sites influence M_s significantly (-140K / mol.%), (3) therefore, the "apparent" M_s change is larger than the "actual" M_s change by Cr only [7]. M_s change by Cr addition only occupying Ti sites is predicted to be about -30 K / mol.%. Then, M_s change by Cr additions should be measured to be near -30 K / mol.% if all Cr atoms occupy Ti sites only.

   M_s change by Cr additions in Ni rich side is found to be -45K / mol.% in this study.   The value is   much smaller than previous experimental value (-120K / mol.%) for stoichiometric NiTi-Cr alloys.   However, the present value is almost comparable but slightly larger compared to the predicted value (-30K / mol.%). Therefore, it is concluded that most Cr occupies Ti sites in Ni rich side, and small amount of both antistructure Ni and Cr occupying Ni sites exist.   The antistructure Ni and Cr at Ni sites reduce M_s values slightly lower than the prediction.   The probability of Cr occupying Ti sites at Ni poor side is smaller than that of stoichiometric compositions, because of smaller deviation from the predicted value than in stoichiometry .

NiTi-Co alloys

Effect of deviation from stoichiometry   Figures 5 (a) and (b) show Co concentration dependence of M_s and A_s for NiTi-Co alloys. The composition dependence of M_s and A_s are quite similar each other. Both M_s and A_s are reduced linearly with increasing Ni concentration when Ti concentration is less than 50.5mol.%. The M_s and A_s do not depend on Ti concentration when Ti concentration is more than 52 mol.%. Then, Ti concentration becomes constant in NiTi-Co alloys containing more than 52mol.%Ti. Phase boundary between single phase (NiTi) and two phase (NiTi and NiTi$_3$) regions is along near 51mol.% Ti, judging from composition dependence of M_s. These features are quite similar to those of NiTi-Cr alloys.

M_s changes by Co additions M_s changes by Co additions are obtained to be -30K / mol.% in Ni rich side and -15K / mol.% in Ni poor side. Both values are similar to the evaluated M_s change of -22K / mol.% for stoichiometric NiTi-Co alloys in which Co is expected to occupy Ni sites. And, both the values are quite different from a predicted value of -120 K / mol.%, where Co occupy Ti sites. Therefore, it is concluded that Co in offstoichiometric NiTi-Co alloys preferably occupies Ni sites regardless of alloy compositions as well as stoichiometric alloys.

There seem to be small deviations ($\pm$7K / mol.) in M_s changes for offstoichiometric NiTi alloys in careful observation, compared to -22K / mol.% for stoichiometric NiTi alloys. The substitution behavior of Co in NiTi alloys has been reported to occupy Ni sites only by both theoretical and experimental works [12, 13]. Then, the deviations cannot be explained by considering atomic configurations only, because M_s change by Co additions is expected to be the same value independent of Ni concentration. Morphology of precipitates such as Ni$_3$Ti may arise internal stress fields in the matrix, which may be effected depending on Ni concentration. Also, the slight deviations in M_s change may be partially explained in terms of driving force (activation energy) to induce the martensite phase. Mobility of dislocations is expected to decrease with increasing Ni concentrations, because NiTi alloys becomes harder with increasing Ni concentration [14, 15]. Therefore, the driving force for martensite formation may become larger by hardening in Ni rich side. If this is the case, M_s change becomes smaller at Ni poor side and larger at Ni rich side than expected. More investigations are necessary to reveal the mechanism of M_s change for offstoichiometric NiTi-Co alloys.

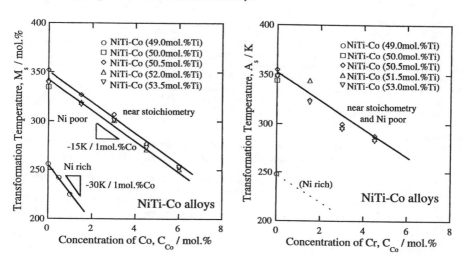

Figure 5   Co concentration dependence of martensite and austenite transformation temperature for NiTi-Co alloys: (a) M_s and (b) A_s.

Effect of Heat Treatment on M_s

The M_s of NiTi is strongly influenced by heat treatment. This is generally explained in terms of precipitation. B2-type binary NiTi is stable above 903K [10]. Then, precipitation occurs during low temperature aging. As the results, (1) M_s is reduced by internal stress fields by forming a number of fine precipitates, and (2) M_s is raised with decreasing Ni concentration of matrix caused by growth of the precipitates. These two factors would complicatedly compete each other.

On the other hand, M_s has been shown to be influenced by high temperature heat treatment above 900K [16]. This cannot be explained in terms of precipitation only. Recently, we have proposed an equation expressing M_s change by taking atomic configuration of B2 phase into account as follows [7],

$$M_{s\ measured} = M_s(0) + \Delta M_s(1) + \Delta M_s(2) + \Delta M_s(3) + \Delta M_s(4) \tag{1},$$

where $M_{s\ measured}$ is a measured M_s, $M_s(0)$ is M_s for the stoichiometric NiTi, $\Delta M_s(1)$ is M_s change by antistructure Ni located at in Ti sites, $\Delta M_s(2)$ is M_s change by antistructure Ti located at Ni sites, $\Delta M_s(3)$ is M_s change by a ternary element located at Ni sites, and $\Delta M_s(4)$ is M_s change by the ternary element located at Ti sites. The M_s change by the ternary element of NiTi alloys are adequately explained by eq.(1) when deviation from the stoichiometry and concentration of ternary elements are small. If the eq.(1) is accepted, M_s should be influenced by the degree of order. The degree of order at an equilibrium state is usually high (close to 1) at low temperature, and low (lower than 1) at high temperature, due to configurational entropy. Therefore, M_s becomes low by a high temperature heat treatment, in comparison with a low temperature heat treatment. This agrees with experimental M_s change found in the literature [16].

CONCLUSIONS

Effects of Cr and Co additions on M_s and A_s for offstoichiometric NiTi alloys are investigated. M_s and A_s obtained are discussed in terms of atomic configurations. The conclusions are as follows.

(1) M_s and A_s are reduced with increasing amount of a constituent element, Ni, and ternary elements, Cr and Co, in single phase regions.
(2) M_s and A_s do not depend on Ti concentration when Ti concentration is more than 52mol.%. NiTi alloys are in two phase region in this case.
(3) M_s and A_s are reduced linearly with increasing Cr and Co concentrations.
(4) M_s changes by Cr addition is evaluated to be -45K / mol.% in Ni rich side. The value is close to our prediction of -30K / mol.% where Cr is expected to occupy Ti sites only.
(5) M_s changes by Cr addition is evaluated to be -65K / mol.% in Ni poor side. The value is in good agreement with -70K / mol.% where Cr is expected to occupy Ti sites only.
(6) M_s changes by Co addition at offstoichiometry are evaluated to be -15 K / mol.% in Ni poor side and -30 K / mol.% in Ni rich side. These are close to the reported values of -22 K / mol.% for stoichiometric NiTi-Co alloys, where Co occupies Ni sites only.

REFERENCES

1. H. Hosoda, K. Inoue, K. Enami and A. Kamio, J. Intelligent Material Systems and Structures, **7**, 312 (1996).
2. K. Otsuka and C. M. Wayman, in Reviews on Deformation Behavior of Metals, ed. by P. Feltham, **2**, 81 (1977).
3. H. Hosoda, Y. Mishima and T. Suzuki, in Intermetallic Compounds, ed. by O. Izumi, (Proc. Sixth JIM Intl. Symp. (JIMIS-6), Japan Institute of Metals, Sendai, Japan, 1991) p.81.
4. H. Hosoda, K. Inoue and Y. Mishima, in High Temperature Ordered Intermetallic Alloys VI, ed. by J. A. Horton, I. Baker, S. Hanada, R. D. Noebe and D. S. Schwartz, (Mater. Res. Soc. Proc. **364**, Pittsburgh, PA, 1995) p.437.
5. H. Hosoda, K. Inoue and Y. Mishima, Gamma Titanium Aluminides, ed. by Y-W. Kim, R. Wagner and M. Yamaguchi (Proc. Intl. Symp. Gamma Titanium Aluminides (ISGTA' 95), TMS, Warrendale, PA, 1995) p.361.
6. R. T. Pascoe and C. W. A Newey, Met. Sci. J., **2**, 138 (1968).
7. H. Hosoda, K. Mizuuchi and K. Inoue, in Proc. Intl. Symp. Microsystems, Intelligent Materials and Robots, ed. by J. Tani and M. Sashi, (7th Sendai Intl. Symp., Sendai, Japan, 1995) p. 231.
8. M. Nishida and T. Honma, J. de Phys., **43**, C4-22 (1982).
9. C. M. Hwang, M. Meichle, M. B. Salmon and C. M. Wayman, Phil. Mag., **47**, 9 (1983); **47**, 117 (1983).

10. J. L. Murray, in <u>Binary Alloy Phase Diagram</u>, ed. by T. B. Massalski, J. L. Murray, L. H. Benett and H. Baker, (**2**, ASM, Metals Park, OH, 1986) p.1763.

11. T. Honma, M. Matsumoto, Y. Shugo, M. Nishida and I. Yamazaki, in <u>TITANIUM'80 Science and Technology</u>, ed. by T. Kimura and O. Izumi, (**2**, Proc. 4th Intl. Conf. Ti, TMS-ASM, Warrendale, PA, 1980) p.1455.

12. H. Hosoda, A. Kamio, T. Suzuki and Y. Mishima, J. Japan Inst. Metals, **60**, 793 (1996) .

13. Y. Nakata, T. Tadaki and K. Shimizu, Mat. Trans. JIM., **33**, 1120 (1991).

14. W. J. Buehler and R. C. Wiley, Trans. ASM, **55**, 269 (1962) .

15. T. Suzuki, J. Japan Inst. Metals, **34**, 337 (1970).

16. T. Saburi, T. Tatsumi and S. Nenno, J. de Phys., **43**, c4-261 (1982).

THE MARTENSITIC TRANSFORMATION AND EXTRA REFLECTIONS APPEARING PRIOR TO THE TRANSFORMATION IN AuCd ALLOY

T. OHBA* and K. OTSUKA**
*Department of Materials Science and Engineering, Teikyo University, Toyosatodai, Utsunomiya 320, Japan, ohba@koala.mse.teikyo-u.ac.jp
**Institute of Materials Science, University of Tsukuba, Tsukuba 305, Japan

ABSTRACT

AuCd alloy is a typical martensitic alloy. The crystal structural of the ζ_2'(trigonal) martensite was recently determined and the analysis indicated the possibility of phonon softening. Despite a large absorption coefficient of Cd phonon softening was actually observed by using an isotope ^{114}Cd. In the present study, precise x-ray scattering studies were performed with a rotating anode x-ray source and synchrotron radiation at Photon Factory in KEK, using small crystals. The structure factors of the parent phase and the diffraction profiles along $<\bar{1}10>^*$ direction were measured with four circle diffractometers. The structure factors decrease when approaching the transformation temperature. Superstructure reflections were observed prior to the onset of the transformation. They are very weak and rather sharp and appear at $q = (1/3, 1/3, 0)$ within the experimental error, i. e. at the commensurate position.

INTRODUCTION

Shape memory alloys are one of the intelligent materials. They have great potential for industrial and medical applications. The shape memory effect is closely related to the martensitic transformation which was known and used for a long time such as steel hardening. The martensitic transformation occurring in the thermoelastic alloys shows the shape memory effect.

The surface relief, shape strain and the existence of the habit plane attracted many researchers. Those characteristic features led to the phenomenological theory[1-3] of martensitic transformation. The phenomenological theory has been applied to several alloys and described well the morphology of the martensite.

The microscopic understanding of the martensitic transformation is also important. The martensitic transformation is a typical first order transformation and is found in many materials these days. The transformation mechanism from the microscopic point of view becomes more important in these situations. Diffraction is the most powerful technique for the microscopic approach. Many efforts using electron microscopy, x-ray and neutron were performed and most of them were concentrated on finding the precursor phenomenon of the martensitic transformation. From a structural consideration, the phonon softening studies of CuAlNi[4], AuCuZn$_2$[5-6] and InTl[7] were performed. However no clear softening was observed in these alloys. Clear phonon softening was reported in the TiNi alloy.[8-10] Diffraction works on TiNi alloys were summarized well in the paper of Shapiro et al.[11] They observed the 'pre-martensitic state' of Ti-Ni alloy and reported that the extra peaks appear at the incommensurate positions and do not have the periodicity of the Brillioun zone. Yamada et al.[12-13] and Gooding and Krumhansl[14] proposed the transformation mechanism following their experimental results. Recently the R-phase was confirmed to be a phase which appears prior to the martensitic transformation by proper heat treatment or by adding ternary elements[15-17].

295

In the past, however, it was not clear that the experiments were carried out on the R-phase or the 'pre-martensitic state'. Furthermore, since the crystal structure of the R-phase was not known, the relationship between the R-phase and the parent phase was not clear.

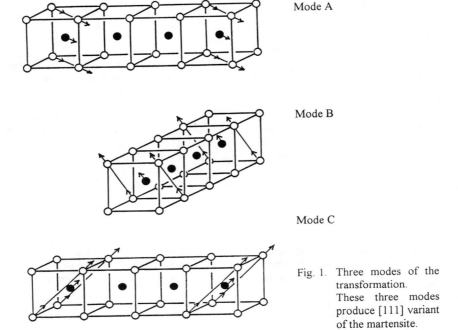

Mode A

Mode B

Mode C

Fig. 1. Three modes of the transformation. These three modes produce [111] variant of the martensite.

AuCd alloy is another typical martensitic alloy. There are two martensite structures depending upon the composition, near the equi-atomic composition. One is the γ_2' phase whose structure is orthorhombic and the other is the ζ_2' phase whose structure is trigonal. The space group of the γ_2' is Pcmm with four atoms in the unit cell. It was determined firstly by Ölander[18] in 1932 and determined precisely by present authors.[19] The ζ_2' phase was known since 1950s, [20] but the crystal structure was not solved for a long time. The crystal structure was first determined by using single crystal x-ray diffraction techniques.[21] The space group is P3 and 18 atoms are present in the unit cell. The c and a axes of the ζ_2' phase correspond to <111> and <211> of the parent phase, respectively. The transformation mechanism was derived from the result of the structure determination. The transformation was described with the superposition of three $< \bar{1}10 >< 110 >$ transverse displacement waves, mode A, B and C, along with their higher harmonics as shown in Fig. 1. They are described as follows:

$$\sum A_n[110]\cos\{2\pi n\frac{(-x+y)}{3}+\delta\}$$

$$\sum A_n[101]\cos\{2\pi n\frac{(-x+z)}{3}+\delta\} \qquad (1)$$

$$\sum A_n[011]\cos(\{2\pi n\frac{(-y+z)}{3}+\delta\}$$

where δ is the phase factor determined uniquely by comparing the structure of martensite with the parent phase.

The phonon softening was expected at 1/3 of <110>* from the above transformation mechanism. The neutron inelastic scattering experiment is the best method to measure the phonon behavior. However the absorption coefficients of Au and Cd are quite large and it is almost impossible to carry out the experiment. Ohba et al. used the isotope [114]Cd to overcome the difficulty of the neutron inelastic experiment and succeeded in the observation of the phonon softening. The phonon softening occurs at the expected position of the reciprocal space[22] shown in Fig. 2.

Elastic scattering study of AuCd was also done by Noda et al. They[23] observed satellite reflections prior to the onset of the transformation and reported a similar behavior which was observed in the 'pre-martensitic state' of TiNi alloys.

In this paper, we study the behavior of the parent phase prior to the transformation. Precise x-ray measurements utilizing synchrotron radiation are reported.

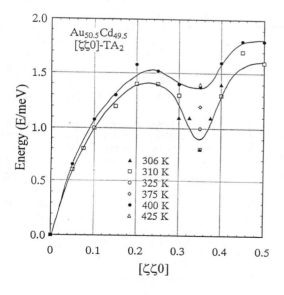

Fig. 2. Phonon softening observed in the $[\xi\xi 0]$-TA$_2$ transverse branch. [114]Cd isotope was utilized to reduce an effect of large absorption.

EXPERIMENTAL

Two kinds of experiments were performed. One is the integrated intensity measurements of the parent phase at various temperatures approaching the transformation temperature. Another is the intensity measurements along $<\bar{1}10>*$ direction. X-ray sources used in the experiments are a rotating anode x-ray diffractometer (Rigaku RU-200) and synchrotron radiation (BL-10A at Photon Factory in KEK). Four circle diffractometers were used for the measurements. The collimator for the rotating anode x-ray source is 1mmφ and receiving slit is 1 degree. The receiving slits at Photon Factory are 3mmφ and 5mmφ. Small fragments of a $Au_{50.5}Cd_{49.5}$ single crystals were picked up and mounted on a glass fiber. The crystal size used in the experiment was approximately 100μm with irregular shape. High temperature apparatus, blowing N_2 gas on the crystal, manufactured by Rigaku was used at Photon Factory and an electrical heater controlled by a personal computer was used at our laboratory. The fluctuation of the temperature was within 0.2K at the monitor for both experiments. The Ms temperature was calibrated with the transformation temperature of bulk sample, that is 310K. The Bragg peak of the parent phase, 200, was observed and it weakened suddenly when the transformation occurred.

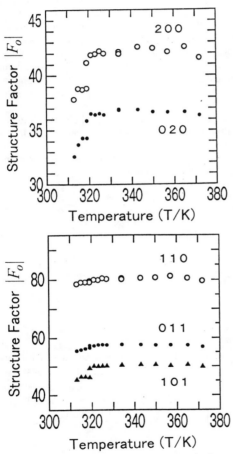

Fig. 3. Observed structure factors ($|F_o|$) of Bragg reflections vs. temperature. They decrease when approaching the transformation temperature.

The orientation matrix[24] which indicate the orientation of the crystal were refined at every temperature using ten or fifteen Bragg reflections. This matrix is important to discuss the position of the reflections, that is the incommensurate problem.

Integrated intensities were measured with CuKα radiation and graphite monochromater at every temperature. The 2θ-ω scan was applied for the integrated intensities measurement. The scan speed for the measurement was 4 degree/min and scan range was followed by the equation of $\Delta\omega = A + B\tan\theta$ with $A = 1.3$ and $B = 0.5$. Five times scans were repeated, if the ratio, $\sigma(|F_o|)/|F_o| < 0.001$, was not satisfied where $\sigma(|F_o|)$ is the standard deviation of the observed Bragg reflection.

The intensity measurements along $<\bar{1}10>*$ direction were done at the Photon

Factory. The interval of the measurements was 0.005 and 0.01 in the reciprocal unit and the measurement time for each step was 5s and 20s. The wavelength utilized was 1.04 Å monochromatized with Si(111) perfect crystal.

Additional measurements along $<\bar{1}10>^*$ direction were performed by the use of CuKα radiation to see the time dependency of the superstructure reflections. The measurements were done with varying time after setting the temperature.

RESULTS AND DISCUSSION

Observed structure factors were obtained from the integrated intensities after the Lorenz-polarization correction. Observed structure factors ($|F_o|$) against temperature are shown in Fig. 3. They decrease when approaching the transformation temperature. In general, structure factors increase with lowering temperature following the Debye-Waller factor. The deviation from the Debye-Waller factor occurs approximately ten degree above the transformation temperature. The FWHMs of the Bragg reflections were measured and do not change with lowering temperature.

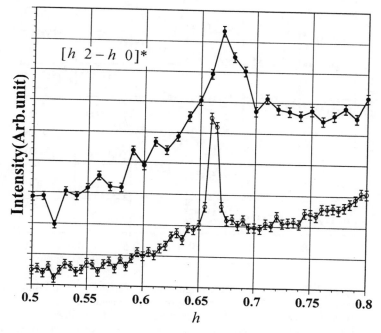

Fig. 4. The step scan measurements along $[h \quad 2-h \quad 0]^*$ around 2/3 4/3 0 reflection. Two profiles were shown in this figure, both of which were observed 2K above the transformation temperatures. A very sharp peak (lower curve in the figure) was observed when the temperature was lowered to 312K from 425K directly. A broad peak (upper curve in the figure) was observed when the temperature was lowered successively from 425K.

The intensity measurements along $<\bar{1}10>$* direction were carried out at 312K and are shown at lower curve in Fig. 4. A sharp superstructure reflection was observed around 2/3 4/3 0. The peak position of the sharp peak was 1/3 of $<\bar{1}10>$*, i. e. at the commensurate position.

The temperature dependence of the intensity was measured by lowering the temperature and is shown in Fig. 5. The measurements were carried out at 425K, 375K, 345K, 320K, 315K and 312K, in succession. The intensity at 312K is drawn in Fig.4 for comparison.

Comparing the reflections observed at 312K shown in Fig. 4, it is obvious that the peak drawn in the upper part is broader than that in the lower part. The full width at half maximum (FWHM) of the broad peak is three times broader than that of the sharp one. Not only was there the peak broadening but also a peak splitting was observed at another equivalent 1/3 positions. The peak splitting indicates that several domains were produced prior to the transformation. The broad peak seems to consist of two or three peaks. The domain size is rather large and grows independently.

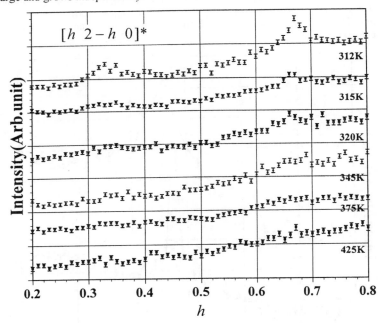

Fig. 5. Temperature dependence of the superstructure reflections. The measurements were performed along $[h\ \ 2-h\ \ 0]$* by lowering the temperature from 425K to 312K in succession. The intensities of the superstructure reflections became strong gradually in this case. The broad peaks were obtained in this process. The profile at 312K around 2/3 4/3 0 is drawn in Fig. 4 for comparison.

The difference between the sharp and broad peaks is depend on the process of the measurements, that is, the sharp peak was measured when the temperature is lowered to 312K from 425K directly. On the other hand, the broad peak was measured at various temperatures in succession as mentioned above.   In the case of broader peaks, the peak position was shifted from the exact 1/3 position which is clearly shown in Fig.4.   The 1/3 superstructure reflections instead of incommensurate reflections must be essential for the martensitic transformation.   For both cases, intensities of superstructure reflections was less than 1/200 of the Bragg reflection, for example 200 reflection.

Additional experiments were performed to see the origin of these differences.   Fig. 6 shows the time dependence of the intensity along $<\bar{1}10>^*$ direction.   When the temperature was lowered to 312K, there are no reflections as indicated "0h" in the figure.   A small change was observed at 6 hours after the temperature being set.   The position of the peak seems to be at 1/3, commensurate position, that is superstructure reflections.   The superstructure reflections appearing at 6 hours split into two or three peaks at 25 hours later shown in the Fig. 6.   This splitting may relate to the growth problem of the martensite.   Roitburd[25] reported the advantage of the multiple domain growth.   This phenomenon seems to relate to the incubation problem recently reported by several researchers.[26-27]

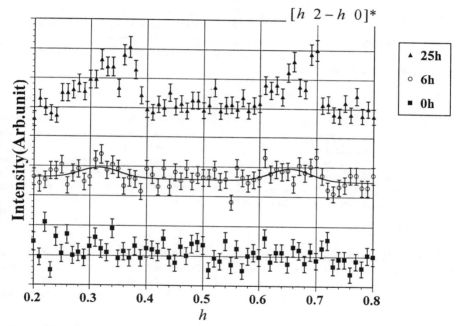

Fig. 6.   Time dependence of the superstructure reflections.   After 6 hours from when the temperature was set, a small change appeared.   The curve fit was applied for the data of 6hours.   The peak at 6hours became broader and split into two or three peaks after 25 hours.   Therefore the curve fit was not applied for them.

The origin of observed sharp peaks must be completely different from dielectric materials. Diffuse scattering from those materials were observed and they are interpreted that the correlation length in a direct space expand when the temperature approaches the transformation temperature.   In other words, the modulation appears in a small region at the beginning and expands over the whole crystal. In dielectric materials, which show a second order transition, a fluctuation in the material varies with temperature and expands over the crystal. The transformation is completed when the fluctuation is spread out. The martensitic transformation is diffusionless and displacive. The mechanism of the transformation is different from the picture shown in a dielectric material.

The correlation between the decrease of the structure factors and extra peaks (superstructure reflections) at 1/3 seems to be strong. The origin of the weakening of the structure factors and sharp peaks prior to the transformation is discussed here. The following possibility can be considered. The small regions whose crystal structure is different from the parent phase appear prior to the transformation. The different structure regions may be called the embryo or the pseudo-martensite. Since the volume of the parent phase decreases, structure factors of the parent phase decrease. On the other hand, the superstructure reflections which come from the embryo or pseudo-martensite increase with lowering temperature.

Upon approaching the transformation temperature, the extra peak (superstructure reflections) becomes broader or splits into several peaks. The superstructure reflections come from the small regions which may be called embryo or pseudo-martensite. They may correspond to the multi-variants in the martensite. The FWHM of the Bragg peaks in the parent phase were not affected. It means that the transformation is heterogeneous. The small regions, embryo or pseudo-martensite, will produce a strain. If the strain is stabilized, hysteresis will appear even in the pre-martensitic state.

CONCLUSIONS

The superstructure reflections prior to the onset of the martensitic transformation were measured. Sharp peaks appeared at $1/3 < \bar{1}10 >^*$ commensurate position when the temperature was set close to the transformation temperature. The broad peaks, which consist of several sharp peaks and slightly shifted from the commensurate position, were observed after one day aging. The integrated intensities of Bragg reflections in the parent phase were measured and they decrease when approaching the transformation temperature. The results indicate that the behavior of the superstructure reflections and the Bragg reflections have correlation.

ACKNOWLEDGMENTS

The authors appreciate Dr. L. Shi of University of Science and Technology of China at Hefei and Mr. Hara of Teikyo University and Dr. M. Tanaka of Photon Factory in KEK for their valuable cooperation. The authors are also grateful to Dr. Y. Murakami, Messrs. S. Morito, Ya Xu, Y. Suzuki, H. Sekido, J. Zhang, and T. Kawano in University of Tsukuba for their helpful assistance in the experiment.

REFERENCES

1. M. S. Wechsler, D. S. Lieberman and T. A. Read, Trans. AIME, 197(1953), p. 1503-1515
2. J. S. Bowles and J. K. Mackenzie, Acta Metall., 2(1954), p. 129-, p. 138-, p. 224-234.
3. D. S. Lieberman, M. S. Wechsler and T. A. Read, J. Appl. Phys., 26(1955), p. 473-484.
4. Y. Morii and M. Iizumi, J. Physical. Soc. Jpn., 54(1985), p. 2948-2954.
5. M. Mori, Y. Yamada and G. Shirane, Solid State Communications, 17(1975), p. 127-130.
6. A. Nagasawa, T, Makita, N. Nakanishi, M. Iizumi and Y. Morii, Metall. Trans. 19A(1988), p. 793-796.

7. T. R. Finlayson and H. G. Smith, Metall. Trans. **19A**(1988), p. 193-198.
8. S. K. Satija, S. M. Shapiro, M. B. Salamon and C. W. Wayman, Physical Rev. **B29**(1984), p. 6031-6035.
9. H. Tietze, M. Mullner, P. Selgert and W. Assmus, J. Phys., **F15**(1985), p. 263-271.
10. P. Moine, J. Allain and B. Renker, J. Phys., **F14**(1984), p. 2517-2523.
11. S. M. Shapiro, Y. Noda, Y. Fujii and Y. Yamada, Phys. Rev. **B30** (1984), p.4314-4321.
12. Y. Yamada, Y. Noda and M. Takimoto, Solid State Communications, **55**(1985), p. 1003-1006.
13. Y. Yamada, Proceedings of the International Conference on Martensitic Transformations, (1986), p. 89-94.
14. R. J. Gooding and J. A. Krumhansl, Phys. Rev. **B39**(1989), p. 1535-1540.
15. T. Tadaki and C. M. Wayman: Metallography, **15**(1982)233-245.
16. M. Meichele, M. B. Salamon and C. M. Wayman: Phil. Mag., **A47**(1983)177
17. C. M. Hwang and C. M. Wayman: Acta Metall., **32**(1984)183.
18. A. Ölander, J. Am. Chem. Soc., **54**(1932), p. 3819.
19. T. Ohba, Y. Emura, S. Miyazaki and K. Otsuka, Materials Trans. JIM, **31**(1990), p. 12-17.
20. W. Köster and A. Schneider, Z. Metallk. **32**(1940), p. 156.
21. T. Ohba, Y. Emura, K. Otsuka, Materials Trans. JIM. **33**(1992), p. 29-37.
22. T. Ohba, S. M. Shapiro, S. Aoki amd K. Otsuka, Jpn. J. Appl. Phys. **33**(1994), p. L1631-L1633.
23. Y. Noda, M. Takimoto, T. Nakagawa and Y. Yamada, Metall. Trans. **19A**(1988), p. 265-271.
24. International Tables for X-ray Crystallography Vol. IV, Kluwer Academic Publishers, Dordrecht(1989), p.275-284.
25. A. L. Roitburd, Mater. Sci. Eng., **A127**(1990), p. 229.
26. H. Abe, K. Ohshima and T. Suzuki, Materials Trans. JIM, **36**(1995), p. 1200-1205.
27. T. Kakeshita, K. Kuroiwa, K. Shimizu, T. Ikeda, A. Yamagishi and M. Date, Mater. Trans. JIM, **34**(1993), p. 423-428.

A STUDY OF THE RUBBER-LIKE BEHAVIOR OF MONO-DOMAIN
Au-Cd MARTENSITE

Xiaobing   REN, Kazuhiro OTSUKA
Institute of Materials Science, University of Tsukuba, Tsukuba, Ibaraki 305, Japan
ren@mat.ims.tsukuba.ac.jp

ABSTRACT

The origin of the rubber-like behavior in mono-domain Au-Cd martensite was explained in terms of a new model that focused attention on the change of long-range elastic interaction energy among vacancies during a domain reversion. Vacancies in martensite, the lower-symmetry phase, produce stress fields with lower symmetry.   During martensite aging, vacancies tend to rearrange themselves to lower elastic interaction energy.   The low-symmetry elastic field results in a low-symmetry vacancy configuration.   When a stabilized martensite domain reverts to a new domain (twin) under external stress, the original vacancy configuration is inherited to the new domain, but such a configuration becomes a high energy configuration because of the lower symmetry of elastic field, and thus it tends to restore the original configuration by reverse twinning. The above vacancy reconfiguration model is consistent with the fact that the rubber-like behavior is closely related to vacancies.

INTRODUCTION

The rubber-like behavior (RLB) of martensite refers to the curious pseudoelasticity in well-aged martensitic state, which was first found in 1932 by Ölander in Au-47.5%Cd martensite[1]. The pseudoelasticity in martensite (i.e., the rubber-like behavior) is quite distinct from the pseudoelasticity in the parent phase. The latter is associated with the reversible movement of martensite/parent interface due to the balance of external stress and internal restoring force arising from the free energy difference between martensite and parent[2]; the former is associated with the reversible movement of martensite domain (twin) boundary [3] due to the balance of external stress and *unknown internal restoring force* developed during aging.   In the past decades there have been quite a few models trying to explain the origin of the unknown restoring force. However, the origin still remains obscure.   These models can be classified into two categories: one is based on the domain boundary effect[4,5], the other is based on the volume effect or bulk effect.   Recent experiment on mono-domain martensite [6,7] has showed that the RLB even occurs in martensites without domain boundary.   This result suggests that domain boundary is not a necessary factor for the RLB, although it promotes the process.   As volume effect is concerned, several recent models have been proposed based on the following possible processes during martensite aging:

- decrease of degree of long-range order (LRO model) [8,9] or increase of degree of short-range order (SRO model) [10,11]
- rearrangement of vacancy-pairs(strain-dipole model) [12]

It should be pointed out that in essentially ordered alloys (most shape memory alloys belong to this category), LRO and SRO are closely related to each other, an increase of SRO will lead to a decrease of LRO.   Thus the SRO model can be considered to be similar to LRO model in nature, and in the following we will indiscriminately refer them as reordering models.

Mat. Res. Soc. Symp. Proc. Vol. 459 © 1997 Materials Research Society

The reordering models (LRO and SRO models) have found certain success in explaining the rubber-like behavior in Cu-based martensites[8-11], but they appear unble to explain the rubber-like behavior in near equi-atomic Au-Cd martensite, in which no appreciable change in order parameters during aging was found by Ohba et al[13]. Besides this, decrease of LRO (or increase of SRO) in ordered alloys would be expected to result in an increase of electrical resistivity, since the disorder disturbs the otherwise perfect lattice and hence increases the resistance to the motion of free electrons. However, this prediction contradicts our recent experimental fact that electrical resistivity of furnace-cooled Au-Cd martensite decreases with aging time[14].

Vacancy-pair model or strain-dipole model tries to overcome the above weakness of reordering models by considering defects in martensite. This model does not require a change in order parameters, thus appears to be able to explain the rubber-like behavior in Au-Cd. However, this model assumes that there exists an excessive amount of vacancies/pairs (0.1 atomic fraction!). Up to now there has been no experimental evidence showing the existence of such a high concentration of vacancies in martensite.

In the following a new model will be proposed which is based on elastic interaction of vacancies. This model assumes that there exists a more realistic concentration of vacancies(0.01 atomic fraction) in martensite.

FREE ENERGY CHANGE DURING DOMAIN REVERSION IN AGED SINGLE-DOMAIN Au-Cd MARTENSITE

Before proposing the model for rubber-like behavior in Au-Cd martensite, it is important to estimate the free energy change during domain reversion (twinning) for aged martensite from available experimental data. Let us consider an aged single-domain martensite (domain 1) changes into another single-domain (twin, or domain 2) under applied tensile stress σ_c as shown below

$$\text{Domain 1} \xrightarrow{\sigma_c} \text{Domain 2}.$$

If neglecting the frictional energy dissipation during domain boundary motion, the applied energy $\sigma_C.\varepsilon_T V$ may be considered to be converted into free energy of domain 2. From energy conservation condition, it follows:

$$\Delta F_0 = F(2) - F(1) = \sigma_C.\varepsilon_T V > 0,$$

where $F(1)$, $F(2)$ and ΔF_0 are free energy of domain 1, domain 2, and their difference, respectively, ε_T is the component of the twinning shear along the tensile axis, V the molar volume of the alloy.

This equation means there is a free energy increase during deforming domain 1 into domain 2 (twin). It is the driving force for domain reversion. This energy increase must come from a chemical origin. For Au-49.5Cd martensite, $\sigma_C \sim 30 \text{MPa}$ and $\varepsilon_T \sim 0.01$[15], $V=11\times10^{-6} \text{m}^3/\text{mol}$, then $\Delta F_0 = 30\times10^6 \times 0.01 \times 11 \times 10^{-6} = 3.3$ J/mol.

Therefore, it is estimated that the change of free energy of Au-49.5Cd martensite(single-domain) during domain reversion should be the order of several joule per mole. It should be of the same order of magnitude as that during martensite ageing. If the free energy of martensite after ageing is decreased by several joule per mole, (say, 3.3J/mol), then there will be an increase of As temperature, $\Delta A_s \sim 3.3/\Delta S$, where ΔS is the entropy of reverse transformation. For Au-49.5Cd martensite, $\Delta S=1.34J/mol.K$[16], then $\Delta A_s \sim 2K$, i.e., A_s will be expected to increase by 2K if single-domain martensite is aged. This result appears to be in good agreement with Murakami's measurement (~2K) on single-domain Au-49.5Cd martensite[7].

For multi-domain Au-Cd martensite, it has been found that ageing in martensite state may result in a remarkable increase in A_s (~15K). This indicates that pinning of defects into domain boundaries can also lower free energy of martensite appreciably.

Any reasonable model for the rubber-like behavior in single-domain Au-Cd martensite should yield a free energy decrease by several joule per mole during aging of martensite. In the next section a model based on vacancy reconfiguration will be proposed.

VACANCY ELASTIC INTERACTION MODEL FOR RUBBER-LIKE BEHAVIOR IN Au-Cd MARTENSITE

Strain Field of a Vacancy in Low-Symmetry Crystal Lattice

A vacancy in a low-symmetry lattice (lower than cubic) produces a strain field with the same symmetry as the lattice, as shown in Fig.1. All martensitic phases are low-symmetry phases, thus it is expected that the strain field of a vacancy in martensite possesses low symmetry, too. The strain field is a long-range field which decays as $1/r^3$.

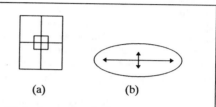

(a) (b)

Fig.1. A vacancy in low-symmetry lattice (a), and its strain field (b).

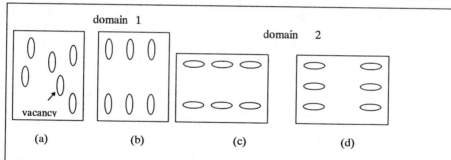

domain 1

domain 2

vacancy

(a) (b) (c) (d)

Fig. 2. Vacancy configuration in single-domain martensite. (a). before ageing, (b) after ageing, (c) formation of domain 2 (twin) under applied stress, (d) rearrangement of vacancies in domain 2 after ageing in domain 2 state.

Vacancy Reconfiguration Due to Their Long-range Elastic Interaction During Aging

Vacancy Reconfiguration Due to Their Long-range Elastic Interaction During Aging

Vacancies in martensite interact with each other through long-range elastic strain field. To lower elastic interaction energy, vacancies tend to rearrange themselves into a low-energy configuration during ageing, as shown in Fig.2a, and 2b. This process is called elastic interaction ordering (EIO)[17]. The low-energy configuration is characterized by a quasi-ordered arrangement of vacancies. Because of the low-symmetry of strain field, the ordered configuration of vacancies should also show a low symmetry, as shown in Fig.2b. When twinning occurs under applied stress, the original lattice is transformed into the new lattice (twin) by a twinning shear. Since no diffusion occurs during this process, the original vacancy configuration is transferred to the new lattice, but with an exchange of symmetry axis. The vacancy configuration after twinning is shown in Fig.2c. The elastic interaction energy of vacancy configuration after twinning (Fig.2c) is clearly different from that before twinning (Fig.2b). Because the configuration in Fig.2b is the low energy configuration, the configuration after twinning (Fig.2c) must be a higher energy state (this can be proved by elastic interaction energy consideration, as will be shown below.). Then, when the external field is removed, there would be a tendency to restore the low energy configuration through reverse twinning. By such a mechanism the martensite shows a pseudoelastic or rubber-like behavior. If the twin (or domain 2) is maintained by external stress for some time, the vacancies will rearrange into a more stable configuration (Fig.2d), and the elastic energy is thus lowered. In this case, there is no driving force for original domain (domain 1) formation, hence original shape cannot be restored after removal of applied stress.

Estimation of the Change of Elastic Interaction Energy During Vacancy Reconfiguration

As shown in Fig.2, both ageing and twin formation involve a change of vacancy configuration. The elastic interaction energy change in either processes should be of the same order of magnitude, thus in the following calculation we do not distinguish them.

Vacancy reconfiguration process results from long-range elastic interactions, thus it is a so-called "elastic interaction ordering (EIO)" process[17]. Elastic interaction energy of vacancies in solid solution has been calculated to be[17]

$$E_{int} = I_0 \varepsilon^2 \eta x_v^2 \ln x_v,$$

where $I_0 = 6\mu V/\gamma$, is related to the elastic properties of the solid solution. μ is the elastic constant, $\gamma = 3(1-\nu)/(1+\nu)$, and ν is the Poisson's ratio. $\varepsilon = (1/a)da/dx_v$ is the compositional expansion coefficient of the lattice with parameter a, and x_v is the atomic fraction of vacancies. η is the EIO order parameter with a maximum value of 1.

If vacancies rearrange themselves from a completely disordered state ($\eta=0$) into an maximum ordered configuration ($\eta=1$), there will be a decrease of total elastic energy given by

$$\Delta E_{int} = I_0 \varepsilon^2 x_v^2 \ln x_v,$$

For Au-49.5Cd mono-domain martensite, assume vacancy or defect concentration is $x_v \sim 0.01$ (atomic fraction), $\varepsilon \sim 0.1$. Take $V \sim 11 \times 10^{-6}$ m^3, $\mu \sim 1.5 \times 10^{10}$ Pa [18], $\gamma = 1.5$, then $I_0 \sim 6.6 \times 10^5$ J/mol. and

$$\Delta E_{int} = 6.6 \times 10^5 \times 10^{-2} \times 10^{-4} \times \ln(10^{-2}) = -3.1 \text{ J/mol}.$$

This result shows that the energy change during vacancy reconfiguration is of the order of a few joule per mole. This value is in agreement with that measured from tensile test as shown in previous section. Thus the vacancy reconfiguration mechanism may be the governing process for the rubber-like behavior in single-domain Au-Cd martensite.

DISCUSSION

Rubber-like Behavior--- A Vacancy Reconfiguration Process

There have been many experimental facts showing that the rubber-like behavior of martensite is closely related with vacancy migration. The above model tries to elucidate the origin of such a process-- elastic interaction induced vacancy migration or reconfiguration. The magnitude of energy change associated with such a process has been shown to be in rough agreement with experimentally measured value.

Since the concentration of vacancy is quite low (~0.01), their reconfiguration is expected to have very small (may be on the order of 10^{-2}) effect on the X-ray diffraction profile. Thus the diffraction spectrum of aged martensite may have little difference from that of fresh martensite, as the case in Au-Cd martensite. However, if the measurement is made accurate enough (say, uncertainty<10^{-2}), it may be possible to observe the small change in diffraction profile, including intensity change and peak shift. Murakami et al's careful X-ray diffraction measurement for Au-Cd single-domain martensite[7] showed that there did exist some difference (although very small) in X-ray profile for aged and unaged martensite. Although their profile change was interpreted as an increase in the asymmetry of the peak, it may also be viewed as a small shift towards higher angle, and a small change of intensity. These changes are so small (10^{-2}) that it can be hardly observed by less accurate measurement.

The vacancy interaction model is based on an assumption that vacancies have a concentration on the order of 10^{-2}. In disordered alloys this value is still too high at room temperature. However, most martensites showing rubber-like behavior are intermetallic compounds, in which there exists a large amount (up to several percent) of "structural vacancies" or "anti-structure defects" even at low temperature[19] If taking this into account, our assumption may not be considered to be unreasonable. Anyway, an experimental measurement of the concentration of vacancy and antistructure defects may provide a critical verification to this model.

CONCLUSIONS

1. A new model based on vacancy rearrangement due to long-range elastic interactions is proposed to explain the rubber-like behavior in single-domain Au-Cd martensite.
2. This model appears to be consistent with known experimental facts and yields correct order of restoring energy.

ACKNOWLEDGMENTS

The authors would like to thank Prof. T. Suzuki of Tsukuba Institute of Science and Technology, Prof. T. Ohba of Teikyo University, Prof. M. Kogachi of University of Osaka Prefecture, and Dr. Y. Murakami of Tohoku University for helpful discussions. The present study is supported by the Grant-in-Aid for General Scientific Research (Ippan B 1995-6) from the

Ministry of Education, Science and Culture of Japan. One of the authors (Xiaobing REN) would also like to acknowledge the financial support by JSPS Fellowship of Japan, and by Grant-in-Aid for Encouragement of Young Scientists (96028; X.R) from the Ministry of Education, Science and Culture of Japan.

REFERENCES

1. A. Ölander, J. Am . Chem. Soc. **56**, p.3819 (1932).
2. K. Otsuka and C.M. Wayman, Review of the Deformation Behavior of Materials, edited by P. Feltham, Freund Publishing House, 1977, vol.2, pp.81-172.
3. H. K. Birnbaum and T.A. Read, Trans. AIME, **218**, p.94 (1960) .
4. J. Janssen, J. Van Humbeeck, M. Chandrasekaran, N. Mwanba and L. Delaey, J. Phys. 43, suppl. **12**, c4-p.715 (1982) .
5. A. Abu Arab and M. Ahlers, J. Phys. 43, suppl, **12**, c2-p.709 (1982) .
6. G. Barcelo, R. Rapacioli, and M. Ahlers, Scripta Metall. **12**, p.1069(1978).
7. Y. Murakami, Y. Nakajima, K. Otsuka and T. Ohba, J. de Phys. IV, **5**, c8-p.1071 (1995) .
8. A. Abu Arab and M. Ahlers, Acta Metall. **36**, p.2627 (1988) .
9. T. Tadaki, H. Okazaki, Y. Nakata, K. Shimizu, Mater. Trans. JIM, **31**, p.941 (1990) .
10. M. Ahlers, G. Barcelo and R. Rapacioli, Scripta Metall. **12**, p.1075 (1978) .
11. K. Marukawa and K. Tsuchiya, Scripta Metall. **32**, p.77 (1995) .
12. T. Suzuki, T. Tonokawa, and T. Ohba, J. de. Phys. **5**, c8-p.1065 (1995) .
13. T. Ohba, K. Otsuka, and S. Sasaki, Mater. Sci. Forum, **56-58**, p.317(1990).
14. Xiaobing Ren and K. Otsuka, to be published.
15. Y. Nakajima, S. Aoki, K. Otsuka and T. Ohba, Mater. Lett. **21**, p.217(1994).
16. K. Morii, S. Miyazaki and K. Otsuka, Proc. ICOMAT-92, 1992, p. 1125.
17. Xiaobing Ren, Xiaotian Wang, K. Shimizu, and T. Tadaki, J. Mater. Sci. & Technol. **12**, p.57 (1996) .
18. T. Tonokawa, S. Morito, Y. Nakajima, A. Ooishi, K. Otsuka and T. Suzuki, Jpn. J. Appl. Phys. **33**, p.2897(1994).
19. R. J. Wasilewski, J. Phys. Chem. Solids, **29**, p.39(1968).

A THEORY OF SHAPE-MEMORY THIN FILMS WITH APPLICATIONS

K. BHATTACHARYA*, R.D. JAMES**
*Division of Engineering and Applied Science, 104-44 California Institute of Technology, Pasadena CA 91125, USA; bhatta@cco.caltech.edu
**Department of Aerospace Engineering and Mechanics, University of Minnesota, Minneapolis, MN 55455, USA; james@aem.umn.edu

INTRODUCTION

Shape-memory alloys have the largest energy output per unit volume per cycle of known actuator systems [1]. Unfortunately, they are temperature activated and hence, their frequency is limited in bulk specimens. However, this is overcome in thin films; and hence shape-memory alloys are ideal actuator materials in micromachines[1]. The heart of the shape-memory effect lies in a martensitic phase transformation and the resulting microstructure. It is well-known that microstructure can be significantly different in thin films as compared to bulk materials. In this paper, we report on a theory of single crystal martensitic thin films. We show that single crystal films of shape memory material offer interesting possibilities for producing very large deformations, at small scales.

A THEORY OF THIN FILMS

Consider a single crystal film with surface S and thickness h: $\Omega_h = S \times (0, h)$. Let $\mathbf{x}$ be a typical point in the film and let $\tilde{\mathbf{y}}(\mathbf{x})$ be any deformation of the film. The total energy stored in the film is given by

$$E_h[\tilde{\mathbf{y}}] = \int_{\Omega_h} \left\{ \alpha |\nabla^2 \tilde{\mathbf{y}}|^2 + W^b(\nabla \tilde{\mathbf{y}}) \right\} d\mathbf{x}. \tag{1}$$

where the first term represents interfacial energy (α is a constant) while the second is the elastic energy. Because the thickness is small, it is important to consider some form of interfacial energy. We choose the above because of simplicity. The elastic energy density W^b has a multi-well structure in martensitic materials as explained below. Minimizing this total energy (1) over all possible deformations subject to suitable boundary conditions gives rise to martensitic microstructure[2]. We are interested in studying this microstructure in very thin films. In [3] we study the behavior of the minimizers of the energy (1) as $h \to 0$ and obtain a limiting theory, which we describe below. It turns out that this is exactly a special Cosserat membrane theory [4].

Consider a film occupying a region $S \in I\!R^2$. Let $\mathbf{z} = (z_1, z_2)$ be a typical point on the film. The deformation of the film is characterized by two three-dimensional vector fields, $\mathbf{y}(\mathbf{z})$ and $\mathbf{b}(\mathbf{z})$ as shown in Figure 1. $\mathbf{y}$ describes the deformation of the base while $\mathbf{b}$ describes the deformation of the film relative to the base. To prevent tearing, $\mathbf{y}$ is assumed to be continuous, but $\mathbf{b}$ can jump at interfaces. The total energy of the film in the limiting theory is given by

$$E[\mathbf{y}, \mathbf{b}] = \int_S W(\mathbf{y}_{,1}|\mathbf{y}_{,2}|\mathbf{b})d\mathbf{z} \tag{2}$$

where $W = W^b/h$ is the stored energy per unit reference area. The notation $\mathbf{A} = (\mathbf{a}_1|\mathbf{a}_2|\mathbf{b})$ means that the columns of the 3×3 matrix $\mathbf{A}$ are the vectors $\mathbf{a}_1, \mathbf{a}_2$ and $\mathbf{b}$; and $\mathbf{y}_{,i} = \partial \mathbf{y}/\partial z_i$.

Figure 1: The deformation of a film is characterized by two vector fields: the point $\mathbf{z}$ in the reference film (left) goes to $\mathbf{y}(\mathbf{z})$ in the deformed film (right) while a vector drawn through the thickness at $\mathbf{z}$ goes to the vector $h\mathbf{b}(\mathbf{z})$.

In martensitic materials, W^b and consequently W has a multi-well structure. Above the transformation temperature, the material is in the *austenite* state. We choose this as the reference, and hence the austenite is described by the identity matrix, $\mathbf{I}$. Below the transformation temperature, the material is in the martensite phase. The *Bain* or *transformation* matrix $\mathbf{U}_1$ describes the distorsion of the martensite lattice relative to the austenite lattice. Typically, the martensite lattice is less symmetric compared to the austenite lattice. Consequently, one can have k symmetry-related variants of martensite characterized by transformation matrices $\mathbf{U}_1, \mathbf{U}_2, \ldots, \mathbf{U}_k$. The number of variants k can be calculated from the change of symmetry, and the transformation matrices can be calculated from the measured lattice parameters.

For example, in a cubic to tetragonal transformation there are three variants of martensite and $\mathbf{U}_1$ is the diagonal matrix with diagonal elements η_1, η_1, η_2. $\mathbf{U}_2$ and $\mathbf{U}_3$ are obtained by permuting the diagonal elements. Ni-36% Al undergoes such a transformation and for this alloy, $\eta_1 = 0.9392, \eta_2 = 1.1302$. Similarly, in a cubic to monoclinic transformation as in Ni-50%Ti, there are twelve variants of martensite[5,6].

Further, a rigid rotation does not change the state of the crystal. Therefore, we define

$$\text{Austenite Well}: \quad \mathcal{A} = \{\mathbf{Q} : \mathbf{Q} \text{ is a rotation matrix}\}$$
$$\text{Martensite Wells}: \quad \mathcal{M} = \bigcup_{i=1}^{k} \mathcal{M}_i; \qquad \mathcal{M}_i = \{\mathbf{Q}\mathbf{U}_i : \mathbf{Q} \text{ is a rotation matrix}\}. \tag{3}$$

W is minimized on $\mathcal{A}$ above the transformation temperature, while it is minimized on $\mathcal{M}$ below the transformation temperature. There is an exchange of stability at the transformation temperature, and W is minimized on both $\mathcal{A}$ and $\mathcal{M}$. We assume that these are the only minimizers of W. Therefore, in order to minimize the total energy, we look for $\mathbf{y}, \mathbf{b}$ such that $(\mathbf{y}_{,1}|\mathbf{y}_{,2}|\mathbf{b})$ remains in the relevant energy wells.

DEFORMATIONS WITH ONE VARIANT

We begin by looking for deformations of the film that involve only one well. In other words we look for vector fields, $(\mathbf{y}(\mathbf{z}), \mathbf{b}(\mathbf{z}))$ such that $(\mathbf{y}_{,1}|\mathbf{y}_{,2}|\mathbf{b}) = \mathbf{Q}(\mathbf{z})\mathbf{F}$ where $\mathbf{F} = \mathbf{I}, \mathbf{U}_1, \ldots,$ or $\mathbf{U}_k$. Let us begin by looking at the case when $\mathbf{F} = \mathbf{I}$. Then, the requirement above describes a very interesting class of deformations: all the deformations that one can perform on a flat sheet of paper without stretching it. The vector $\mathbf{b}$ remains normal to the deformed film. Typical examples are deformations that take a flat sheet into a cone or cylinder. If $\mathbf{F} \neq \mathbf{I}$, we first stretch the sheet uniformly by $\mathbf{F}$ and then subject it to the paper-folding deformations described above. This is shown in Figure 2.

Let us constrast this with the behavior of bulk bodies. A one-well deformation would require that $(\tilde{\mathbf{y}}_{,1}|\tilde{\mathbf{y}}_{,2}|\tilde{\mathbf{y}}_{,3}) = \mathbf{Q}(\mathbf{x})\mathbf{U}$. But, since all three columns are derivatives, this implies that $\mathbf{Q}(\mathbf{x})$ is a constant and consequently, the only possible deformation is a rigid body rotation of a pure variant.

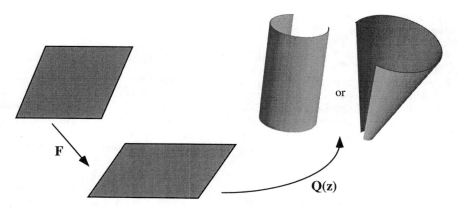

Figure 2: Deformation involving just one well involves a stretch and a fold

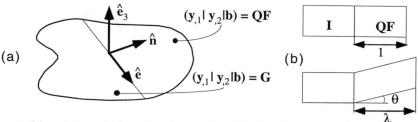

Figure 3: (a) A deformation involving two wells. (b) The deformation across an interface (undeformed above and deformed below) consists of shear θ and an elongation λ.

We understand this result as follows. Because the films are thin, it is possible to roll them up with very little energy. In contrast, bending a bulk body requires quite a lot of energy. Thus, we see that our theory captures quite well the "floppiness" of very thin films. We also see that the range of possible deformations of a thin film are much richer than bulk bodies even with a single variant.

DEFORMATIONS WITH TWO VARIANTS: INTERFACE CONDITIONS

Consider a deformation involving two variants in a film with normal $\hat{\mathbf{e}}_3$ shown schematically in Figure 3a (Here we would choose $\mathbf{F}, \mathbf{G}$ to be equal to $\mathbf{I}, \mathbf{U}_1, \ldots,$ or $\mathbf{U}_k$ depending on the interface we wish to study). In order for the film to be unbroken, it is necessary that $\mathbf{y}$ be continuous across the interface; however, there is no such restriction on $\mathbf{b}$. Therefore, we need to satisfy an *invariant line* condition across the interface:

$$(\mathbf{QF} - \mathbf{G})\hat{\mathbf{e}} = 0 \qquad \text{for some } \hat{\mathbf{e}} \text{ such that } \hat{\mathbf{e}} \cdot \hat{\mathbf{e}}_3 = 0 \qquad (4)$$

where $\hat{\mathbf{e}}_3$ is the film normal. Here, $\hat{\mathbf{e}}$ is the invariant line since $\mathbf{QF}\hat{\mathbf{e}} = \mathbf{G}\hat{\mathbf{e}}$. Our energetic argument assures us that a thin film can overcome any incoherence in the thickness direction at the cost of a small elastic energy. Therefore, all that we need is to satisfy coherence along the plane of the film and this is exactly the invariant line condition.

In [3] we show that given $\mathbf{F}$ and $\mathbf{G}$ as above, we can find a rotation $\mathbf{Q}$ and a direction $\hat{\mathbf{e}}$ that satisfy (4) if and only if

$$\hat{\mathbf{e}}_3 \cdot \left(\text{cof} \left(\mathbf{F}^T \mathbf{F} - \mathbf{G}^T \mathbf{G} \right) \right) \hat{\mathbf{e}}_3 \leq 0. \tag{5}$$

Above, cof denotes the cofactor of a matrix. If this condition is true, then one can find (i) two independent directions $\hat{\mathbf{e}}$ and (ii) a one-parameter family of $\mathbf{Q}$ for each $\hat{\mathbf{e}}$.

To understand (5) consider a square and stretch it by a quantity λ_1 along one side and a quantity λ_3 on another. Then, it is possible to find an invariant line if and only if one of the two stretches is greater that one while the other is less that one, i.e., if and only if $(\lambda_1 - 1)(\lambda_3 - 1) \leq 0$. The equation (5) expresses this condition in terms of the matrices $\mathbf{F}$ and $\mathbf{G}$ and the film normal $\hat{\mathbf{e}}_3$.

Let us now contrast the condition for formation of an interface in thin films and bulk. In thin films the condition (5) is equivalent to the condition $\lambda_1 \leq 1 \leq \lambda_3$ on the eigenvalues $(\lambda_1, \lambda_2, \lambda_3)$ of the matrix $\mathbf{G}^{-T} \mathbf{F}^T \mathbf{F} \mathbf{G}^{-1}$; if this eigenvalue condition holds, then there is a set of film orientations $\hat{\mathbf{e}}_3$ that exhibit coherent interfaces. In bulk, the eigenvalues have to satisfy a much stricter condition: $\lambda_1 \leq \lambda_2 = 1 \leq \lambda_3$. Therefore, it is possible to form many *more* interfaces in thin films than in bulk. On the other hand, if we can form an interface in the bulk, we can form an interface in the thin film; further, the interface in the film is the trace of the bulk interface on the plane of the film.

Therefore, all twins (interfaces between two variants) that are possible in bulk are possible in thin films. However, there are more possibilities in thin film: for example in contrast to bulk, it is possible to construct interfaces between all pairs of variants in $(110)_c$ film of NiTi.

For future use, let us look near one of the interfaces as shown in Figure 3b. For simplicity, assume $\mathbf{G} = \mathbf{I}$. After deformation, one side of the film shears by an angle θ and stretches by a quantity λ relative to the other. It turns out that $\sin \theta = \hat{\mathbf{e}} \cdot \mathbf{F}^T \mathbf{F} \mathbf{n}$ while $\lambda = |\mathbf{F} \hat{\mathbf{n}}|$.

AUSTENITE-MARTENSITE INTERFACE

We now examine if it is possible to form an *exact* austenite-martensite interface: an interface between the austenite and a *single* variant of martensite. Recall that it is almost never possible to form such an interface in the bulk. Instead the austenite forms an interface with fine twins of martensite. In fact, this is the basis of the crystallographic theory of martensite. In contrast, in thin films, we can form a wide variety of exact austenite-martensite interfaces. We give two examples.

$(111)_c$ films of Cubic to Tetragonal materials.

We substitute $\mathbf{F} = \mathbf{U}_1$, $\mathbf{G} = \mathbf{I}$ and $\hat{\mathbf{e}}_3 = \frac{1}{\sqrt{3}}(1, 1, 1)$ in (5) and conclude that it is possible to find an interface between the austenite and a single variant of martensite if and only if the material parameters η_1, η_2 satisfy the condition $(1 - \eta_1^2)(2\eta_2^2 + \eta_1^2 - 3) \geq 0$. The normals to the two possible interfaces in a cubic basis are

$$\begin{pmatrix} \pm 3(1 - \eta_1^2) - \sqrt{(1 - \eta_1^2)(2\eta_2^2 + \eta_1^2 - 3)} \\ \pm 3(\eta_1^2 - 1) - \sqrt{(1 - \eta_1^2)(2\eta_2^2 + \eta_1^2 - 3)} \\ 2\sqrt{(1 - \eta_1^2)(2\eta_2^2 + \eta_1^2 - 3)} \end{pmatrix}. \tag{6}$$

Table 1: Exact austenite-martensite interfaces in a $(110)_c$ Nickel-Titanium film

Interface with Variant	Normal to the interface
1	(-0.623295, 0.623295, -0.472236) or (-0.498218, 0.498218, 0.709619)
1'	(-0.694088, 0.694088, -0.191005) or (0.179439, -0.179439, 0.967266)
2	(0.623295, -0.623295, -0.472236) or (0.498218, -0.498218, 0.709619)
2'	(0.694088, -0.694088, -0.191005) or (-0.179439, 0.179439, 0.967266)
3	(-0.694088, 0.694088, 0.191005) or (0.179439, -0.179439, -0.967266)
3'	(-0.498218, 0.498218, -0.709619) or (-0.623295, 0.623295, 0.472236)
4	(0.623295, -0.623295, 0.472236) or (0.498218, -0.498218, -0.709619)
4'	(-0.179439, 0.179439, -0.967266) or (0.694088, -0.694088, 0.191005)
5	(-0.360979, 0.360979, -0.859877) or (0.681337, -0.681337, -0.267507)
5'	(-0.681337, 0.681337, -0.267507) or (0.360979, -0.360979, -0.859877)
6	No interface possible with this variant
6'	No interface possible with this variant
The variants are labeled according to Matsumoto et al [5]	

$(110)_c$ films of Nickel Titanium.

Using (5) to check the various possibilities, we find that the austenite can form an exact interface with 10 out of the 12 variants of martensite. These variants are identified in Table 1 along with the calculated normals.

TUNNELS AND TENTS

We now examine if it is possible to form tunnels and tents as shown in Figure 4. The idea is to deposit a film on a substrate, then release it in some region and look for the following behavior: the film is flat in the austenite state, while it bulges up to a tunnel or a tent as it transforms to the martensite (perhaps under some back pressure). If this is possible, then these structures possibly can be exploited to make micropumps, microvalves and other micromachine actuators.

It is possible to form such a tunnel[3] if the relative shear $\theta = 0$ and the normal stretch $\lambda > 1$; this is equivalent to

$$\hat{e}_3 \cdot \mathrm{cof}(\mathbf{F}^T\mathbf{F} - \mathbf{I})\hat{e}_3 = 0 \qquad \text{and} \qquad \mathrm{trace}(\mathbf{F}^T\mathbf{F}) - \hat{e}_3 \cdot \mathbf{F}^T\mathbf{F}\hat{e}_3 - 2 > 0. \qquad (7)$$

This turns out to be a rather severe restriction on the matrix $\mathbf{F} = \mathbf{U}_i\mathbf{U}_{sub}^{-1}$ for the ith variant of martensite where $\mathbf{U}_{sub}$ is the matrix that takes the stress-free austenite to the stress-free substrate. We have assumed above that the substrate is relatively thick. Perhaps it is possible to satisfy (7) with a clever choice of film and substrate. Similarly, it is possible to form an n-sided pyramidal tent[3] if the film is a plane of n-fold symmetry of the austenite and we satisfy (7). The orientation of the pyramid is governed by (4). In a material with cubic austenite, we expect a 4-sided pyramid if we release a suitably oriented square in a $(100)_c$ film or a 3-sided pyramid if we release a suitably oriented equilateral triangle in a $(111)_c$ film.

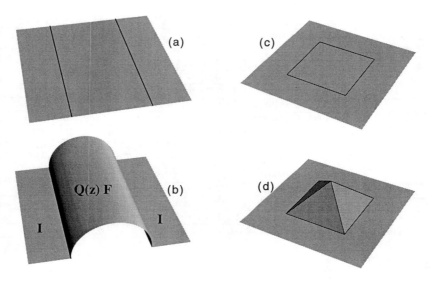

Figure 4: Proposed tunnel and tent microstructure

COMMENTS

We can easily extend our theory to the case of a film bonded to a deformable substrate, or to multilayer structures. However, note that the theory we have studied so far is suitable for the situation when the lateral size of the film (dimension of S) is much larger than the thickness h. Consequently, our theory does not contain any "bending" energy as is clear from the solutions shown in Figure 2. However, in many micromechanical devices of interest, shape memory film multilayers are used as cantilevers where the lateral size is comparable (a few times) to the total thickness. In these situations, it is critical to consider bending. The higher-order corrections of our theory that include bending are discussed in [3].

ACKNOWLEDGMENTS

We gratefully acknowledge the partial financial support of the AFOSR (KB:F49620-95-1-0109 and RDJ:AF/49620-96-1-0057), NSF (KB:CMS-9457573 and RDJ:NSF/DMS-9505077), ONR (RDJ:N/N00014-91-J-4034) and the TRW Foundation (KB).

REFERENCES

1. P. Krulevitch, A.P. Lee, P.B. Ramsey, J. Trevino, J. Hamilton and M.A. Northrup, to appear in J. MEMS.
2. J.M. Ball and R.D. James, Arch. Rat. Mech. Anal. **100**, 13, (1987).
3. K. Bhattacharya and R.D. James, In preparation.
4. P.M. Naghdi, Handbuch der Physik, Vol VI a/3 (1972).
5. O. Matsumoto, S. Miyazaki, K. Otsuka and H. Tamura, Acta Metall. **35**, 2137, (1987).
6. K. Bhattacharya, In preparation.

CONSTITUTIVE EQUATION OF NiTi SUPERELASTIC WIRE

F. Yang, Z.J.Pu, and K.H.Wu
Department of Mechanical Engineering
Florida International University
Miami, FL 33199

ABSTRACT

During the superelastic deformation process, because of the involvement of the austenite-martensite phase transformation, a superelastic wire will experience a self-heating or self-cooling process due primarily to the latent heat of the material. As the strain rate increases, and conditions become more adiabatic, the self-heating and self-cooling will cause a temperature rise upon loading and will drop upon unloading. As a consequence, an apparent effect of the strain rate on the superelastic behavior in the shape-memory alloys with a large diameter or more adiabatic conditions will be noticed. In the present paper, a constitutive stress-strain-strain rate equation is proposed to describe the self-heating behavior. In order to verify the model, a series of experiments have been conducted to study the effect of the strain rate, wire diameter, and adiabatic condition on the superelastic behavior of the shape-memory alloy wire. As will be shown later, the proposed equation can predict the behavior of the superelastic wire accurately, and the prediction is in good agreement with the experimental data.

INTRODUCTION

Shape-memory alloys (SMAs) are among the most exciting of the intelligent materials. Their actual and potential mechanical applications have created a demand for research that barely keeps pace with the consumption of reported results [1-3]. Superelasticity is one of the most important aspects of the shape-memory alloys. To date, there are numerous studies available on superelastic behavior. However, most of the previous work focused on the study of the influence of temperature on the stress-strain curves [4-5]. Little attention has been paid to the influence of the strain rate on the superelastic hysteresis behavior of the NiTi SMAs.

Rodriguez et al. [6] studied the effect of the strain rate on the superelastic behavior of Cu-based shape-memory alloys. It was reported that the threshold stress for the stress-induced austenite-to-martensite phase transformation slightly increases with the increasing strain rate. Conversely, the threshold stress for the reverse transformation, corresponding to the lower plateau of the stress-strain curve, decreases with the decrease of the strain rate. In the superelastic deformation process, because the involvement of the austenite-martensite transformation, the release of absorption of the latent heat within the material will cause the self-heating and self-cooling of the specimen. Rodriguez et al. pointed out that the strain rate itself does not impose any influence on the superelastic behavior of the Cu-based alloys. On the contrary, the apparent influence of the strain rate is originated from the self-heating or self-cooling effect because of the involvement of the latent heat. Mostly recently, Witting [7] had conducted a series of experiments to study the strain-rate sensitivity of superelastic behavior of NiTi-based alloys. It was noticed in his study that the strain rate has a pronounced effect on the stress-strain curve of NiTi superelastic wires. The upper plateau stress, which corresponds to the loading cycle, significantly increases and the lower plateau stress, corresponding to the unloading, noticeably decreases as the strain rate increases. In Witting's work, a phenomenological constitutive equation is proposed to describe the strain-rate effect of the superelastic behavior.

317

Mat. Res. Soc. Symp. Proc. Vol. 459 ® 1997 Materials Research Society

Wu et al. [8] studied the strain-rate effect of the NiTi superelastic wires. In their study, two conditions are carefully maintained to avoid the self-heating (or self-cooling) effect. First, the finest NiTi wire available in the commercial market (d=0.0254 mm) was used to conduct the tests. Secondly, all the tests were conducted in water. The experimental results clearly show that the strain rate has no effect on the superelastic behavior of the tested shape-memory wires when these two conditions are fulfilled. Their results agree with the Rodriguez's viewpoint, i.e., the apparent strain-rate effect, commonly observed in the large diameter wire tested in air, is caused by the self-heating or self-cooling effect rather than by the strain rate itself.

This paper presents the results of a continuous effort on the mechanical behavior of superelastic wire in our group. A constitutive equation has been proposed to describe the apparent stress- strain-strain rate-temperature relationship. An attempt has been made to conduct experiments to verify the model and to provide a database for the design of the superelastic devices and applications using superelastic wires.

CONSTITUTIVE STRESS-STRAIN-STRAIN RATE-TEMPERATURE EQUATION

As stated above, the commonly observed strong influence of the strain and strain rate on the threshold stress for the stress-induced austenite-to-martensite transformation is an apparent phenomenon. The nature of this phenomenon is associated with the latent heat of the material and the thus-induced self-heating effect. Here, it was assumed that the stress of the superelastic wire during the stress-induced austenite-to-martensite transformation is independent of the strain and strain rate, but dependent on the real-time temperature of the wire. The stress-strain-strain rate-temperature constitutive equation of the shape-memory alloy during the stress-induced austenite-to-martensite transformation could be easily described by the Clausius-Clapeyron relationship as follows:

$$\sigma = \sigma_0 + a_p T_s \tag{1}$$

where T_s is the real-time temperature of the superelastic wire. σ is the threshold stress for the stress-induced austenite-to-martensite transformation at the real temperature of the wire, T_s. a_p is a coefficient in the Clausius-Clapeyron relationship. If the test is conducted in an environment with a temperature of T, the real-time temperature of the wire, including the effect of self-heating, can be described by the following equation:

$$T_s = T + \Delta T (\varepsilon, \dot{\varepsilon}) \tag{2}$$

Combining Eqs. (1) and (2), the threshold stress for the stress-induced austenite-to-martensite transformation at the environment, T, strain, ε, and strain rate, $\dot{\varepsilon}$, can be expressed as follows:

$$\sigma = \sigma_0 + A T_s = \sigma_0 + A[T + \Delta T (\varepsilon, \dot{\varepsilon})] = \sigma_0 + a_p T + a_p \Delta T (\varepsilon, \dot{\varepsilon}) \tag{3}$$

Equation (3) describes the stress-strain-strain rate-temperature relationship of the superelastic wire. Here, the key is to determine the $\Delta T(\varepsilon, \dot{\varepsilon})$ as a function of strain, ε, and strain rate, $\dot{\varepsilon}$.

Assume at a certain time, t, the temperature of the sample is T. For any time change, dt, the temperature change is (T_s-T) during the stress-induced austenite to martensite transformation. The energy equilibrium equation during the stress-induced transformation then is as follows:

$$\frac{1}{2} (\frac{d\varepsilon}{dt})\sigma\pi r^2 l_0 dt + (\frac{d\varepsilon}{dt})dt\pi r^2 l_0 \rho \frac{\Delta H}{\varepsilon_u} - 2h\pi rl_0 (T_s - T)dt = C_p \pi r^2 l_0 \rho dt \tag{4}$$

where, the first term on the left-hand side is the energy produced by the applied stress. The second term on the left-hand side is the energy involved in the phase transformation. ε_u is the total strain taken from the beginning till the end of the plateau of a stress-strain curve of a wire, in

318

which the transformation is fully completed. ΔH is the latent heat of the phase transformation. The third term is the energy exchanged between the specimen and the surrounding environment. Here h is the convection coefficient between the specimen and the environment. The term on the right side designates the energy the sample absorbs due to the change of temperature. And r is the radius of the sample, l_0 is the length of the sample. ρ and C_p are the density and the specific heat of the wire, respectively.

The solution to Eq. (4) is given by Eq. (5). Substituting Eq. (5) into Eq. (3), the threshold stress for the stress-induced austenite-to-martensite transformation at environment, T, strain, ε, and strain rate, $\dot{\varepsilon}$, can be obtained and is shown in Eq. (6). Equation (6) is the constitutive stress-strain-strain rate-temperature equation that is capable to describe the behavior of the superelastic wires.

$$\Delta T(\varepsilon) = \frac{r\dot{\varepsilon}\sigma_0 + 2r\dot{\varepsilon}\rho\,\dfrac{\Delta H}{\varepsilon_u}}{r\dot{\varepsilon}a_p - 4h}\,(e^{\frac{r\dot{\varepsilon}a_p - 4h}{2r\dot{\varepsilon}C_p\rho}\varepsilon} - 1) \qquad (5)$$

$$\sigma = \sigma_0 + a_p T + a_p\,\frac{r\dot{\varepsilon}\sigma_0 + 2r\dot{\varepsilon}\rho\,\dfrac{\Delta H}{\varepsilon_u}}{r\dot{\varepsilon}a_p - 4h}\,(e^{\frac{r\dot{\varepsilon}a_p - 4h}{2r\dot{\varepsilon}C_p\rho}\varepsilon} - 1) \qquad (6)$$

NUMERICAL RESULTS

Figures 1a and 1b show the variation of ΔT and threshold stress, σ, as a function of strain ε and strain rate $\dot{\varepsilon}$, respectively. As can be readily seen in Figure 1a, ΔT increases with increasing strain and strain rate. When the strain rate is relatively low, ΔT is nearly zero, and appears to be independent of strain variation. This implies that if the strain rate is relatively low, the wire has sufficient time to exchange heat with the environment and the wire is able to maintain the same temperature with the environment. Inasmuch, the stress remains constant with the strain as can be seen in Figure 1b. As the strain rate progressively increases, the ΔT increases and its dependence on the strain magnitude becomes more pronounced (see Fig. 1a). Similarly, the threshold stress also increases correspondingly with the increasing strain rate and displays a dependence to the strain magnitude (see Fig. 1b). When the strain rate is sufficiently high, both ΔT and the threshold stress reach a saturated value and are independent of the further increase of the strain rate.

Figures 1c and 1d show the variation of ΔT and threshold stress, σ, as a function of strain, ε, and wire diameter, d, respectively. Based on Figure 1c, ΔT increases with increasing strain and wire diameter. For a very small diameter wire, ΔT is almost near zero, and, again, is independent to the variation of the strain. This means that if the wire diameter is very small, heat induced by the phase transformation can dissipate rapidly and the wire can easily maintain a uniform temperature with its environment. For this reason, the stress remains constant with the variation of the strain and can be clearly seen in Figure 1d. From Figures 1c and 1d, it can be noticed that both ΔT and the threshold stress increase correspondingly with the increasing diameter, then reach to saturated values, and their dependence on the magnitude of strain also becomes more pronounced.

Figures 1e and 1f show the variation of ΔT and threshold stress, σ, as a function of strain, ε, and the heat convection coefficient, h, respectively. As expected, increasing the heat convection

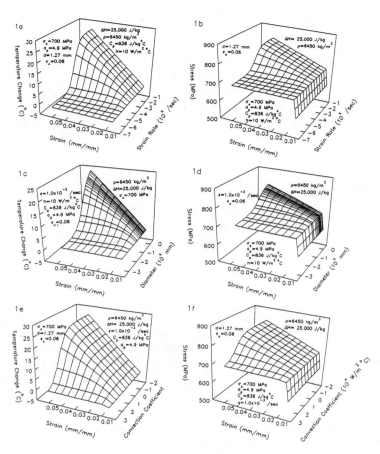

Figure 1. The variation of ΔT and σ as a function of (a) (b) strain and strain rate, (c) (d) strain and wire diameter, (e) (f) strain and heat convection coefficient, respectively.

coefficient, h, tends to suppress the ΔT value, to the extent that it is almost near zero, and the strain appears to have no influence on the temperature difference (see Fig. 1e). This suggests that if the heat-exchange rate is very high, the sample has sufficient time to exchange heat with the environment and the wire is able to achieve the same temperature as the environment. For the same reason, the stress remains almost constant with the variation of the strain for a very fine wire, as noticed in Figure 1f. A high value of h has a tendency of increasing both ΔT and the threshold stress, as well as their dependence on the strain magnitude. However, both ΔT and the threshold stress will eventually reach a saturated state as h continues increasing.

COMPARISON OF THEORETICAL AND EXPERIMENTAL DATA

A series of experiments have been conducted to characterize the stress-strain curves of the superelastic wire. NiTi superelastic wires with different wire diameters were used in this study.

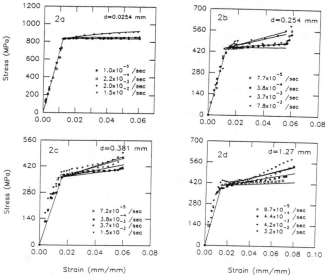

Figure 2. The comparison of the theoretical prediction stress-strain curves based on Constitutive Equation and experimental stress-strain curves with four different wires.

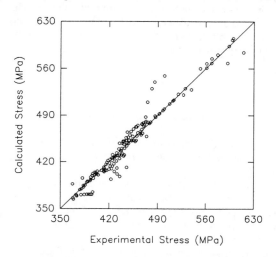

Figure 3. The comparison of theoretical prediction stress and experimental stress with different strains, strain rates and wire diameters.

Wires were tested on an INSTRON machine and in air at room temperature. Figure 2 shows the experimental data of the stress-strain curves with four different wires at different strain rates. In this figure, the theoretical prediction curves are also plotted as a reference. Figure 2 clearly demonstrates that all the theoretic prediction curves agree well with the experimental data. The experimental data of the threshold stress at different strain, strain rate, and wire diameter versus the theoretic prediction value are plotted in Figure 3. The experiments were carefully designed and conducted and more than 300 data are generated in Figure 3. The results shown in Figure 3 further confirm the fact that the proposed stress-strain-strain rate-temperature constitutive equation is capable of predicting the behavior of the superelastic wires.

LIMITATION OF THE CONSTITUTIVE EQUATION

The development of the constitutive equation is based on an assumption that the temperature of the wire in the cross section is uniform, and does not vary with the spatial location. This is only true for the wires with a relatively small diameter in comparison with its length. If the wire diameter is relatively large, the assumption is violated and the prediction is no longer accurate. Practically, it is difficult to define an upper bound for the validity of the theory. Nevertheless, in our experiments, the maximum diameter of the wire is 1.27 mm and the data comply with the model quite well. It is therefore safe to say that the diameter's upper bound is at least beyond 1.27 mm.

CONCLUSIONS

In superelastic wire, it is noticed that the threshold for the stress-induced austenite-to-martensite transformation depends on the strain, strain rate, wire diameter, and heat-transfer coefficient. This behavior is associated with a self-heating phenomenon caused by the latent heat of the material. In the present paper, a constitutive equation has been proposed to describe the apparent stress-strain-strain rate-temperature relationship through the consideration of the self-heating effect. Experimental data are provided to verify the mode and the results demonstrate that the proposed model can predict the response of the superelastic wire very well and the prediction is in good agreement with the experimental data.

REFERENCES

1. C. Liang and C.A. Rogers, *Journal of Intelligent Material Systems and Structure* **1** (1990) 207.
2. T. Takagi, *Journal of Intelligent Materials and Structure* **1** (1990) 149.
3. B. K. Wada, J. L. Fanson, and E.F. Crawley, *Journal of Intelligent Materials and Structure* **1** (1990) 157.
4. T.W. Duegin, K.N. Melton, D. Stockel, and C.M.Wayman, in Engineering Aspect of Shape Memory Alloys (Butterworth-Heinemann, 1990).
5. J. Perkins, in Shape Memory Effects in Alloys, ed. by J. Perkins (Plenum Press, 1975).
6. C. Rodriguez and L.C.Brown, in Shape Memory Effects in Alloys, ed. by J. Perkins (Plenum Press, 1975), p. 27.
7. P.R. Witting, Ph.D. Dissertation, University of New York at Buffalo, 1994.
8. K.H. Wu, F. Yang, Z. Pu, and J. Shi, *Journal of Intelligent Materials and Structures*, **7** (1996) 138.

THE TWO-WAY MEMORY EFFECT BY PRE-STRAINING IN A 45Ni50Ti5Cu ALLOY

G. Airoldi*, T.Ranucci**
*Dipartimento di Fisica, Universita' di Milano, Via Celoria 16, 20133 Milano,I,
airoldi@mvmidi.mi.infn.it
**ITM-C.N.R., Area della Ricerca di Milano, Via Bassini 15,20133 Milano, I

ABSTRACT

The 45Ni50Ti5Cu shape memory alloy exhibits one single thermoelastic martensitic transformation from the parent B2 phase to monoclinic B19'and looks as an appealing system to be used for the two way memory effect (TWME). First results, appeared in literature, on the TWME in 45Ni50Ti5Cu indicate that in this alloy high figures for the two way strain can be obtained. The prestrain method is here investigated to imprint the TWME in a 45Ni50Ti5Cu alloy: a pseudoplastic deformation higher than the reorientation strain is applied in martensite, followed, after unloading, by a recovery of the one-way strain by heating to 110°C. The two way memory strain is detected in the following thermal cycle. Results show the two way memory strain can be imprinted in just one cycle, but it can even considerably be increased by repeating the procedure: the highest figure for the two way memory strain is however obtained whenever a plastic unrecoverable strain is accumulated in the specimen in some prestraining cycles. Often the maximum two way strain is of the same order as the plastic strain and the former decreases with the increase of the latter.

INTRODUCTION

The pre-strain training procedure looks an appealing way to develop the two-way memory effect (TWME). The results previously obtained on 40Ni50Ti10Cu alloys [1] have however shown that two deformation strain contributions are present in the TWM strain (ε_{TW}): one connected with the B2 parent phase - orthorombic B19 transformation, the other with the B19 - monoclinic B19' transformation. Moreover the temperature range to exploit the whole ε_{TW} is rather wide, approximately 200°C. On the other hand the ε_{TW} figure connected with just the B2-B19 transformation, although confined to a narrower temperature range of the order of 25°C, appears low in absolute [2].

The 45Ni50Ti5Cu alloy, due to the presence of one single direct thermoelastic martensitic transformation from the B2 parent phase to the monoclinic B19' looks a good candidate to reach a high ε_{TW} in a fairly narrow temperature range.

The procedure: deformation in B19' + unloading + recovery to B2, is defined a pre-straining step. The effect of repeated pre-straining steps on ε_{TW} are here investigated. During either each pre-straining step or the two way memory effect, both deformation and electrical resistance are detected in order to investigate the potential of the electrical resistance to serve as an internal variable of an actuator.

EXPERIMENT

Commercial 45Ni50Ti5Cu (at%) wires, 1mm in diameter, supplied by Furukawa Electric Co. (J), have been examined. Specimens 150mm in lenght have been solution treated in vacuum at 850°C (3.6ks) and quenched in water at room temperature.

Electrical Resistance (R) vs. temperature (T) measurements in the range -80°C÷+110°C were carried out by a conventional four terminal DC method using a digital micro-ohmmeter ESI 1701B.

Mechanical tensile tests were carried out by a 1455 Zwick testing machine (load cell 2kN), equipped with a thermostatic chamber operated between -80°C and 110°C; strain was detected by a Zwick extensometer (gauge lenght 10mm, range ±2mm). The specimen temperature was detected by three T-type thermocouples, spot welded one inside the gauge region, the others outside the extensometer region.
Pre-strains were applied in strain control mode (strain rate 10^{-4} s^{-1}) at room temperatures where the martensitic B19' phase is almost thoroughly settled. After unloading the pseudoplastic strain connected with the one way memory strain is recovered by heating to 110°C. ε_{TW} is detected during the following thermal cycle between -70°C and +90°C.

RESULTS

The R vs. T, detected before training, of a typical specimen is plotted in figure 1: it can be appreciated the presence of one single transformation, almost completed at room temperature with M_f, M_s, A_s, A_f respectively 34°C, 49°C, 45°C, 70°C.
In each pre-straining step the specimen is submitted to a fixed total applied strain ε_t, and unloaded; a heating to 110°C recovers the one-way strain ε_{ow}. The strain left unrecovered is the plastic strain ε_p. In figure 2 one pre-straining step and the related ε_t, ε_{ow} and ε_p are exemplified.

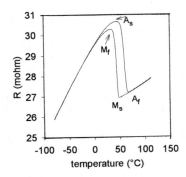

Fig.1 - R vs. T for a typical specimen

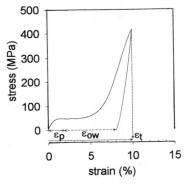

Fig. 2 - Stress-strain during one pre-straining step up to ε_t=10%. ε_{ow} recovered during heating and ε_p left unrecovered are shown.

ε_{TW} is evaluated during the following thermal cycle between -70°C and +90°C.
For each selected total applied strain, the above sequence can be repeated and ε_{TW} is evaluated each time.
The total applied strains investigated were 8, 10, 12, 13, 15%.
A typical sequence of stress vs. strain curves performed on one specimen, with

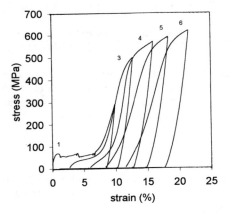

Fig. 3 - Stress vs strain during several (1,3,4,5,6) prestrain steps in sequence each with ε_t=10%.

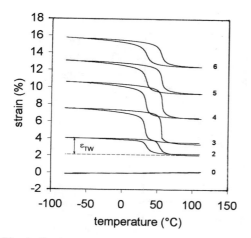

Fig. 4 - Strain vs temperature during the termal cycles before (0) and after 2,...,6 prestraining steps, each with ε_t=10%.

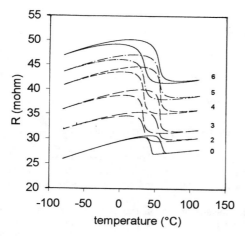

Fig. 5 - R vs T during the thermal cycles before (0) and after the 2,...,6 pre-straining steps.

pre-strain steps $\varepsilon_t = 10\%$, is plotted in figure 3.

A recovery cycle to 110°C, is applied between one curve and the following one.

An example of the thermal cycles following several pre-straining steps, each corresponding to the above described procedure, are given in figure 4: in each pre-straining step the specimen was submitted to a $\varepsilon_t = 10\%$. Each ε vs. T curve is shifted upward of a value corresponding to the unrecovered ε_p at T=110°C in the previous pre-strain steps. The R vs. T curves related to the same thermal cycles are given in figure 5.

The ε_{TW} values, deduced as shown in figure 4, are plotted in figure 6a as a function of the pre-strain cycle number and for each pre-strain value; in figure 6b the plastic strain ε_p, left unrecovered at T=110°C is plotted for each step sequence. It can be seen that except for the maximum pre-strain steps of 15%, both the ε_{TW} values and ε_p, left after recovery, increase with the pre-strain cycle number, though with different rates.

Finally ε_{TW} as a function of the unrecovered plastic strain, left after the previous pre-strain cycle, is given for all the performed tests in figure 7.

DISCUSSION

A minimum total pre-strain $\varepsilon_t = 8\%$ has been selected in order to be sure the variant reorientation process is completed and some unrecovered ε_p is left after high T recovery.

In all the performed tests, the first stress-strain curve, after the elastic range, always shows some pop-in, in the strain range related to the variant reorientation process (see

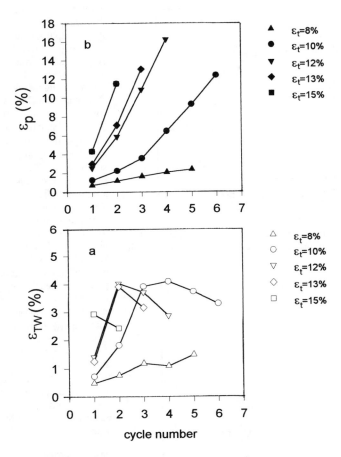

Fig. 6 - a) ε_{TW} and b) ε_p as a function of the prestrain cycle number;

figure 3), which almost disappears in the curve related to the following pre-strain cycle. The load instabilities correspond to the knees found in the correspondent curve R vs ε, detected altogether during the prestrain step in martensite, as shown in figure 8, and already put into evidence in a previous paper[3].

For clarity in figure 8 each curve is shifted upward of 0.5 on the R/R_0 axis. It can there be appreciated two sharp steps, present in the first pre-strain loading, which modify into one in the second pre-strain step and disappear in the following ones. That is most probably due to the detwinning process which in the first pre-strain cycle is different in comparison to the following ones. Actually this behaviour testifies the effect of the training procedure itself: the accumulated plastic strain has teached the specimen the way which best accomodates the new applied strain in the following pre-strain step.

ε_{TW} generally increases in few pre-strain steps, reaches a maximum and decreases after some pre-strain cycles: that is clearly shown either in figure 4, where ε vs T is given during all the thermal cycles performed after each pre-strain step with $\varepsilon_t =$ 10%, or in figure 6a where ε_{TW} is given for all the performed tests.

In the case of low pre-strain steps, as for $\varepsilon_t = 8\%$ in the present investigation, the maximum ε_{TW} value is not reached even after six prestraining steps. For higher applied pre-strains, as for $\varepsilon_t =10, 12, 13\%$, the higher is ε_t ,the lower is the prestrain cycle number to reach the maximum ε_{TW}. For $\varepsilon_t=15\%$, after the first pre-strain cycle, the maximum ε_{TW} appears already overcome.

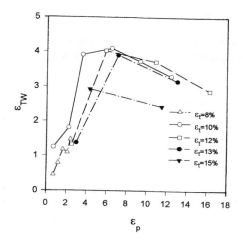

Fig. 7 - ε_{TW} as a function of ε_p for all tests.

Each pre-strain cycle builds a plastic strain ε_p, left unrecovered at high temperature: as well shown in figure 6b, ε_p plainly increases for the pre-strain steps at $\varepsilon_t =8\%$, with higher increasing rates for higher pre-strains ε_t. The highest ε_{TW} is reached before the higher increasing rate for ε_p steps in: this appears for an ε_p value comparable to the ε_{TW} itself.

The ε_{TW} values here given, reliably checked on several specimens, appear lower in absolute when compared to the data obtained in literature[4] on similar 45Ni50Ti5Cu alloys. The difference cannot be explained at the light of present investigation, though it might be attributed to a different start thermal treatment.

As it concerns the relationship between strain and resistance

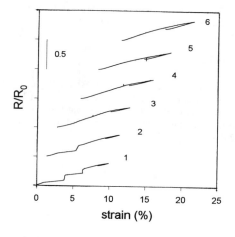

Fig.- 8 - R/R$_0$ vs ε during six prestraining with $\varepsilon = 10\%$. R$_0$ is the electrical R at $\varepsilon=0$

during the TWME, it can be pointed out the similarity between ε and R, as shown in figures 4, 5, across the transformation range.

Actually, in well trained specimens, as shown in figure 9, the R/R_0 vs ε curve, across the transformation range exhibits an almost linear behavoiur, worth to be considered in applications.

CONCLUSIONS

The pre-strain training appears an appealing way to imprint the TWME. In the 45Ni50Ti5Cu alloy here investigated the highest ε_{TW} values are found for plastic strains unrecovered comparable or even higher than ε_{TW}.

The highest ε_{TW} is settled after at least two pre-strain cycles for a total applied strain step higher than 10%.

The steps found in the R vs. ε in martensite during pre-straining evidence the electrical resistivity modifies with variant reorientation.

In trained specimens, where the electrical resistivity change connected with band structure change has the same sign of the resistance change induced by deformation, R can act as a control variable of deformation.

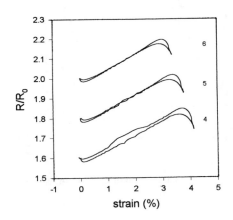

Fig. - 9 - R/R_0 vs ε during thermal cycles after the 4th, 5th, 6th pre-strain step at $\varepsilon_t = 10\%$.

ACKNOWLEDGMENTS

The specimens have been supplied by Furukawa Electric Co., J., here gratefully acknowledged.

REFERENCES

1) G.Airoldi, T.Ranucci, G.Riva, A.Sciacca, J.Phys.:Condens.Matter 7,3709 (1995)
2) G.Airoldi, T.Ranucci, G.Riva, A.Sciacca, Scripta Mater. 34,287, (1995)
3) G.Airoldi, T.Ranucci, G.Riva, Trans.Mat.Res. Soc. Jpn., vol 18B,1109 (1994)
4) S.Miyazaki,S.Chujo, Proc.1st Int.Conf.on Shape Memory and Superelastic Tecnologies, SMST-94,1994,pg.73

Part VI

Shape-Memory Alloys II

THERMOTRACTIVE TITANIUM-NICKEL THIN FILMS FOR MICROELECTROMECHANICAL SYSTEMS AND ACTIVE COMPOSITES

D. S. GRUMMON AND T. J. PENCE
Department of Materials Science and Mechanics
Michigan State University, East Lansing, MI 48824

ABSTRACT

Thin films of thermoelastic titanium-nickel are of interest as a material basis for force-producing elements in microelectromechanical systems, and for active phases in mechanically-adaptive composite materials. The successful introduction of this material system into such application areas will depend on development of reliable thin film deposition protocols, together with the refinement of analytical models which successfully predict the response of active microstructures to a variety of dynamic thermal and mechanical stimuli. In the present paper we review some of our recent experimental and theoretical work which bear on these problems. With respect to thin film fabrication techniques we focus on problems of composition control and the manipulation of microstructure, with particular emphasis on opportunities afforded by amorphous precursor phases formed during low temperature processing, and the fine-grained, thermally stable crystalline microstructures obtainable using hot-substrate deposition. The films resulting from either approach retain the important thermomechanical response features of the well-known bulk-alloy system: shape memory and transformational superelasticity. The response can be modeled in terms of a continuum description augmented with internal variables that track fractional partitioning of the material between austenite and variants of the martensite.

INTRODUCTION

Equiatomic titanium-nickel forms the basis of an important class of shape-memory and superelastic alloys whose unique constitutive behavior may play a role in enabling materials technologies for smart material systems, microelectromechanical actuators, and mechanically active composite materials systems. They may also find application to problems in fatigue, damping of small systems, and functionally graded interfaces for joining dissimilar materials. The highly reversible and energetic displacive transformation exercised by titanium-nickel has several interesting technical implications: For example, ohmic electrical excitation can alter elastic compliance by a factor of four or more, of potential interest for control of impedance for sensors and transducers. Joule excitation can also alter internal friction and damping capacity, the latter being particularly high for the low-temperature martensite phase. Of particular interest to microelectromechanical systems, however, is the fact that the martensite-to-austenite transformation can generate large displacements and very high force output: the equivalent of one gram of force can, in principal, be generated from a 10 μm wide film that is only 3 μm thick. This corresponds to very high actuator energy-densities (exceeding 10 MJ-m^{-3}) which is several orders of magnitude greater than can be achieved using other available microactuator materials [1]. The most serious disadvantage, apart from general thin film process sensitivities, is the relatively slow cyclic response in comparison to alternative materials, though the latter drawback diminishes to some degree at smaller dimensional scale. The design of real actuators capable of long-term cyclic displacement will require thorough understanding of the role of intrinsic and extrinsic stress evolution during processing and micromachining, applied to reliable analytical models of thermotractive response to indirect and ohmic heating, and may require the development of novel approaches to the incorporation of bias-force elements in self-assembled structures[1].

Thin films of TiNi may also be processed to achieve classic transformational superelasticity [2, 3] This effect allows completely recoverable anelastic strain excursions to over 5%, at low stress, while generating few dislocations or other damage artifacts. This characteristic may find

---

[1] For additional discussion of stress-effects and thin film displacive behavior, see Zhang, et al., in this symposium.

Mat. Res. Soc. Symp. Proc. Vol. 459 © 1997 Materials Research Society

use in the design of functionally graded materials (FGMs) for joining problems complicated by widely differing thermal expansion coefficients, and may also be applicable to certain problems in low-cycle fatigue [4]. Patterning techniques for titanium-nickel films have been shown to be compatible with microelectronic lithography procedures, and the films have been successfully deposited and micromachined on polyimide substrates [5, 6]. New microscale interconnect schemes and passive/active dimensional adjustment capabilities may thus be possible.

The present paper will begin with a discussion of the difficulties associated with the preparation of thermotractive thin films in view of the complications imposed by the high sensitivity of critical phase-transformation temperatures to deviations from stoichiometry. We will then provide some examples from the range of microstructures which can be achieved via selection of deposition temperature and post-deposition annealing. Finally, we will present some mechanical behavior data for uniaxial loading over a range of test temperatures which compare reasonably well with results of recent analytical modeling efforts. The latter utilizes a minimal number of material parameters: transformation temperatures, transformation strain, elastic modulus, and transformation latent heat. The martensite flow and transformation finish stresses may also be specified in refined treatments. Together, these permit calculation of ongoing strain accumulation and annihilation under general histories of temperature and uniaxial-stress.

EXPERIMENTAL METHODS

All sputtering targets used in the experiments reviewed here were 50.8 mm diameter x 3 mm thick spark cut disks from a single 76 mm-diameter hot-rolled bar having a measured composition of 49.4 at% Ti - 50.6 at% Ni, a nominal A_f temperature of 263 K, and an approximate grain size of 100 μm. Sputtering was done with a horizontally mounted Torus-2C gun in a chamber pumped to a base-pressure of 3×10^{-5} Pa, with $P_{O2} < 2 \times 10^{-6}$ Pa. Planar flint glass substrates were used from which free-standing films could be readily detached.

For the composition studies, argon working-gas pressure was set at various levels between 0.4 and 1.1 Pa. New targets were brought to nominal steady-state conditions by wearing-in for 0.5 hrs at a cathode power ~500 watts. Composition of these films was measured on specimens with a minimum thickness of 3 μm using energy-dispersive X-Ray microanalysis in a Hitachi S-2500 scanning electron microscope with a LaB$_6$ filament and a Link AN-1000 system using a silicon detector and applying ZAF-4 correction codes. Spectra were collected at an accelerating potential of 20 keV, working distance of 15 mm, and at constant specimen angle and dead-time percentage, and were corrected using identical pure-Ni and known $Ni_{49.9}Ti_{50.1}$ calibration standards. Deposition rates were inferred from film thickness measurements on a Dektak-II profilometer.

For the hot-substrate deposition microstructure studies, substrates were kept at elevated temperature during deposition using resistive heating elements both in front of and behind the substrates. After outgassing, and with the substrate at temperature, the diffusion-pumped chamber maintained a base-pressure of $<3 \times 10^{-4}$ Pa. High-purity argon was gettered over pure titanium at 1023 K to remove trace oxygen and nitrogen, and was maintained at a working pressure of 0.73 Pa. Both as-sputtered films, and films subsequently given solutionizing and aging heat treatments were studied. The heat treatment of as-sputtered films were conducted in a turbo-pumped quartz-tube vacuum furnace. During the entire annealing process, the base pressure was maintained under 1.33×10^{-4} Pa.

Displacive transformation temperatures were evaluated by electrical resistimetry at cooling and heating rates of 5~10 K/min, and were examined in a Hitachi H-800 TEM operated at 200 kV. Tensile tests were carried out in a screw-driven microtensile machine at various temperatures, on specimens configured as straight-sided strips having no reduced-section, with gauge length and width of 6 mm and 1.6 mm, respectively. For this reason, the ductilities and tensile strengths observed reflect a severe specimen geometry in which failure in the grips was sometimes observed. The series of stress-strain curves shown here were made with a single specimen, with the first of the tests conducted at the lowest of the indicated test temperatures, such that the last of the curves represented behavior after ten to twelve cycles at strain amplitudes on the order of four percent. In each case, after completion of a stress-strain loop, the specimen was heated at 373 K to fully austenitize it, and then cooled to the temperature selected for the subsequent test.

RESULTS AND DISCUSSION

Effect of Process Parameters on Film Composition

Table I below summarizes work by Jiun-Chung Lee [7] showing deposition conditions and the results of composition measurements for a variety of sputtering conditions in which cathode power, working distance, working pressure and take-off angle were systematically varied. The measured deviation of the sputtered film composition from that of the binary TiNi sputter cathode are given in the last column.

Table-I
Conditions and Results for Composition-Dependence Study

Test	Initial Cathode Voltage V	Final Cathode Voltage V	Initial Cathode Current mA	Final Cathode Current mA	Average Cathode Power W	Working Pressure Pa	Deposit Incidence Angle Deg.	Average Deposition Rate nm/min	Total Deposit Depth nm	Film Composition a/o Ni ± std. dev.	Δ Ti %
Effect of Cathode Wear↓											
1a	790	739	314	336	250	0.41	0	153	9168	51.73 ± 0.14	-1.14
1b	738	689	340	365	250	0.41	0	146	8797	50.94 ± 0.22	-0.35
1c	687	643	367	. 387	250	0.41	0	143	8608	51.10 ± 0.20	-0.51
1d	642	607	388	400	250	0.41	0	137	8251	50.25 ± 0.12	+0.34
Effect of Cathode Potential↓											
2a	500	500	63	70	36	0.53	0	21	3139	52.79 ± 0.11	-2.20
2b	625	625	151	184	108	0.53	0	65	7797	56.02 ± 0.11	-5.43
2c	750	750	165	211	145	0.53	0	82	7415	56.12 ± 0.07	-5.53
Effect of Cathode Power and Deposit Incidence Angle↓											
3a	947	830	153	175	150	0.44	0	111	8896	53.32 ± 0.05	-2.73
3b	1008	918	294	324	300	0.44	0	196	7829	49.72 ± 0.30	+0.87
3c	932	983	482	465	450	0.44	0	294	5876	51.15 ± 0.12	-0.56
3a1	947	830	153	175	150	0.44	26	76	6098	51.59 ± 0.26	-1.00
3b1	1008	918	294	324	300	0.44	26	141	5636	50.64 ± 0.19	+0.05
3c1	932	983	482	465	450	0.44	26	282	5643	51.55 ± 0.16	-0.96
Effect of Argon Pressure and Deposit Incidence Angle ↓											
4a	600	600	106	141	78	0.44	0	51	6933	51.76 ± 0.05	-1.17
4b	600	600	260	320	177	0.64	0	108	8641	50.66 ± 0.17	+0.07
4c	600	600	400	475	262	1.01	0	162	12970	52.41 ± 0.18	-1.82
4a1	600	600	106	141	78	0.44	26	40	5524	51.05 ± 0.23	-0.46
4b1	600	600	260	320	177	0.64	26	87	6974	51.68 ± 0.28	-1.09
4c1	600	600	400	475	262	1.01	26	114	9186	52.59 ± 0.25	-2.00
Effect of Working Distance↓											
5a	600	600	399	431	251	0.61	0	158	5348	51.47 ± 0.33	-0.88
5b	600	600	467	474	283	0.61	0**	82	4538	51.79 ± 0.05	-1.20

**Working distance increased to 76 mm

The results from tests 1 (a-d) are plotted in Figure 1 to illustrate the effect of sputter-cathode erosion. This plot shows measured thin film composition from deposits collected during the first, second, third and fourth hours of sputtering from a single cathode at a constant power level of 250 watts. Dashed horizontal lines indicate, for reference, the approximate sputter cathode composition. Three regimes are apparent in the data: in Stage-I the nickel-content of the deposit de-

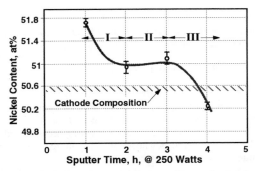

Figure 1. Change in sputtered film composition as a function of sputtering time indicating effect of cathode wear.

clines rapidly from its initially high level. In Stage-II a brief steady-state interval occurs in which the composition is constant for a period of approximately two hours. This is followed in Stage-III by another steady decline in nickel content which eventually results in compositions richer in titanium than the target.

The data shows that cathode wear over an interval of only four hours can result in substantial variations in resultant thin film composition. Since it may be presumed that the well-cooled sputter cathode has reached steady-state surface composition conditions prior to the collection of this data (i.e., that the overall composition of the sputtered flux must match the bulk composition of the cathode), it is believed that the observed variation in film composition is connected with complex effects arising from development of cathode geometry and local surface topography.

Erosion resulted in the usual 'race-track' profile expected from magnetically-enhanced sources, and also produced a microscopic topography similar to that shown in Figure 2, which was taken from a target sputtered for 15 hours at an average power of 93 watts (giving a total ion fluence on the order of 10^{20} ions/cm^2 on the 50 cm diameter target). The topography consisted of ridged mounds, 5 to 30 mm in diameter (Figure 2a), generally containing sharp 1-5 μm diameter cup-and-cone features (Figure 2b) with cone-angles generally between 50 and 60 degrees.

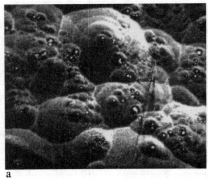
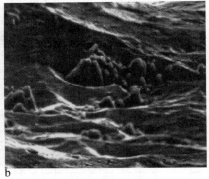

a   b

Figure 2. Target surface topographies after severe target erosion. (a) Ridged mound formations; (b) Details of cups and cones at the apex of the mound features [these correspond to the bright spot features in (a)]. [7]

The development of such surface features on the sputter cathode alters the ion incidence angle distribution as well as the sputtered-atom ejection-angle distribution. The idea that changes in these angular distributions might affect film composition is reinforced by the data from test runs 3 and 4, in which specimens were acquired at positions both on-axis and at a location 26 degrees away from the cathode normal. The general trend in this data is for slightly higher nickel content to evolve in deposits made away from the cathode axis at the higher cathode powers and sputtering rates, suggesting that the nickel flux is somewhat 'under-cosine' in relation to the titanium flux [8], i.e., that the Ti-flux is concentrated slightly more in the direction of the target normal. This is consistent with sputtering in the shallow-cascade regime where relatively few collisions occur within the target and insufficient momentum randomization occurs to produce a perfect cosine distribution. Under these conditions, there is a slight tendency for the less massive species to be more forward-sputtered.

The observed sensitivity of sputtered thin film composition to target erosion complicates the interpretation of the results of tests 2, 3, 4 and 5 (see Table I), which were designed to respectively explore the effects of varying cathode potential, power, working gas pressure, and working distance on resulting composition, and further analysis of the latter data is not presented here. Consideration of the data set as a whole, however, reveals a tendency for the sputtered film composition to approach that of the target as the sputtering rate increases. This is apparent in Figure 3, in which all of the composition data from Table-1 are plotted in terms of the ratio of the film composition to that of the sputter cathode, as a function of the average deposition rate for each test. Here it can be seen that as deposition rates approach approximately 250 nm/min (15 μm/hr) the composition ratio between film and target approaches 1.0, that is, the loss of titanium from the growing film approaches zero. This suggests that capture of Ti by reactive contaminants in the vacuum may play a significant role in shifting composition, and argues for both high deposition rates and special attention to reduction of N_2 and O_2 in the vacuum environment.

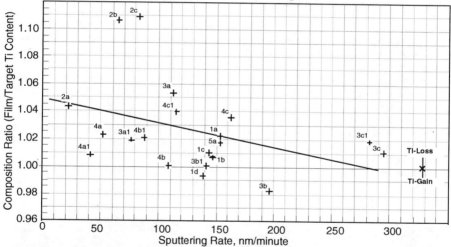

Figure 3. Compilation of data shown in Table I with thin film composition ratio (calculated as the titanium content of the sputter-cathode divided by that of the thin film) as a function of the deposition rate used in the test.

<u>Dependence of Microstructure on Substrate Temperature</u>

Low-Temperature Deposition   The preparation of austenitic titanium-nickel thin films by sputter deposition may follow one of two possible routes. In the most widely-used approach, the film is deposited to an un-heated substrate, in which case the ambient temperature[2] typically rises to ~400 K. If a reasonably low working-gas pressure is used, the result is a strong, homogenous and fully dense amorphous phase whose Cu-K$_\alpha$ X-Ray diffraction spectrum displays a broad rise between 2-θ angles of 40 and 50 degrees. To obtain the required crystalline structure, these films must be vacuum-annealed at a minimum temperature of ~700K for 1 to 2 ks, or for corre-

---

[2] Essentially identical results follow from substrate temperatures as high as ~600K.

spondingly shorter times at higher temperatures. This annealing step is not generally a problem in the case of silicon substrates, but may pose some difficulty for temperature-sensitive materials such as polymers, or for metals which would tend to rapidly diffuse into the film. One approach to the minimization of thermal damage during the crystallization anneal on dielectric substrates is to monitor the electrical resistivity of the film. As shown in Figure 4, the temperature coefficient of resistivity is negative for the amorphous phase, but positive for the B2 austenite phase. Thus, a probe for crystalline structure (at temperatures well into the austenite stability range) may consist of simply monitoring resistivity change in response to a small transient temperature change.

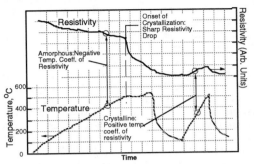

Figure 4 Plot of electrical resistivity and temperature vs time for annealing of amorphous titanium-nickel. The amorphous phase shows a negative temperature coefficient, and a sharp drop in resistivity associated with the onset of crystallization. Once crystallized the temperature coefficient becomes positive.

The end result of the cool-substrate approach is usually a relatively coarse grained (1 - 3 μm) microstructure which, if the annealing step is carefully controlled, may possess a substantial supersaturation of either nickel *or* titanium [9, 10]. The latter point is of some significance, and marks one of the principal differences between thin films and bulk alloys in regard to the microstructures which can be engineered. In both bulk and thin film alloys, the titanium solvus line rises very steeply up to high temperatures, such that alloys conventionally solidified from the melt cannot easily support a supersaturation of titanium. As a consequence, Ti_2Ni precipitates form rapidly at the grain boundaries, where they generally degrade mechanical properties. The amorphous precursor phase in cool-substrate sputtered deposits, on the other hand, has been shown to support substantial Ti supersaturation even after the crystallization anneal, as illustrated by the TEM micrographs made by Chang [11] in Figure 5a. This allows the application of careful post-crystallization isothermal anneals to generate fine, transgranular precipitate distributions such as are seen in Figure 5b, which may be exploited to control transformation characteristics in much the same way as in the well-established protocols for nickel-rich precipitates [10]. In the thin film setting this fact offers the added advantage that the composition-dependency of martensite transformation temperature is somewhat lower on the Ti-rich side of stoichiometry. It may thus be expected that future advances in TiNi thin film microactuator materials will involve development of Ti-rich compositions which have been less fully exploited in ingot-metallurgy settings.

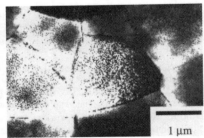

| a | b |

Figure 5 (a) Microstructure of a 51 at.% Ti thin film following a crystallization anneal at 773K showing a homogenous precipitate-free microstructure. (b) The same film after isothermal annealing at 923 K for 3.6 ks at 2.6×10^{-4} Pa showing both transgranular and intergranular precipitates.

Elevated-Temperature Deposition   Raising the substrate temperature during deposition to greater than approximately 620 K produces crystalline TiNi films directly, and can obviate the need for post-deposition crystallization annealing. The resultant microstructures are quite different than those produced by cool-substrate deposition in several respects. First, the structures are generally closer to thermodynamic equilibrium, meaning that intermetallic precipitate structures are more likely to be fully developed, and in many cases, are relatively coarse and incoherent. Secondly, and somewhat counter-intuitively, the B2 grain size in hot-substrate deposits can be made much finer than is usually the case for austenites crystallized from the glassy phase. This can provide excellent austenite strength which is more-or-less independent of precipitate distribution and dislocation substructure, and without recourse to cold-work/annealing steps (which are problematic in the thin film setting). Finally, though the implications for mechanical properties are not clear, within certain ranges of elevated substrate temperature, TiNi films on silicon can show pronounced crystallographic texture.

An example of the fine grain size achievable is shown in the bright-field TEM micrograph and corresponding electron diffraction pattern made by Li Hou [5], shown in Figure 6. The foil for these images was made from material near the film-substrate interface, and shows a polycrystalline B2 structure with a 25 - 50 nm grain size, though it is possibly imbedded in a matrix of amorphous material. Tilting experiments indicate that the polycrystal is not textured.

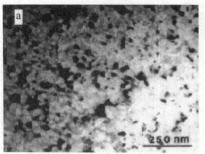 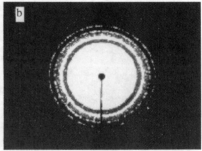

Figure 6. (a) Nanoscale grain size at the film/substrate interface in a film deposited at 723 K (b) Associated SADP

However, in films deposited at 723 K, the grains tended to grow as the deposit thickened, and the near-surface structure of the film was characterized by a grain size of approximately 200 nm, no texture, and mature, incoherent Ni_4Ti_3 precipitates, as shown in Figure 7. The film from which micrographs in Figures 6 and 7 were taken had excellent mechanical properties including austenite ductility greater than 4%, with tensile strength approaching 1 GPa. It also displayed electrical resistivity behavior characteristic of two-step B2→R→B19' transformation on cooling, with M_s and A_f temperatures of 250 and 303 K respectively, and exhibited nearly perfect transformational superelasticity at temperatures between 313 and 353 K.

 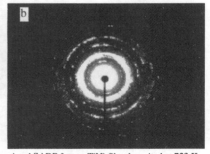

Figure 7. (a) Near-surface microstructure and (b) Associated SADP from a TiNi film deposited at 723 K

337

When deposition temperature is reduced to 673 K, two principal effects are notable. First, as is apparent in Figure 8a, the near-substrate grain size is larger, presumably reflecting a reduced crystalline grain nucleation rate. Secondly, though the initial near-substrate microstructure in this 7-micron thick film is not textured, the films develop strong <110> fiber texture at the mid-plane and near surface, as indicated by the 0 - and 30-degree tilted-foil electron diffraction patterns shown in Figure 8e and 8f. Small precipitates of Ni_4Ti_3 are present throughout, and the 150-200 nm grain size is roughly constant from substrate to free-surface. This film also displayed classical two-step martensite transformation and very good transformational superelasticity in which an energy-storage density in excess of 20 MJm^{-2} was measured [12]. The microstructure at the near-surface of this was similar that that shown for the mid-plane.

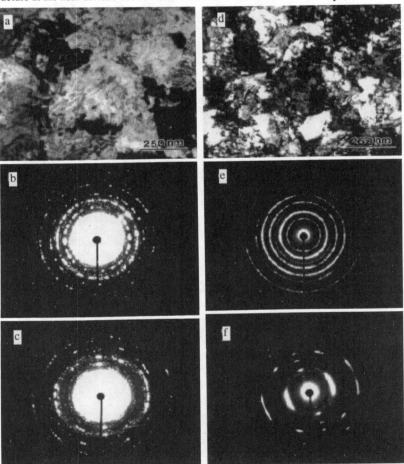

Figure 8. Brightfield TEM images and diffraction patterns from a film deposited at 673 K. (a) is the structure at a location near the substrate interface. Corresponding SADPs in (b) and (c) are taken from this region at zero and 30 degree tilts, and indicate an absence of texture. (d) is the midplane structure, and corresponding zero and 30 degree tilted SADPs indicate a strong (110) fiber texture.

The very fine grain sizes observed in hot-substrate deposits are very stable with respect to coarsening at elevated temperature. Figure 9 is a TEM bright-field image of a film deposited at 723 K and chamber-cooled, producing the microstructure shown previously in Fig. 7a. The film was then isothermally annealed for 3.6 ks at 858 K, and furnace cooled. This annealing temperature is below that which would be needed to dissolve excess nickel in this film, whose measured composition was approximately 51.1 at% Ni. The grain size is ~150 nm, essentially unchanged from the post-deposition condition. However, the precipitate phase has dramatically coarsened, producing large blocky, incoherent grains which have been indexed as monoclinic Ni_3Ti_2. As can be seen in the inset in Figure 9, this high-temperature anneal has not significantly altered the two-step transformation characteristics of this film from the as-deposited condition. It has, however, resulted in a slight increase in M_s and A_f from 266 and 303 K, respectively, for the as-deposited condition, to 271 and 307 K for the annealed film, consistent with a possible slight reduction in nickel concentration in the matrix.

If the annealing temperature is raised so as to lie in the region of the phase diagram where Ni solubility in TiNi increases to measurable levels, the precipitates formed during deposition can be dissolved, and new second-phase distributions can be produced on cooling. An example is illustrated in Figure 10, which shows the microstructure of a film deposited at 673 K and then annealed at 1073 K for 7.2 ks. The grain size has now coarsened appreciably, but is still well within the sub-micron range at ~400-500 nm. The precipitate distribution has, however, become much finer and more complex, with small intergranular particles coexisting with coarser precipitates that show some evidence of interface strain fields suggesting partial coherence. The inset, showing electrical resistivity behavior, indicates that the M_s temperature has been substantially depressed. This film was mechanically brittle, and did not display superelastic behavior.

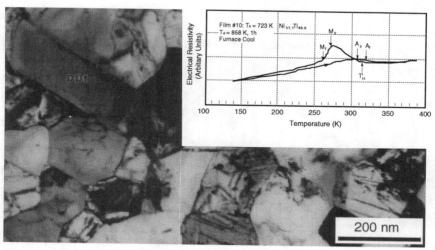

Figure 9. TEM micrograph from a film deposited at 723 K and subsequently annealed for 3.6 ks at 858 K, with an inset showing electrical resistivity as a function of temperature.

Thermomechanical Response Modeling   From a continuum viewpoint, the accumulation and annihilation of transformation strain is central to both superelasticity[3] and shape-memory.

---

[3] As detailed in [4] and [5], and as can be seen in Fig. 11, fine-grained thin films produced by hot-substrate deposition display excellent austenite strength which permits expression of classical transformational superelasticity in the as-deposited condition.

Accordingly, our mathematical models have developed to track fractional partitioning of the material between the austenite and martensite phases, the latter of which is allowed, by virtue of reduced crystal symmetry, numerous isostructural shear-variants. The development of such theories is currently an active area of constitutive modeling [13]. It requires an internal variable description of the underlying phase and variant fractions that, in view of transformation hysteresis, is not a direct function of stress τ, temperature T, or any other (intrinsic) thermodynamic variable that is capable of driving the transformation. Thus, a primary concern then is proper determination of the phase-fraction internal variables on the basis of a specified history of τ and T. Related issues involve this same determination when the transformation is driven indirectly by convective heating and/or constraint reaction forces.

The algorithms that we have developed involve an internal variable description for partitioning between austenite and two martensite variants, M^+ and M^-, whose strains are complementary shears from the base austenite state, A. Between any 2 of these 3 species the transformation may proceed in either a forward or a reverse direction, giving, in principle, 8 combined transformation possibilities. Of these, two ($A \rightarrow M^+ \rightarrow M^- \leftarrow A$ and $A \rightarrow M^+ \leftarrow M^- \leftarrow A$) are associated with conditions of strong temperature decrease in conjunction with weak stress variation.

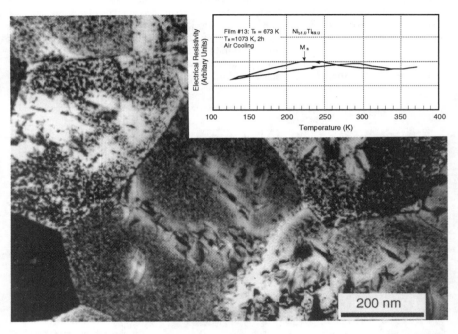

Figure 10. TEM micrograph from a film deposited at 673 K and subsequently annealed for 7.2 ks at 1073 K, with an inset showing electrical resistivity as a function of temperature.

The other 6 possibilities correspond to the other permutations of the directional indicators, $\rightarrow$ and $\leftarrow$, in a way that is dependent on the current value of τ and T, and on the direction of the increments $d\tau$ and dT. In general, a total of 6 possibilities require consideration, since the 2 combined transformation possibilities $A \rightarrow M^+ \rightarrow M^- \rightarrow A$ and $A \leftarrow M^+ \leftarrow M^- \leftarrow A$ are not associated with any reasonable thermodynamic process. Dependence on the current value of τ and T involves zones in (τ, T)-space that are logically organized by the four transformation temperatures M_f, M_s, A_s, A_f. Dependence on the increments dτ and dT is organized in terms of threshold slopes in (τ, T)-space that follow from Clausius-Clapeyron type relations. Thermodynamic consistency with respect to this slope determination necessitates refining the energy balance argument for conventional first-order phase transformations, in order to account for hysteresis and phase-coexistence over regions in (τ, T)-space [14].

Elsewhere in this proceedings [15] we present details of this modeling procedure as it applies to a restricted set of combined transformation possibilities. Models of various sophistication are possible. At minimum, 7 material parameters are required to account for arbitrary trajectories in (τ, T)-space: the 4 transformation temperatures, the transformation strain, an overall elastic modulus, and the transformation latent heat. Additional refinements involve specification of differing or temperature-dependent moduli in the austenite and martensite, coefficients of thermal expansion, and low temperature stresses associated with martensite flow due to the initiation and completion of detwinning reactions. Figure 11 exhibits the use of this type of model by Xiaochang Wu for calculating stress-strain in a 7 micron thick sputtered Ti-50.6% Ni fine-grained polycrystalline thin film, loaded in uniaxial tension at the test temperatures indicated. Squares represent experiment while the solid lines indicate simulation. For this alloy, M_f, M_s, A_s and A_f were 235, 263, 295 and 308 Kelvins, respectively, as determined by electrical resistimetry. Model

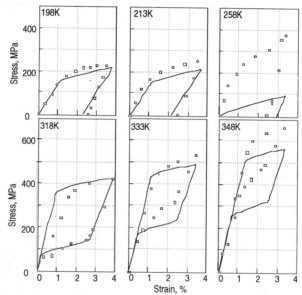

Figure 11. Experimentally determined (open squares) and simulated (lines) stress-strain curves for a 7 micron thick sputtered Ti-50.6% Ni thin film, loaded in uniaxial tension at the test temperatures indicated. For $T<A_f$, the initial value of the internal phase-fraction variables was calculated from stress-free cooldown starting at $T<A_f$ Good fits are generally obtained except at temperatures near M_s.

calculations employed a density of 6.5 g/cm^3 and a constant entropy change for the displacive transformation of 414.7 kJ/(m^3-K), based upon a calorimetric measurement made on the film itself. Young's moduli were taken as 14 and 38 GPa for the martensite and austenite phases respectively. Good fits are generally obtained except at temperatures near M_s, where this version of the model may suffer distortion associated with the assumption of temperature-independent elastic moduli, with errors in measurement of the true M_s temperature, or other factors [15]. Modeling results are also sensitive to assumptions regarding the initial microstructure. Figure 12 shows results for additional simulations of behavior at 258 K, with curve A corresponding to Fig. 11 (cooling from above A_f to just slightly below M_s, giving a small martensite fraction), and curves B and C in which an initial microstructure is assumed with 45 and 50 percent martensite respectively. These assumptions produce somewhat improved fits to the experimental data, which may have involved a degree of undercooling prior to the start of the test.

SUMMARY

Sputter-deposited titanium-nickel thin films display thermotractive characteristics of shape-memory and transformational superelasticity if processed so as to have chemical composition near equiatomic stoichiometry and precipitate distributions that are not so fine as to inhibit the martensite transformation process. Composition control in the sputter-deposition process is complicated by the tendency for composition drift with cathode, though a close match between cathode and film composition seems to be favored by high deposition rates. Deposition onto unheated substrates produces amorphous deposits which require post-deposition annealing, but may be processed to produce precipitate distributions not readily generated in melt-solidified alloys, especially on the Ti-rich side of stoichiometry. Deposition onto heated substrates can produce extremely fine-grained structures with mature, incoherent precipitates. These films display shape-memory and transformational superelasticity and have very good austenite strength. Both the grain size and precipitate structure have excellent thermal stability at temperatures below the deposition temperature, but both grain size and precipitate distribution can be controlled by choice of suitable post-deposition annealing treatment.

The very fine grain size of films deposited on heated substrates produces high austenite strength without recourse to cold work and annealing treatments, and irrespective of composition near stoichiometry. This results in excellent strain-recovery in transformational superelasticity experiments. These characteristics can be successfully modeled using a continuum description which maintains internal variables to track fractional partitioning of the structure between austenite and the possible martensitic variants, such that a simple two-variant description is able to adequately describe the basic constitutive behavior of the system.

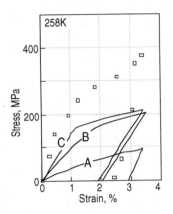

Figure 12. Experimental (squares) and simulated stress-strain curves at 258 K, for which alternative assumptions regarding the initial martensite-austenite phase-fractions were used. Curve A (also shown in Fig. 11) assumed cooling from above A_f; Curves B and C assumed 45 and 50 percent martensite, respectively, in the starting microstructure.

ACKNOWLEDGMENTS

This work has been funded by the National Science Foundation under grants #MSS8821755 and MSS9302270, by the Ford Motor Company, and by the Composite Materials and Structures Center and the Center for Fundamental Materials Research at Michigan State University.

REFERENCES

[1] R. H. Wolf and A. H. Heuer, "TiNi (Shape-Memory) Films on Silicon for MEMS Applications", J. Microelectromenchanical Systems **4**, 1057 (1995).

[2] Li Hou and D. S. Grummon, "Transformational Superelasticity in Sputtered Titanium-Nickel Thin Films", *Scripta Metallurgica* **33**, 989-995 (1995).

[3] S. Miyazaki, T. Hashinaga, K. Yumikura, H. Horikawa, T. Ueki and A., Ishida, Proc. 1995 N. Amer. Conf. On Smart Structures and Materials, (1995).

[4] D. S. Grummon, S. Nam and L. Chang, "Effect of Superelastically Deforming NiTi Surface Microalloys on Fatigue Crack Nucleation in Copper", Proc. Mat. Res. Soc. **246**, pp. 259-264 (1992).

[5] D. S. Grummon, Li Hou and T. J. Pence, "Progress on Sputter-Deposited Thermotractive Titanium-Nickel Films", *J. de Physique* IV, **C8** pp 665-670 (1995).

[6] Li Hou, T. J. Pence and D. S. Grummon, "Structure and Thermal Stability in Titanium-Nickel Thin Films Sputtered at Elevated Temperature on Organic and Polymeric Substrates", Mat. Res. Soc. Proc. **360**, pp 369-374 (1995).

[7] Jiun-Chung Lee, Master's Thesis, Michigan State University, 1994.

[8] E. Kay, Adv. Electronics and Electron Physics, **17** p245 (1962).

[9] L. Chang and D. S. Grummon, "Structure Evolution in Sputtered Thin Films of Titanium-Nickel, Part I - Diffusive Transformations", *Philosophical Magazine A*, 1996, in press.

[10] L. Chang and D. S. Grummon, "Structure Evolution in Sputtered Thin Films of Titanium-Nickel, Part-II - Displacive Transformations", submitted to *Philosophical Magazine A*, 1996, in press.

[11] L. Chang, Ph. D. Thesis, Michigan State University, 1993.

[12] Hou Li, D. S. Grummon and T. J. Pence, "Transformational Superelasticity in Nanophase Sputtered Deposits of Equiatomic Titanium-Nickel", 9th Annual University/Industry Symposium, MSU Center for Fundamental Materials Research, Apr. 10, 1995.

[13] L.C. Brinson and M.S. Huang, *J. Intell. Mater. Syst. and Struct.* **7**, 108-114 (1996).

[14] T.J. Pence, D.S. Grummon and Y. Ivshin in Mechanics of Phase Transformations and Shape Memory Alloys, ed. L.C. Brinson and B. Moran, (ASME AMD vol. 189, 1994), p. 45-58.

[15] X. Wu, T.J. Pence and D.S. Grummon, this symposium.

THE R-PHASE TRANSFORMATION IN THE Ti-Ni SHAPE MEMORY ALLOY AND ITS APPLICATION

I. Ohkata*, H. Tamura**
*Medical Division, PIOLAX Inc., Yokohama, 240 Japan
**Yokohama Research Laboratories, The Furukawa Electric Co., Ltd., Yokohama, 220 Japan

ABSTRACT

We discuss a comprehensive design approach of Ti-Ni alloy coil springs and introduce a new application of the R-phase transformation. In order to attain high cyclic performance, one must understand the two relationships between design parameters and material characteristics and between material characteristics and cyclic performance. Metallurgical parameters and coil spring dimensions play an important role as design parameters in the former relationship. High cyclic performance of an actuator is closely related to the suppression of the monoclinic martensite. Transformation temperatures and their stress dependence is of primary importance as material characteristics in the latter relationship. A thermostatic mixing valve, which is the latest application of the R-phase transformation in Japan is then discussed as a new type of a shape memory alloy actuator. The R-phase transformation is employed to achieve not only a long cycle life but a linear operation with the set temperature to continuously control the mixing ratio of hot and cold water. This is achieved by changing the total length of the two-way actuator in a linear manner with the set temperature. The linear characteristic is satisfied between 35-50°C by optimizing thermomechanical treatment and the dimensions of Ti-Ni and biasing coil springs.

INTRODUCTION

The R-phase transformation in the Ti-Ni shape memory alloy exhibits a small temperature hysteresis and excellent fatigue property[1]. The transformation also completes in a narrow temperature range, which makes it suitable for a fairly high rate actuation. However, the R-phase transformation is sensitive to thermomechanical treatment, alloy composition including the addition of a third element and repeating thermal cycles[2]. In order to utilize the R-phase transformation efficiently, one must start an actuator design from controlling material characteristics including transformation behavior and stress dependence of transformation temperatures. Figure 1 shows how coil spring dimensions and metallurgical parameters including the composition and heat treatment play an important role in determining the cyclic performance of an actuator. The relationships have been intensively but separately studied between design parameters and material characteristics and between material characteristics and cyclic performance.

One purpose of this study is to comprehend how the design parameters are determined in order to attain a large number of cycles by discussing the two relationships. Since coil springs have been mostly employed in SMA applications, we focus our attention on the cyclic performance of Ti-Ni shape memory coil springs. Another purpose is to introduce a new water mixing valve which utilizes the R-phase transformation. Water mixing valves require continuous and gradual temperature control. We have succeeded in controlling the temperature range of the R-phase transformation and optimizing an actuator operation in terms of the operation temperature, linear temperature characteristic and long cycle life.

Mat. Res. Soc. Symp. Proc. Vol. 459 © 1997 Materials Research Society

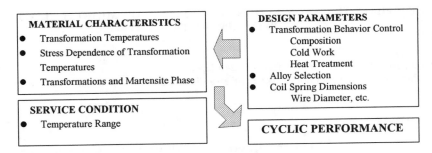

MATERIAL CHARACTERISTICS	DESIGN PARAMETERS
● Transformation Temperatures	● Transformation Behavior Control
● Stress Dependence of Transformation Temperatures	Composition
● Transformations and Martensite Phase	Cold Work

Figure 1 Comprehensive actuator design approach to achieve high cyclic performance. The approach links the two relationships between design parameters and material characteristics and between material characteristics and cyclic performance.

RELATIONSHIP BETWEEN MATERIAL CHARACTERISTICS AND CYCLIC PERFORMANCE

We consider intrinsic cyclic performance of different transformations, effect of the temperature range in a repeating thermal cycle and effect of strain.

Intrinsic Cyclic Performance of Different Transformations

Different cyclic performance is obtained depending on a martensitic phase and an alloy system. A coil spring exhibits poor cyclic performance as shown in Figure 2 when only the B2-monoclinic martensite transformation takes place in a repeating thermal cycle. On the other hand, a half million operation is reported for an actuator which utilizes the R-phase transformation[3]. However, the shape recovery strain is much smaller and about 1% for the latter case.   Ti-Ni-Cu alloys with more than 8at% Cu substituted for Ni have an in-between shape memory property. The cyclic performance is typically over one million, less than one hundred and ten thousand cycles for the R-phase(Ti-Ni), B2-monoclinic martensite(Ti-Ni) and B2-orthorhombic martensite(Ti-Ni-Cu) transformations, respectively[4].   A required number of cycles and actuation stroke and degree of allowable temperature hysteresis in service usually determine which alloy system and transformation should be utilized.

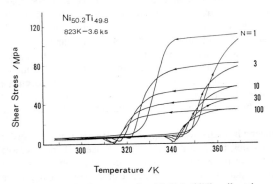

Figure 2   Cyclic performance of a Ti-50.2at%Ni coil spring which utilizes the B2-monoclinic martensite transformation. The spring is heat-treated at 823K for 3.6ks, and the shear stress is measured under 0.75% strain. N indicates the number of cycles.

Effect of Temperature Range in Repeating Thermal Cycle

The temperature range in service has primary importance because if service temperature decreases below the Ms temperature of a Ti-Ni alloy determined under zero stress, one cannot avoid the generation of the monoclinic martensite in service. Figure 3 compares the cyclic performance of a Ti-50.2at%Ni coil spring in three temperature ranges[5]. The Ms and Mf temperatures of the alloy is 294K and 282K, respectively. It is clear that the temperature-shear stress curve changes more remarkably with a decreasing lower temperature limit below Ms. The shape recovery force decreased by about 35% after 10^4 cycles when the lowest temperature was 4K below Ms((b)) and 65% when 4K below Mf((c)).

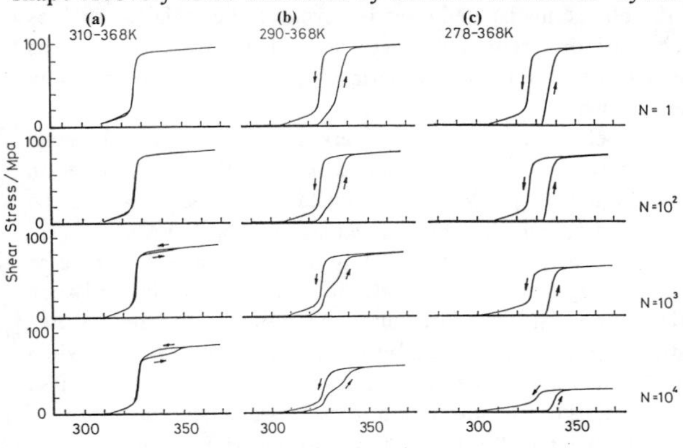

A large separation of the R-phase and R-monoclinic martensite transformations is favored in order to assure that the lowest service temperature lies between Mf' and Ms. One can attain such control by appropriately choosing the alloy composition and thermomechanical treatment condition.

Effect of Strain

One must take account of the change in transformation temperatures under stress and

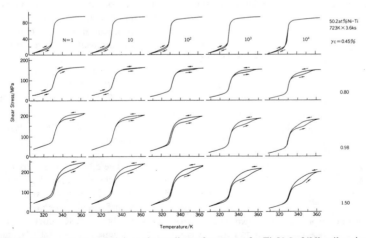

Figure 3   Cyclic performance of a Ti-50.2at%Ni coil spring in three different temperature ranges under 0.4% strain. The lowest temperature in cycles (b) and (c) is below Ms, which is 294K. N indicates the number of cycles.

Figure 4   Effect of the constant strain(γ_c) on the cyclic performance of a Ti-50.2at%Ni coil spring. Springs were subjected to a repeating thermal cycle under four different γ_cs between 308K and 363K, which is above Ms=283K.   N indicates the number of cycles.

strain. The operation temperature of an actuator can be therefore controlled by using a biasing spring. However, the cyclic performance remarkably degrades if the Ms temperature is increased above the lowest service temperature by stress. When a Ti-Ni coil spring is subjected to a repeating thermal cycle under a constant strain, its cyclic performance can be predicted from its initial temperature-stress characteristic[6]. Figure 4 shows the effect of the constant strain on the cyclic performance of a Ti-50.2at%Ni coil spring[7]. When the strain increases, a small hysteresis loop appears above the apparent Ms' temperatures in the first cycle. While the temperature-stress characteristic remains almost unchanged after 10^4 cycles if the initial curve does not exhibit the small hysteresis, the recovery force remarkably decreases when the small hysteresis appears in the initial curve. This indicates that a Ti-Ni coil spring has the critical strain below which the cyclic performance remains excellent.

We consider the critical strain is closely related to the generation of the stress-induced monoclinic martensite from the B2 phase. The stress dependence of the transformation temperatures is schematically shown in Figure 5[7]. When the constant strain is small and the stress at high temperatures does not exceed τ_X, which is the critical stress at which Ms and Ms' coincide, temperature-stress curve t_u-a-b-t_1-c-d-t_u is obtained. Only the R-phase and its reverse transformations are involved in this case. Segment t_u-b is sloped due to the change in the lattice parameters of the R-phase. When the constant strain is large and the stress at high temperatures becomes above τ_X, the temperature-stress curve is represented by T_u-A-B-C-T_1-D-E-F-T_u. When a coil spring is cooled down from T_u, the first transformation to occur is not the R-phase transformation but the B2-monoclinic martensite transformation at point A. The generation of the martensite then relaxes the strain, and when the stress reaches τ_X at point B, the remaining B2 phase transforms to the R-phase. We believe that the R-phase and monoclinic martensite coexist at point T_1. In the heating cycle, the R-phase and monoclinic martensite start reversely transforming to the B2 phase at point D and F, respectively.

The maximum stress in service should not exceed τ_X. This condition is usually achieved by appropriately determining coil spring dimensions including the wire diameter, mean diameter and number of effective turns. τ_X is a function of the alloy composition and thermomechanical treatment[8].

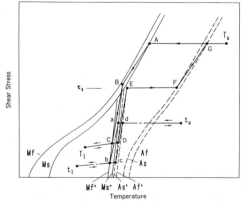

RELATIONSHIP BETWEEN DESIGN PARAMETERS AND MATERIAL CHARACTERISTICS

Transformation Behavior of Ti-Ni Alloys

The effect of the alloy composition and thermomechanical treatment has been intensively studied on the transformation behavior[9]. Figure 6 shows DSC curves of a work-hardened Ti-49.7at%Ni alloy as a function of the heat treatment temperature[10]. Each transformation temperature

Figure 5   Schematic representation of two types of temperature-shear stress curves, t_u-a-b-t_1-c-d-t_u and T_u-A-B-C-T_1-D-E-F-G-T_u, with the stress dependence of transformation temperatures. The Ms' and Ms temperatures coincide at stress τ_x.

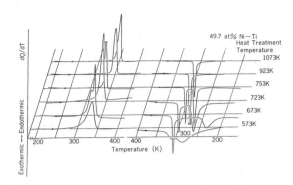

Figure 6 DSC curves for a Ti-49.7at%Ni alloy heat-treated at different temperatures indicated in the figure. The scan rate is 20K/min.

determined from the DSC curves is plotted in Figure 7. Work-hardened Ti-Ni alloys exhibit two transformations in cooling, that is, the R-phase and R-monoclinic martensite transformations when it is heat-treated below 773K[11]. The separation of the two transformations becomes larger for higher Ni-content alloys and largest for the heat treatment at around 673K. The addition of a third element to Ti-Ni alloys usually decreases the transformation temperatures[12] and brings about a larger separation of the two transformations[13].

Ms' and Mf' are not sensitive to the heat treatment temperature. The Ms' temperature gradually decreases with increasing heat treatment temperature, while Ms increases more rapidly when heat-treated above 673K. This implies that the separation of the two transformations becomes smaller for higher heat treatment temperature, which can result in the generation of the stress induced martensite and poor cyclic performance.

The transformation temperature range is also a function of the alloy composition and thermomechanical treatment condition. A wide temperature range implies a gentle operation of an actuator. Both the R-phase and R-monoclinic martensite transformations exhibit a wider transformation temperature range for lower heat treatment temperature. However, the dependence is much more significant for the R-monoclinic martensite transformation.

Stress Dependence of Transformation Temperatures

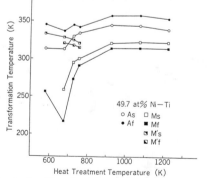

49.7 at% Ni—Ti
o As □ Ms
• Af ■ Mf
◨ M's
◩ M'f

Figure 7 Transformation temperatures of a Ti-49.7at%Ni alloy determined from Figure 6 as a function of the heat treatment temperature.

It is important to know the change in the transformation temperatures with stress to secure the separation of the R-phase and R-monoclinic martensite transformations in service as well as to control the operation temperature of an actuator. The transformation temperatures increase under stress according to the Clausius-Clapeyron relationship. The stress dependence of the R-phase transformation temperatures is much smaller than that of the R-monoclinic martensite transformation reflecting smaller transformation strain and transformation heat of the former[8].

Ms and Ms' coincide at τ_x(Figure 5), and above this stress the generation of the monoclinic martensite accelerates the degradation of the recovery force. Coil spring dimensions should be so determined that the maximum stress in service does not exceed τ_X. The transformation temperatures and τ_X depend on the alloy composition and thermomechanical treatment. Therefore, the compilation of material characteristic data is indispensable for the actuator design as a function of the design parameters.

A NEW APPLICATION OF R-PHASE TRANSFORMATION - WATER MIXING VALVE

Background of Development

Figure 8 shows three conventional types of water mixing valves. In the double valve type((a)), water temperature is controlled by manually adjusting two valves for hot and cold water. The water mixing ratio can be controlled with one lever in the single lever type((b)). However, it is not always easy to realize desired water temperature in these two types. The water temperature fluctuates even during an apparent steady state use. The fluctuation is caused by the variation in the water pressure resulting from water discharge elsewhere in the waterline. The pressure increase and decrease brings about the temperature drop and rise, respectively. The thermostat type((c)) automatically controls the water mixing ratio by sensing the water temperature. The market share of the thermostat type has been increasing to about 40% in Japan.

The conventional thermostat type mostly employs a wax actuator. Figure 9 shows the structure of a thermostat type valve. When the water temperature exceeds the set temperature, the wax expands and the actuator moves the spool to left resulting in a decreased mixing ratio of hot and cold water. When the water temperature becomes lower, the spring contracts the wax, which increases the mixing ratio. The temperature control is thus automatically done.

The most critical problem with wax actuators is a slow thermal response. When one interrupt the water and open the valve again or when one quickly increases the set temperature, too hot water is sometimes discharged. This is caused by a poor thermal response and called overshoot. The structure of a wax actuator is shown in Figure 10. Paraffin wax is enclosed in a copper vessel with a rubber diaphragm fixed at one end. The actuator utilizes the volume expansion accompanying the solid-to-liquid transition of wax in heating. The overshoot is caused by slow thermal conduction and the delay in actual volume expansion of the paraffin wax.

Thermostatic Operation with SMA Actuator

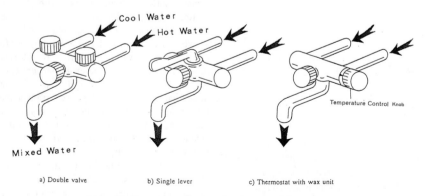

Figure 8   Three types of conventional water mixing valves

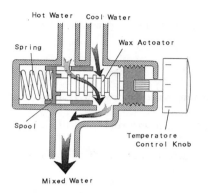

Figure 9   Structure of a conventional thermostat type valve using a wax actuator.

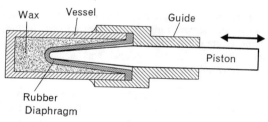

Figure 10   Cross section of a wax actuator

An excellent thermal response is expected for an SMA coil spring by directly exposing it to water. Figure 11 schematically compares the temperature-force characteristics under a constant deflection of a Ti-Ni coil spring and a wax actuator with an identical outer diameter. The force is much stronger for the Ti-Ni coil spring. A simpler mixing valve with a Ti-Ni coil spring can be designed with a better thermal response if one can attain small hysteresis and control the force with temperature.

Figure 12 schematically shows the new mixing valve with an SMA two-way actuator. The temperature control knob varies the total length of the actuator through the equi-pitched screw. Therefore, the total length of the actuator has a linear relationship with the set temperature. In order to control the water temperature, the spool must be shifted in a linear manner with the set temperature under equilibrium. We can prove that the R-phase transformation is well suited for such control as follows.

The linearity between the total actuator length(L_{tot}) and set temperature(T_{set}) is given by:

$$\Delta L_{tot} = \alpha \bullet \Delta T_{set} \tag{1}$$

where Δ and α denote a change in an amount from an initial state and a constant, respectively. With spring constant K_b and deflection δ_b, the biasing force is given as $K_b \bullet \delta_b$. We approximate the force of a Ti-Ni coil spring at temperatures between Ms' and Mf' as $(K_{s0}+\beta \bullet \Delta T_{act}) \bullet \delta_s$, where K_{s0} and δ_s denote a constant and the deflection of a Ti-Ni coil spring, respectively. β represents the temperature dependence of the spring constant. ΔT_{act} is the change in actual water temperature. At equilibrium, the two forces must be equal:

$$K_b \bullet \delta_b = \left(K_{s0} + \beta \bullet \Delta T_{act}\right) \bullet \delta_s \tag{2}$$

or

$$K_b \bullet \Delta \delta_b = K_{s0} \bullet \Delta \delta_s + \beta \bullet \Delta T_{act} \bullet \delta_s \tag{2'}$$

L_{tot}, δ_b and δ_s follow:

$$\Delta L_{tot} + \Delta\delta_b + \Delta\delta_s = 0 \tag{3}$$

The relationship between $\Delta\delta_s$ and ΔT_{act} is determined by a valve structure and, with γ being a constant, given by

$$\Delta T_{act} = \gamma \bullet \Delta\delta_s \tag{4}$$

From (1), (2'), (3) and (4), the following relationship is derived between ΔT_{set} and $\Delta\delta_s$:

$$K_b \bullet \alpha \bullet \Delta T_{set} + \left(K_b + K_{s0} + \beta \bullet \gamma \bullet \delta_s\right) \bullet \Delta\delta_s = 0 \tag{5}$$

When $\Delta\delta_s$ is much smaller than δ_{s0}, which is the deflection of a Ti-Ni coil spring in the initial state, the linearity between ΔT_{set} and $\Delta\delta_s$ is established as:

$$K_b \bullet \alpha \bullet \Delta T_{set} + \left(K_b + K_{s0} + \beta \bullet \gamma \bullet \delta_{s0}\right) \bullet \Delta\delta_s = 0 \tag{6}$$

Note that $\Delta\delta_s$ represents the spool displacement in Figure 12. Therefore, the spool displacement can be controlled in a linear manner with the set temperature by varying the total actuator length as in (1).

It is essential to control the temperature dependence of the spring constant during the R-phase transformation. If the transformation is gentler with temperature, the above expression for the force of a Ti-Ni coil spring holds in a wider temperature range, and a linear spool displacement and water temperature control can be obtained therein. In conventional wax actuators, the temperature dependence of force is controlled by mixing two waxes with different melting temperatures.

SMA actuators so far utilize the difference in deflections or forces below M_f' and above A_f'. The new mixing valve utilizes the linear relationship between the temperature and force during the R-phase transformation for the first time. The temperature dependence is controlled by the design parameters described earlier, that is, the alloy composition, thermomechanical treatment conditions including heat treatment temperature and cold working ratio and dimensions of Ti-Ni and biasing coil springs. Based upon fundamental data on the relationships between these parameters and the temperature dependence, we have succeeded in obtaining linear control in the 35 to 50°C range as well as excellent cyclic performance utilizing the R-phase transformation (Figure 13).

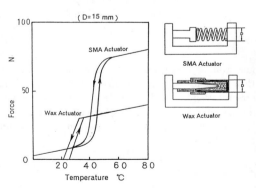

Figure 11 Temperature-force characteristics of SMA and wax actuators with an identical outer diameter, D(schematic).

New Water Mixing Valve[14]

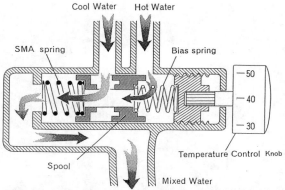

Cool Water   Hot Water

SMA spring   Bias spring

— 50

— 40

— 30

Temperature Control Knob

Spool

Mixed Water

Figure 12 Structure of the new mixing valve using SMA and biasing coil springs. The spool position and the mixed water temperature is controlled by varying the total length of the actuator linearly with the set temperature by the control knob. A linear temperature-deflection(force) characteristic in a wide temperature range is imparted to the SMA spring by optimizing the alloy composition and thermomechanical treatment condition.

Figure 14 exhibits the new water mixing valve. The Ti-Ni coil spring directly contacts the mixed water in Figure 12 and has a good thermal response. The water temperature is set by the temperature control knob as in the conventional thermostat valve. The knob changes the total length of the actuator, the biasing force and the spool position. The temperature fluctuation can be compensated as follows. When the actual water temperature becomes higher than the set temperature, the SMA spring pushes the spool to right, which in turn results in a decreased fraction of hot water and the recovery to the set temperature. When the actual temperature fluctuates lower, the opposite mechanism works.

Figure 15 compares the thermal response of the new and conventional valves. The set temperature is 42°C, and the temperature fluctuation was measured after one minute interrupt of discharge. For the conventional valve with a wax actuator, the water temperature fluctuates by about ten degrees from 38°C to 47°C and is stabilized at 42°C. The fluctuation range is smaller than two degrees for the new valve, which is not distinctly sensed by people.

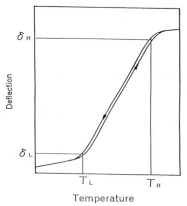

Temperature

Figure 13 Schematic representation of the temperature-deflection characteristic of the linear shape memory component used in the new mixing valve. The characteristic exhibits linearity between (temperature, deflection)=(T_L, δ_L) and (T_H, δ_H).

CONCLUSION

We have demonstrated that the R-phase transformation in the Ti-Ni alloy can be utilized for linear temperature control with a large number of operation cycles. The optimization of SMA actuators requires the link of the two relationships between design parameters and

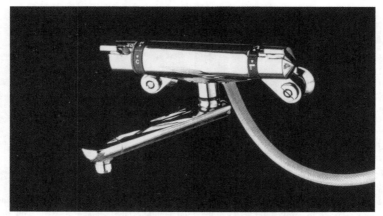

Figure 14　Appearance of the new mixing valve using the linear shape memory component (courtesy of TOTO Ltd.).

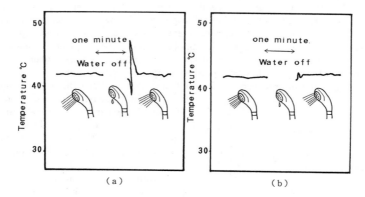

Figure 15　Comparison of the thermal response of a conventional thermostat type valve(a) and the new mixing valve(b).　42°C water is interrupted for one minute, and then the valve is re-opened.

material characteristics and between material characteristics and cyclic performance.　An addition of a new function like the linear temperature control also becomes possible by the same approach.

REFERENCES

1.　H. Tamura, Y. Suzuki and T. Todoroki, Proc. Int. Conf. on Martensitic Transformations, Japan Inst. Metals, p.736(1986).

2.　S. Miyazaki and K. Otsuka, Phil. Mag. A, **50**, p.393(1984).

3. Y. Suzuki, J. Rob. Mech., **1**, p.240(1989).

4. Y. Tsuzuki and H. Horikawa, Furukawa Rev., **9**, p.18(1991).

5. H. Tamura, K. Mitose and Y. Suzuki, J. DE PHYS. IV, Colloque C8, supplément au J. de Phys. III, **5**, C8-617(1995).

6. Y. Suzuki and H. Horikawa, Mat. Res. Soc. Symp. Proc., **246**, p.389(1992).

7. H. Tamura, Furukawa Rev., **6**, p.123(1988).

8. T. Todoroki and H. Tamura, J. Japan. Inst. Metals, **50**, p.538(1986) (in Japanese).

9. T. Todoroki and H. Tamura, Trans. Japan Inst. Metals, **28**, p.83(1987).

10. H. Tamura and Y. Suzuki, Furukawa Rev., **4**, p.33(1986).

11. S. Miyazaki, Y. Ohmi, K. Otsuka and Y. Suzuki, J. DE PHYS., Colloque C4, supplément au n° 12, Tome 43, C4-255(1982).

12. K.H. Eckelmeyer, Scripta Met., 10,p.667(1976).

13. C.M. Hwang, M. Meichle, M.B. Salamon and C.M. Wayman, Phil. Mag. A, **47**, p.9(1983).

14. H. Matsui, M. Enoki, T. Kato, J. DE PHYS. IV, Colloque C8, supplément au J. de Phys. III, **5**, C8-1253(1995).

EFFECT OF NITROGEN ON SHAPE MEMORY
BEHAVIOUR OF FE-MN-SI-CR-NI ALLOYS

A. ARIAPOUR, D. D. PEROVIC, A. McLEAN
Department of Metallurgy and Materials Science, University of Toronto, Toronto, Ontario, Canada, M5S 3E4

ABSTRACT

The composition of an Fe-Mn-Si-Cr-Ni stainless steel with shape memory effect was altered in this work in order to increase the strength of the alloy. The alloy possessed a low yield strength which is a major draw back for structural applications.

Nitrogen alloying, using nitrogen pressurized melting (P=1-10 atm), was employed to introduce a nitrogen concentration of up to 0.36 wt%. The effect of nitrogen alloying on shape memory effect was studied through mechanical testing. It was found that nitrogen alloying increased the hardness; however, nitrogen as an interstitial alloying element suppressed the $\gamma \Rightarrow \varepsilon$ transformation and therefore decreased the shape memory effect.

Introducing small amount of Nb (e. g., 0.36 wt%) to the nitrogen containing alloys caused formation of NbN. The NbN compound was in the form of globular dispersed particles (200 nm) which increased the strength of the alloy without significantly changing the shape memory effect.

INTRODUCTION

A weak shape memory effect in an Fe-Mn alloy was reported for the first time in 1975 (1). Upon adding Si to an Fe-Mn binary alloy Sato et. al., (2) obtained a perfect shape memory effect (SME) in single crystalline material. An optimum range of Mn and Si concentrations were determined by Murakami et. al., (3) in order to have a good shape memory effect in Fe-Mn-Si alloys. This range was 28-34% of Mn and 4-6.5% of Si. The maximum shape memory effect which has been reported for polycrystalline Fe-Mn-Si alloys is only 2-4% (4). However, the lower price of these alloys compared to Ni-Ti and better corrosion resistance compared to Cu-based alloys has made them worthwhile for further investigation.

Since certain application requires high corrosion resistance (such as fasteners and couplings) Otsuka et. al., (5) studied the effect of Cr and Ni addition on Fe-Mn-Si- based alloys. They found that alloys with the composition of Fe-28%Mn-6%Si-5%Cr, Fe-20%Mn-5%Si-8%Cr, and Fe-16%Mn-5%Si-12%Cr-5%Ni (wt%) have good corrosion resistance without showing any degradation in SME.

Improvement of corrosion resistant Fe-Mn-Si-based alloys has provided a great potential for employing these alloys in structural applications such as tendon rods in prestressed concrete and armour materials. The main disadvantage of the existing alloys is their low strength. The objective of this work involves altering the composition of

Mat. Res. Soc. Symp. Proc. Vol. 459 © 1997 Materials Research Society

existing Fe-16%Mn-6%Si-9%Cr-6%Ni (wt%) shape memory alloy in order to improve the strength of this alloys without degrading the shape memory effect.

Nitrogen as an interstitially dissolved element is known to improve the strength of steels. The strengthening effect of nitrogen is explained by Pickering's equation (6):

$$\sigma_{yield}=15.4(4.4+23C+1.3Si+0.24Cr+0.94Mo+1.2V+0.29W$$
$$+2.6Nb+1.7Ti+0.82Al+32N+0.16\delta \text{ ferrite}+0.46/d^{1/2}$$

where d is the average grain size.

Corrosion studies on nitrogen containing steels also imply the benefits of presence of nitrogen (7-8). These studies show that nitrogen containing stainless steels are less prone to sensitization than carbon containing alloys (9). Therefore, nitrogen is a good candidate if improvement in the strength and corrosion resistance for a stainless steel is required.

Increasing strength is also profitable to the shape memory effect. Presently, it is well accepted that shape memory effect in Fe-Mn-Si-based alloys is due to the reversible martensitic transformation of fcc⇔hcp (10). By increasing the strength of austenite, permanent slip in austenite is inhibited. Permanent slip is a barrier to the reverse transformation process. Nitrogen has also been shown to lower the stacking fault energy of the iron-based alloys (11). Therefore, the nucleation barrier for ε-martensite formation should be decreased by addition of nitrogen.

Although, there are many advantages in nitrogen alloying of Fe-Mn-Si-based alloys, one should bare in mind that nitrogen is a strong austenite stabilizing element. It depresses the Neel temperature (T_N) in iron alloys (12). Excessive concentration of nitrogen provides the condition for stabilizing austenite with $T_N=280\text{-}500^\circ$K against ε-martensite with $T_N=250^\circ$K (13).

The results of this work, presented in the following sections, reveal the effect of nitrogen alloying on strengthening and SME for an Fe-Mn-Si-based alloy.

EXPERIMENTAL PROCEDURES

Materials Preparation

In this study an alloy with chemical composition close to Fe-16%Mn-9%Cr-6%Si-6%Ni (wt%) was used as reference alloy (alloy Ref), see Table (I). This alloy has been formulated for coupling applications by NKK Steel Corporation.

Nitrogen alloying in this work was done by pressurized melting under nitrogen overpressure. In this way the solubility of nitrogen in the iron melts was increased (i.e., 0.044% in pure iron at one atmosphere and 1600 °C).

Alloy development involved three different procedures. The first attempt was introducing nitrogen to the reference alloy by melting it at 1 atm and 10 atm nitrogen overpressure in an isostatic pressure furnace (i.e., alloys R-1atm and R-10 atm).

Since Mn, Ni, and N are all austenite stabilizing elements, the second attempt involved replacing Ni by N. For this reason an alloy with the composition close to Fe-16%Mn-9%Cr-6%Si (wt%) was first produced in an induction furnace in argon atmosphere and then it was remelted at 1 atm and 10 atm of nitrogen overpressure in an isostatic pressure furnace (i.e., alloys AA-1 atm and AA-10 atm).

In the third series some of Ni was replaced by N in the processed alloys (i.e., Cr 8 and Cr 13). Niobium was added to the composition of the Cr 8 alloy. It has been shown that small amount of Nb (i.e., 0.3 wt%) in steels produces interstitial-free alloys. However, the strength of the alloy is improved drastically through dispersion hardening by formation of NbN (14).

Hardness and Shape Recovery Measurements

In order to evaluate the strength of the alloys, Rockwell hardness measurements (HRA) were performed on the samples which were austenized at 700 °C for 30 minutes and then cooled in air. For each data point at least three independent measurements were performed.

The shape recovery was determined from compression and recovery tests on the alloys. The change in the shape of the samples was traced by the change in the distance of two indentation marks on a polished surface of samples. This distance was set along the axis of applied stress during the compression test. The distance of the indentation marks were measured before compression test (L_0), after the test (L_1), and after recovery treatment at 500 °C for 30 minutes (L_2). The shape memory effect (SME) was determined by the following formula.

$$SME = \frac{L_2 - L_1}{L_0 - L_1} x100$$

In addition, strain recovery percentage was determined using the following definition:

$$\varepsilon_r = \frac{L_2 - L_1}{L_0} x100$$

Phase Identification and Microstructural Studies

Phase identification was performed by X-ray diffraction analysis using Cu-Kα radiation. The analysis was performed on the polished surface of the materials.

High resolution scanning electron microscopy, and transmission electron microscopy were used for microstructural studies. TEM samples were prepared by electro-thinning and ion milling of 3 mm disks. All the samples were perforated in 10% perchloric acid and 90% methanol at -20°C. For the nitrogen containing alloys final thinning was obtained by ion milling of perforated samples.

RESULTS AND DISCUSSION

Table (I) shows the chemical composition of the various alloy systems. It is seen that by increasing the nitrogen overpressure in each alloying system the amount of dissolved nitrogen has been increased in the system.

Table (II) contains the hardness values, shape memory effect and the strain recovery values for various alloys.

Table (I)- Chemical Composition of the Alloys

Alloy Type	Mn	Si	Cr	Ni	P	S	N	Nb	Fe
Ref	15.75	5.93	9.15	5.74	0.012	0.012	0.010	-	balance
R-1atm	15.72	5.92	9.13	5.73	0.011	0.004	0.175	-	"
R-10atm	15.69	5.60	9.11	5.72	0.011	0.004	0.362	-	"
AA	16.10	5.65	9.22	0	-	0.16	0.017	-	"
AA-1atm	16.05	5.63	9.16	0	-	0.16	0.152	-	"
AA-10atm	16.03	5.62	9.16	0	-	0.11	0.325	-	"
Cr 8	18.25	4.18	8.05	2.04	-	0.001	0.36	0.36	"
Cr 13	17.85	3.99	12.67	2.39	-	0.001	-	-	"

Table (II) - Hardness (HRA) and Shape Memory Behaviour Values

Alloy Type	HRA	SME	ε_r
Ref	50.6	43	2.1
R-1atm	53.5	16	0.9
R-10 atm	55.2	11	0.6
AA-1atm	49.6	25	1.4
AA-10 atm	60.7	18	1.0
Cr 8	64.9	35	1.7
Cr 13	61.2	8	0.4

In general, the hardness values increased with increasing nitrogen content. However, different mechanisms were involved in strengthening of these alloys. Microstructural studies revealed that in alloys R-10atm, AA-10atm, and Cr 13 some precipitated particles were produced. These precipitates were not dissolved by solutionizing at 1100 °C. The precipitates in these alloys were complex nitrides with the composition of MN_2 (M represents the metallic elements) which was determined by X-ray diffraction analysis. Alloy Cr 8 showed the highest level of hardness. High resolution scanning electron microscopy revealed two types of particles in this alloy. The majority of the particles were globular dispersed particles of 200 nm size. These particles were NbN particles as determined by EDX analysis. The second type of particles were similar to the other alloys. Therefore, nitrogen alloying of the Fe-Mn-Si-Cr-Ni alloys in conjunction with Nb was a more effective way in strengthening than just addition of N.

The results of mechanical tests show that shape memory effect is drastically decreased by increasing the amount of nitrogen in solid solution. This is evident from samples R-1 atm and AA-1 atm and Cr 13. However, when part of the dissolved nitrogen participated in the composition of the nitrides the shape memory effect was improved. This was evident especially in the behaviour of the Cr 8 alloy. In this alloy almost all the dissolved nitrogen was in conjunction with Nb.

The validity of above statements was proved by studying the X-ray diffraction data. These results revealed that all the nitrogen containing alloys, except Cr 8, had little ε-martensite phase either in samples quenched in liquid nitrogen or in samples mechanically deformed. This implies that in those alloys excess nitrogen is responsible for the absence of the ε-martensite phase. In alloy Cr 8, the action of Nb to make an interstitial-free alloys suppressed the effect of nitrogen.

Figure (1) represents the typical TEM microstructures of the reference alloy (Ref) deformed by 5%. A well developed structure, containing thick plates of ε-martensite, isolated and overlapped stacking faults are observed in this micrograph.

Figure (2) is typical representation of the microstructure for nitrogen containing alloys which showed weak shape memory effect for deformation of 5%. As can be seen in these alloys, some ε-martensite plates were present, however the X-ray diffraction analysis could not detect them. Therefore, the poor shape memory effect in these alloys is the poor development of the martensite phase and not irreversibility of martensite.

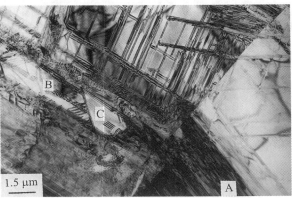

Figure I -   TEM micrograph of Ref sample, showing well developed stress induced ε-martensite phase (A), isolated (B) and overlapped stacking faults (C)

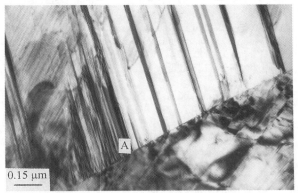

Figure II-   TEM micrograph of nitrogen containing alloy (R-1atm), showing poor development of stress induced ε-martensite (A)

SUMMARY

The Fe-Mn-Si-Cr-Ni shape memory alloys are stainless steel alloys with great potential for structural applications. However, they suffer from low strength. In order to increase the strength of these alloys, nitrogen alloying was employed in this work.

It is found that addition of nitrogen as an interstitial element suppresses the $\gamma \Rightarrow \varepsilon$ transformation. Therefore, the SME deteriorates.

Another approach in strengthening these alloys is using dispersed forming elements such as Nb. In this way the strength was increased while the shape memory effect is not affected drastically.

Among the various alloys in this study, alloy AA-10 atm and Cr 8 had a combination of high strength and reasonable SME. The advantage of one of these alloys against the other depends on the area of application. For example for tendon rods in prestressed concrete, the requirements are hardness of HRA=68-74, strain recovery of 0.2%, and good corrosion resistance. Therefore, other properties of these alloys must be studied and compared before finalizing the proposed area of applications.

ACKNOWLEDGMENTS

The authors would like to express their appreciation towards Dr. L. Mac. Schetky, from Memry Corporation, for providing the reference alloy, and also Dr. J. S. Dunning from Department of Energy, Albany Research Center, for providing alloys Cr 8 and Cr 13 which were used in this work.

REFERENCES

1. K. Enami, A. Nagasawa, and S. Nenno; Scripta Metall., 9, 1975, 941.
2. A. Sato, E. Chishiama, K. Soma, and T. Mori; Acta Metall., 30, 1982, 1177.
3. M. Murakami, H. Otsuka, H. Suzuki, and Sh. Matsuda; Trans. ISIJ, 27, 1987, B 88-B89.
4. Q. Gu, J. Van Humbeeck, and L. Deleay; Scripta Metall., 30(2), 1994, 1587.
5. H. Otsuka, H. Yamada, T. Maruyama, H. Tanahashi, Sh. Matsuda, and M. Murakami, ISIJ Int., 30, 1990, 674.
6. E. B. Pickering; "Physical Metallurgy and the Design of Steels", Appl. Sci. Pub., 1978, p. 23.
7. J. E. Truman, M. J. Coleman and K. K. Pirt, Corrosion J., 12, 1977, p. 236.
8. J. J. Eckenrod, C. W. Konach; ASTM, 679, 1979, p. 17.
9. J. W. Simmons, D. G. Atteridge, and J. C. Rawers; Corrosion, 45, 1994, p. 491.
10. L. Jian, C. M. Wayman; Scripta Metall., 27, 1992, p. 279.
11. R. E. Schramm and R. P. Reed; Met. Trans., 6A, 1975, p.1345.
12. E. Jones, T. Datta, C. Almasan, D. Edwards, H. M. Ledbetter; Mat. Sci and Eng., 91, 1987, p. 181.
13. E. Gartstein and A. Rabinkin; Acta Met., 27, 1979, p. 1053.
14. W. C. Leslie, in "The Physical Metallurgy of Steels", McGraw-Hill, 1981, 316.

IMPROVEMENT OF THE CORROSION RESISTANCE OF NITI STENTS BY SURFACE TREATMENTS

C. TRÉPANIER*, M. TABRIZIAN*, L'H. YAHIA*, L. BILODEAU**, D.L. PIRON***
*GRBB, Biomedical Engineering Institute, École Polytechnique de Montréal, CANADA
**Institut de Cardiologie de Montréal, CANADA
***Department of Material Engineering, École Polytechnique de Montréal, CANADA

ABSTRACT

Because of its optimal radiopacity, superelasticity and shape memory properties Nickel-Titanium (NiTi) is an ideal material for the fabrication of stents. Indeed, these properties can facilitate the implantation and precise positioning of those devices. However, *in vitro* studies on NiTi report the dependency of the alloy biocompatibility and corrosion behavior to surface treatments. Oxidation of the surface seems to be very promising to improve both the corrosion resistance and the biocompatibility of NiTi. The present study investigate the effect of electropolishing, heat treatment (in air and in a salt bath) and nitric acid passivation to modify the oxide layer on NiTi stents. Techniques such as potentiodynamic polarization tests, Scanning Electron Microscopy (SEM) and Auger Electron Spectroscopy (AES) have been used to develop relationships between corrosion behavior, surface characteristics and surface treatment. Results show that all surface treatments improve the corrosion behavior of the alloy. SEM results indicate that treated stents which exhibit a smooth and uniform surface show a higher corrosion resistance than non treated stents which possess a very porous oxide layer. AES results, indicate that the best corrosion behavior was observed for the stents which exhibit the thinnest oxide layer (electropolished and passivated samples).

INTRODUCTION

Since 1986, implantation of tubular endoprostheses (stents) in blood vessels is performed to prevent occlusion and restenosis of coronary arteries after a percutaneous intervention e.g. angioplasty[1]. These devices constitute, nowadays, the most effective way to treat coronary occlusions. Clinical results demonstrate a decrease of at least 30% of the restenosis rate after their implantation[2]. However, clinically used metallic stents, stainless steel and Tantalum based, have limitations because of their suboptimal radiopacity and mechanical properties that may complicate the insertion and positioning of stents at the site of obstruction[3]. In contrast, optimal radiopacity, superelasticity and shape memory properties make Nickel-Titanium (NiTi) an ideal material for such an application. These properties can facilitate the implantation and precise positioning of stents.

However, before any new alloy can be approved for implantation in the human body, the biocompatibility of the material must be established. *In vitro* studies of NiTi, in contact with different cell cultures, report the dependency of the alloy biocompatibility behavior to surface treatments[4-6]. Furthermore, in a recent study, Yahia et al. have shown the toxic effect of NiTi by *in vivo* implantation of non-treated NiTi screws in the tibia of rabbits[7].

Indeed, among the factors which determine the biocompatibility of an implant, surface properties and corrosion resistance of the material are the most important. Passive metals, like NiTi, have a stable oxide layer on their surface which renders them corrosion resistant and relatively inert in physiological conditions. This passivity may be enhanced by modifying the thickness, topography and chemical composition of this oxide layer by different surface treatments[6,8,9]. The oxide layer is very protective and may be very promising to improve both corrosion resistance and biocompatibility of NiTi[9-10]. Also, it is able to sustain large deformations induced by the shape memory effect[9]. It has been shown that laser surface melting treatment promotes the oxidation of NiTi and improve the corrosion resistance of the alloy[8]. However, this technique is very expensive and not very appropriate for implants of complex geometry such as stents. Modification of the oxide layer can also be achieved by more conventional methods such as chemical polishing and heat treatments in salt bath and in air[11]. Also, a study on Ti based implant materials demonstrated a reduction in the dissolution of ions by aging the surface oxide or by thermal oxidation[12]. Electropolishing and nitric acid passivation are two other techniques recommended for surface treatments of medical devices (ASTM-F86)[13]. These methods are simple, non expensive and effective to treat implants of different shapes.

The present study investigate the effect of electropolishing, heat treatment (in air and in a salt bath) and nitric acid passivation to modify oxide layer properties on NiTi stents. Techniques such as potentiodynamic polarization tests, Scanning Electron Microscopy (SEM) and Auger Electron Spectroscopy (AES) have been used to develop relationships between corrosion behavior, surface characteristics and surface treatment.

EXPERIMENT

Material

NiTi stents (50.8 at% Ni) have been manufactured by Nitinol Devices and Components (California, USA) by laser cutting diamond shaped apertures in NiTi tubings. The stents are 14 mm long and have a diameter of 4 mm in the expanded state. Five different groups of samples were prepared:

1. Non treated (NT)
2. Electropolished (EP)
3. Air Aged (AA)
4. Heat Treated (HT)
5. Passivated (PA)

The first group of stents are supplied directly after the machining and expansion of the stents with an heavy oxide layer. Electropolished stents (EP) have been first, micro-abraded to remove mechanically the primary oxide layer, then, chemically polished at room temperature and finally, electropolished. After electropolishing, a mirror-like surface finish is obtained. Air aged samples (AA) are electropolished stents that have been air aged at 450°C to produce a light yellow oxide layer. The heat treated stents (HT) are electropolished stents that were heat treated in salt at 500°C to produce a dark blue oxide layer. The passivated stents (PA) are electropolished stents that were passivated in a nitric acid solution at room temperature. They optically exhibit the same surface finish as the electropolished stents. All surface treatments were performed by Nitinol Devices and Components.

Method

The corrosion resistance of 15 stents (3 for each surface treatment) was evaluated by **anodic polarization tests**. The potentiodynamic experiments were carried out using a computer controlled potentiostat (EG&G Princeton Applied Research, model 273). The tests were conducted in 37°C Hank's physiological solution of the following composition: NaCl 8 g/l, KCl 0.4 g/l, NaHCO$_3$ 0.35 g/l, KH$_2$PO$_4$ 0.06 g/l, Na$_2$HPO$_4$ 0.0475 g/l and C$_6$H$_{12}$O$_6$ (glucose) 1 g/l.

This solution is buffered with HEPES (3.5745 g/l) at normal physiological pH of 7.4. A saturated calomel electrode (SCE) was used as the reference and a platinum plate as the counter electrode. The solution was deoxygenated with nitrogen gas during 1 hour before starting the experiment. The samples were immersed 30 minutes prior to the test to stabilize the open circuit potential (OCP). A scan rate of 10 mV/min was applied in the range between the OCP to the pitting potential of the sample.

Scanning Electron Microscopy (SEM), Jeol JSM 840 microscope, was used to study the topography of one sample of each surface treatment (before and after corrosion test). Micrographs were taken at areas located on the external and lateral faces of the stents in the second electron imaging mode (SEI).

Auger Electron Spectroscopy (AES) analyses were performed with a Jeol JAMP 30 microscope on 3 samples (2 stents before corrosion experiment and 1 stent after corrosion experiment) for each surface treatment. AES survey spectra (100-1200 eV) have been recorded from two different spots (5 μm^2) on external faces of each sample to identify the surface composition. Then, depth profiles have been measured by combining AES analysis and Argon ion sputter etching to evaluate the oxide layer thickness and the distribution of each element in the sample.

RESULTS

Anodic polarization tests

Typical potentiodynamic scanning curves for each surface condition are illustrated in figure 1.

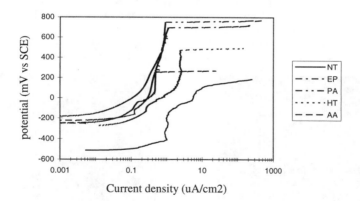

Figure 1 Potentiodynamic scanning curves for stents with different surface treatments

From these results, it can be noticed that all surface treatments improve the corrosion resistance of the NiTi stents in comparison to the non-treated stents (NT). Indeed, all stents present a passivation current density one order of magnitude lower than the NT stents, indicating a decrease in the corrosion rate of the treated samples. Also, among different surface treatments, an increase of the pitting potential (from 100 to 600 mV) and in the passivation range (from 50 to 500 mV) can be observed which denote a higher corrosion resistance of treated samples.

SEM study

Microscopic examination of treated and non treated stents before corrosion indicates an improvement of the surface topography after treatments. As shown on figure 2, NT stents exhibit a very porous and non-uniform surface topography. After treatment, the stent surface appears smoother and more uniform (figure 3). Still, the topography is not uniform on all faces of devices for all different surface treatments. Indeed, the oxide layer on lateral faces of HT and AA treated stents is irregular compared to external faces.

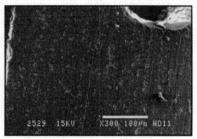 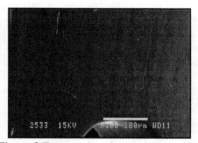

Figure 2 Topography of non treated stents   **Figure 3** Topography of treated stents (EP)

Microscopic analysis of treated stents after corrosion showed a difference in the corrosion process on different faces of the devices. Indeed, lateral faces of stents corrode faster than the external faces. This localized attack is more pronounced for AA and HT treated stents. Micrographs shown on figure 4 and 5 demonstrate the local attack of the external faces of stent AA compared to the lateral faces. This non-uniformity of corrosion may result from the geometry of the samples or from an non-homogeneity of the surface treatment.

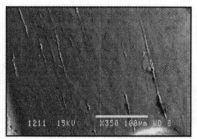 

Figure 4 External faces of stent (AA)   **Figure 5** Lateral face of stent (AA)

AES study

The AES survey spectra demonstrate the presence of C, Ti and O on the devices before and after corrosion. Presence of C on the devices can be attributed to contamination of the material. Small signal from Ni could also be randomly recorded on NT, EP, HT and PA stents. This Ni signal may be due to segregation of Ni that can occur in NiTi alloys under certain conditions[14]. The oxide layer thickness measured by AES for the non treated samples was of the order of 10 000 Å. AES depth profiling of treated samples indicate not only a variation of oxide thickness among the different treatments, as reported in Figure 6, but also a significant difference between the non treated and treated stents. No significative difference in oxide layer thickness could be recorded for stents with the same surface treatment before and after corrosion experiment.

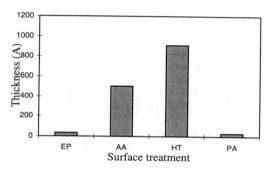

Figure 6 Variation of oxide layer thickness for the different surface treatments from AES study

CONCLUSION

The objective of this work was to study the influence of different surface treatments on the corrosion resistance of NiTi stents. Results showed that electropolishing (EP), heat treatment (HT), air aging (AA) and nitric acid passivation (PA) of NiTi improve the corrosion behavior of the alloy. Surface topography analyses by SEM and oxide layer thickness measurements by AES provided relations between surface physicochemical properties and corrosion behaviors. SEM micrographs indicated that the surface condition plays an important role in NiTi stents corrosion resistance. Treated samples with smooth and uniform surface demontrated a higher corrosion resistance than non treated ones which possess a very porous oxide layer. From the AES results, it seems that corrosion resistance of NiTi stents is not directly related to oxide layer thickness. Indeed, the best corrosion behavior was observed for stents with the thinnest oxide layer (EP and PA). Similar results were obtained by Sohmura on the reliability of thin oxide for improvement of corrosion resistance of NiTi[9]. Also, it has been demonstrated that improvement of corrosion resistance of NiTi by laser melting surface treatment could be attributed, in part, to homogenization of the surface[8].

In an undergoing project, determination of the nature of the oxide by X-ray Photoelectron Spectroscopy (XPS) may explain the differences between their different level of protection against corrosion.

The present study has important implications for the final steps of fabrication of NiTi implants. It seems that electropolishing of NiTi increases significantly the corrosion resistance of the alloy and that subsequent treatments (air aging, heat treatment and passivation) seem to have a detrimental or small effect on this improvement. Moreover, the additional steps may increase the cost of production of stents.

ACKNOWLEDGMENTS

The authors gratefully acknowledge the financial and material support of NSERC (Canada) and Nitinol Devices and Components Inc, (USA).

REFERENCES

1. U. Sigwart, J. Puel, V. Mirkovitch, F. Joffre, L. Kappenberger, N. Eng. J. Med., **316**, 12, p. 701, (1987).

2. P.W. Serruys et al., N. Eng. J. Med., **331**, p.489, (1994).

3. K.W. Lau, U. Sigwart, Ind. Heart J., **43**, 3, p. 127, (1991).

4. J.L.M. Putters, D.M.K.S. Kaulesar Sukul, G.R. de Zeeuw, A. Bijma, P.A. Besselink, Eur. Surg. Res., **24**, p. 378, (1992).

5. M. Assad, L'H. Yahia, E.A. Desrosiers, S. Lombardi, C.H. Rivard in Shape Memory and Superelastic Technologies, edited by A. Pelton, D. Hodgson and T. Duerig (Shap. Mem. Super. Tech. Proc., Pacific Grove, CA, 1994), p. 215.

6. S.Shabalovskaya, J. Cunnick, J. Anderegg, B. Harmon, R. Sachdeva in Shape Memory and Superelastic Technologies, edited by A. Pelton, D. Hodgson and T. Duerig (Shap. Mem. Super. Tech. Proc., Pacific Grove, CA, 1994), p. 209.

7. M. Berger-Gorbet, B. Broxup, C.H. Rivard, L'H. Yahia, J. Biomed. Mat. Res., **32**, p. 243, (1996).

8. F. Villermaux, M. Tabrizian, L'H. Yahia, M. Meunier, D.L. Piron, Appl. Surf. Sci., (in press).

9. T. Sohmura, (World Biomat. Congress Proc., Kyoto, Japan, 1988), p. 574.

10. J.P. Espinos, A. Fernandez, A.R. Gonzalez-Elipe, Surf. Sci., 295, p.402, (1993).

11. ASM Handbook, 5, ASM International, p.835, (1994).

12. A. Wisbey, P.J. Peter, M. Tuke, Biomat., **12**, p.470, (1990).

13. Annual Boof ASTM Standards, 13.01, ASTM, Philadelphia, (1996).

14. C. Chan, S. Trigwell, T. Duerig, Surf. Interf. Anal., **15**, p.349, (1990).

ACTUATOR APPLICATIONS WITH SINGLE-CRYSTALS OF Cu-Zn-Al SHAPE MEMORY ALLOY

A.A. YAWNY, M. SADE, F.C. LOVEY
División Metales, Centro Atómico Bariloche (CNEA). (8400) S.C. de Bariloche, Argentina.
yawny@cab.cnea.edu.ar

ABSTRACT

In the present work the use of single-crystals of Cu-Zn-Al Shape Memory Alloys (SMA) in actuator applications is analyzed. The actuator considered here, a device capable of doing work in response to temperature changes, is based on a single-crystal nucleus of a Cu-Zn-Al SMA coupled to a conventional spring that represents the load to be displaced. A special experimental stage was designed for performing controlled thermal cycles under load. In this way the effects of different parameters (cycle number, friction, temperature range, load level) on the actuator behavior can be studied. From the results obtained, the use of a single-crystal of an SMA in a thermostatic device is analyzed and compared with the commercial wax actuator performance.

INTRODUCTION

Shape Memory Alloys (SMA) are characterized by strong shape changes which are due to diffusionless transformations, the so-called martensitic transitions [1]. The characteristic effects shown by SMA (one-way shape memory effect, pseudoelasticity, trained two-way shape memory effect) suggest the possibility of applications in different areas like medicine, electric industry or mechanical couplings [2]. However, few products have reached industrial success. Two common points among different trials can be remarked:

- Most of the intended applications to date are based on the use of *polycrystalline* material of different alloys (TiNi based alloys, Cu-Al-Ni, Cu-Zn-Al).
- The *lack of a design tool* for an accurate SMA device. In some cases a procedure by trial and error is adopted. In others, results from beam theory are used to make CAD-programs and design-charts. The main difficulty is due to the complex behavior of SMA elements in the stress-strain-temperature space (σ-ε-T): non-linearity, hysteresis, path dependence and time evolution of behavior.

In addition, loading geometries different from pure uniaxial tension are commonly found. Bending and torsion arrays are used for displacement magnification. The resulting non-homogeneous distribution of stress diminishes the material usage efficiency.

Shape Memory Alloy (SMA) applications are classified according to the function performed by the SMA element: free recovery, constrained recovery, actuator or work production devices, pseudoelastic and damping applications [2-3]. In this work the use of Cu-Zn-Al *single-crystals* in *actuator* applications is analyzed. A general *conceptual design procedure* is discussed for uniaxial stress loading geometry. This procedure is directly related to basic aspects of Cu-Zn-Al SMA behavior. We call *actuator* a device in which an SMA element performs mechanical work in response to an external stimulus, i.e. an electric current (electrical actuator) or a temperature change (thermal actuator). In the following only a thermal actuator device will be considered. The conceptual design tool is finally applied to the analysis of a *thermostatic device*.

EXPERIMENT

Single-crystals of Cu-Zn-Al with a composition of Zn=16.486 wt%, Al =7.822 wt%, Cu rest were used. The measured spontaneous (stress free) transformation temperature Ms was 285 K. Flat samples were obtained by spark machining (sample S1a: length 20 mm, section: 1 mm x 1 mm; S1b: length 10 mm, section: 1mm x 1 mm). The tensile axis orientation was determined using the X-ray Laue technique, the Schmid factor for the induced variant being $\mu = 0.48$. The samples were kept 20 minutes at 1103 K and air-cooled to 303 K, being kept at this temperature at least seven days in order to reduce subsequent martensite stabilization and to increase thermal cycle reproducibility. Finally, an electrolytic surface-polishing procedure with HNO_3 15 vol % in methanol was applied.

An equipment was developed for sample characterization. It allowed the performing of thermal cycles under constant load or under spring-loading conditions (elongation-dependent force). Temperature cycling has been done with a Peltier thermobattery Melcor CP-1.4-17-10L in the 273 - 353 K range through a computer-controlled current source. The load was measured with an Omega LCF 25 load cell (maximum load ±12 lb). The elongation was recorded with an LVDT Schaevitz 050 HR linear differential transformer (±1.25 mm range). The temperature was measured with a thermocouple in direct contact with the sample. The interaction of the SMA element with different loads and with springs of different elastic constants was studied by simulating the working conditions of a real device. Friction of different magnitudes was simulated by changing the load applied. The temperature stability of the cycle has been studied by continuous cycling through the transformation range. A maximum temperature rate of 5 K/min^{-1} was obtained for near-room temperature cycling. In addition, the surface production of defects was followed by optical and scanning electron microscopy observations.

RESULTS

In the following part, results related to a basic characterization of a commercial wax thermostatic nucleus are shown. Then similar conditions are imposed to a basic SMA actuator and its response is analyzed.

Commercial wax actuator

A typical commercial wax thermostatic device for regulating the flow of hot water in radiators in water central heating systems is composed basically by two elements: the nucleus and the valve. Diagrams of the wax nucleus and its elongation-temperature (ΔL-T) response coupled to the matched water valve measured in our laboratory are shown in Figures 1A and 1B respectively. The force-elongation characteristic of the matched commercial water valve is shown in Figure 1C. From the above characterization the following informations can be extracted. The valve displacement from completely shut to completely open position is 3 mm, over a temperature range of 15K. Thermal hysteresis of the stress-free wax nucleus is 0.5 K. These values are essentially unmodified for different load situations due to the low dependence of the wax thermal expansion coefficient ($\alpha \cong 0.3\ 10^{-4}\ K^{-1}$) on applied stress. The maximum force necessary to completely close the commercial valve is near 55 N. A change in the desired room temperature level is obtained through the displacement of the wax nucleus relative to the valve.

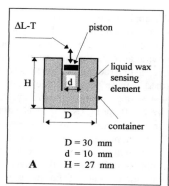
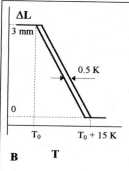
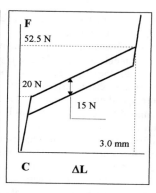

D = 30 mm
d = 10 mm
H = 27 mm

FIGURE 1: **A**- Schematic representation of a wax nucleus device. The volume increase due to a positive temperature change produces the piston displacement. **B**- Displacement (ΔL) -Temperature (T) response for the wax nucleus coupled to the water valve.
C- Force (F) - displacement (ΔL) characteristic of a commercial valve for thermostatic wax actuator.

SMA Basic Actuator Device

We define a Basic Actuator Device as one composed by the SMA in interaction with an elastic material as shown in Figure 2. In this way a real actuator device can be simplified. In fact in most cases the actuator nucleus interacts with total external forces that can be adequately represented by an ideally elastic material like a spring. Friction can be simulated in a second step by changing the spring length in the turning points of transformation, thus increasing the hysteresis width. Based on the actuator performance requirements, the single-crystal element and the spring material can be designed taking into account the mutual interactions.

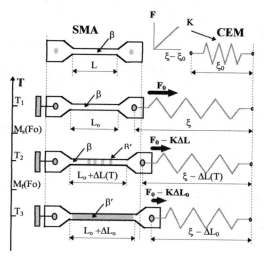

FIGURE 2: Shape memory alloy coupled to a spring in an actuator application. **Fo** is the initial applied stress on the SMA and **KΔL** is the stress relaxation due to the length change during transformation.
The transformation takes place in a continuous way as the temperature is lowered from Ms to Mf while the parent β phase transforms to the martensitic phase β′. The characteristics transformation temperatures are stress dependent following a Clausius- Clapeyron relation [1].

In order to analyze the feasibility of single-crystal SMA applications we considered the informations extracted from the wax nucleus characteristics as basic requirements to be imposed on the SMA nucleus performance. However, if another type of valve were considered the requirements on the SMA could be different. In this way, taking into account the maximum force involved, the required displacements and the temperature range, a flat monocrystalline sample (S1a) was characterized in the equipment previously described. Figure 3 shows the results corresponding to sample S1 in interaction with a conventional spring. In Fig. 3A the cycles obtained at three different levels of force (Fig. 3B) for a relatively soft spring (K_1= 0.37 N) are shown. They were obtained by changing the initial length of the spring applied to the β phase before the start of the transformation. As the force F_0 on the SMA nucleus is increased, the transformation is shifted to higher temperatures. During the transformation, the force decreases due to the relaxing effect associated to the progressive sample elongation. In Fig. 3C the effect of changing the spring stiffness can be appreciated. The higher the elastic constant of the spring, the higher the temperature range of transformation due to the higher stress relaxation. The martensitic transformation temperature (Ms) depends only on the Fo value. It should be emphasized that in all cases the hysteresis of the transformation ranged between 1.5 K to 2.0 K.

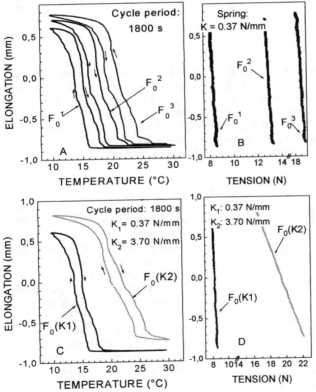

FIGURE 3: Thermal cycles for a Cu-Zn-Al single-crystal of Ms=285 K obtained in controlled experiments. **A**: three different load values (Fo) are obtained by changing the initial length of a spring of K_1=0.37 N/mm. Transformation temperature shifts upwards as the stress increases. **B**: stress variation during transformation for the cycles of Fig. 2A. **C**: effect of a different spring K value on the temperature width of the transformation thermal cycle. The higher stress change for the K_2=3.7 N/mm spring, shown in Fig. **2D**, explains the increase in the temperature width.

In order to study the reproducibility of the material behavior with the number of cycles, several similar samples were cycled up to different amounts of cycles (having obtained a maximum N = 14000). It was found that the cycles are reproducible within a maximum shift of 1 K.

On the other hand, surface defects appeared after a large number of cycles (N=1500), in agreement with the results of Sade et al. [4] obtained from pseudoelastic cycled samples. It has been shown that the defects, aligned parallel to the habit plane of transformation are the origin of fatigue cracks leading to fracture of the sample.

DISCUSSION AND CONCLUSIONS

Our results show that the SMA nucleus has a transformation hysteresis of about 1.5 - 2 K. This value is higher than the 0.5 K value obtained for the wax nucleus. The elongation-temperature slope of the wax nucleus is 3 mm/15 K, being a fixed value. By changing the spring stiffness in the SMA basic actuator device the slope of transformation can be tailored, as observed in fig. 3C. The maximum slope value is intrinsic to the SMA element [5] and can be obtained with a sufficiently soft spring in order to avoid strong relaxation during transformation. As the spring stiffness increases, the magnitude of the slope of the transformation decreases, the minimum value being determined by the maximum stress which can be applied to the SMA element and the maximum elongation required. By changing the slope of transformation, a SMA based actuator may behave in the range between nearly *on-off and proportional* controllers. Similar force values can be reached with wax and SMA nucleus.

As was stated above most applications with SMA until the present are based on the use of polycrystalline material. This approach is reasonable when complex geometries are necessary. However, in actuator applications the use of single-crystals of SMA in simple uniaxial tension shows advantages. Polycrystalline SMA materials show typical intrinsic hysteresis values between 15 K and 20 K [2]. These values are at least a factor of ten higher than the 1.5 K to 2 K obtained for the Cu-Zn-Al single-crystal samples (Fig. 3). On the other hand, the presence of grain boundaries has a strong deleterious effect on the fatigue life during cycling due to the high degree of elastic anisotropy inherent to these materials. The fatigue life of pseudoelastically cycled single-crystals of Cu-Zn-Al is higher than N = 1 E5 cycles for an applied stress level below 60 MPa [6].The observed surface defects in all the thermally cycled samples are of the same type as those obtained after pseudoelastic cycling. Thus, it is accepted that pseudoelastic fatigue results are valid for thermal cycling experiments. A Cu-Zn-Al single-crystal with 1 mm^2 transverse section can reach a total life of N = 1 E5 cycles for an applied load of 60 N. This result is satisfactory for the valve considered.

It is worthwhile to consider that the design of actuator devices with SMA single-crystals has additional advantages due to the simple shape of the elongation - temperature behavior which makes easier the prediction of the interactions with the elastic material or with an additional friction element. The elongation-temperature behavior of a single-crystal can be considered nearly independent of stress. A simple translation relates response for different stress levels. On the other hand, the total elongation is dependent on the applied stress level if polycrystalline samples are used.

The definition of the basic actuator device enhances the design process. As an example a conceptual analysis of a thermostatic application is made in Fig. 4. The SMA single-crystal elongation-temperature behavior can be well approximated by a linear relation and an hysteresis. For simplicity, we assume that the heat rate delivered by the radiator to the room (**P**) is proportional to the hot water flow wich in turn is also proportional to the valve aperture. Since the valve aperture is directly controlled by the SMA elongation

(ΔL), a linear relation with an hysteresis results between (**P**) and (**T**). The relation between the room temperature (**T**) and the corresponding room heat power (**P$_A$**) (room characteristic) is assumed as linear too. In Fig. 4A the effect of the spring stiffness in the temperature control precision ΔT_C can be analyzed. As the spring stiffness is increased ($K_2 > K_1$) the **P-T** slope magnitude decreases and the temperature precision increases ($\Delta T_C^{K2} < \Delta T_C^{K1}$). In Fig. 4B the effect of the slope on the accuracy of temperature control is shown. A shift of the room characteristic **P$_A$-T** due to an increase in the external-room temperature ($T_E \rightarrow T_{E'}$) induces a change δTc in the controlled temperature. This temperature change is smaller for the softer spring ($\delta T_C^{K1} < \delta T_C^{K2}$). The optimum choice of the spring K value should consider both effects.

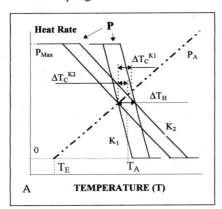 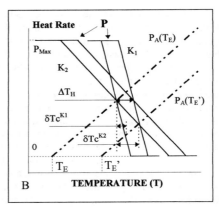

A   **TEMPERATURE (T)**   B   **TEMPERATURE (T)**

FIGURE 4: Schematic analysis of the behavior of an actuator of single-crystal SMA applied to a thermostatic controller. Two cases are considered corresponding to springs with stiffness K_1 and K_2 respectively ($K_2 > K_1$) giving hysteresis cycles with different slopes. T_E is the external temperature and T_A is the desired room temperature. The effect of the intrinsic hysteresis ΔT_H on the effective controlled width ΔTc can be decreased choosing a 'hard' spring as can be seen in Fig. **3A**. A perturbation in the heat rate - temperature characteristic of the room from $P_A(T_E) \rightarrow P_A(T_{E'})$ produces a higher shift δT_C in the equilibrium temperature for the harder spring as can be seen in Fig. **3B**.

REFERENCES

1. M. Ahlers, Prog. Mater. Sci. **30**, 135 (1986).

2. T.W. Duerig, K.N. Melton, D.Stockel and C.M.Wayman (Editors), Engineering Aspects of Shape Memory Alloys, (Butterworth - Heinemann, London, 1990)

3. T.W. Duerig, Materials Science Forum **56-58** Pt II, edited by B.C. Muddle (Trans Tech Publications, Zurich, 1990) pp. 679 - 691.

4 . M. Sade, R. Rapacioli and M. Ahlers, Acta Metall. **33**, 487 (1985).

5. F.C. Lovey, A.Amengual, V.Torra and M.Ahlers, Phil. Mag. A **61**, 161 (1990).

PHASE TRANSFORMATIONS AND CRYSTALLOGRAPHY OF TWINS IN MARTENSITE IN Ti-Pd ALLOYS

M. Nishida*, Y. Morizono*, H. Kijima**, A. Ikeya**, H. Iwashita** and K. Hiraga***

*Department of Materials Science, Kumamoto University, Kurokami, Kumamoto 860, Japan, nishida@gpo.kumamoto-u.ac.jp
**Graduate School, Kumamoto University, Kurokami, Kumamoto 860, Japan
***Institute for Materials Research, Tohoku University, Katahira, Aoba, Sendai 980-77, Japan

ABSTRACT

Phase transformations and crystallography of twins in near-equiatomic Ti-Pd alloys have been studied. In the first half we have found that the transformation temperature decreases with decreasing the Pd contents and the successive transformations take place in the Ti-rich alloys. While the transformation temperature is nearly constant with the composition and a single transformation takes place in the equiatomic and Pd-rich alloys. The solubility limit of TiPd compound in Ti-rich side is extended to about 55 at%Ti at 900°C and abruptly decreased with decreasing temperature. In the latter half we have found three twinning modes, i.e., {111} Type I, <121> Type II and {101} compound twins, in the martensite. The {111} Type I and <121> Type II twinnings which are conjugate to each other coexist in the same variant. The {111} Type I twins are dominantly observed and the <121> Type II twins are less frequently observed. The former twinning is considered to be a lattice invariant shear. There is no martensite variant consisting wholly of the <121> Type II twins throughout the present observation. The <121> Type II twinning is considered to be a deformation twin due to the elastic interaction during the transformation. The {101} compound twinning is also considered to be a deformation twin, since the twin has an isolated fashion in the martensite variant. The boundary structure of the above three twinning modes was also discussed on the basis of lattice image.

INTRODUCTION

Ti-Pd alloys of the near-equiatomic composition have been one of promising high temperature shape memory materials since they undergo a thermoelastic martensitic transformation from the B2 to a B19 type structure around 500°C upon cooling [1-2]. However, some fundamental characteristics of the alloy have not been clarified yet. In the first half of the present study we discuss the compositional dependence of the transformation behaviors. The effect of heat treatment on the transformation behaviors is also investigated and then the phase boundary of TiPd compound is estimated. In the latter half the crystallography of twins in the martensite is studied by conventional (CTEM) and high resolution electron microscopy (HREM).

EXPERIMENTAL PROCEDURES

Ti-Pd alloys of near-equiatomic compositions were prepared from 99.7 mass% Ti and 99.8 mass% Pd by arc melting in an argon atmosphere. The ingots were hot-forged, and then homogenized at 1000°C for 36 ks in vacuum. The rods of 3 mm in diameter were spark cut from the ingot. The rods were cut into disks of about 1.5 mm and 0.2 mm in thickness. The former and the latter disks were used for DSC measurements and TEM observations, respectively. They were solution-treated in vacuum at 1273 K for 3.6 ks, and then quenched into ice water. Some of thicker disks were then aged from 600 to 1000°C with 100°C interval for 360 ks. Specimens for TEM studies were electropolished using the twin jet method in an electrolyte of H_2SO_4 and methanol, 1 : 4 in volume, around 273 K. CTEM observations were carried out in a JEOL-2000FX microscope. HREM observations were carried out in a JEOL-200CX and 4000EX microscopes. The crystal structure of the B19 martensite is of MgCd type with an orthorhombic unit cell. The following lattice parameters were used for the analysis; a = 0.489 nm, b = 0.281 nm, c = 0.456 nm [3].

RESULTS AND DISCUSSION

Transformation Behavior

Figure 1 shows typical DSC curves of near-equiatomic alloys. Two types of curves are obtained with the composition. Two exothermic and endothermic reactions are observed during heating and cooling, respectively, in Ti-rich alloys as seen in Ti-45 at%Pd alloy. We define that the first and the second peak temperatures during heating are A2* and A1* as indicated in Fig. 1, respectively. In the same way the first and the second peak temperatures during cooling are defined as M1* and M2*, respectively. On the other hand, a single reaction is only measured in equiatomic and Pd-rich alloys as seen in Ti-50 at%Pd alloy. The compositional dependence of A2* and A1* is summarized in Fig. 2. The A2* is increased with Pd content up to the equiatomic composition and then overlapped to A1* which is nearly constant with the composition. It is apparent from these results that the successive transformations take place in Ti-rich alloys. The crystal structure of Ti-rich and equiatomic alloys at room temperature was determined to be B19 by X-ray and electron diffraction experiments. Pd-rich alloys consisted of the B19 martensite and Ti_2Pd_3 compound. To identify the intermediate products between B2 parent and B19 martensite phases *in-situ* TEM observation was carried out. However, since large amount of Ti_2Pd precipitates was formed during heating, the intermediate phase could not be detected by both the electron diffraction experiments and image observations. The detailed nature of the intermediate phase is now under study and will be reported in due course. Therefore, the intermediate phase is designated as X in the present study as shown in Fig. 2.

Phase Boundary of TiPd Compound

From the compositional dependence of transformation temperatures in Fig. 2, the phase boundary of TiPd compound in Pd-rich side is deduced to be near the equiatomic composition. This was supported by aging experiments. In other words, transformation temperatures were not changed in Pd-rich alloys by aging. While the phase boundary of TiPd compound in Ti-rich side is likely extended to 55 at%Ti at 1000°C at least since the A2* is decreased with Ti content. Figure 3 shows the effect of aging temperature on the transformation peak temperatures in Ti-45 and

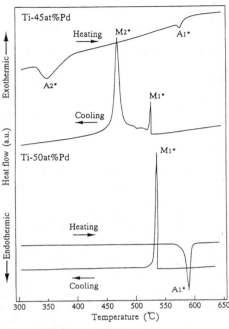

Figure 1 Typical DSC curves of near-equiatomic Ti-Pd alloys

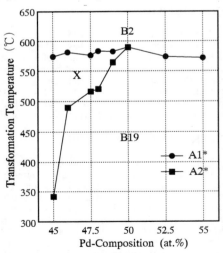

Figure 2 Compositional dependence of transformation peak temperatures during heating in near-equiatomic Ti-Pd alloys quenched from 1000°C.

47.5 at%Pd alloys. The all specimens are aged for 360 ks at each temperature. The A2* in Ti-45 at%Pd alloy is increased in the specimens aged below 800°C. That in Ti-47.5 at%Pd alloy is increased in the specimens aged below 700°C. The double peaks, i. e., A1* and A2*, are overlapped to the single one in the both alloys aged below 700°C. These results indicate that the phase boundary of TiPd compound between 800 and 900°C is around 55 at%Ti and that between 700 and 800°C is around 52.5 at%Ti. Consequently, the solubility limit of TiPd compound in Ti-rich side is extended to about 55 at%Ti above 900°C and abruptly decreased with decreasing temperature.

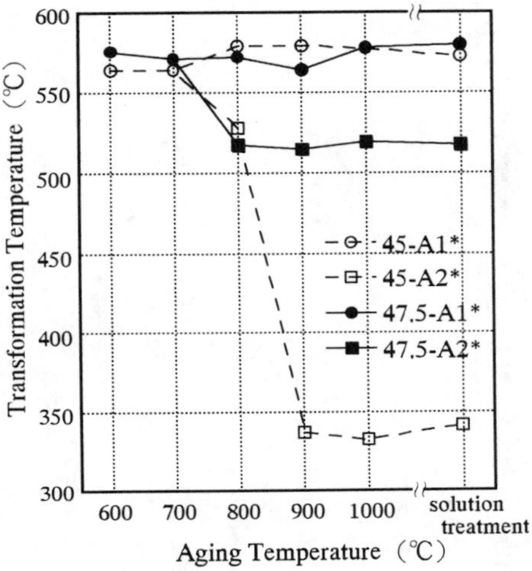

Figure 3 Effect of aging on the transformation peak temperatures during heating in Ti-rich Ti-Pd alloys.

Crystallography of Twins in Martensite

Since the martensite structure of near-equiatomic Ti-Pd alloys used in the present study was B19 as mentioned above, the Ti-50 at%Pd alloy is adopted for the investigation of crystallography of twins. In the present study, three twinning modes were confirmed as listed in Table 1. The twinning elements were calculated by the Bilby-Crocker theory [4]. The {111} Type I and <121> Type II twinnings which are conjugate to each other coexisted in the same variant as described below. Most of twinning modes observed were {111} Type I twins which is considered to be a lattice invariant shear the same as in the previous reports [2,5] and other modes were less frequently observed.

Figure 4 (a) shows a bright field image of twins taken along the [121] which is parallel to the K_1 plane of $(1\bar{1}1)$ Type I and $(10\bar{1})$ compound twinnings and η_1 direction of [121] Type II twinning. That is, the three twinning modes are observed in the edge-on state. Figures 4 (b) and (c) are selected area electron diffraction patterns taken from areas B and C in (a), respectively. The pattern in (b) consists of two sets of reflections of [121] zone axis which are in mirror symmetry with respect to the $(1\bar{1}1)$ plane. The alternating platelets in the area B can be concluded to be the $(1\bar{1}1)$ Type I twins. On the other hand, the pattern in (c) shows a single pattern of [121] zone axis. The single pattern taken from alternating platelets is characteristic of Type II twin obtained along η_1 direction [6,7]. In other words, the patterns from the matrix and the twin are related by the rotation of 180° around the η_1 = [121] axis. The inclined platelets and matrix in the area C are determined to be in the [121] Type II twin relation. The inclined angle between the trace of K_1 planes for the $(1\bar{1}1)$ Type I and [121] Type II twins is about 15° in average and compatible with the calculated value from the indices of each K_1 plane listed in Table 1. The Bilby-Crocker theory predicts that the conjugate twinning mode is always geometrically possible with the same twinning shear [4]. However, there was no martensite variant consisting wholly

Table 1 Twinning elements in B19 martensite in Ti-Pd alloy.

	K_1	η_1	K_2	η_2	s
Type I Twin	{111}	$<0.671\ 1\ \overline{0.329}>$	$\{\bar{1}\ 0.678\ \overline{0.356}\}$	<121>	0.361
Type II Twin	$\{\bar{1}\ 0.678\ \overline{0.356}\}$	<121>	{111}	$[\overline{0.671}\ 1\ \overline{0.329}]$	0.361
Compound Twin	{101}	$<\bar{1}0\bar{1}>$	$\{\bar{1}0\bar{1}\}$	<101>	0.140

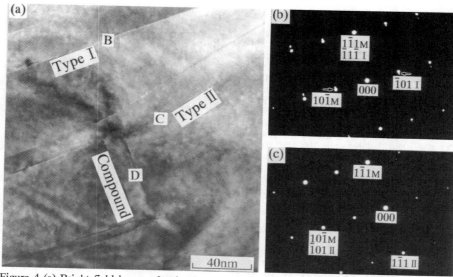

Figure 4 (a) Bright field image of twins in martensite. (b) and (c) Electron diffraction patterns taken from areas B and in (a), showing ($1\bar{1}1$) Type I and [121] Type II twin relations, respectively.

of the <121> Type II twins throughout the present observation. Therefore, the <121> Type II twin is considered to be a deformation twin due to the elastic interaction during the transformation rather than a lattice invariant shear. The electron diffraction pattern taken from area D consisted of two sets of reflections of [121] zone axis which were in mirror symmetry with respect to the ($10\bar{1}$) plane, although the pattern is not reproduced here. The isolated platelet and matrix in the area D can be concluded to be in the ($10\bar{1}$) compound twin relation. Since the {101} twinning always has an isolated fashion as shown in (a), this twinning may be induced as a result of elastic interaction during the martensitic transformation. This assumption is supported that {101} compound twins with banded morphology were frequently observed in the specimen applied thermal stress cycles, where the Ti-Pd alloy was adopted as a stress relaxation layer of ceramics/metal joint [8]. The <121> Type II and {101} compound twins are firstly observed in the present study.

Boundary Structure of Twins

Figures 5 (a) and (b) show two-dimensional lattice images of the ($1\bar{1}1$) Type I and [121] Type II twin boundaries taken along [121] direction, respectively. The boundary of the ($1\bar{1}1$) Type I twin in Fig. 5 (a) is straight and has mirror symmetry with respect to the K_1 plane, which is characteristic of the Type I twinning. The irrational boundary of the [121] Type II twin in Fig. 4 (b) is gradually and randomly curved with broad strain contrast as compared with that of the ($1\bar{1}1$) Type I twin. There is neither ledge nor step structure at the boundary as reported by Knowles in the <011> Type II twin boundary in Ti-Ni martensite [9]. This observation suggests that a strain around the [121] Type II twin boundary in the B19 martensite of Ti-Pd alloy is elastically relaxed by gradual displacement of the atoms instead of ledge and step structures the same as that around <011> Type II twin boundary in the B19' martensite of Ti-Ni alloy [10].

Figure 6 shows a lattice image of the (101) compound twin boundary taken along [010] direction. The boundary of the (101) compound twin is straight and the two twin crystals have mirror symmetry with respect to the (101) K_1 plane, although a broad strain contrast can be seen in places. The strain contrast may be introduced as a elastic interaction during the transformation.

378

CONCLUDING REMARKS

Phase transformations and crystallography of twins in near-equiatomic Ti-Pd alloys have been studied. The obtained results are summarized as follows.

(1) The transformation temperature decreases with decreasing Pd contents and the successive transformations take place in the Ti-rich alloys. While the transformation temperature is nearly constant with the composition and a single transformation takes place in the equiatomic and Pd-rich alloys.

(2) The phase boundary of TiPd compound in Pd-rich side is deduced to be near the equiatomic composition. While that in Ti-rich side is extended to about 55 at%Ti at 900°C and abruptly decreased with decreasing temperature.

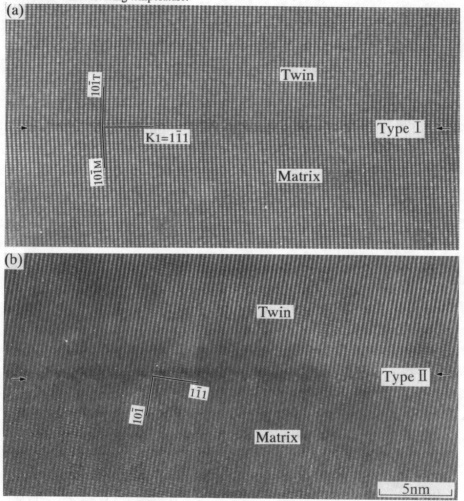

Figure 5 Lattice images of ($1\bar{1}1$) Type I (a) and [121] Type II (b) twin boundaries. Electron beam // [121].

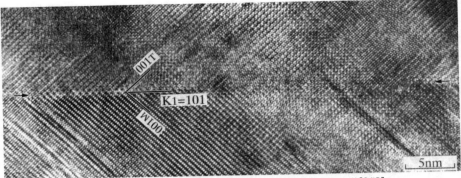

Figure 6 Lattice image of (101) compound twin boundary. Electron beam // [010].

(3) There are three twinning modes, i.e., {111} Type I, <121> Type II and {101} compound twins, in the martensite. The Type I twin is considered to be a lattice invariant shear in the present transformation. The Type II and the compound twins are considered to be a deformation twin.

(4) Lattice images of the Type I and the compound twins exhibit the well-defined crystallographic characteristics of those boundaries. There is neither ledge nor step structure at the <121> Type II twin boundary. The lattice image suggests that a strain around the boundary is elastically relaxed by gradual displacement of the atoms.

ACKNOWLEDGMENTS

The present work was partly supported by the Grant-in-Aid for Fundamental Scientific Research (C, 08650772, 1996) from the Ministry of Education, Science, Sports and Culture, Japan. The latter half of the study was carried out under the Visiting Researcher's Program of the Institute for Materials Research, Tohoku University.

REFERENCES

1. H. C. Donkersloot and J. H. N. van Vucht, J. Less Common Met. **20**, p.63 (1970).

2. K. Enami, H. Seki and S. Nenno, Tetsu-to-Hagane. **72**, p. 563 (1986).

3. P. Krautwasser, S. Bhan and K. Shcubert, Z. Metallkd. **59**, S. 724 (1968).

4. B. A. Bilby and A. G. Crocker, Proc. Roy. Soc. Ser. A. **288**, p. 240 (1965).

5. P. G. Lindquist and C. M. Wayman, MRS Int'l. Mtg. on Adv. Mats. **9**, p. 123 (1989).

6. K. M. Knowles and D. A. Smith, Acta Metall. **29**, p. 101 (1981).

7. T. Hara, T. Ohba, S. Miyazaki and K. Otsuka, Mater. Trans. JIM. **33**, p. 1105 (1992).

8. Y. Morizono, M. Nishida and A. Chiba, Trans. Mat. Res. Soc. Japan. **16B**, p. 1151 (1994).

9. K. M. Knowles, Phil. Mag. **A45**, p.357 (1982).

10. M. Nishida, K. Yamauchi, I. Itai, H. Ohgi and A. Chiba, Acta Metall. et Mater. **43**, p. 1229 (1995)

COOLING RATE DEPENDENCE OF FATIGUE LIFE TIME (N_f) OF Cu-Al-Ni ALLOY PREPARED BY LIQUID-QUENCHING

K. MORI, Y. TANAKA, S. FURUYA, T. OKADA, K. KOMATSUZAKI and Y. NISHI
Department of Materials Science, Tokai University, 1117 Kitakaname, Hiratsuka, Kanagawa,
259-12 Japan, am026429@keyaki.cc.u-tokai.ac.jp

ABSTRACT

Since the fatigue life time is a serious problem of the Cu-Al-Ni alloy, it is difficult to apply for practical use. The authors have investigated the influence of the cooling condition of liquid-quenching on fatigue property. The liquid-quenching is performed by a piston-anvil apparatus. It is easy to control the solidification condition. The fatigue life time increases with increasing the cooling rate. The long life time is due to the small grain size, the small volume of γ_1' phase, the low density of lattice defects and the low transformation temperature.

INTRODUCTION

The Cu-Al-Ni alloy is generally known as smart materials which show the shape and color memory properties. The Cu-Al-Ni alloy is commercially cheaper than the Ni-Ti alloy. However, the Cu-Al-Ni alloy has a cumbrous problem of resistance to fatigue. Sakamoto has suggested that the resistance to fatigue of the Cu-Al-Ni alloy depends on the deformation modes. The psedoelastic deformation is the most harmful mode for fatigue life time [1]. If we conquer the fatigue problem, the required process is the controlling the grain size. The small grain size prevents to concentrate the exccess stress. Liquid-quenching is one of useful processes for grain refining. Various properties have been reported for the liquid-quenched crystalline alloys [2-5]. We have reported that the liquid-quenching enhances the grain refining [6,7]. Thus it is possible for the liquid-quenching to improve the fatigue resistance of Cu-Al-Ni alloy. The aim of present work is to investigate effects of the cooling condition of liquid-quenching on resistance to fatigue in the pseudoelastic deformation region of the Cu-Al-Ni alloy.

EXPERIMENTAL PROCEDURE

A Cu-26.6at%Al-3.7at%Ni alloy was prepared by melting pure Cu (99.9%), Al (99.9%) and Ni (99.9%). The homogenization of the master alloy was performed for 3 hours at 1073K. Sheet specimens were prepared by melting in an infrared furnace at 1400K under a protective atmosphere of Ar-5vol%H_2 and then liquid-quenching by the use of a piston-anvil apparatus [8]. The chemical composition of the samples as listed in Table I   was analyzed in by ICP. The cooling rate (R ; K/s) at 1000K was varied by controlling the sample thickness (D ; m) as a following equation [9], the R value was determined by the D value.

$$\log R = - \log D + 0.9 \tag{1}$$

TABLE I   Chemical analysis of master alloy and liquid-quenched alloy

	Cu (at %)	Al (at %)	Ni (at %)
Nominal composition	69.7	26.6	3.7
Master alloy	69.9	26.4	3.7
Liquid-quenched alloy	70.0	26.4	3.6

The microstructure of the samples was observed by an optical microscope. The structure changes were investigated by means of X-ray diffraction. The transformation temperature was determined by using a differential scanning calorimeter. The heating rate is 20K/min. The fatigue test was carried out as follows. The shape of the specimen was a small rectanglar parallelepiped. The thickness was controlled by electro-polishing for the fatigue test. The size of the specimen was cut into 9mm in length, 5mm in width and 0.04mm in thickness. The specimen was bent by the use of a cylinder of 3mm in diameter at room tenperature and then heated in furnace at 503K (see Fig. 1). The test is repeated upto the fracture. The fractured surface was observed by a scanning electron microscope.

RESULTS AND DISCUSSION

Figure 2 shows the relationship between logarithmic number (N_f) of cycles to failure and logarithmic cooling rate (R). The long fatigue time is obtained in the specimen with faster cooling rate. Figure 3 shows the SEM photograph of fatigue fractured surface of the liquid-quenched Cu-Al-Ni alloy. The SEM photograph in Fig. 3 (E) indicates the intergranular cracking of the slow cooled alloy (brittle type). The intergranular cracking is generated by the stress-induced martensitic transformation. The stress concentration along the grain boundary is to maintain the coherency between the matrix and the stress-induced martensite. On the other hand, a dimple-like pattern is observed on the fractured surface of the fast cooled alloy in Fig. 3 (B). It indicates that the fast cooled alloy has a ductile fracture mode. Both the slow (large columnar like grain) and the fast (small granular like grain + small columnar like grain) cooled alloys show the pseudoelastic deformation due to a stress-induced martensitic transformation in the matrix state.

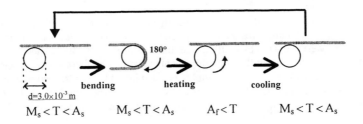

Figure 1   Cycle of fatigue fracture test.

Figure 4 shows the optical microscope photographs of the cross section for the liquid-quenched Cu-Al-Ni alloy. The large size of columnar is observed in Fig. 4 (E) for the slow cooled alloy. It grows from surface. The small grain size of granular is observed in Fig. 4 (B) for the fast cooled alloy. Figure 5 shows the relationship between logarithmic diameter of grain size (d) and logarithmic cooling rate (R). The grain size decreases with increasing cooling rate. Since the small grain size prevents to concentrate the excess residual stress, the stress concentration of the fast cooled alloy is smaller than that of the slow cooled alloy. Thus, it is easy for the slow cooled alloy to fracture by an intergranular cracking. It is origin of small fatigue life time.

The fracture type of the fast cooled alloy is a transgranular cracking. It shows large fatigue life time. Figure 6 shows X-ray diffraction patterns for the liquid-quenched Cu-Al-Ni alloy. The γ_1' martensite phase gradually disappears with the increasing the cooling rate. When the volume is small of γ_1' martensite phase, the strain is small in the β_1 parent phase. Figure 7 shows relationship between full width of quarter maximum (FWQM) of β_1 (220) phase and logarithmic cooling rate R. The fast cooled alloy shows the small FWQM value. It shows that it is difficult to generate the dislocation for the fast cooled sample. The X-ray study indicates the small strain of β_1 parent phase for the fast cooled alloy. It enhances the fatigue life time. Although the fast liquid-quenching generally enhances the dislocation density, the high defects density can be obtained for the slow cooled Cu-Al-Ni alloy.

Figure 8 shows models of stress-induced martensite depending on the cooling rate. Since the small grain prevents to concentrate the excees redidual stress, (see in Fig. 8 (a) and (b)), the stress distribution of the fast cooled alloy expands downward (see in Fig. 8 (c) and (d)). Therefore, it is difficult to occur the stress-induced martensite transformation, which is annoyance for fatigue life time. If it is easy to generate the γ_1' phase, the transformation temperature should be high for the slow cooled sample. Figure 9 shows the relationship between A_s temperature and logarithmic cooling rate (R). The low A_s temperature is observed in the specimen with faster cooling rate. It shows that the increasing in the cooling rate restrains the stress-induced transformation.

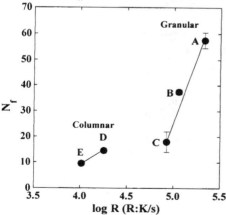

(B) Fast cooled alloy   (E) Slow cooled alloy

Figure 2   Relationship between logarithmic number (N_f) of cycles to failure and logarithmic cooling rate (R).

Figure 3   SEM photograph of fatigue fractured surface of liquid-quenched Cu-Al-Ni alloy for fast (B) and slow (E) cooled samples.

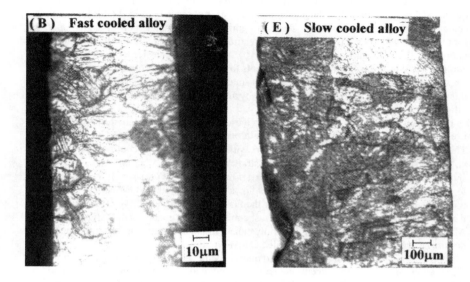

Figure 4   Optical microscope photograph of cross section for liquid quenched
Cu-Al-Ni alloy : (B) for fast cooled alloy; (E) for slow cooled alloy.

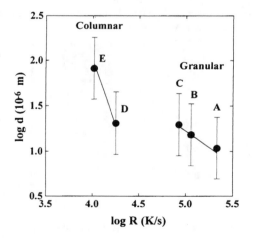

Figure 5   Relationship between logarithmic diameter of grain (d)
and logarithmic cooling rate (R).

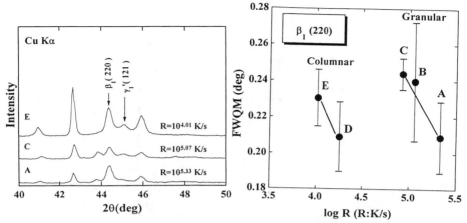

Figure 6   X-ray diffraction patterns for liquid-quenched Cu-Al-Ni alloy at different cooling rate.

Figure 7   Relationship between full width of quarter maximum (FWQM) of $\beta_1(220)$ phase and logarithmic cooling rate (R).

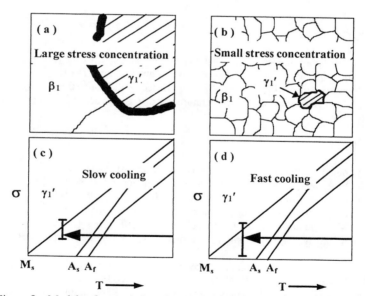

Figure 8   Models of stress-induced martensite depending on cooling rate. (a) & (c) are for slow cooled sample. (b) & (d) for fast cooled sample. (a) & (b) are morphorogy. (c) & (d) are for relation between resistance to stress and temperature of stress-induced martensite depending on cooling rate.

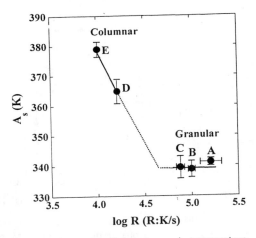

Figure 9   Relationship between A_s temperature and logarithmic cooling rate (R).

CONCLUSION

The effects of liquid-quenching is investigated on the fatigue life time for the Cu-Al-Ni alloy. The liquid-quenching process enhances the fatigue life time. The long fatigue time is obtained in the specimen with faster cooling rate. Based on the grain size, density of lattice defects, γ_1' phase generation and A_s temperature, we discuss the cooling rate dependence of fatigue life time.

REFFERENCES

1. H. Sakamoto, Y. Kijima and K. Shimizu, Trans. Japan Inst. Metals, **23**, p585 (1982).
2. Y. Nishi, H. Aoyagi, K. Suzuki and E. Yajima, Trans. Japan Inst. Metals, **23**, p.703 (1982).
3. Y. Nishi, H. Aoyagi and E. Yajima, J. Japan Inst. Metals, **47**, p964 (1983).
4. Y. Nishi, M. Tachi and E. Yajima, Scripta Met.,**19**, p865 (1985).
5. Y. Nishi, M. Tachi and E. Yajima, Scripta Met., **19**, p289 (1985).
6. Y. Nishi, Y. Miyagawa, N. Suketomo, T. Morishita and E. Yajima, Scripta Met., **19**, p1273 (1985)
7. Y. Nishi, M. Tachi and E. Yajima in proceeding of the Fifth International Conference on Rapidly Quenched Metals, 1985, edited by S. Steeb and H. Warlimont, p. 1435-1438.
8. Y. Nishi, K. Suzuki and T. Masumoto in proceeding of the Fourth International Conference on Rapidly Quenched Metals, 1981, edited by T. Masumoto and K. Suzuki, p. 217-220.
9. Y. Nishi, K. Suzuki and T. Masumoto, J. Japan Inst. Metals, **45**, p1300 (1981).

AGING INFLUENCE ON THE SHAPE MEMORY EFFECTS AND SUPERELASTICITY IN TITANIUM-NICKEL SINGLE CRYSTALS

YU.I. CHUMLYAKOV, I.V. KIREEVA, V.N. LINEYTSEV, A.G. LWISYUK
Siberian Physical Technical Institute, Revolution sq., 1, Tomsk 634050, Russia,
chum@phys.tsu.tomsk.su

ABSTRACT

On the Ti-50.8%Ni, Ti-51%Ni and Ti-51.3%Ni (at.%) single crystals, the investigations of the shape memory effects (SME) and superelasticity (SE) have been carried out in order to find out the dependence on a size and a volume fraction of the Ti_3Ni_4 dispersive particles, a crystal orientation and a sign of applied stress. It is shown, that in crystals without particles the strong dependence of SME on an orientation and a sign of applied stress takes place and SE is not observed. Precipitation of the coherent particles of Ti_3Ni_4 of 50-100 nm in size leads to the reduction of SME, weakening of orientation dependence of σ_{cr} in interval of stress-induced martensitic transformations (MT).

INTRODUCTION

The precipitation of the Ti_3Ni_4 dispersive particles in Ti-(50.6-52)at.%Ni alloys leads, as it is shown by electron-microscopic structural investigations and a studying of physical, mechanical and functional properties of SME and SE, to principal change in the fine structure of martensite, the MT temperatures and strength properties [1-3].

The Ti_3Ni_4 dispersive particles, which are non-isomorphous to matrix, have the rombohedrical structure, appear during aging in the interval 673-773 K. They have dimensions of 50-100 nm, don't undergo MT and lead to the change in the fine structure of martensite crystals from twinning II to compound twinning. This change in the martensite fine structure is accompanied by the decrease of SME as compared with monophase materials and by the appearance of SE which is not observed in quenched poly and single crystals. Also there is the increase of the MT temperatures. It have been found that there is the significant decrease of $\sigma_{cr}(M_n)$ as compared with specimens before aging [1-3].

For the deformation by slip of poly and single crystals containing the volume fraction f>1-3% of the plastically non-deforming particles BeO, Al_2O_3, HfO_2 it's shown that in single crystals the orientation dependence of the strain-hardening coefficient Θ disappears and in polycrystals the dependence of $\sigma_{0.1}$ and Θ on grain size does not observed [4]. For such gradient materials, the compatibility of plastic deformation of matrix and particles is achieved due to creation of "geometrically-necessary" dislocations, which determine plastic behaviour of materials. During the MT development in the geterophase TiNi single crystals containing the Ti_3Ni_4 particles, which don't undergo transformaions, one can expect the qualitatively analogous peculiarities of the mechanical behaviour.

The single crystals were grown by Bridgman technique in the inert gas atmosphere and then were solution-treated at 1173 K for 24-30 hr. Then the specimens for mechanical tests were spark-cut, mechanically grinded, chemically etched, and electropolished.

Mat. Res. Soc. Symp. Proc. Vol. 459 © 1997 Materials Research Society

EXPERIMENTAL RESULTS AND DISCUSSION

It was experimentally found that in the homogeneous state the orientation dependence and asymmetry of the σ_{cr} depend on the Ni concentration in crystals. Such the orientation dependence and asymmetry in Ti-50.45at.%Ni crystals are significantly stronger than in the Ti-50.8at.% Ni (I), Ti-51 at.%Ni (II) and Ti-51.3at.%Ni (III) (fig.1) ones.

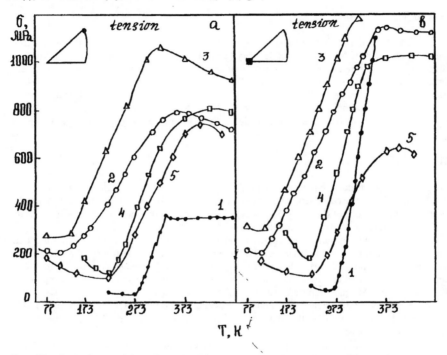

Fig. 1 The dependence of the $\sigma_{0.1}$ on the test temperature of the TiNi single crystals: 1-Ti-50.45 at.%Ni; 2-Ti-51at.%Ni-quench; 3-Ti-51at.%Ni aged at 573 Kfor 1hr.; 4-Ti-51at.%Ni aged at 773K for 1hr.; 5-Ti-51.3at.%Ni aged 773 K for 1hr.

Figure 2 shows the temperature dependence of the electrical resistance $\rho(T)$ in the Ti-51at.% Ni crystals aged at 573 K for 1 hr. This heat treatment leads to the increase of the deforming stresses σ_{cr} as compared with the quenched state. For the specimens aged at 773 K for 1 hr., the decrease in deforming stress is observed which is particularly appreciable in the stress-induced MT interval (fig.1). M_s shifts to the high temperature range as compared with the quenched crystals. In the temperature range of the stress-induced martensite formation the increase of the angle $\alpha=d\sigma/dT$ on the dependence $\sigma_{cr}(T)$ is observed. In this temperature interval at $T<M_d$, the orientation dependence of the σ_{cr} and their asymmetry-the dependence of the σ_{cr} on sign of the applied stresses-practically disappears (fig.3). In the case of Ti-50.45at.%Ni crystals at $T<M_d$, the strong σ_{cr} dependence on orientation and applied stress sign is observed (fig.1).

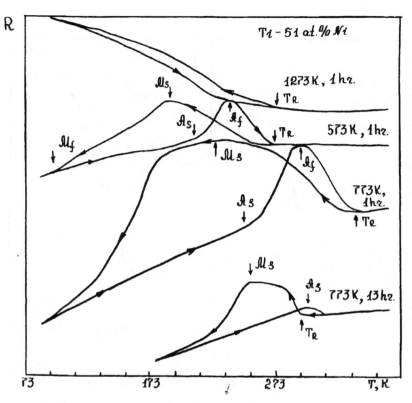

R

$Ti - 51$ at.%Ni

M_s

$1273K$, $1hr.$
↓ T_R

A_{s}

A_f

$573K$, $1hr.$
↓ T_R

M_f

↑ M_s

A_f

$773K$, $1hr.$

A_s

↑ T_R

M_s

A_s $773K$, $13hr.$

↑ T_R

73   173   273   T, K

Fig.2 The temperature dependence of the electrical resistance in the Ti-51 at.%Ni aged for some temperatures for some hours.

In monophase and after aging at 573 K for 1 hr. I-III crystals, the SE is not observed. The magnitude of SME appears to be close to the value found earlier on the single crystals without particles and these values are in the good agreement with theoretical predictions for B2-B19' transformations [3-6]. After aging at 773 K for 1 hr., at 673 K for 1.5 hr. of the I, II, III crystals, the value of SME decreases as in [3,6] and the effects of SE appear (fig.4). The loops of mechanical hysteresis at $T>A_f$ possess the peculiarities: 1) in the crystals I (fig.4) during loading, the yielding point appears and then the unstable plastic flow takes place with the rising of ε; at unloading, the deviation of the σ(ε) curves from the "seeming" elastic unloading is observed which was previously noted for polycrystals [8,9]. This non-symmetry of the loading and unloading curves more obviously appears in the III crystals with the high fraction of particles. In this case the decrease of the mechanical hysteresis magnitude with the increase of the test temperature is observed.

So, it was experimentally shown in the titanium-nickel single crystals enriched by nickel, that the precipitation of the d<5 nm dispersive particles during aging at 573 K for 1 hr. does not

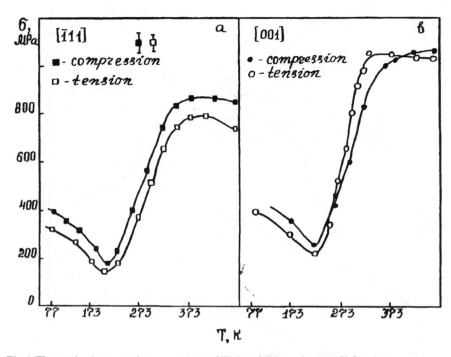

Fig.3 The tension/compression asymmetry of Ti-51at.%Ni aged at 773 K for 1hr.

change the functional properties of the SME and SE as compared with monophase crystals, in the case of d>50-100 nm (aging at 673-773 K for 1÷1.5 hr.) the weakening of the orientation dependence and asymmetry of the σ_{cr} takes place and SE appears. The SME magnitude decreases and its orientation dependence weakens (table).

The coherent Ti_3Ni_4 particles are the sources of internal stresses and, consequently, they are the places of the nucleation of the martensite crystals. Several the crystallographic variants of the "particle-matrix" correlation were experimentally observed in [1,2]. Therefore the local fields from the precipitates play the main role in the martensitic crystal nucleation and these fields provide the nucleation of the several crystallographic variants in crystal during the homogeneous external loading. It's obviously that in monophase single crystals these local fields of stresses are absent. Therefore, both the nucleation and the growth of the martensitic crystals occur in accordance with the Boas-Schmid's law in the crystallographic systems with the maximum shear stresses [3,7].

Therefore the development of MT in the heterophase crystals containing homogeneously distributed stress concentrators leads to degeneration of the SME dependence on crystal orientation and applied stress sign, providing large quantity of the martensite variants.

The plastic incompatibilities appearing during the martensite growth in the structural non-homogeneous crystals with particles (which don't undergo MT) lead to the storing of elastic energy in the crystal [4]. Consequently, in this case the elastic energy will be stored both due to the interaction of martensite variants with each other and due to the plastic deformation incompatibility of the matrix and particles lattices. This leads to formation of the conditions for the appearance of SE. Peculiarities of the mechanical hysteresis loops, stages of the unloading curves are evidence of small friction forces and the high mobility of the twinning boundaries [9].

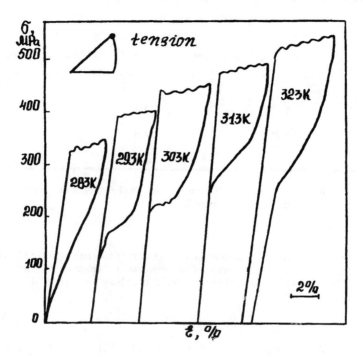

Fig.4 The SE dependence of [111] Ti-50.8at.% Ni single crystals aged at 673 K for 1.5 hr. (tension).

It is supposed that this peculiarity connected with the change in the martensite crystal fine structure during the precipitation of the Ti_3Ni_4 particles from twinning II to compound twinning in aging crystals [2]. The physical cause of the twinning type changing during precipitation connected with the following: the compound twinning have smaller Burgers vector than the twinning II. This leads to the decrease of the elastic interaction of the twinned martensite with precipitates both during the overcoming of their elastic fields and during the by-passing of particles which don't undergo MT. The increase of the aging time at 773 K to 15 hr. leads to the growth of the particles size till some microns, SME appears to be close to monophase results (table), SE

The dependence of maximum magnitudes of the reversible deformation in Ti-50.8at.%Ni single crystals in tension, * - [3,5]

Orientation		Test temperature				
	ε_0, %	quenching	673K-1hr.	673K-1.5hr.	773K-1hr.	773K-15hr.
<111>		9.0	8.8	5.4	4.6	8.0
<110>		8.7*	-	5.5	-	-
<001>		2.7*	-	4.6	-	-
<112>		-	-	4.6	-	-

is not observed. This have be connected with the twinning type changing of the martensite crystals from the compound twinning to the twinning II.

ACKNOWLEDGMENTS

The research described in this publication was made due to the financial supports of Grant - 95-02-03500 from Russian Fond on Fundamental Investigations and Grants from State Committee of Higher Educational Scool (UGTU-UPP, Ekaterinburg, Grant from MGTU named Bauman, Moscow).

REFERENCES

1. M. Nishida, C.M. Wayman and H. Honma, Met.Trans. A17 (7-12), 1505-1515 (1986).
2. M. Nishida, C.M. Wayman and A. Chiba, Metallography 21, 273-291 (1988).
3. K. Otsuka, K. Shimizu, Y. Suzuki, Y. Sekiguchi, T. Tadaki, T. Honma, S. Miyazaki, Shape memory alloys (Metallurgic Publisher, Moscow, 1990), p. 224.
4. M.F. Ashby, Phyl.Mag., 21, 399-424 (1970).
5. T. Saburi, S. Nenno in Solid-Solid Phase Transformations, edited by H.I. Aaronson, E.E. Laughlin, R.F. Sekerka, and C.W.Wayman, (The Metallurgical Society of AIME, Warrendale, PA, 1982), pp.1455-1479.
6. S. Miyazaki, S. Kimura, K. Otsuka and Y. Suzuki, Scripta Met. 18, 883-888 (1984).
7. T.E. Buchheit, J.A. Wert, Metallurgical and Materials Transactions A, 25A (11), 2383-2389 (1994).
8. L. Orgeas, D. Favier, Journal de Physique IV 5, C8-605-C8-610 (1995).
9. Y. Liu, , D. Favier and L. Orgeas, Journal de Physique IV 5, C8-593-C8-598 (1995).

Part VII

Poster Session II

PSEUDOELASTICITY IN HIGH STRENGTHENING FCC SINGLE CRYSTALS

YU.I. CHUMLYAKOV, I.V. KIREEVA, G.S. KAPASOVA, E.I. LITVINOVA
Siberian Physical Technical Institute, 1, Revolution sq., Tomsk 634050, Russia,
chum@phys.tsu.tomsk.su

ABSTRACT

It was experimentally shown that the achievement of a high deforming stress level due to dispersion hardening and solid solution strengthening of FCC single crystals with a low stacking-fault energy leads to the deformation mechanism changing from slip to twinning, the dependence of mechanical properties on a crystal orientation and a sign of applied stresses. During deformation by twinning at T<150-300K effects of pseudoelasticity associated with elastic twinning is observed.

INTRODUCTION

In alloys with thermoelastic martensitic transformations the pseudoelasticity (PE) at $T>A_f$ is usually observed, when the stress-induced martensite during unloading is thermodynamically unstable one and martensite crystals formed under loading disappear during removal of latter [1]. In absence of martensitic transformations PE is very seldom observed. The appearance of PE is usually associated with twinning deformation [2-7]. Analysis of experimental and theoretical works over the development of the elastic twinning and twinning PE, associated with this phenomenon, permits to distinguish a number of necessary conditions:

1) it is necessary to suppress the development of the slip deformation by a crystal structure variation, a concentration changing of interstitial and substitution alloying elements, a precipitation of different nature particles, a dislocation density increase, i.e., it is necessary to create conditions for twinning deformation: $\tau_{cr}^{tw} < \tau_{cr}^{sl}$, where τ_{cr}^{tw}, τ_{cr}^{sl} are critical shear stresses of twinning and slip, respectively [2];

2) it is necessary to provide an easy twin nucleation and to create a high density of twinning sources homogeneously distributed over crystal. This is possible due to the low stacking-fault γ_{sf} energy, the precipitation of plastically non-deforming particles of second phase, which are able to provide local places of twin nucleation over incoherent "particle-matrix" boundaries [2-9];

3) it is necessary to create conditions for a stable twin growth on account of an inhomogeneous external stress field or a strong varying internal stress fields, since in the homogeneous external stress field an elastic twin loses its stability, when it achieves of critical size like crack and pile up of dislocations [2];

4) it is necessary to provide moving forces for the reversible movement of the twins in the "loading-unloading" cycle on account of an elastic interaction of twinning dislocations with each other, an accumulation of an elastic energy, when the twinning dislocations by-pass plastically non-deforming particles of a second phase [8-10], a destruction both short-rang and long-range orders and a twin lattice symmetry on account of an interstitial atoms position changing during the movement of twinning dislocations [10,11].

For the experimental search of PE FCC single crystals with the low stacking-fault energy (γ_{sf} =0.004-0.02 J/m^2) and the high level of deforming stresses due to a precipitation of plastically non-deforming particles have been selected: a) alloy I: Cu-15at.%Al -2at.%Co; 2) alloy II: Cu-12at.% Al-2at.%Co; c) alloy III: austenitic stainless steels with composition close to 316 steel alloyed by 10 wt.% Mn and nitrogen C_N=0-0.7 wt.%.

The combination of the low stacking-fault γ_{sf} energy and the high level of applied stresses due to dispersion hardening (alloy I, II) and solid solution hardening (alloy III) was supposed to provide the deformation mechanism changing from slip to twinning. PE will be associated with the twinning deformation. Moving forces for the reversible movement of twins in alloys I, II will be back long range stress fields arising at a by-passing of plastically non-deforming particles by twinning dislocations. The twinning deformation will be the new mechanism of removal of local deformation gradients forming near particles. Twinning dislocations will be "geometrically necessary" ones and will be able to the reversible movement in such gradient materials [8,10]. In crystals III twinning will lead to the interstitial atoms position changing from octahedral places to tetrahedral ones [10,11].

RESULTS AND DISCUSSION

The achievement of the high level of deforming stresses leads to the deformation mechanism

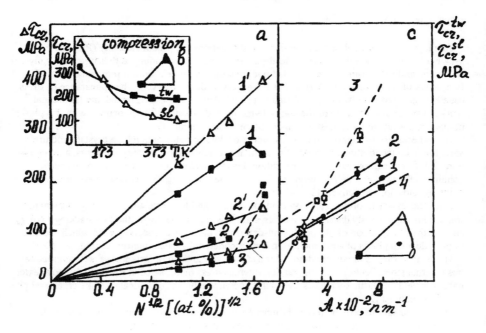

Fig.1 (a) The concentration dependence of solid solution hardening $\Delta\tau_{cr} = (\tau_{cr}^{2}(N) - \Delta\tau_M^{2})^{1/2}$ (where $\tau_{cr}(N)$, τ_M are critical shear stresses for crystals with and whithout nitrogen, respectively) for single crystals of alloy III; 1, 1'-T=77 K; 2, 2'-T=300 K; 3, 3'-T=573 K; (b) the temperature dependence of τ_{cr} at C_N=0.7 wt.%; (c) the dependence of critical shear stresses for slip (τ_{cr}^{sl}) and twinning (τ_{cr}^{tw}) on size of second phase particles; 1, 3 - tension (alloy I); 2 - tension (alloy II); 4 - compression (alloy I); A=1/(L-D), where L-the particle spacing, D-the particle diameter in glide plane. Open triangles, circles and squares-τ_{cr} for slip, filled triangles, circles and squares-τ_{cr} for twinning.

changing from slip to twinning. At tension of crystals I, II the orientation dependence of deforming stresses was found (Fig.1c). [111], [123] crystals have smaller critical shear stresses τ_{cr} than [001] ones and this difference increases with the growth of the deforming stress level. Metallographic, electron microscopic and X-rays investigations have shown, that at $\tau_{cr}>140$ MPa in "soft" [111], [123] orientations deformation is result of twinning (Fig.2a), the parabolic stage with the high strainhardening coefficient Θ at T<300 K is associated with the elastic twinning. With the increase of the CoAl particles density the increase of stage length with the high Θ and the growth of PE magnitude are observed (Fig.2a). In polycrystals of these alloys PE was found (Fig.3b), it does not dependent on grain size (d~16-140 mm), PE keeps up to T<300 K. In [001] crystals deformation occurs by slip, PE does not observe (Fig.2b).

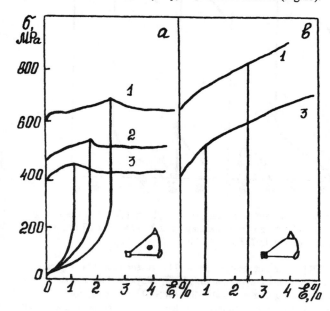

Fig.2 The stress-strain and mechanical hysteresis curves of single crystals Cu-15at.%Al-2at.%Co at 77 K in dependence on the particle size. 1a, 1b-aging at 723 K, 10 hr., 4*4*30 nm - the particles size, L=24 nm - the particle spacing; 2a-aging at 773 K, 67 hr., 10*15*60 nm, L=50 nm; 3a, 3b- aging at 773 K, 1 hr.+ 873 K, 1 hr., 12.5*25 *100 nm, L=78 nm.

Solid solution hardening by nitrogen of austenitic stainless steel ctystals (alloy III) leads to the appearance of the orientation dependence and asymmetry of τ_{cr} (Fig.1a,b). At tension of [111] crystals and at compression of [001] ones crystals are "soft", they have τ_{cr} significantly lower than "hard" [001] crystals at tension and [111] ones at compression. At $C_N<0.5$ wt.% deformation of [111] crystals at tension and [001] ones at compression by splited dislocations is realized and at $C_N>0.7$ wt.% deformation occurs by twinning from the beginning of the plastic flow. Abrupt fall of τ_{cr} at $C_N>0.5$ wt.% (Fig.1a) at T<173 K and vice versa hardening at T>173 K are connected with the deformation mechanism changing from slip to twinning. These peculiarities of solid solution hardening are connected with the difference of the temperature dependence of critical shear stress for twinning and slip: $\tau_{cr}^{tw}{\sim}G(T)$ and, consequently, the athermal slip is observed, while at the slip deformation $\tau_{cr}^{sl}(T)$ has the thermally activated part at T<473 K (Fig.1b).

At the twinning deformation at compression of [001] crystals with the high content of nitrogen $C_N=0.7$ wt.% at T<173 K, effects of PE (Fig.4a) were found. The PE crystallographic resource of such crystals was defined in experiments over cycling with the consistent increase of degree of plastic deformation. At set plastic deformation ε_{pd}=15% the maximum reversible deformation is ε_{rev}=9%. At tension of [111] crystals at T<173 K to define the PE crystallographic resource very difficult because of brittle fracture of specimens at ε<1.5-2%.

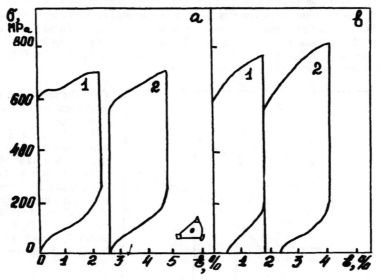

Fig.3 The mechanical hysteresis loops in single crystals (a) and polycrystals (b) of Cu-15at.%Al-2at.%Co alloy at 77 K; a-tension of [135] single crystals, aging at T=773 K for 67 hr.; b-polycrystals, grain size - 30 mm, aging at T=773 K for 27 hr.

At the slip deformation in [001] crystals at tension and in [111] orientations at compression PE was not found (Fig.4b). Investigations shown, that in I, II, III alloys the total reversibility of deformation is not observed even during first cycles at ε<3%. In [001] crystals at compression and [111] orientation at tension PE is observed. In this case electron microscopic investigations of dislocation structure shown the existence of stacking faults and microtwins [12,13]. Therefore in second cycle of loading the plastic flow begin at lower stresses than in first cycle. It is connected with the easy nucleation of deformation twins on residual stacking faults and microtwins.

And so, it experimentally shown, that the combination of a low stacking-fault energy γ_{sf} and a high level of stresses leads to the strong orientation dependence of deformation mechanism of slip and twinning. At the fixed orientation deformation mechanism is determined by sign of applied stresses (tension/compression). Physical nature of such dependence of deformation mechanism from an orientation and a sign of stresses may be connected, firstly, with the influence of the external stress field on the magnitude of the a/2<110> dislocation splitting into the Shockley partials according to the known reaction:

$$a/2<011>=a/6<112>+a/6<121> \qquad (1)$$

According to the theory of dislocations [14] in the external stresses field in dependence on an orientation and a sign of applied stresses the equilibrium stacking-fault energy γ_0 must be changed by the effective stacking fault-energy γ_{eff}

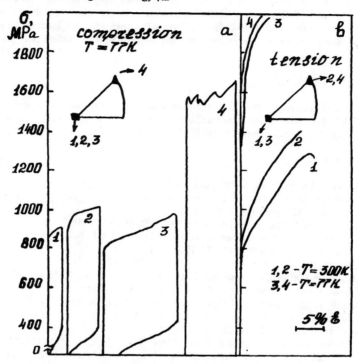

Fig.4 The stress-strain and mechanical hysteresis curves of single crystals of austenitic stainless steel with nitrogen $C_N=0.7$wt.% at compression (a) and tension (b).

$$\gamma_{eff}=\gamma_0\pm0.5*(m_2-m)*\sigma*b_1 \qquad (2)$$

where m_1 and m_2 are the Schmids factors for the leading $a/6<211>$ and trailing $a/6<121>$ Shockley dislocations; b_1 is the Burgers vector module of the partial dislocations. At tension of [111] and at compression of [001] crystals the external stresses field promotes the increase of individual dislocation splitting into partials [14]. This was experimentally determined on alloys III at $C_N<0.5$ wt.% [12,13], and at achievement of critical stresses $\tau_{cr}=1000$ MPa slip is changed by twinning and γ_{eff} is close to zero, i.e., individual dislocations are unstable ones in the external stresses field. They split into Shockley partials, stacking faults and twins form by the "slip source" mechanism [4], which does not require of the development of the preliminary slip deformation, as the "pole source" [3].

Secondly, twinning dislocations get over interstitial atoms and particles at lower stresses than individual dislocations because of the Burgers vectors difference $b_1=a/6<211>$ and $b=a/2<110>$.

Elastic interactions of dislocations with interstitial atoms and a by-passing of particles decrease with the decrease of the Burgers vector, in that case twinning is preferable of slip.

Thirdly, at twinning in single crystals with plastically non-deforming particles back long range stresses arise due to the formation of Orowan loops from partial dislocations around particles [8-11]. At twinning of FCC crystals with the high content of interstitial atoms the interstitial atoms position changing from octahedral places to tetrahedral ones occurs, it leads to the accumulation of additional elastic energy in twins and promotes the reversible movement of twins at unloading [10].

ACKNOWLEDGMENTS

The research described in this publication was made due to the financial supports of Grant - 95-02-03500 from Russian Fond on Fundamental Investigations and Grant from State Committee of Higher Educational Scool (UGTU-UPP, Ekaterinburg).

REFERENCES

1. K. Otsuka and K. Shimizu, Metals Forum 4 (4), 142-152 (1981).
2. V.S. Boiko, R.I.Garber and A.M. Kosevich, Reversible plasicity of crystals, (Fizmatlit of Nauka Publishers, Moscow, 1991), p. 237.
3. S. Mahajan, D.F.Williams, Int. Met. Rev. 18 (179), 43-61 (1973).
4. J.A. Venables, J. Phys. and Chem. Solids, 25 (7), 693-700 (1964).
5. J.A. Venables, Phyl. Mag. 6 (63), 379-396 (1961).
6. M.L.Green, M.Cohen, Acta Met. 27 (9), 1523-1538 (1979).
7. L.P. Kubin, A.Fourdeu, I.Y. Guedou, I. Rien, Phil. Mag. A. 46, 357-378 (1982).
8. L.M. Brown, W.M. Stobbs, Phyl. Mag. 23 (185), 1185 (1971).
9. M.F.Ashby in Strengthening Methods in Crystals, edited by A. Kelly, and P. Nickolson (Elsevier Science Publishers, New York, 1971), pp. 137-152.
10. J.W. Chan, Acta Met. 25, 1021-1026 (1977).
11. P.H. Adler, G.B. Olson and W.S. Owen, Metal. Trans. 17A, 1725-1737 (1986).
12. Yu.I. Chumlyakov, A.D. Korotaev, Izvestiya Vuz. Fizika (Russian Physics Journal) 35 (9), 783-779 (1992).
13. Yu.I. Chymlyakov, I.V.Kireeva, A.D. Korotaev, E.I. Litvinova, Yu.L. Zuev, Izvestiya Vuz. Fizika (Russian Physics Journal) 39 (3), 189-210 (1996).
14. S.M. Copley, B.N. Kear, Acta Met. 16 (2), 227-231 (1968).

SHAPE MEMORY EFFECT AND PSEUDOELASTICITY IN Ti-40%Ni-10%Cu (At.%) SINGLE CRYSTALS

YU.I. CHUMLYAKOV, I.V. KIREEVA, YU.I. ZUEV, A.G. LYISYUK
Siberian Physical Technical Institute, 1, Revolution sq., Tomsk 634050, Russia,
chum@phys.tsu.tomsk.su

ABSTRACT

The dependence of shape memory effect (SME) and pseudoelasticity (PE) on crystal orientation and on sign of applied stresses (tension/compression) has been investigated on Ti-40%Ni-10%Cu (at.%) single crystals. It have been shown that both SME and PE depend on an orientation and an applied stress sign. Asymmetry effects were found at deformation in high-temperature B2 phase and in the range of stress-induced martensite transformations.

INTRODUCTION

For the development of physical and micromechanical models of SME and PE, of plastic deformation of alloys undergoing MT at cooling and under load systematic investigations of mechanical properties of single crystals, including Ti-40%Ni-10%Cu, are required, and first of all investigations of the orientation dependence and asymmetry of deforming stresses are important [1-7].

Single crystals Ti-40%Ni-10%Cu were grown using Bridgman technique without seeds in inert gas atmosphere. Homogenization was held at 1123-1193 K during 24 hr. Then crystals were cut into specimens of certain orientation, specimens for tension being $12 \times 1 \times 2.5$ mm^3 gauge size, and for compression $2.5 \times 2.5 \times 5$ mm^3. Damaged layer was removed by fine grinding, chemical etching and electropolishing. Heat treatments A, B were performed in gas atmosphere: A-annealing at 1093 K during 1 hr, with subsequent water quenching; B-annealing at 1193 K during 24 hr., fast cooling in furnace.

RESULTS AND DISCUSSION

It has been experimentally shown by electric resistivity $\rho(T)$ measurements that for crystals B (fig.1b) for the first transformation B2-B19 M_s^1=293 K, M_f^1=268 K, A_s^1=275 K, A_f^1=313 K; for the second transformation B19-B19' M_s^2=263 K, M_f^2=161 K, A_s^2=175 K, A_f^2=270 K. The use of this method for precise determination of M_s, M_f, A_s, A_f is limited by low sensitivity of electric resistance to the formation of orthorhombic phase B19, while B19-B19' transformation is accompanied by the significant change of resistance. For crystals A the distinguishing of two transformations through the change of $\rho(T)$ is difficult (fig.1b). MT temperatures in single crystals appears to be shifted to the lower temperature range as compared with data over polycrystals [1-3], which is related with the change of titanium concentration during growth.

The dependence $\sigma_{cr}(T)$ in single crystals is typical for materials undergoing MT and appears to be analogous to that found earlier [1] on polycrystals (fig.1). In high temperature B2 phase σ_{cr} depends on crystal orientation, sign of applied stresses (tension/compression) and heat treatment. Crystals A in any orientation have lower σ_{cr}, than B and this is crucial, as will be proved below, for creation of proper conditions for PE. Under the same heat treatment, for example B, (fig.1) orientations <111> are "soft", possessing lower σ_{cr}, while <001>, <011> ones are "hard".

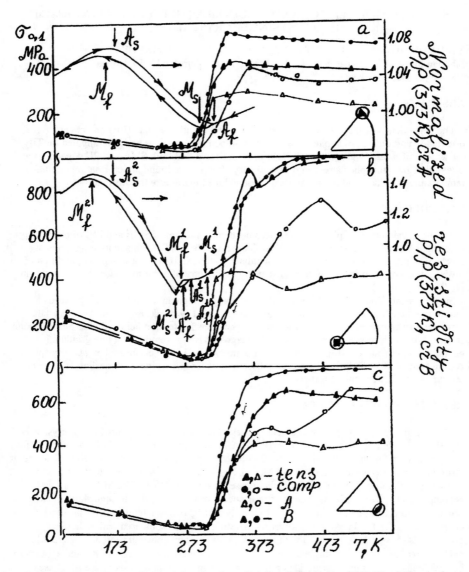

Fig. 1 The dependence of σ_{cr} and the electric resistivity of the Ti–40at.%Ni–10at.%Cu single crystals on the test temperature.

Regardless of the way of heat treatment the phenomena of asymmetry have been found in B2 phase: dependence of σ_{cr} on the sign of applied stress (tension/compression). Crystals of "hard" orientation <001> in tension are characterized by low plasticity and brittle fracture, in

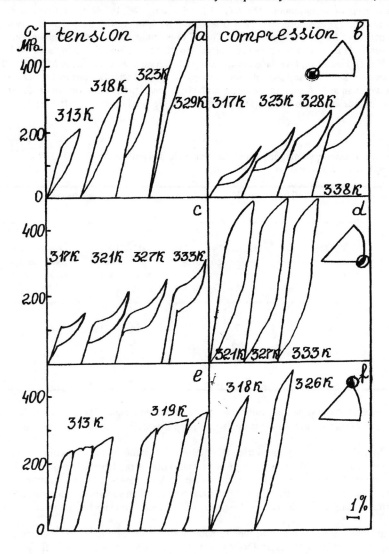

Fig.2 The tension/compression flow curves of the Ti-40at.%Ni-10at.%Cu single crystals at the different test temperatures (crystal B).

compression deformation occurs by kinking bands. "Soft" orientations <111> are plastic, <011> being intermediate ones.

The strong orientation dependence and asymmetry of σ_{cr} in B2 phase determine the temperature interval of stress-induced MT ΔT_σ, dependence σ_{cr}, PE. In low strength crystals after treatment A no PE was observed in any orientation, while in crystals treated by B way high strength properties of B2 phase promote PE in "hard" orientations <011>, <001> in tension and compression at $T>A_f$ in narrow temperature interval, no PE having been found in "soft" <111> orientation (Fig.2). Rise of σ_{cr} level leads to widening of interval of stress induced MT ΔT_σ. The SME depends on orientation and sign of applied stress (table) and after B treatment appears to be slightly lower than after A one. In compression SME is lower than in tension, and reaches maximum magnitude in tension <111>, <O11> crystals, being smaller in <100> (table).

TABLE

The dependence of maximum magnitudes of the reversible deformation in Ti-40at.%Ni-10at.%Cu single crystals (crystal B)

Orienta tion	Theory [1,7]				Experiment			
	B2-B19'		B2-B19		B2-B19'		$\alpha=d\sigma_{0.1}/dT$	
	ten-sion α_0, %	compre-ssion α_0, %	ten-sion α_0, %	compre-ssion α_0, %	ten-sion α_0, %	compre-ssion α_0, %	ten-sion	compre-ssion MPa*K^{-1}
<111>	9.8	3.6	2.7	-	6.4	2.3	4.2	7.4
<110>	8.4	5.2	5.3	-	6.8	2.5	6.0	14.2
<100>	2.7	4.2	2.7	-	2.8	3.4	12.0	6.9

In M_s^1-M_s^2 interval σ_{cr} does not depend on T, crystal orientation, sign of applied stress, while at $T<M_s^2$ the rise of σ_{cr} with the temperature decrease and orientation dependence of σ_{cr} is observed.

In Ti-40%Ni-10%Cu single crystals during B2-B19-B19' transformations PE is only observed at high strength properties of B2 phase, which are reached due to σ_{cr} dependence on crystal orientation and heat treatment. In "soft" orientations at $T=A_f$ necessary stresses for martensite formation make up about 1/3 of σ_{cr}(B2) and the martensite lamellae growth is accompanied by plastic deformation depressing the reversible motion of interface boundaries "B2-martensite" in unloading.

The experimental magnitudes of SME with respect to orientation of the crystals and sign of applied stress are compiled to the table. One can see, that theoretical estimates of SME for B2-B19 transformation are considerably lower than experimental ones, while for B2-B19', on the contrary, accounts give SME magnitudes higher than experimental ones. On the other hand, for B2-B19' MT in quenching Ti-Ni single crystals theoretical accounts are in satisfactory agreement with experimental values. So, SME magnitudes are determined not only by eventual structure of martensite, but also depend on concrete mechanism and SME of B2-B19' MT appears to be different from that of B2-B19-B19' MT. In advanced analysis it should be noticed, that B19-B19' MT is unperfected in Ti-40%Ni-10%Cu alloys even at T < 77 K.

Unlike the deformation by slip in B2 phase which is bidirectional: same slip system is effective both in compression and in tension, martensite deformation does not possess such feature. Therefore the asymmetry of deforming stresses σ_{cr}, the dependence of SME magnitude on the way of loading (tension/compression) are related to unidirectional nature of martensite shear for each habit plane and are the result of limited number of correspondence variant pair. Experimentally asymmetry of σ_{cr} and dependence of SME magnitude on tension/compression, difference of flow curves in tension and compression were observed in TiNi polycrystals [4,5] and was described fairly well in assumption of independence of mechanical work, related to stress-induced MT, on the sign of load. Such an approach can be applied to analysis of the orientation dependence of $\alpha = d\sigma/dT$ in tension and compression in temperature interval of stress-induced MT. In Clapeyron-Clausius equation for stress-induced MT [1,5,7] only ε_0 turns out to depend on orientation and sign of load. From (fig.1) and table one can see that increase of ε_0 leads to the decrease in a in the same way of deformation, for instance in tension. In tension ε_0 is greater than in compression, and $\alpha_t < \alpha_{comp}$. That is why temperature interval of stress-induced MT ΔT_{σ} is determined not only by level of $\sigma_{cr}(B2)$, but also by the value of α, which is dependent on orientation, and well-known scheme [1] for determination of PE and SME window can be supplemented by orientation dependence and asymmetry of $\sigma_{cr}(B2)$ and α. Connection of high $\sigma_{cr}(B2)$ with low values of α (i.e. high magnitudes of ε_0), which is experimentally realized in tension of <011>, causes the broadening of the interval ΔT_{σ}, promote the appearance of PE, having maximum crystallographic resource in this orientation.

CONCLUSION

1. In Ti-40%Ni-10%Cu single crystals (at.%) the non-schmid's phenomena are observed in high temperature B2 phase: orientation dependence and asymmetry of deforming stresses.

2. Experimentally found the asymmetry between deforming stresses in tension and compression and their orientation dependence within the temperature interval of stress-induced MT. Transformation stresses are higher in compression than in tension in <111>, <110> orientations, while the magnitude of SME, on the contrary, is smaller in compression, than in tension.

3. The dependence of PE on crystal orientation, sign of applied stresses, strength properties of B2 phase has been found. In crystals of "soft" orientations <111> no PE was observed, while in "hard" <100>, <110> ones the conditions for reversible motion of martensite crystals are realized.

ACKNOWLEDGMENTS

The research described in this publication was made due to the financial supports of Grant - 95-02-03500 from Russian Fond on Fundamental Investigations, Grants from State Committee of Higher Educational Scool (UGTU-UPP, Ekaterinburg, Grant from MGTU named Bauman, Moscow).

REFERENCES

1. K. Otsuka, K. Shimiza, Y. Suzuki et.al., Shape Memory Alloys, (Metallurgic Publisher, Moscow, 1990), p. 224.
2. T.H. Nam, T. Saburi, Y. Kawamura, K. Shimizu, Materials Transations, JIM 31 (4), 262-269 (1990).
3. T.N. Nam, T. Saburi, Y. Nakata, K. Shimiza, Materials Transacions, JIM 31 (12), 1050-1056 (1990).
4. L. Orgeas, D. Favier, Journal de Physique IV 5, C8-605-C8-610 (1995).
5. T.E. Buchheit, J.A. Wert, Metallurgical and Materials Transactions A, 25A (11), 2383-2389 (1994).
6. K. Bhattacharya, R.V. Kohn, Acta Mat. 44 (2), 529-542 (1996).
7. T. Saburi, S. Nenno in Solid-Solid Phase Transformations, edited by H.I. Aaronson, E.E. Laughlin, R.F. Sekerka, and C.W.Wayman, (The Metallurgical Society of AIME, Warrendale, PA, 1982), pp.1455-1479.

NONTHERMOELASTIC AND THERMOELASTIC MARTENSITIC TRANSFORMATION BEHAVIOR CHARACTERIZED BY ACOUSTIC EMISSION IN AN Fe-Pt ALLOY

H. OHTSUKA*, K. TAKASHIMA** and G. B. OLSON***
* National Research Institute for Metals, 1-2-1 Sengen, Tsukuba, Ibaraki 305, Japan.
  ohtsuka@nrim.go.jp
** Department of Materials Science and Mechanical Engineering
  Kumamoto University, 2-39-1 Kurokami, Kumamoto 860, Japan.
*** Department of Materials Science and Engineering,
  Northwestern University, 2225 N. Campus Dr., Evanston IL, 60208-3108, U.S.A.

ABSTRACT

Acoustic emission (AE) signals generated during the martensitic transformation in an Fe-24at%Pt alloy with various ordering treatments have been measured, and the AE parameters have been correlated with phase transformation events. AE events are generated during cooling of specimens, which is the first detection of the AE events associated with the martensitic transformation in Fe-Pt alloys. In all specimens, AE events were observed at higher temperature than the Ms temperatures determined by electrical resistivity measurement. AE events started to increase abruptly for the weakly ordered specimen with decreasing temperature. In contrast, AE events increased gradually for the highly ordered specimen with decreasing temperature. This is consistent with the nature of the transformation behavior in this alloy. The critical defect size of weakly ordered and highly ordered specimens were calculated using Ms temperature determined by AE technique. Results obtained in this investigation strongly suggests that the AE technique is useful for analysis of transformation kinetics in this alloy.

INTRODUCTION

An alloy in the Fe-Pt system is a typical Fe-based shape memory alloy and effects of ordering on martensitic transformation have been investigated in detail [1-8] because martensitic transformation behavior changes from nonthermoelastic to thermoelastic with increasing degree of order. As for the nucleation sites of martensite, they are classified into two types[9]; one is preexisting nucleation sites and the other is autocatalytic nucleation sites. The former include grain boundaries, preexisting defects in the parent phase and interface between inclusion and parent phase. The latter mean the newly developed nucleation sites produced by the first nucleation event through the stress field, strain and interface of product phase. The number of nucleation events at preexisting and autocatalytic nucleation sites were qualitatively estimated and compared with each other in the same Fe-Pt alloy by Olson and Ohtsuka[8], but the transformation temperatures were measured by electrical resistivity change in this case. The acoustic emission (AE) technique will be useful to precisely determine the density of two kinds of nucleation sites in Fe-Pt alloys. Recently, the AE technique has been applied to the characterization of martensitic transformations in various alloys[10-20]. However, most studies of AE during martensitic transformation have been carried out on Fe-Ni alloys[11,12], Cu-Al-Ni alloys[13,14], Au-Cd[15,16] alloys and Ti-Ni alloys[17-19], and to the author's knowledge there are no studies to date which have been applied to Fe-Pt alloys. The purpose of the present study is to examine the applicability of the AE technique for the detection of the martensitic transformations in an Fe-Pt alloy. AE signals were measured in the Fe-24at%Pt alloy with various degrees of order and correlated with martensitic transformation behavior and the critical defect size of weakly and highly ordered Fe-Pt specimens in this study.

EXPERIMENTAL PROCEDURE

The material studied is an Fe-24.0(at%)Pt alloy. The alloy was homogenization treated for 7 days at 1473K, and solution treated for 1 hr at 1573K and ordered at 823K for 4, 22 and 108hrs.

The austenite grain size was measured as 150μm. Specimens for AE measurement and electrical resistivity of 0.5 X 1.5 X 20 mm³ in size were cut form the heat-treated alloy. The specimen was attached directly to an AE transducer using a silicone vacuum grease. The AE transducer used was an M53 (Fuji Ceramics Co.), and its resonant frequency is approximately 300kHz. The specimen was cooled slowly at a cooling rate of approximately 2K/min. AE signals detected were amplified by a pre-amplifier and analyzed by an NF 9502 AE measurement system. The gain of the pre-amplifier was 40dB, and the threshold level was 50dB with reference to 0.1 μV at the input of the preamplifier (or 31.6 μV at the input of the preamplifier). Transformation behavior was also monitored in specimens of 3mm square by 0.5mm thickness by video-recorded observations of surface relief contrast in a cold-stage optical microscope.

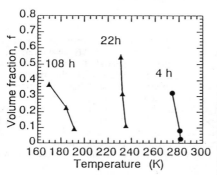

Fig. 1 volume fraction of martensite formed during cooling as a function of temperature.

RESULTS AND DISCUSSION

Transformation Behavior Observed by Optical Microscope

Figure 1 shows the volume fraction of martensite as a function of temperature in the specimens ordered for 4, 22 and 108hrs. The volume fraction was measured on the printed images of each specimen at various temperatures during cooling. In the specimens ordered for 4 and 22hrs, volume fraction increases very abruptly with decreasing temperature and one nucleation event triggers other nucleation events by the mechanism of autocatalytic nucleation.but it increases more gradually in the specimens ordered for 108 hrs.

Acoustic Emission Characteristics During Cooling

Figures 2(a) shows the relation between total AE event counts, event rate and temperature during cooling in 4hrs ordered specimen. The Ms temperature measured by electrical resistivity change is also shown on the abscissa. In the 4 hr ordered specimen, no event is observed above 282K and AE events begin to be observed sporadically with decreasing temperature. AE event counts start to increase just below 278K and increase abruptly during further cooling. The Ms temperature determined by electrical resistivity change is 281K, and the generation of AE event counts corresponds to the martensitic transformation in this specimen. The test was stopped when the number of events exceeds approximately 6,000 events which is the limitation of the measurement system used in this study. Figures 2(b) and (c) show amplitude distribution of AE events observed in the temperature range between 282 and 278K, and that between 278 and 273K, respectively. AE event amplitude values are shown with reference to 0.1 μV at the output of the sensor used. Amplitude values are distributed from 50 to 85 dB above 278K, while amplitude values below 278K are distributed from 50 to 106 dB. These results indicate that higher amplitude events are generated below 278K. Since the amplitude of AE event is related to the size of the AE source, high amplitude AE events are associated with the formation of larger martensite plates. Figures 3(a)-(c) show the same sets of experimental data in the 22 hr ordered specimen. As shown in Fig. 3(a), AE event counts start to increase just below 257K and increases abruptly during further cooling as observed in the 4 hr ordered specimen. Although AE events are observed sporadically at the temperature rage between 268 and 257K, their amplitude is less than 80 dB as shown in Fig. 3(b). Higher amplitude events are observed below 257K as shown in Fig. 3(c). The Ms temperature determined by electrical resistivity is 235K, and this temperature is 22K lower than the temperature at which AE events starts to increase abruptly. Figures 4(a)-(c) show the same sets of experimental data in the 108 hr ordered specimen. AE

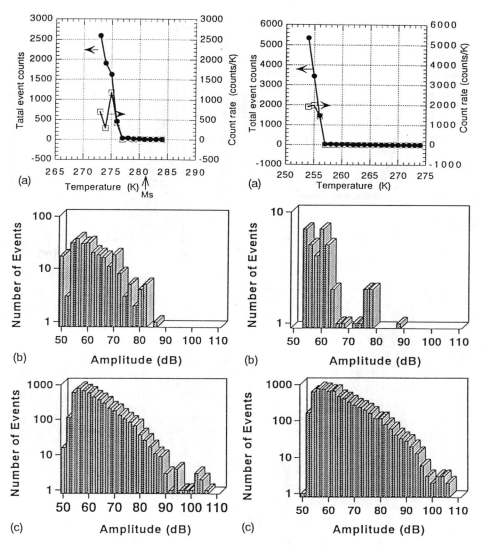

Fig.2 Number of AE events as a function of temperature and amplitude. (a) Total AE event counts and event rate as a function of temperature in 4hr-ordered specimen. The Ms temperature determined by electrical resistivity change is shown on the abscissa. (b) Amplitude distribution of AE events observed in the temperature range between 282 and 278K, and (c) that below 278K and 273K in 4hr-ordered specimen.

Fig.3 Number of AE events as a function of temperature and amplitude. (a) Total AE event counts and event rate as a function of temperature in 22hr-ordered specimen. (b) Amplitude distribution of AE events observed in the temperature range between 268 and 257K, and (c) that between 257 and 273K in 22hr-ordered specimen.

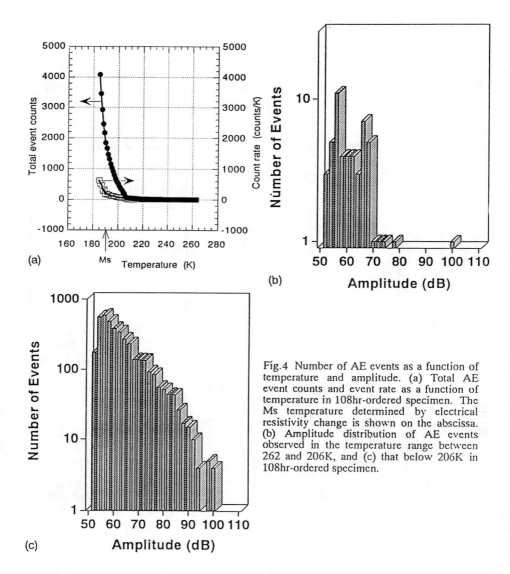

(a)

(b)

(c)

Fig.4 Number of AE events as a function of temperature and amplitude. (a) Total AE event counts and event rate as a function of temperature in 108hr-ordered specimen. The Ms temperature determined by electrical resistivity change is shown on the abscissa. (b) Amplitude distribution of AE events observed in the temperature range between 262 and 206K, and (c) that below 206K in 108hr-ordered specimen.

events start to increase at 206K, but AE events increases gradually during further cooling. This AE behavior is consistent with the characteristics of martensitic transformation in the 108 hr ordered specimen observed by optical microscope. In this specimen, AE events start to be observed at 262K and AE events are generated sporadically during cooling to 206K. AE event amplitude of this temperature range is quite low (less than 70 dB) as shown in Fig. 4(b). Higher amplitude events are observed below 206K(Fig. 4(c)), but the highest amplitude is 100 dB and a proportion of higher amplitude events is less compared with the 4 and 22 hrs ordered specimens. In all specimens, the AE events started to be observed sporadically at higher temperature where AE events begin to generate abruptly. This suggests that the martensitic transformation may occur sporadically as the temperature decreases or In the 22 and 108 hrs ordered specimens, the temperatures at which AE events begin to increase abruptly are much higher than the Ms temperatures determined by electrical resistivity change. This may be due to the nucleation of martensite plates which are too small to be observed by optical microscope and to be detected by electrical resistivity change, but it is not clear yet.

DISCUSSION

The thickness of critical martensite nucleus at Ms(AE) temperature is calculated for 4h and 108h ordered specimens and compared with each other.

The transformation driving force for the weakly ordered 4h specimen can be estimated from the reported transformation hysteresis[1], using the transformation latent heat measured for nominally disordered Fe-24.5at%Pt[21], and assuming the equilibrium T_0 temperature is midway between the Ms and As (reversion start) temperatures for nonthermoelastic behavior. The $\gamma \rightarrow \alpha$ free energy change (J/mol) vs. T(K) is approximated as:

$$\Delta G^{\gamma \rightarrow \alpha}(T) = -2435 + 4.97T \tag{1}$$

For transformation in highly ordered Fe-24.5at%Pt, calorimetry has indicated a constant $\gamma \rightarrow \alpha$ heat capacity difference of $\Delta Cp=-6.75$ (J/molK) with $\Delta H=0$ at $T=Mf$[2]. Following Salzbrenner and Cohen[22] we assume $T_0=Af$ for thermoelastic behavior. Applying these conditions to the 108h specimen where Af=273K and Mf=123K[1] gives ΔG (J/mol) vs. T(K) in the form:

$$\Delta G^{\gamma \rightarrow \alpha}(T) = 830 - 40.9T + 6.75T\ln T \tag{2}$$

Using volume free energy change $\Delta g = \Delta G/Vm$ with $Vm=7.75 \times 10^{-6}$ m^3, the critical driving forces can be related to nucleating defect sizes through heterogeneous nucleation theory. For barrierless heterogeneous nucleation in which the elastic self energy of the nucleus is compensated by the elastic interaction with a defect, the potency of a defect can be described by the thickness of critical nucleus stabilized by the interaction. Measuring this thickness as nd where d is the interplanar spacing of closest-packed planes, modeling of the interaction with linear defects gives a critical driving force for nucleation of the form[23,24]

$$\Delta g_c(n) = -\{(2\gamma/nd) + g_R\} \tag{3}$$

where γ is the nucleus specific interfacial energy, and g_R is a net resistive term attributed to work of interfacial motion and an elastic energy contribution associated with possible distortions in the nucleus interface. The g_R is expected to scale with the austenite isotropic shear modulus μ, and modeling of Fe-Ni alloys gives $g_R=1.03 \times 10^{-3}\mu$. In contrast to the gR term, the interfacial energy of γ of a semicoherent γ-α interface is sensitive to elastic anisotropy as the local shear distortions occur on the soft (011)[011]A system of the C' elastic constant. Scaling interfacial energy estimates for Fe-Ni alloys to C' gives $\gamma/C'=4.5 \times 10^{-12}$ m for the temperature range of interest for the weakly ordered 4h specimen, the elastic constant measurements of Ling and Owen[25] give $\mu=5.9 \times 10^{10}$ N/m^2 and $C'=3 \times 10^{10}$ N/m^2 for the disordered condition. Measured lattice parameters give d=2.15Å. In the case of ordered specimen, based on the elastic constant measurements of Ling and Owen[25] on Fe-25at%Pt, applying a 14K temperature shift for the difference in Tc, the temperature dependence of the C' shear constant near 200K in Fe-24at%Pt is estimated as C' (N/m^2)=8.13$\times 10^7$T-2.9$\times 10^9$. from the assumed scaling of the interfacial energy with C', this then defines a temperature-dependent γ. As it is well established that order strengthening greatly diminishes the degree of slip accommodation accompanying transformation in Fe-Pt alloys, we will here adopt the approximation that $g_R=0$ in the highly ordered condition. The Ms temperatures at which the very first event is detected by AE technique were 281 and 262 K for 4h and 108 h ordered specimens, respectively. Applying these estimated parameters and experimentally measured data, defect size n can be determined as 17 and 156 for 4h and 108h ordered specimens, respectively. This derivation is based on many assumptions on thermodynamics, but it means the defects responsible for the very initial transformation are about nine times bigger in ordered state compared to those in the disordered state. This may originate from various reasons, such as the strain energy (caused by elastic modulus), interfacial energy, etc, but mainly because the critical volume free energy change is much larger in disordered state.

CONCLUDING REMARKS

AE signals generated during martensitic transformation in an Fe-24at%Pt alloy with various ordering treatments have been measured. AE events were observed at significantly higher temperature than the Ms temperatures determined by electrical resistivity change. AE events started to increase abruptly during cooling for weakly ordered specimen, but increases gradually for highly ordered specimen. The results obtained by the AE measurements were consistent with the characteristics of the martensitic transformation in this alloy. More detailed AE waveform analysis (including source wave analysis) would appear to be promising for analyzing the dynamics of the martensitic transformation that occur in this alloy. The critical defect size for the very initial nucleation of martensite was calculated for the disordered and ordered Fe-Pt specimens using the Ms temperature determined by AE measurement, and it was found that the defect size is much larger for the ordered specimen.

REFERENCES

[1] D.P.Dunne and C.M.Wayman: Metall. Trans., **4,**, p.137(1973).
[2] M.Umemoto and C.M.Wayman: Metall. Trans., **9A**, p.891(1978).
[3] T.Tadaki and K.Shimizu: Scripta Metall., **9**, p.771(1975).
[4] T.Tadaki and K.Shimizu:Trans. JIM, **11**, p.44(1970).
[5] T.Tadaki:Trans. JIM, **18**, p.864(1977).
[6] S.Muto, R.Oshima and F.E.Fujita:Metall Trans., **19A**, p.2723.
[7] S.Muto, R.Oshima and F.E.Fujita:Metall Trans., **19A**, p.2931(1988).
[8] G.B.Olson and H.Ohtsuka: Proc. of Int. Symposium and Exhibition on Shape Memory Materials, Beijing (1994), Sept.
[9] G.B.Olson and M.Cohen:Ann. Rev. Mater. Sci., **11**, p.1(1981).
[10] K.Takashima, Y.Higo and S.Nunomura:Phil Mag., **A49**, p.231(1984).
[11] K.Takashima, M.Moriguchi and H.Tonda: Progress in AE V, (1990), 105.
[12] Z-Z. Yu and P.C.Clapp:Metall. Trans., **20A**, p.1601(1989).
[13] K.Yoshida, A.Takahashi, K.Sakamaki and H.Takagi: Progress in AE V, (1990), 83.
[14] K.Yoshida, A.Takahashi, K.Sakamaki and H.Takagi: Progress in AE VI (1992), 497.
[15] J.Bram and M.Rosen:Phil. Mag.,**44**, p.895(1981).
[16] J.Bram and M.Rosen:Acta Met., **30**, p.655(1982).
[17] J.Bram and M.Rosen:ScriptaMet.,**13**, p.565(1979).
[18] K.Takashima and M.Nishida: Progress in AE VII (1994), 155.
[19] K.Takashima and M.Nishida: Journal de Physique IV (1995), C8-735
[20] G.B.Olson, K.Tsuzaki and M.Cohen: Mat. Res. Soc. Symp. Proc. **57**, p.129(1987).
[21]. H.C. Tong and C.M. Wayman, Metall. Trans., **5**, p.1945(1974).
[22] R.J. Salzbrenner and M. Cohen, Acta Metall., **27**, p.739(1979).
[23] G.B. Olson and M. Cohen, Metall. Trans., **7A**, p.1925(1976).
[24] G.B. Olson and M. Cohen in Dislocation in Solids, vol.7, North-Holland (1986), p.295.
[25] H.C. Ling and W.S. Owen, Acta Metall., **29**, p.1721(1981).

THE EFFECT OF REVERSE TRANSFORMATION ON RECOVERY · RECRYSTALLIZATION PROCESS IN Ti-Pd-Ni HIGH TEMPERATURE SHAPE MEMORY ALLOYS

YA XU*, KAZUHIRO OTSUKA*, TATSUHIKO UEKI**, KENGO MITOSE**
*Institute of Materials Science, University of Tsukuba, Tsukuba, Ibaraki 305, Japan
**Yokohama R & D Lab.,The Furukawa Electric Company Ltd., 2-4-3 Okano, Nishi-ku, Yokohama   220, Japan

ABSTRACT

The effect of martensitic reverse transformation on recovery·recrystallization process in cold rolled Ti-Pd-Ni high temperature shape memory alloys has been investigated systematically by flash heating treatment, micro-Vickers hardness test, differential scanning calorimetry and transmission electron microscopy. It was found that the temperatures of softening in hardness after flash heating treatments agree well with the reverse transformation temperatures in the present alloys, and most of the softening occurs within 60 seconds when annealing temperature is raised to above the reverse transformation temperature. We conclude that the recovery· recrystallization process is controlled by the reverse transformation. The reasons are considered based on the large difference in atomic diffusion rate in the parent phase and in the martensite.

INTRODUCTION

Ti-Pd-Ni alloys have attracted considerable attention as high temperature shape memory alloys in recent years, since their martensitic transformation start temperatue(Ms) can be varied from about 813K to the ambient temperature by substituting Pd by Ni in $Ti_{50}Pd_{50}$[1-2]. It was found that the alloys with high Pd content transform from B2 parent phase to the B19 (orthorhombic) martensite[2-3]. Mach research has been done on the shape memory characteristics of this alloy system by torsion and compression[2,4], tensile tests at room temperature[5] and at high temperature[6-8]. It was found that fairly good shape memory characteristics appear at room temperature, but the shape memory effect performance becomes bad with increasing the tensile test temperature. The reason was found to be due to the low critical stress for slip at high temperature by Otsuka et al.[7-8]. Thereafter, an improvement of the shape memory effect by thermomechanical treatment was made by Golberg et al.[9-11], and it was found that this improvement is closely related to the recovery·recrystallization behavior during the heat treatment. Therefore, it is important to understand the recovery·recrystallization behavior in Ti-Pd-Ni alloys, and a systematic research on recovery·recrystallization process of several Ti-Pd-Ni alloys was made by the present authors[12]. As a result, a close relationship was found between the recovery·recrystallization and the reverse transformation of the Ti-Pd-Ni alloys. But the effect of the reverse transformation on recovery·recrystallization is not clear yet. The purpose of the present research is to investigate this effect in detial and to understand its mechanism by flash heating treatment, differential scanning calorimetry(DSC), micro-Vickers hardness measurements and transmission electron microscopy(TEM).

EXPERIMENTAL PROCEDURE

Ingots of $Ti_{50}Pd_{50-x}Ni_x$ (x=0,5,10,15,20,30) alloys were plasma melted and homogenized at

Mat. Res. Soc. Symp. Proc. Vol. 459 © 1997 Materials Research Society

1273K for 5 hours. Then the ingots were hot rolled at 1023K into 0.6~1.25mm-thick sheets. These sheets were rolled at room temperature into 0.5~1.0mm-thick with a 24.5~26.4% reduction in thickness. The martensitic transformation temperatures were determined by DSC measurement. Flash heating and isothermal annealing treatments at various temperatures were made to investigate the effect of reverse transformation on the recovery•recrystallization behaviour. The recovery and recrystallization processes were assessed by micro-Vickers measurement. The structural changes after flash heating or isothermally annealing at various temperatures were observed by TEM.

RESULTS AND DISCUSSION

Figure 1 shows the results of micro-Vickers measurements of cold rolled $Ti_{50}Pd_{50-x}Ni_x$ alloys

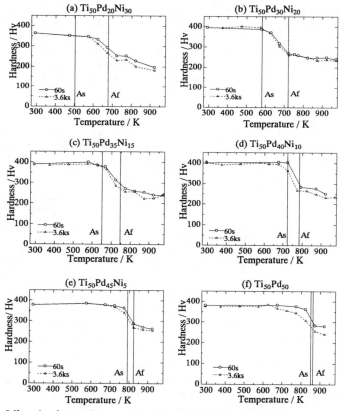

Figure 1   Micro-hardness changes of cold rolled $Ti_{50}Pd_{50-x}Ni_x$ alloys after flash heating at various temperatures for 60s (solid lines), and annealed for 3.6ks (dotted lines). Reverse transformation temperatures (As and Af) determined by DSC are also shown.

after flash heating at various temperatures for 60 seconds. The reverse transformation start temperature(A_s) and finish temperature(A_f) of cold rolled specimens were determined by DSC at a heating rate of 0.05K/sec., and these results are also shown in Fig. 1. For comparison, the results of isothermal annealing at various temperatures for 3.6 ks of our previous work[12] are also shown in Fig. 1 as dotted lines. We find that most of the softening in hardness after a flash heating is almost the same as that after isothermal annealing for 3.6 ks(first softening stage), and the temperature ranges of the softening in hardness after a flash heating treatment agree well with the reverse transformation temperatures except for the $Ti_{50}Pd_{20}Ni_{30}$ alloy.

The A_s, A_f and softening start temperature(T_s), softening finish temperature(T_f) obtained from the hardness curves of flash heated specimens(Fig. 1) are summarised and plotted as a function of Pd content in Fig. 2. We see that the T_f agrees well with A_f except for some deviation at low Pd

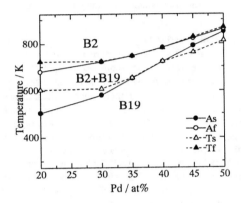

Figure 2   Plots of reverse transformation temperature and softening temperature vs. composition for $Ti_{50}Pd_{50-x}Ni_x$ alloys. Ts—start temperature and Tf—finish temperature , of the first softening stage for flash heated specimens.

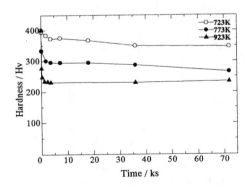

Figure 3   Micro-hardness changes of $Ti_{50}Pd_{40}Ni_{10}$ alloy after isothermally annealing at 723K, 773K and 923K, respectively.

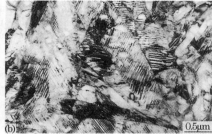

Figure 4 TEM micrographs of $Ti_{50}Pd_{40}Ni_{10}$ alloy after isothermally annealing at (a) 723K for 72ks, (b) 773K for 72ks, respectively.

content(<30at%). Since it has been proved that the softening in hardness is due to the recovery • recrystallization process[12], we conclude that the recovery • recrystallization process is controlled by the reverse transformation. The reasons are considered as follows. 1) The recovery • recrystallization process is much slower in martensitic state than that in the parent state because of a large difference in atomic diffusion rate in the parent(B2) and in the martensite with a close-packed structure(B19), which was proposed in Ref.[12]; 2) the recovery • recrystallization process occurs dominantly at the parent-martensite interfaces with the progress of the reverse transformation, where atoms are unstable and mobile. Thus, the hardness shows a sharp decline as the reverse transformation goes on. On the other hand, since the reverse transformation temperature of $Ti_{50}Pd_{20}Ni_{30}$ alloy is relatively low(A_s=503K, A_f=687.5K), the atomic diffusion is still diffcult at this temperature range, even in B2 structure. Therefore the reverse transformation is of little assistance to the recovery and recrystallization process. Thus the deviation between the reverse transformation and the softening temperatures at low Pd content can also be interpreted.

In order to prove the above proposition more strictly, the changes of micro-hardness of cold rolled $Ti_{50}Pd_{40}Ni_{10}$ alloy were measured during isothermal annealing at 723K, 773K and 923K which correspond to martensite, parent+martensite and parent regions, respectively. These results are shown in Fig. 3. We see that there is only a slight softening even after annealing at 723K for 72ks. In contrast, most of the softening occurred within 60 seconds when annealed at 923K which is higher than A_f, and no further softening in hardness was observed as annealing time goes on afterwards. This indicates that the softening in the parent phase is much faster than that in martensite. The microstructures after annealing at 723K and 773K for 72ks, respectively, are shown in Fig. 4. No recrystallized structure was observed after annealing at 723K for 72ks; partially recrystallized structure was observed after annealing at 773K for 72ks. These results give strong support to the above proposition that the recovery • recrystallization process occurs in the parent phase, dominantly at the parent-martensite interfaces with the progress of the reverse transformation.

The microstructural changes after the flash heating at various temperatures for the present alloys were observed by TEM and these results are shown in Fig. 5. No subgrain structures were observed after flash heating at 723K which just corresponds to the T_f for $Ti_{50}Pd_{20}Ni_{30}$ and $Ti_{50}Pd_{30}Ni_{20}$ alloys(Fig.5a, 5b). We consider that the softening in hardness for these two alloys is due to the recovery involving the annihilation and rearrengment of dislocations for these two alloys. In addition, a twin structure was observed for these specimens flash heated above their A_f temperature. Figure 6a shows the twin structure of $Ti_{50}Pd_{20}Ni_{30}$ alloy after flash heating at 773K. These twins were determined as Type I $(111)_{B19}$ twins(Fig. 6b), which are considered to be due to

Figure 5   TEM micrographs of $Ti_{50}Pd_{50-x}Ni_x$ alloys after flash heating at their Tf temperatures; (a) $Ti_{50}Pd_{20}Ni_{30}$ alloy after flash heating at 723K; (b) $Ti_{50}Pd_{30}Ni_{20}$ alloy after flash heating at 723K; (c) $Ti_{50}Pd_{40}Ni_{10}$ alloy after flash heating at 793K; (d) $Ti_{50}Pd_{50}$ alloy after flash heating at 873K.

Figure 6 (a)   TEM micrograph of $Ti_{50}Pd_{20}Ni_{30}$ alloy after flash heating at 773K; (b) Selected area diffraction pattern, taken from the marked area in (a), indicating a Type I $(111)_{B19}$ twinning pattern in $[\bar{1}2\bar{1}]_{B19}$ zone.

a martensitic transformation during subsequent quenching. Figure 5c shows the microstructure of $Ti_{50}Pd_{40}Ni_{10}$ alloy after flash heating at 793K which is just above its A_f. Subgrains and variants containing internal twins were observed. Figure 5d shows the microstructure of $Ti_{50}Pd_{50}$ alloy after flash heating at 873K which is just above its A_f. Recrystallized grains with internal twins were observed, which indicates that the primary recrystallization took place. According to the above microstructure observations, we see that the softening in hardness corresponds to different recovery•recrystallization stages with the increase in the reverse transformation temperature of $Ti_{50}Pd_{50-x}Ni_x$ alloys. These results are summarised as follows. 1) $Pd \leq 35at\%$, the softening in hardness represents the recovery which involves annihilation and rearrangement of dislocations and other lattice defects introduced by cold rolling; 2) $Pd = 40at\%$, the softening in hardness represents the recovery which involves formation and growth of subgrains; 3) $Pd \geqq 45at\%$, the softening in hardness represents the recovery and primary recrystallization.

CONCLUSION

1. The recovery·recrystallization process in $Ti_{50}Pd_{50-x}Ni_x(x \leqq 20at\%)$ alloys with high reverse transformation temperatures is controlled by the reverse transformation.
2. The reasons are considered as follows. 1) The recovery · recrystallization process is much slower in martensitic state than that in parent state because of a large difference in atomic diffusion rate in the parent phase(B2) and in the martensite with a close-packed structure(B19); 2) the recovery · recrystallization process occurs dominantly at the parent-martensite interfaces with the progress of the reverse transformation, where atoms are unstable and mobile.
3. The softening in hardness corresponds to different recovery·recrystallization stages with the increase of the reverse transformation temperature in $Ti_{50}Pd_{50-x}Ni_x$ alloys, that is, recovery only for $Ti_{50}Pd_{50-x}Ni_x(x \geqq 10\ at\%)$, and recovery and primary recrystallization for $Ti_{50}Pd_{50-x}Ni_x(x \leqq 5\ at\%)$ alloys.

ACKNOWLEDGMENTS

The authors benefitted very much by the discussions with Prof. E. Furubayashi at Waseda University and express their sincere appreciation for his critical comments. They also thank Dr. T. Kainuma at the National Research Institute for Metals for the use of a micro-Vickers hardness tester. The present work was supported by the Grant-in-Aid for General Scientific Research (Shiken B, 1995-7) from the Ministry of Education, Science and Culture of Japan.

REFERENCES

1. H.C. Donkersloot and J.H.N. Van Vucht, J. Less-Common Metals, 20, 83 (1970).
2. V.N. Khachin, N.A. Matveeva, V.P. Sivokha and D.V. Chernov, Dokl. Acad. Nauk SSSR, 257, 167 (1981).
3. K. Enami, K. Horii and J. Takakura, ISIJ International, 29, 430 (1989).
4. W.S. Yang and D.E. Mikkola, Scr. Metall. 28, 161 (1993).
5. P.G.Lindquist & C.M. Wayman, Proc. of the Int. Conf. on Engineering Aspects of Shape Memory Alloys (Butterworth-Heineman, London, 1990), p. 58.
6. K. Otsuka, K. Oda, Y. Ueno, Min Piao, T. Ueki and H. Horikawa, Scr. Metall. 29, 1355 (1993).
7. Y. Ueno, M. Piao, K. Oda, K. Otsuka, T. Ueki and H. Horikawa, Proc. 3rd International SAMPE Symposium (Chiba, Japan, 1993), p. 1274.
8. K. Enami and T. Hoshiya, Proc. 3rd IUMRS Int. Conf. on Advanced Materials, edited by C. T. Liu, K. Otsuka, K. Shimizu, Y. Suzuki and J. Van Humbeeck, (Tokyo, Japan, 1993), p.1013.
9. D. Golberg, Ya Xu, Y. Murakami, S. Morito, K. Otsuka, T. Ueki and H. Horikawa, Scr. Metall. 30, 1349 (1994).
10. D. Golberg, Ya Xu, Y. Murakami, S. Morito, K. Otsuka, T. Ueki and H. Horikawa, Intermetallics 3, 35 (1995).
11. D. Golberg, Ya Xu, Y. Murakami, K. Otsuka, T. Ueki, H. Horikawa, Materials Lett. 22, 241

MICROSTRUCTURES AND MECHANICAL PROPERTIES OF 6061 Al MATRIX SMART COMPOSITE CONTAINING TiNi SHAPE MEMORY FIBER

J. H. LEE, K. HAMADA*, K. MIZIUUCHI**, M. TAYA* and K. INOUE
Materials Science and Engineering, University of Washington, Box 352120, Seattle, WA 98195-2120, inoue@u.washington.edu
*Mechanical Engineering, University of Washington, Box 352600, Seattle, WA 98195-2600, U.S.A.
**Osaka Municipal Research Institute, Jyoutou-ku, Osaka 536, Japan

ABSTRACT

6061 Al-matrix composite with TiNi shape memory fiber as reinforcement has been fabricated by vacuum hot pressing to investigate the microstructure and mechanical properties. The yield stress of this composite increases with increasing amount of prestrain, and it also depends on the volume fraction of fiber and heat treatment. The smartness of the composite is given due to the shape memory effect of the TiNi fiber which generates compressive residual stresses in the matrix material when heated after being prestrained. Microstructual observations have revealed that interfacial reactions occur between the matrix and fiber, creating two intermetallic layers. The flow strength of the composite at elevated temperatures is significantly higher than that of the matrix alloy without TiNi fiber.

INTRODUCTION

The characteristics of shape memory alloys (SMAs) are associated with martensitic and its reverse transformations. A martensitic transformation in a SMA is induced not only by changing temperature but also by applying stress [1-5]. The martensite transformation temperature depends on the chemical composition of alloy, method of fabrication and heat treatment [6,7]. After being prestrained at temperatures below martensite finish temperature, M_f, following shape memorization, SMA fiber can shrink back to its original length when heated to and above its austenite finish temperature, A_f. If such shrinkable SMA fiber is embedded in metals and alloys to form metal-matrix composite (MMC), compressive residual stress is induced in the matrix when heated up at least to austenite start temperature, A_s. This results in enhanced tensile flow stress of MMC.

The concept of such a smart composite with TiNi fiber is shown in figure 1 [1], which is described as follows. TiNi fiber is first heat treated to shape memorize their initial length at above A_f temperature. It is then quenched to room temperature (around M_s in the present case), given prestrain and then embedded in the matrix to form a composite by, for instance, vacuum hot pressing. Composite thus produced is then heated to temperatures higher than A_f where the TiNi fiber tends to shrink back to its initial length (i.e., shape recovery). The recovered strain is as much as the amount of prestrain if the constraint of the matrix material does not prohibit the shape recovery of the fiber. This shape recovery induces residual tensile stress in the TiNi fiber, while the matrix is subjected to compressive stress. This compressive stress generated in the matrix contributes to the enhancement of the tensile properties of the composite. The purposes of this work are to study the microstructure formed upon the fabrication of the 6061 Al-matrix composite containing TiNi fiber and to study effects of prestraining and volume fraction of fiber on the strength of the composite with and without aging heat treatment.

EXPERIMENT

TiNi fiber (Ti-50.3 at. % Ni) of 200 μm diameter (made by Kantoc Ltd., Fujisawa, Japan) and sheet metal of 6061 Al alloy (0.5 mm thick, Alcoa made) were used in this study to produce smart composite. The sheet metal was sheared off using a shear cutter into a rectangular shape of 10 mm width and then slits of 0.3 mm width and 1 mm length were made in both narrow sides with an interval of 1 mm using a low-speed diamond wheel cutter. After removal of oxide layer, TiNi fiber was wound around such 6061 Al alloy sheets through slits, and several prepreg sheets thus made were stacked on a pair of hot press dies. Then, composite specimens were fabricated by a custom-made vacuum hot press system at selected holding temperatures and pressures. After holding specimens under a set of holding pressure and temperature for a various period of time, composite specimens were cooled down to room temperature within a furnace. Two volume fractions were chosen in the study and they were 2.7 and 5.3 %. Figure 2 is a schematic illustration showing the hot pressing procedure of wire wound prepreg employed in the present study.

Composite specimens were cut to produce tensile specimens, whose dimensions are 10 mm wide and 90 mm long with a gauge length of 26 mm. To obtain mechanical properties and transformation temperatures of TiNi fiber used, tensile tests were performed at temperatures between 297 and 373 K after shape memorizing treatment at 773 K for 30 min. in air followed by quenching into water. The transformation starting and finishing stresses can be obtained from tensile stress-strain curves obtained each test temperature. DSC was also utilized for the measurement of transformation temperatures without stressing. In order to see the effects of T6 aging treatment on mechanical properties of the composite, the heat treatment was performed by following a standard procedure of 6061 Al alloy. That is, the solution treatment was first made at 803 K for 1 hr and then water quenched followed by the second heat treatment at 443 K for 6 hr and quenching. To investigate prestrain effects, various amounts of prestrain were given to some composite specimens by deforming them in tension at 293 K. Tensile tests were performed on an Instron testing machine at a constant strain rate, 1×10^{-4}/s. The displacement of tensile specimens used in the study was measured by using an extensometer.

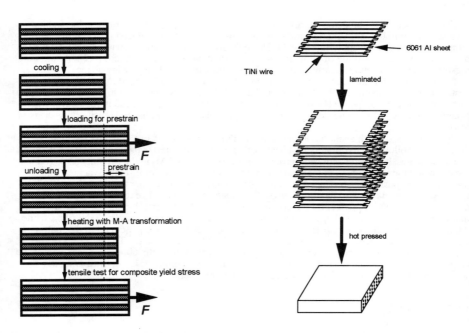

Fig. 1 Concept of smart composite with TiNi fibers.

Fig. 2 Hot pressing procedure of wire wound prepreg

RESULTS

Processing and Microstructures

The transformation temperatures of TiNi fiber used in the present study were measured by a differential scanning calorimeter (DSC), which are tabulated in Table 1. This table also contains the elastic moduli of the fiber for both martensite and austenite phases, which were measured from data obtained in tension tests using an extensometer. The table also contains the elastic modulus of 6061 Al alloy.

Table 1. Transformation temperatures and elastic moduli of Ti-50.3 at % Ni SMA fiber and 6061 Al alloy.

	M_s	M_f	A_s	A_f	Elastic Modulus
TiNi	288 K	280 K	318 K	329 K	26.3 GPa (Martensite)
					67.0 GPa (Austenite)
6061 Al					70.0 GPa

Figure 3 shows the EPMA results of a 6061 Al composite sample containing TiNi SMA fiber fabricated at a vacuum hot pressing of 7 MPa at 823 K for 30 min. The elements of regions in the composite obtained by EPMA are tabulated in Table 2. EPMA analysis revealed that there was a considerable interfacial reaction between the fiber and the matrix under this fabrication condition. As seen in region 1 of figure 3, only TiNi is observed. On the other side, in region 2, Al_3Ti was

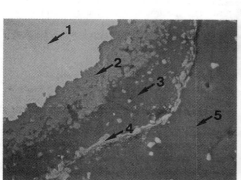

Fig. 3. EPMA semi-quantitative analysis of boundary reaction layer.

Table 2. EPMA results obtained from regions in composite (at %).

	Al	Si	Cr	Cu	Ni	Ti
1					51.22	48.78
2	64.65	4.20			6.05	25.11
3	75.46				23.18	1.36
4	98.45		0.15	0.47	0.21	0.72

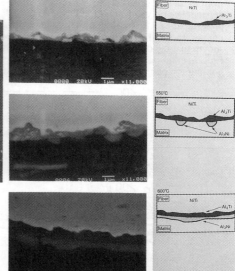

Fig. 4 SEM Micrographs of Smart Composite by Hot Pressing at Various Temperatures

formed and Al₃Ni in the region 3. Such dual phase structures in interfacial regions has also been observed in Ti-matrix composite containing TiPd SMA fiber. Figure 4 shows typical SEM micrographs of the composite made by vacuum hot pressing at 773, 823 and 873 K. It should be noted that the Al_3Ti phase is the only intermetallic phase formed in reaction areas when the composite was made at 773 K. On the other hand, both Al_3Ti and Al_3Ni were formed both at 823 and 873 K. At this moment it is not sure which interfacial structure has the strongest bonding.

Mechanical Behavior

Figure 5 shows yield stress - prestrain curves obtained in tension at 293 K for the composite fabricated at 773 K for 30 min. followed by T6 aging treatment. This figure also contains data for the composite without T6 aging treatment. In figure 5, prestrain was given at 293 K. From this figure, the yield stress is found to increase with increasing both the amount of prestrain and the volume fraction. It should be noted that T6 treated samples all show higher yield stress than that without T6 treatment. As seen, the yield stress after T6 treatment becomes about three times as high as that without T6 treatment. Another feature to be pointed is that the yield stress obtained by testing at 373 K is also higher than that tested at 293 K. This stress increase is associated with the annihilation of stress-induced martensite formed upon prestaining. The following typical equations can be obtained by a least square fit method for the composite having a 2.7 % volume fraction with and without T6 aging treatment.

$$Y = 15.677 \ X \ + \ 283.15 \quad \text{with T6.}$$
$$Y = 18.719 \ X \ + \ 77.599 \quad \text{without T6.}$$

where X is the amount of prestrain and Y is the yield stress.

Figure 6 shows the prestrain dependence on yield stress for the TiNi/6061Al -F composite (where F stands for no heat treatment). The hot press condition used was at 823 K for 30 min, and composite specimens made under this condition were tested at 375 K. The results obtained from these specimens are very similar to those seen in figure 5. Figure 7 shows the prestrain dependence on yield stress for TiNi / Al 6061-F composite specimens fabricated by vacuum hot pressing at 873 K for 30 min. This figure show that the yield stress is higher when tested at higher temperature. The figure also shows that the yield stress is higher for higher prestrain. It is also shown that the composite becomes stronger when made at higher hot pressure. From the experimental results obtained, the optimum vacuum hot press conditions are considered to be a holding temperature of 773 K, a holding time of 30 min and a maximum pressure of 54 MPa.

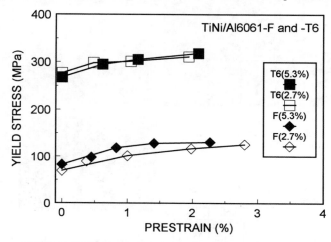

Fig. 5 Yield Stress vs. Prestrain Curves Tested at 375K

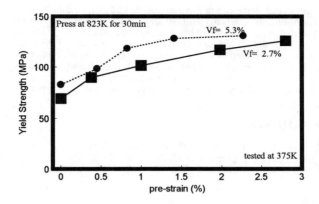

Fig. 6 Dependence of Composite Yield Strength on Pre-Strain

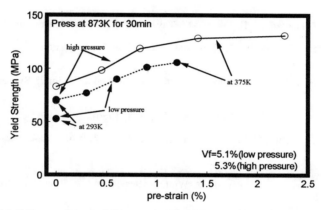

Fig. 7 Dependence of Composite Yield Strength on Pre-Strain

CONCLUSIONS

6061 Al - matrix composite containing TiNi SMA fiber was fabricated by vacuum hot pressing under various conditions, and the microstructure and mechanical properties have been investigated. The results obtained are summarized as follows.

1. The optimum vacuum hot press conditions were at 773 K for 30 min. at 54 MPa.

2. The prestrain effects were clearly shown for the composite with and without aging treatment of 6061 Al alloy matrix composite containing TiNi SMA fiber.

3. The composite with TiNi SMA fiber used as reinforcement becomes stronger at high temperatures due to the generation of residual stress caused by a shape memory effect of TiNi fiber at above As or/and A_f.

4. The yield stress of the composite increases with increasing prestrain, fiber volume fraction and test temperature at temperatures above the A_f temperature.

5. To improve mechanical properties in this composite, T6 aging treatment is needed.
6. Interfacial reactions occur between TiNi fiber and 6061 Al matrix upon hot pressing at conditions employed, and the interfacial phases are intermetallics of Al_3Ti and Al_3Ni.
7. The yield stress (Y) of the composite with a 2.7 volume fraction of TiNi fiber and the amount of prestrain (X) is roughly expressed to have a linear relationship and the relationship can be expressed by the following forms for the composite with and without T6 aging treatment.

$$Y = 15.677 \ X \ + \ 283.15 \quad \text{with T6 aging treatment}$$
$$Y = 18.719 \ X \ + \ 77.599 \quad \text{without T6 aging treatment}$$

ACKNOWLEDGMENTS

One of the authors (J. H. L.) would like to thank the Dong-A University, Pusan, Korea for support during his stay in the University of Washington. The authors also would like to thank the National Science Foundation, Grant No. CMS - 9414696 for partial support of the work.

REFERENCES

1. M. Taya, A. Shimamoto and Y. Furuya, Proc. ICCM-10, Whistler, B.C., Canada, 1995, p. V275 - V282.

2. K. Inoue, K. Hamada, J. H. Lee and M. Taya, Proc. TMS, Cincinnati, OH, 1996.

3. C. Lexcellent, H. Tobushi, A. Ziolkowski and K. Tanaka, Int. J. Pres. Ves. and Piping 58, p. 51 (1994).

4. P. H. Lin, H. Tobushi, K. Tanaka, T. Hattori and M. Makita, J. Intell. Sys. and Str. 5, p. 694 (1994).

5. W. D. Armstrong, J. Intell. Sys. and Str. 7, p. 448 (1996).

6. J. H. Lee, J. B. Park, G. F. Andreasen and R. S. Lakes, J. Biom. Mater. Res. 22, p. 573 (1988).

7. Shape memory materials and phenomena - Fundamental aspects and applications, edited by C. T. Liu, H. Kunsmann, K. Otsuka and M. Wuttig (Mater. Res. Soc. Proc. 246, Boston, MA 1991), p. 13 -55.

ENHANCED MECHANICAL PROPERTIES OF SMART POLYMER-MATRIX HYBRID COMPOSITE BY SHAPE MEMORY EFFECT

H. IZUI*, K. HAMADA**, K. OGAWA**, M. TAYA**, and K. INOUE*
*Dept. of Material Science and Engineering, University of Washington, Seattle, WA 98195
**Dept. of Mechanical Engineering, University of Washington, Seattle, WA 98195

ABSTRACT

New smart hybrid composite with TiNi shape memory alloy (SMA) fiber as reinforcement is designed to be composed of TiNi fiber and carbon/epoxy (CFRP) composite. One-dimensional composite modeling is utilized to evaluate residual stresses of the smart composite generated by shape memory effects. The modeling indicated that the tensile strength of TiNi/CFRP composite is improved due to the shape memory effect of TiNi fiber for both parallel and perpendicular configurations of TiNi fiber with respect to the direction of carbon fiber. Modeling also shows the tensile strength is increased with increasing both volume fraction of TiNi fiber and amount of prestrain.

INTRODUCTION

Two types of smart composite with TiNi SMA fiber as reinforcement have recently been designed. The composite designed includes metal-matrix and polymer-matrix composite, such as TiNi fiber/6061Al matrix(TiNi/Al) and TiNi fiber/epoxy matrix (TiNi/epoxy) composites [1,2,3]. Embedding SMA fiber in these matrix materials produces composites with improved mechanical properties such as stiffness, tensile yield stress, fracture toughness, and fatigue resistance. This improvement is attributed to residual stresses formed due to shape memory effects.

We are currently developing carbon fiber reinforced polymer (CFRP) composite containing continuous fiber of TiNi SMA both experimentally and theoretically. Here CFRP used is a Toray product and made of Carbon Toray T-300 fiber and bis-phenol epoxy. The objective of the present study is to evaluate the effects of prestrain and volume fraction of SMA fiber on tensile strength of the composite. From the results obtained, fracture properties can be predicted.

ANALYTICAL MODEL

One-Dimensional Composite Model

Assuming that the absolute strain of the composite is so small that no sliding occurs between TiNi fiber and CFRP composite, the compatibility equations for the smart hybrid TiNi/CFRP composite can be expressed as

$$\varepsilon_{sc} = \varepsilon_f = \varepsilon_c \tag{1}$$

where ε_{sc}, ε_f and ε_c are the total strain of the smart hybrid composite, TiNi fiber, and CFRP composite, respectively. In this case, the applied stress of the smart composite, σ_{sc}, can be expressed by the force equilibrium equation given by the stress of TiNi fiber, σ_f, and that of the composite, σ_c. The relationship is given in form (2),

$$\sigma_{sc} = V_f \times \sigma_f + (1 - V_f) \times \sigma_c \tag{2}$$

where V_f is the volume fraction of TiNi fiber. The total strain of TiNi fiber, ε_f, is given in form (3),

$$\varepsilon_f = \varepsilon_f^{el} + \varepsilon_f^{CTE} + \varepsilon_f^{trans} \tag{3}$$

where 'el', 'CTE', and 'trans' stand for elastic strain, thermal strain, and transformation strain, respectively. In the case of CFRP composite, its total strain ε_c is given in form (4),

$$\varepsilon_c = \varepsilon_c^{el} + \varepsilon_c^{CTE} \tag{4}$$

The constitutive equations of TiNi fiber and CFRP composite are given in normal forms,

$$\sigma_f = E_f \times \varepsilon_f^{el} \tag{5}$$

$$\sigma_c = E_c \times \varepsilon_c^{el} \tag{6}$$

where E_f and E_c are the elastic moduli of TiNi fiber and CFRP composite, respectively. The thermal strain of TiNi fiber and CFRP composite for a temperature change of dT is given in the following forms,

$$\varepsilon_f^{CTE} = \alpha_f \times dT \tag{7}$$

$$\varepsilon_c^{CTE} = \alpha_c \times dT \tag{8}$$

where α_f and α_c are the coefficients of thermal expansion of TiNi fiber and CFRP composite, respectively.

Phase Transformation of SMA Fiber

The volume fraction of martensite in a SMA depends on applied stress and temperature. Various types of equations have been proposed to describe the thermomechanical behavior of SMA as a function of the volume fraction of martensite at a given temperature and at given applied stress, and in this paper the volume fraction of martensite, ε_M, during martensitic transformation is assumed to be expressed in the following form,

$$\xi_M (T, \sigma_f) = 1 - \exp \left[a^M \times (M_s - T) + b^M \times \sigma_f \right] \tag{9}$$

$$a^M = \ln(0.01) / (M_s - M_f), \quad b^M = a^M / C_M$$

where M_s and M_f are the martensite start and finish temperatures without applied stress, respectively. C_M is the slope of the temperature-stress phase diagram and is given by knowing a material constants a^M and b^M. Here, transformation finishing is defined by a value of ξ_M being

426

99 %. By using form (9), the ξ_M increment with the increment of T and σ_f is expressed in the following form,

$$d\xi_M (T, \sigma_f, dT, d\sigma_f)=$$

$$\exp[a^M \times (M_s - T) + b^M \times \sigma_f] - \exp[a^M \times (M_s - T - dT) + b^M \times (\sigma_f + d\sigma_f)] \quad (10)$$

In the case of the reverse transformation, where martensite goes back to austenite, the volume fraction of austenite, ξ_A, is expressed in a form similar to that of form (9), as given in form (11),

$$\xi_A (T, \sigma_f) = \exp[a^A \times (A_s - T) + b^A \times \sigma_f] \quad (11)$$

$$a^A = \ln(0.01) / (A_s - A_f), \quad b^A = a^A / C_A \quad (12)$$

where A_s and A_f are the austenite transformation start and finish temperatures without applied stress, respectively. Here transformation finishing is defined by a value of ξ_M being 1 %. By using this equation, the ξ_A increment with the increment of T and σ_f is given in form (13),

$$d\xi_A (T, \sigma_f, dT, d\sigma_f) =$$

$$\exp[a^A \times (A_s - T - dT) + b^A \times (\sigma_f + d\sigma_f)] - \exp[a^A \times (M_s - T) + b^A \times \sigma_f] \quad (13)$$

It has been accepted that the elastic modulus and coefficient of thermal expansion of a SMAs significantly depend on the martensite volume fraction. Thus, in this work the elastic modulus and the thermal expansion coefficient are assumed to have linear relationships with the volume fraction and it is shown in forms (14) and (15),

$$\alpha_f(\xi) = \xi \times \alpha_{fMAR} + (1 - \xi) \times \alpha_{fAUS} = \alpha_{fAUS} + \xi \times (\alpha_{fMAR} - \alpha_{fAUS}) \quad (14)$$

$$E_f(\xi) = \xi \times E_{fMAR} + (1 - \xi) \times E_{fAUS} = E_{fAUS} + \xi \times (E_{fMAR} - E_{fAUS}) \quad (15)$$

where α_{fMAR} and α_{fAUS} are the coefficients of thermal expansion for martensite and austenite, respectively, and E_{fMAR} and E_{fAUS} are the elastic moduli of martensite and austenite, respectively.

The dimensional change (i.e. strain of SM fiber upon phase transformation) is also assumed to be expressed by a linear function of the change in the martensite volume fraction as shown in form (16),

$$\varepsilon_f^{trans} = d\xi \times \varepsilon^{t\,max} \quad (16)$$

where ε^{tmax} is the maximum transformation strain obtainable when 100 % of martensite becomes 100 % of austenite, and vice versa.

MATERIAL PROPERTIES

Carbon/epoxy prepreg used as a matrix was produced by Toray, consisting of Carbon Toray

Table 1   Properties of CFRP and TiNi SMA fiber

	Composite		TiNi SMA fiber	
	Longitudinal (0 deg)	Transverse (90 deg)	Martensite	Austenite
Tensile strength	1760 MPa	80 MPa	900 MPa (at RT)	600 MPa (at 373 K)
Elastic modulus	125 GPa	8.8 GPa	26.3 GPa	67 GPa
Ultimate strain	1.3 %	1.0 %	-------	-------
Coefficient of thermal expansion	0.3×10^{-6} 1/K	36.5×10^{-6} 1/K	6.6×10^{-6} 1/K	11.0×10^{-6} 1/K
Density	-------		6.5 g/cm^3	

T-300 fiber and bis-phenol epoxy (CFRP)matrix. The volume fraction of the carbon fiber is 55 %. The nominal thickness of prepreg sheet is 0.15 mm. TiNi SMA fibers of 200 μ m diameter were used to fabricate a hybrid composite for improving mechanical properties. The TiNi fiber is composed of 49.7 at % Ti and 50.3 at % Ni. Tave interfacial strength between the matrix and the TiNi fiber, the fiber was chemically polished for removing the oxide film using a chemical solution of 30 % nitric acid, 4 % hydrofluoric acid and 66 % distilled water. The transformation temperatures of the TiNi fiber were analyzed by DSC, and they are martensite start temperature (M_s) being 288 K, martensite finish temperature (M_f) being 280 K, austenite start temperature (A_s) being 318 K, and austenite finish temperature (A_f) being 329 K. Plastic deformation of TiNi fiber occurs without stress induced martensitic transformation at 368 K. Table 1 shows the properties of both the CFRP composite[4] and TiNi SMA fiber.

During the fabrication of the hybrid composite, the TiNi fiber was prestrained to a desired value, followed by hot pressing. The hybrid TiNi/CFRP composite is then heated up above the austenite finish temperature of TiNi fiber, where the TiNi fiber tends to shrink back to its initial length.

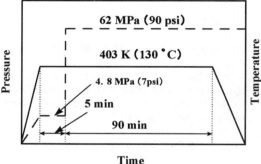

Fig. 1 Pressure-temperature-time diagram for hot pressing

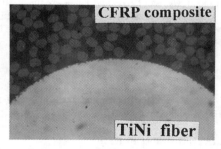

Fig. 2 Cross-sectional view of a hybrid composite

Hybrid composite samples were fabricated by hot pressing at a holding pressure of 62 MPa and at a holding temperature of 403 K for 90 min after preheating, as shown in Fig.1. Figure 2 is a cross-sectional view of a hybrid sample produced at the conditions given in Fig. 1, revealing both carbon fibers and one TiNi fiber.

RESULTS

Analytical results for smart, hybrid TiNi/CFRP composite were obtained based on one-dimensional modeling for both parallel and perpendicular configurations. In the parallel configuration, TiNi fiber is placed parallel to T-300 carbon fibers, whereas in the perpendicular configuration, TiNi fiber is perpendicular to T-300 fiber. The results obtained are shown in Fig. 3. Figure 3 represents the effects of volume fraction and prestrain on residual stress of the TiNi/CFRP composite. As seen, these two configurations of the composite have shown a similar tendency to each other. That is, the residual stress in composite increases with both increasing prestrain and volume fraction. Comparing the hybrid composite with TiNi fiber having the parallel configuration, the composite with TiNi fiber having the perpendicular configuration is found to show higher residual stress when compared at the same amount of prestrain (which is 5 vol%). This difference is caused by the difference in elastic modulus of carbon fiber in the 0 deg direction and in the 90 deg direction. The elastic modulus of the former case is 125 GPa, whereas that of the letter is 8.8 GPa. The increase in residual stress is found to be proportional to the volume fraction of TiNi fiber. This is caused by linear increase in transformation strain associated with shape memory effects.

The increment of tensile strength is

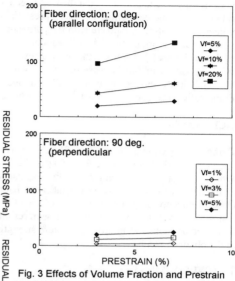

Fig. 3 Effects of Volume Fraction and Prestrain on residual stress of the hybrid composite

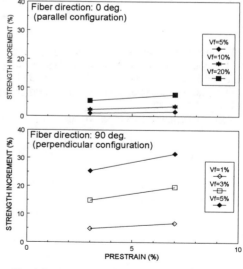

Fig. 4 Dependence of Hybrid Composite Strength Increment on Volume Fraction and Prestrain

replotted in Fig. 4 in order to show the percentage increment of strength as a function of prestrain for various volume fractions of TiNi fiber. As seen in these figures, the tensile strength of the hybrid composite increases with increasing prestrain and with volume fraction. The strength increment of the hybrid composite with the perpendicular configuration was higher than that with the parallel configuration. Since this CFRP composite is brittle, its tensile strength and fracture strength are identical. Therefore, the present results indicate that the fracture toughness of the hybrid composite would be increase substantially in the perpendicular configuration.

The analytically obtained results described above are being compared with those obtained experimentally. The comparison will be reported elsewhere.

CONCLUSIONS

Carbon/epoxy composite and TiNi SMA fiber were used to form smart, hybrid CFRP/TiNi composite. The shape memory effect of TiNi fiber on the hybrid TiNi/CFRP composite has been analytically investigated by using a one-dimensional composite model..

The modeling results obtained have indicated that the tensile strength of the hybrid composite is increased with increasing volume fraction of TiNi fiber and with increasing prestrain of TiNi fiber. This stress increase is caused by the increase in residual stress associated with shape memory effects. The possibility of improving fracture strength of the smart hybrid composite is suggested from the obtained results.

ACKNOWLEDGMENTS

The authors would like to thank the National Science Foundation, Grant No. CMS-9414696 for partial support of the work. One author (H.I.) also would like to thank Nihon University for the support to stay at University of Washington.

REFERENCES

1. Y. Furuya, A. Sasaki and M. Taya, Mater. Trans. JIM, Vol.34, No.3, 1993, p.224-227.
2. M. Taya, Y. Furuya, Y. Yamada, R. Watanabe, S. Shibata, and T. Mori, Proc. Smart Mater., edited by V.K. Varadan (SPIE, 1993), p.373-383.
3. M. Taya, A. Shimamoto, and Y. Furuya, Proceedings of ICCM-10 (Whistler, B.C., Canada, 1995), p.V-275-282.
4. Torayca Technical Reference Manual(Toray Marketing & Sales, Kirkland, WA, USA., 1988)

MODEL TRACKING OF STRESS AND TEMPERATURE INDUCED MARTENSITIC TRANSFORMATIONS FOR ASSESSING SUPERELASTICITY AND SHAPE MEMORY ACTUATION

X. WU, T. J. PENCE AND D. S. GRUMMON
Department of Materials Science and Mechanics
Michigan State University, East Lansing, MI 48824-1226

ABSTRACT

Both austenite/martensite transformations, and martensite/martensite variant reorientation are central to superelasticity and shape memory actuation. Here we discuss a simple model for tracking these transformations that is appropriate for device modeling. The approach is to augment conventional continuum mechanical descriptions with internal variables that track fractional partitioning of the material between austenite and the various martensite variants. A three-species model involving austenite and two complementary martensite variants provides sufficient generality to capture the martensite variant distributions that underlie shape memory, and the strain-accommodation associated with superelasticity. Transformations between all of these species can be tracked on the basis of triggering algorithms that reflect both transformation hysteresis and the variation of triggering stress with temperature. The algorithm described here is for temperature-dependent response resolved in a single direction. It requires only the following experimentally determined parameters: the four transformation temperatures M_f, M_s, A_s, A_f, the transformation strain, the Young's modulus, and the transformation latent heat.

INTRODUCTION

Austenite/martensite transformations, and transformations between different martensite variants, underlie shape memory and superelastic response. The role of temperature and stress in triggering these transformations are central to various models for the description of thermomechanical response at a level that is useful for the elementary treatment of device actuation. Here we present a model that tracks phase fraction internal variables which is conceptually similar to [1]-[7]. These models differ in various ways based upon: the definition of the phase fraction internal variables; the linking of these internal variables to thermomechanical response; and the algorithms given for the evolution of these variables. Specifically, we indicate how a model presented in [7] for high temperature austenite/martensite transformations can be extended to the low temperature regime so as to treat direct martensite/martensite transformations. The most important of these is a generic stress-assisted detwinning transformation starting from a strain-free "random" martensite, *i.e.*, martensite consisting of balanced opposing variant fractions, typically in some self-accommodated arrangement. Under load this then transforms to a "biased" martensite in which certain variants predominate. We consider a relatively simple version of the model for single direction response in which the material is characterized by only two martensite variants. Seven parameters then characterize the material: the four transformation temperatures M_f, M_s, A_s, A_f, the austenite/martensite transformation strain γ^*, the elastic modulus E, and the latent heat of the austenite/martensite transformation ΔH.

An interesting aspect of this model is that the stresses associated with detwinning start and detwinning finish are determined in terms of the other material parameters. This is due to the simplified form of the model. A more sophisticated treatment is possible in which: these stresses are specified independently; the martensite and austenite have different moduli; and conventional thermal expansion is present. However, these extensions require a discussion of various thermodynamic issues with respect to the construction of thermodynamic phase diagrams in stress-temperature space. The version of the model presented here is free of these details, at the expense of increased generality. The nature of the complications associated with these additional extensions

431

is briefly discussed, although a complete development will be given elsewhere. The theory is presented in the context of uniaxial tension, and material parameters are chosen on the basis of experimental work on TiNi as given in [8]. These values are: $M_f = 235$ K, $M_s = 263$ K, $A_s = 295$ K, $A_f = 308$ K, $\gamma^* = 0.06$, E = 38 GPa, $\Delta H = 116 \times 10^6$ J/m^3. After presenting the underlying theory in the following section, we apply it to various isothermal loading situations. We also examine path dependence as it pertains to the production of biased martensite.

THEORY

Following [7], consider two martensitic variants M$^+$ and M$^-$ whose crystallographic lattice strains are complementary shears from the base austenite state A. Transformations are triggered by changes in temperature T and stress τ, and may involve: M$^-\leftrightarrow$A, A$\leftrightarrow$M$^+$, or M$^-\leftrightarrow$M$^+$. At each instant there are 3 possibilities for A$\leftrightarrow$M$^+$: no transformation, A$\rightarrow$M$^+$, or M$^+\rightarrow$A. Similar considerations hold with respect to M$^-\leftrightarrow$A and M$^-\leftrightarrow$M$^+$. Transformations at the macroscopic level are associated with stress-free strains: γ^*, -γ^*, and $2\gamma^*$ for A$\rightarrow$M$^+$, A$\rightarrow$M$^-$, and M$^-\rightarrow$M$^+$, respectively. Let the triple $\xi = \{ \xi_-, \xi_A, \xi_+ \}$ give the phase fractions of M$^-$, A, M$^+$ respectively. Overall material balance then requires: $\xi_- + \xi_A + \xi_+ = 1$. Hookean behavior is assumed in each individual phase, so that pure-phase macroscopic strains are the following functions of stress: $\{\gamma_-, \gamma_A, \gamma_+\} = \tau \{1, 1, 1\}/E + \gamma^* \{-1, 0, 1\}$. Conventional thermal expansion is neglected in the version presented here. The overall macroscopic strain γ at each instant is given by a rule of mixtures:

$$\gamma = \{\xi_-, \xi_A, \xi_+\} \cdot \{\gamma_-, \gamma_A, \gamma_+\} = \gamma^*(\xi_+ - \xi_-) + \tau/E.$$

As the shape memory effect shows, the strain γ is not a direct function of stress and temperature. This is accounted for by ξ not being a direct function of τ and T. Instead ξ is determined from the process path in (τ,T)-space starting from presumed known initial conditions for ξ. Thus the algorithm for determining ξ is of central concern. To this end, temporarily consider a more fundamental point of view in which the four (stress-free) transformation temperatures: M_f, M_s, A_s, A_f are regarded as entities due to the splitting of a single underlying base transformation temperature T* = ($M_f + M_s + A_s + A_f$)/4. This base transformation temperature is characteristic of standard first order transformations in the absence of both hysteresis and phase coexistence. The presence of hysteresis and phase coexistence in the actual austenite/martensite system then generates the splitting of T* into M_f, M_s, A_s, A_f. The underlying phase diagram in (τ,T)-space corresponding to the single transformation temperature T* is shown in Fig. 1a. Here (τ,T) = (0,T*) is a standard triple point from which emanate 3 straight neutrality curves associated with transformations: M$^-\leftrightarrow$A, A$\leftrightarrow$M$^+$ and M$^-\leftrightarrow$M$^+$. Their slopes are given by the Clausius-Clapeyron relation, where the entropy jumps $\Delta\eta$ are determined from the latent heat of transformation:

$$\frac{dT}{d\tau}_{A\leftrightarrow M+} = \frac{\gamma^* T^*}{\Delta H}; \qquad \frac{dT}{d\tau}_{M-\leftrightarrow A} = -\frac{\gamma^* T^*}{\Delta H}; \qquad \frac{dT}{d\tau}_{M-\leftrightarrow M+} = \infty.$$

Crossing these 3 neutrality curves causes abrupt changes in ξ between {1,0,0}, {0,1,0} and {0,0,1}.

Unfolding this simple phase diagram allows for gradual changes in ξ that also exhibit hysteresis. Such an unfolding splits T* into the 4 transformation temperatures, which obey $M_f < M_s$, $A_s < A_f$. We restrict attention here to the numerical values given above so that $M_s < A_s$. For describing stress-free transformations that are purely temperature driven, we introduce the envelope functions $\alpha_{min}(T)$ and $\alpha_{max}(T)$ which respectively give the fraction of austenite obtained by increasing T from below M_f where $\xi = \{0.5, 0, 0.5\}$ and by decreasing T from above A_f where $\xi = \{0,1,0\}$. Thus $\alpha_{min}(T) \le \alpha_{max}(T)$ and $\alpha_{min}(T)=0$, $\alpha_{max}(T)=0$ for T< A_s, T< M_f respectively, whereas $\alpha_{min}(T)=1$, $\alpha_{max}(T)=1$ for T> A_f, T> M_s respectively. In general, there will be a sigmoidal transition between 0 and 1, which can be determined from experiment [9]. In the absence of detailed data, these envelope functions can be well-estimated by:

$$\alpha_{min}(T) = \frac{1}{2}\left\{1 - \cos\left(\left(\frac{T-A_s}{A_f-A_s}\right)\pi\right)\right\}, \qquad \alpha_{max}(T) = \frac{1}{2}\left\{1 - \cos\left(\left(\frac{T-M_f}{M_s-M_f}\right)\pi\right)\right\}.$$

Stress-free thermal transformations involving unbiased martensite then give phase fractions obeying $\xi = \{0.5(1-\xi_A), \xi_A, 0.5(1-\xi_A)\}$ where $\alpha_{min}(T) \le \xi_A \le \alpha_{max}(T)$. Hysteresis algorithms for the determination of ξ_A in this type of setting are discussed in [9]-[11].

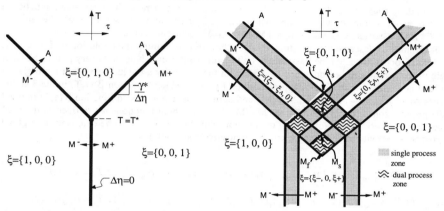

Figure 1. Simple phase diagram (left) and unfolded phase diagram (right). The unfolded diagram involves 6 single process zones, of which 4 involve 2 unconnected parts. Their 4 overlapping regions define the dual process zone.

Transformations due to both temperature and stress require accounting for the effect of the four-fold splitting $T^* \to M_f, M_s, A_s, A_f$ on the simple phase diagram away from the temperature axis. In particular, each of the base neutrality curves $A \leftrightarrow M^+$, $M^- \leftrightarrow A$ and $M^- \leftrightarrow M^+$ unfolds into 4 separate curves. Figure 1b presents the unfolding considered here, it involves no slope change to the original neutrality curves. Note that the four $A \leftrightarrow M^+$ curves that split from the original curve are extended into the $\tau < 0$ region until they meet the $M^- \leftrightarrow A$ curve passing through M_f. Similar considerations apply to the $M^- \leftrightarrow A$ curves. This was the extension considered in [7] and is all that is needed for treating high temperature processes, *i.e.*, those not involving direct $M^- \leftrightarrow M^+$ transformations. In order to treat low temperature detwinning transformations it is necessary to account for the fourfold splitting of the original $M^- \leftrightarrow M^+$ neutrality curve that is on the T-axis in the simple phase diagram. In the unfolding of Figure 1b, two of the four unfolded $M^- \leftrightarrow M^+$ curves connect to the high T construction at the termination of the two $M^- \leftrightarrow A$ curves that pass through A_s and through A_f. Hence these connections occur at stress values

$$\tau_s = \frac{(A_s - M_f)\Delta\eta}{2\gamma^*} \qquad \text{and} \qquad \tau_f = \frac{(A_f - M_f)\Delta\eta}{2\gamma^*}.$$

The other two unfolded $M^- \leftrightarrow M^+$ curves connect to the high T construction at the termination of the $A \leftrightarrow M^+$ curves, *i.e.*, at stresses $-\tau_s$ and $-\tau_f$.

This construction gives rise to seven stable zones in which no transformation takes place: three are the pure phase zones inherited from the original construction, three are stable dual phase zones inherited from the unfolding of the 3 neutrality curves well away from the original triple point at $(\tau, T) = (0, T^*)$ and the final zone is a stable triple phase zone inherited from the unfolding of the original triple point. These are separated by 7 transformation process zones; 6 are single process zones and 1 is a dual process zone. The 6 single process zones all proceed away from the stable triple phase zone and involve one of the single transformations: $A \to M^+$, $M^+ \to A$, $A \to M^-$, $M^- \to A$, M^-

$\rightarrow M^+$, $M^+ \rightarrow M^-$ as indicated in Figure 1b. In particular, the $M^- \rightarrow M^+$ zone is the low temperature continuation of the $M^- \rightarrow A$ zone, a view which is consistent with that of [12]. A similar continuation holds with respect to the $M^+ \rightarrow M^-$ and $M^+ \rightarrow A$ zones.

The dual process zone consists of four regions of overlap in the four $A \leftrightarrow M$ single process zones. They are clustered near the stable triple phase zone. In the dual process zone, one of the following four dual simultaneous transformations may take place depending on the orientation of the (τ,T)-path: $\{M^- \rightarrow A, \ M^+ \rightarrow A\}$, $\{A \rightarrow M^-, \ A \rightarrow M^+\}$, $\{A \rightarrow M^-, \ M^+ \rightarrow A\}$, $\{M^- \rightarrow A, \ A \rightarrow M^+\}$, corresponding, roughly, to: T increase, T decrease, τ decrease, τ increase. The demarcation between any two process alternatives occurs when the (τ,T)-path slope passes through the values: $\pm \gamma * T * / \Delta H$ associated with the original $M \leftrightarrow A$, $A \leftrightarrow M^+$ neutrality curves.

It remains to describe the quantitative algorithm for determining the evolution of $\xi = \{\xi_-, \xi_A, \xi_+\}$ in the process zones on the basis of the (τ,T)-path. To this end, note that every point (τ,T) in a single process zone can be connected to a point on the T-axis by a unique curve that parallels the unfolded neutrality curves. This mapping defines a positive-sloped function $\beta^+(\tau,T)$ in the 3 single process zones for: $M^+ \rightarrow A$, $A \rightarrow M^+$, $M^+ \rightarrow M^-$, and defines a negative-sloped function $\beta^-(\tau,T)$ in the 3 single process zones for: $M^- \rightarrow A$, $A \rightarrow M^-$, $M^- \rightarrow M^+$. Note that both functions are defined in the dual process zone. These β functions allow the previous envelope functions to be extended throughout the process zones via the following 6 process zone envelope functions:

$$\alpha_{A \rightarrow M^+}(\tau,T) = \alpha_{\max}(\beta^+(\tau,T)), \quad \alpha_{M^+ \rightarrow A}(\tau,T) = \alpha_{\min}(\beta^+(\tau,T)), \quad \alpha_{M^+ \rightarrow M^-}(\tau,T) = \alpha_{\min}(\beta^+(\tau,T)),$$

$$\alpha_{A \rightarrow M^-}(\tau,T) = \alpha_{\max}(\beta^-(\tau,T)), \quad \alpha_{M^- \rightarrow A}(\tau,T) = \alpha_{\min}(\beta^-(\tau,T)), \quad \alpha_{M^- \rightarrow M^+}(\tau,T) = \alpha_{\min}(\beta^-(\tau,T)).$$

This, in turn, enables the hysteresis algorithm to be extended to the process zones in terms of the following evolution equations for ξ:

$$d\xi_+ = \frac{-\xi_A}{2\alpha_{A \rightarrow M^+}} d\alpha_{A \rightarrow M^+} \quad \text{for } A \rightarrow M^+, \qquad d\xi_- = \frac{-\xi_A}{2\alpha_{A \rightarrow M^-}} d\alpha_{A \rightarrow M^-} \quad \text{for } A \rightarrow M^-,$$

$$d\xi_+ = \frac{-\xi_+}{1 - \alpha_{M^+ \rightarrow A}} d\alpha_{M^+ \rightarrow A} \quad \text{for } M^+ \rightarrow A, \qquad d\xi_- = \frac{-\xi_-}{1 - \alpha_{M^- \rightarrow A}} d\alpha_{M^- \rightarrow A} \quad \text{for } M^- \rightarrow A,$$

$$d\xi_+ = \frac{-\xi_+}{1 - \alpha_{M^+ \rightarrow M^-}} d\alpha_{M^+ \rightarrow M^-} \quad \text{for } M^+ \rightarrow M^-, \qquad d\xi_- = \frac{-\xi_-}{1 - \alpha_{M^- \rightarrow M^+}} d\alpha_{M^- \rightarrow M^+} \quad \text{for } M^- \rightarrow M^+.$$

The equations outlined above, in concert, define a well-posed algorithm for the determination of ξ on any (τ,T)-path beginning from specified initial values for ξ (viz. [7]). In particular, it delivers the stated envelope function behavior on $\tau = 0$ paths.

RESULTS AND DISCUSSION

Figure 2 illustrates the use of this model for isothermal loading and unloading at various temperatures. For temperatures below A_f it is necessary to specify the initial condition for ξ. Here we consider initial conditions corresponding to maximal austenite and minimal unbiased martensite, *i.e.*, $\xi = \{ 0.5(1- \xi_A), \xi_A, 0.5(1- \xi_A)\}$ where $\xi_A = \alpha_{max}(T)$. These simulations show superelasticity due to stress-induced martensite at high temperatures. They also show retained biased martensite after unloading at low temperatures, which can be randomized by subsequent stress-free heating and cooling (the shape memory effect). As noted by Brinson [4], simpler martensite models do not deliver unbiased martensite on cooling and so reaccumulate transformation strain below M_s.

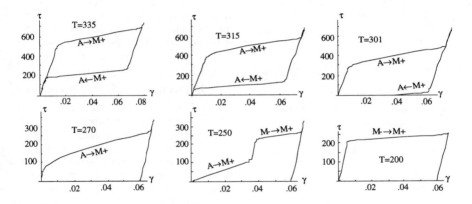

Figure 2. Isothermal loading and unloading at various temperatures. Initial conditions for ξ involved maximum austenite and random martensite, as obtained from stress-free cool down beginning at $T > A_f$. The irregular form of these curves is due to discretization and rendering.

Figure 3 illustrates the use of the current model in predicting production of biased martensite from pure austenite due to two separate types of (τ,T)-paths, both of which begin from $(\tau,T)=(0,T_{init})$ where $T_{init} > A_f$ so that the initial phase fraction is pure austenite. The path endpoints are of the form $(\tau,T)=(\tau_i,T_{end})$ where $T_{end}< M_f$ for various stress values τ_i. We take $T_{init}= 315$ K, $T_{end}= 220$ K. For each stress τ_i, the first type of path (load-cool) involves first loading to τ_i at $T = T_{init}$ followed by cooling to T_{end}, whereas the second type of path (cool-load) involves first cooling to T_{end} at $\tau=0$ followed by loading to τ_i.

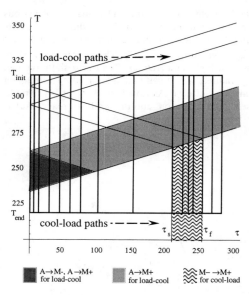

For both paths, the final state will be one of pure martensite. As indicated in [13], the values of ξ_+ will generally differ. The cool-load paths involve $\{A\to M^-,\ A\to M^+\}$ transformations on cooling followed by an $M^-\to M^+$ transformation on loading (provided that $\tau_i > \tau_s$). As shown in Figure 3, this gives $\xi = \{0.5, 0, 0.5\}$ for $\tau_i < \tau_s = 211$ MPa and gives $\xi = \{0, 0, 1\}$ for $\tau_i > \tau_f = 256$ MPa. Intermediate values τ_i on the cool-load paths give a fractional partitioning between M^- and M^+ which is inherited from the function $\alpha_{min}(\beta^-)$. In contrast, the load-cool paths involve no transformations on loading since this path does not intersect the $A\leftrightarrow M^+$ curve passing through M_s (provided $\tau_i < (T_{init} - M_s)\Delta\eta / \gamma^* = 365$ MPa). However, this curve is crossed on cooling, whereupon $A\to M^+$ transformations are triggered in the absence of $A\to M^-$ transformations. $A\to M^-$ transformations are only initiated if the path intersects the $M^-\leftrightarrow A$ curve passing through M_s. This will occur on cooling only if $\tau_i < (M_s - M_f)\Delta\eta /(2\gamma^*) = 98$

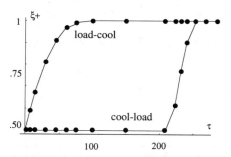

Figure 3. Production of M^+ on load-cool and cool-load paths. The transformation processes depicted in the top frame on the load-cool paths are for $d\tau = 0$, $dT < 0$.

MPa. Conversely, if $\tau_i > 98$ MPa then the $A\to M^+$ process goes to completion with $\xi_- = 0$, so that $\xi = \{0, 0, 1\}$. In particular, the load-cool paths yield a larger phase fraction of M^+ than the cool-load path for all $0 < \tau_i < \tau_f$. Intermediate paths involving simultaneous cooling/loading will yield corresponding intermediate results.

The form of the model presented here involves an $M^-\to M^+$ single process zone bounded by stresses τ_s and τ_f that follow directly from the geometry of the unfolded phase diagram. For the specified numerical values of the transformation temperatures, latent heat and transformation strain, these triggering stresses were $\tau_s = 211$ MPa and $\tau_f = 256$ MPa. In general, values suggested by physical measurement for these stresses are somewhat lower. This can be traced to the nature of the triple point unfolding presented here, where an unfolded A↔M$^+$ and M$^-$↔A curve was extended completely through the unfolded triple point until its connection to the unfolded M$^-$↔M$^+$ curve. At this connection point the combined curve exhibited an abrupt slope change. In contrast, an unfolding in which this connection involves a gradual slope change will cause these connections to occur at lower stress values. Extensions to the present model that incorporate this type of feature will be presented elsewhere. Such extensions impact the phase diagram energy balances that lead to Clausius-Clapeyron relationships prior to unfolding. A related issue involves smoothing the abrupt slope change in the A→M curves at the T-axis, which would presumably tend to delay completion of a stress-driven A→M transformation for $M_f < T < M_s$. In addition, allowing for different elastic moduli in the martensite and austenite phases causes the strain discontinuity on transformation to be stress-dependent. This also introduces curvature into the neutrality curves on the (τ,T)-plane [7], although such an effect is of order $|(E_A-E_M)\tau|/(E_A E_M)$, and is therefore small. Even so, by maintaining overall thermodynamic consistency in the (τ,T)-plane, in conjunction with the specification of specific heats and convective heat transfer coefficients, it is possible to use the present type of model to treat heat flow effects, both passively (as output) or actively (as input), in the full spectrum of isothermal to adiabatic conditions [14].

ACKNOWLEDGMENTS

This work has been funded by the MSU Composite Materials and Structures Center and by the National Science Foundation under grant #MSS8821755.

REFERENCES

1. K. Tanaka, *Res Mechanica* **18**, p. 251-263 (1986).

2. C. Liang and C.A. Rogers, *J. of Intell. Mater. Syst. and Struct.* **1**, p. 207-234 (1990).

3. H. Tobushi, H. Iwanaga, K. Tanaka, T. Hori and T. Sawada, *JSME International Journal-ser. 1* **35**, p. 271-277 (1992).

4. L.C. Brinson, *J. Intell. Mater. Syst. and Struct.* **4**, p. 229-242 (1993).

5. J.G. Boyd and D.C. Lagoudas, *J. Intell. Mater. Syst. and Struct.* **5**, p. 333-346 (1994).

6. L.C. Brinson and M.S. Huang, *J. Intell. Mater. Syst. and Struct.* **7**, 108-114 (1996).

7. T.J. Pence, D.S. Grummon and Y. Ivshin in *Mechanics of Phase Transformations and Shape Memory Alloys*, ed. L.C. Brinson and B. Moran, (ASME AMD vol. 189, 1994), p. 45-58.

8. L. Hou and D.S. Grummon, *Scripta Met.* **33**, p. 989-995 (1995).

9. R. Loloee, T.J. Pence and D.S. Grummon in *Proceedings of ICOMAT-95* (*J. de Physique* IV C8, 1995), p. 545-550.

10. A.A. Likhachev and Y.N. Koval, *Scripta Met.* **27**, p. 223-227 (1992).

11. Y. Ivshin and T.J. Pence, *Int. J. Engng. Sci.* **32**, p. 681-704 (1994).

12. R. Wasilewski, *Scripta Met.* **5**, p. 127-130, 131-135 (1971).

13. T. Duerig, D. Stockel and A. Keeley in *Engineering Aspects of Shape Memory Alloys,* (Butterworth-Heinemann, 1990), p. 181-194.

14. Y. Ivshin and T.J. Pence, *J. of Intell. Mater. Syst. and Struct.* **5**, p. 455-473 (1994).

COMPRESSION-INDUCED MARTENSITIC TRANSFORMATIONS IN
Cu-Zn-Al ALLOYS WITH e/a > 1.50

A.Cuniberti [(1)] and **R.Romero** [(1)] [(2)]
(1) Instituto de Física de Materiales Tandil, Facultad de Ciencias Exactas, Universidad Nacional del Centro de la Provincia de Buenos Aires. Pinto 399, 7000 Tandil, ARGENTINA.
Tel/fax +54-293-42821
(2) Comisión de Investigaciones Científicas de la Provincia de Buenos Aires.

ABSTRACT

Results are reported about the pseudoelastic cycles β-martensite obtained by compression of β Cu-Zn-Al single crystals with electronic concentration 1.52. Stress-strain curves at different temperatures and *in situ* observations were carried out. The results are compared with those obtained from tension test.

INTRODUCTION

In Cu-Zn-Al alloys the β phase is metastable at room temperature and can be retained by quenching, suffering two successive orderings $A_2 \rightarrow B2 \rightarrow L2_1$. The martensitic phases can be either thermally-induced, spontaneous transformation, or stress-induced, cooling or stressing the β phase. Because of the non-diffusive character of the transformation, the martensite inherits the atomic order from the β phase. The martensitic structures are based on the stacking sequence of the close-packed planes, they are 9R, 18R and 2H where the number indicates the stacking sequence period and the letter the structure simetry: R rhomboedric and H hexagonal. Each martensite can be obtained from the β phase depending on the electronic concentration, temperature and stress conditions [1].

β Cu-Zn-Al single crystals with e/a higher than 1.50 transform spontaneously to 2H martensite, by tensile load it is possible to obtain 18R and 2H martensite depending on temperature conditions [2] [3]. At temperatures above about Ms+20K the 18R phase is induced and the pseudoelastic cycles are similar to those for alloys with e/a 1.48, with a small stress hysteresis equivalent to 3-4K. Below Ms+20K the martensite induced is 2H, the stress-strain curves show a marked yield point followed by a plateau at a lower stress and a stress increase on further loading, an important stress hysteresis is observed and commonly no reverse transformation takes place. Although the martensitic transformations induced by tensile load had already been studied in detail, up to now, there was not any information about the phase transformations that take place when the applied load is compressive. In this work, the martensitic transformations occurring in β Cu-Zn-Al single crystals with e/a>1.50 are studied by compressive pseudoelastic cycles at different temperatures and *in situ* observations.

EXPERIMENTAL

The composition of Cu-Zn-Al single crystals used in the present study is Cu-8.54at%Zn-21.75at%Al, e/a=1.52 and Ms=264K, determined by electrical resistivity [2]. They were grown starting from alloys prepared with high purity materials (5N).

The load axes orientation of each single crystal was determined in β phase by Laue technique. Cylindrical samples of 3mm diameter and 10mm length for compression tests were spark machined. After machining, the samples were kept for an hour at 1100K, quenched in iced water and kept for a minimum of 6 hours at room temperature to anneal out all quenched in short range disorder [1] [3]. Compression tests were carried out in an Universal Testing Machine Shimadzu Autograph DSS-10T-S at $8 \cdot 10^{-4}s^{-1}$.

RESULTS

Typical stress-strain curves for the compression induced transformations obtained from the β phase are shown in Figure 1. As can be seen, three types of curves are obtained depending on the temperature test and the strain reached. At the lower temperature a discontinuous stress-strain curve with strong stress serrations is observed and on unloading the martensitic phase is retained, Figure 1 (a). At the higher temperature, Figure 1 (b), the transformation advances continuously and, if the loading is stopped before a critical strain is reached, the reverse transformation is observed on unloading but the stress hysteresis is about three times higher than the typical for β→18R phase transformation (~3MPa). At a critical strain an abrupt increase in stress appears accompanied by an audible *click* and the martensitic phase is retained on unloading, Figure 1 (c). This type of pseudoelastic cycles were also observed for other Cu-Zn-Al single crystals with 1.534 and 1.51 of electronic concentration [4].

The change in the stress-strain cycles occurs at a temperature of about (284±5)K, i.e. Ms+20K. This transition temperature is like that obtained from tensile tests [3], notwithstanding there are differences between pseudoelastic cycles induced by tension and compression.

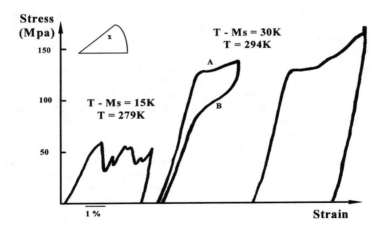

Figure 1: Typical stress-strain curves for the compression-induced martensitic transformations from the β phase.

The plateau stress on loading after the yield stress observed in tension tests at low temperatures does not appear in compression tests, successive stress serrations are observed in its place. The load path at temperatures above Ms+20K is similar between both load directions but, by compression, there exists a considerable increase in the stress hysteresis on unloading.

In situ observations were carried out and optical micrographs obtained at room temperature, T>Ms+20K, are shown in Figure 2, corresponding to points A and B in Figure 1 (b). Fine and parallel bands appear with the compressive stress increase and they disappear on unloading, this behaviour is typical of 18R martensite. Besides 18R bands, other bands crossing them are formed and these ones remain when the load is removed. In this case the crystallographic orientation was not possible to determine because of their discontinuous interface. The apparition of the second phase could be the reason for the increase in the stress hysteresis of the pseudoelastic cycles. Once reached a critical strain ε_c and before the $\beta \rightarrow 18R$ transformation has concluded a burst phase transformation takes place and the surface marks are no longer seen, this process coincides with the sudden stress increase described above and would be associated with a sample elongation. The measured critical strain ε_c was 4.2%, this value is smaller than the strain associated with the transformation to 18R martensite for this alloy [5] and axe orientation, about 6.8%.

A burst transformation, accompanied by a contraction of the sample length and an audible acoustic emission, was also observed in 1.48 18R single crystals by cooling under constant tensile load and the resulting phase was 2H martensite [6].

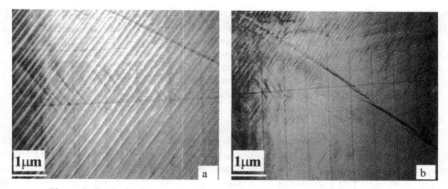

Figure 2: Optical micrograph during a) compressive loading, b) unloading.

According to the experimental observations, high stress hysteresis in the stress-strain cycles and the burst type transformation, the martensitic phase obtained by compressive load would be 2H, even at high temperatures at which 18R martensite is induced by tension. *In situ* observations suggest a 2H plates growth absorbing a lamellar 18R structure, this transformation mechanism has been observed in 1.53 Cu-Zn-Al single crystals by tension but at temperatures below Ms+20K [3].

Figure 3 shows the martensitic shear systems for the phase transformations observed by tension for axe orientations inside the T β phase unit triangle; this triangle is equivalent to compressive axe orientations inside the C triangle taking into account the crystallographic relationship between tension-compression for the β↔18R phase transformation [1].

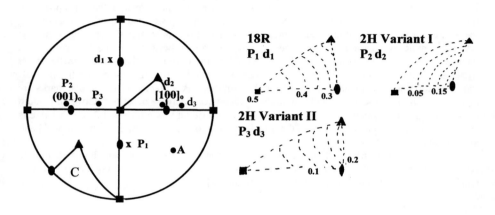

Figure 3: Stereographic projection showing the main directions and planes corresponding to 18R and 2H phase transformations observed by tension.
Equal Schmid factor lines for compressive axes inside the C β triangle.
$P_1 d_1$: habit plane and macroscopic shear direction for the 18R transformation.
$P_2 d_2$: habit plane and macroscopic shear direction for the 2H variant I [7].
$P_3 d_3$: habit plane and macroscopic shear direction for the 2H variant II [7].
A: habit plane for a 2H transformation [3], shear direction unknown.

According to equal Schmid factor lines for compressive axe orientations in the C triangle, the β→18R transformation is the most favoured. Two 2H martensite variants have been observed having a different orientation relationship with the 18R phase [6] [7], variant 1: $(001)_{2H}//(001)_{18R}$, $[010]_{2H}//[010]_{18R}$, $[100]_{2H}//[100]_{18R}$, variant 2: $(100)_{2H}//(001)_{18R}$, $[010]_{2H}//[010]_{18R}$, $[001]_{2H}//[100]_{18R}$. The Schmid factor is not so different for both 2H variants and, althoug it remains low, it becomes higher when the axe orientation is not near the [001] of β, as in this case. The system with pole denoted by A is only unfavourable for orientations near the $[111]_β$.

The Schmid factor indicates that it is possible the occurrence of a transformation to 2H phase even though this is not the most favourable.

CONCLUSIONS

The martensitic transformation induced by compressive load from β Cu-Zn-Al single crystals with e/a higher than 1.50 presents differences with respect to that induced by tensile load.

- At temperatures below Ms+20K the stress-strain cycles show great stress serrations by compression instead of a stress plateau observed by tension, the martensitic phase is retained on unloading and according tensile tests the resulting phase would be 2H martensite.

- At temperatures above Ms+20K a smooth pseudoelastic cycle is observed and 18R bands are formed; once reached a critical strain, about 4%, a sudden phase transformation takes place and this martensitic phase does not disappear on unloading. This behaviour differs from that observed in tension tests which lead to a β-18R transformation and the reverse transformation with small stress hysteresis.

X-rays and electronic microscopy analyses are needed to determine the induced phases, but the stress-strain curves and the optical observations suggest that the final martensitic phase obtained by compression tests would be 2H phase, independently the temperature test between Ms+10K and Ms+53K.

Acknowledgement. The authors wish to thank Dr.J.L.Pelegrina, Centro Atómico Bariloche, who gave us the single crystals used in this work.

REFERENCES

1. M.Ahlers, Prog.Mater.Sci. 30, 135 (1986).
2. J.L.Pelegrina. Doctor Thesis. Universidad Nacional de Cuyo, Argentina (1990).
3 J.L.Pelegrina and M.Ahlers, Acta Metall.Mater. 38, 293 (1990).
4. A.Cuniberti and R.Romero, Proc. Asociación Física Argentina Vol.7, in press.
5. R.Romero and J.L.Pelegrina, Phys.Rev.B 50, 9046 (1994).
6. A.Tolley, D.Ríos Jara, F.C.Lovey, Acta Metall. 37, 1099 (1989).
7. J.-E.Bidaux and M.Ahlers, Z.Metallk.83, 310 (1992).

STRUCTURAL DEFECTS IN THE 18R MARTENSITE OF CuZnAl SHAPE MEMORY ALLOYS

A.M. CONDO, F.C. LOVEY
Centro Atómico Bariloche, CNEA, 8400 San Carlos de Bariloche, Argentina.

ABSTRACT

The lattice relaxation at the type I faults in the 18R martensite was studied by high resolution TEM. Local atomic plane rotations and small changes of the interplanar distances were observed over a distance of 1.5 nm. It is shown that this distorted region can be rationalized as an intermediate structure between the martensite and the second variant of the parent phase.

INTRODUCTION

A large density of non-basal plane defects, called type I faults, are commonly observed in the 18R type of martensites. They have been studied by several authors either with conventional transmission electron microscopy (TEM) [1,2] or high resolution TEM [3,4]. Since the type I faults are connected to the basal plane stacking faults they would have the same displacement vector. However, it has been found [2] that a further relaxation occurs along the non-basal faults because of the mismatch of the atomic planes joining at the defects. Such relaxation could have important consequences on the mechanical properties of the 18R martensites serving as obstacles for dislocation gliding [5], or could be related in some way to the rubber-like behavior [6].
In this work the lattice relaxation at the type I faults was studied by high resolution TEM. Atomic plane rotations and local changes of the lattice parameters are reported. The results suggest that the atomic relaxation gives an intermediate structure approaching the second variant of the parent β phase.

EXPERIMENT

A single crystal of 18R martensite with an alloy composition of Cu -17.52 at% Al -12.96 at% Zn has been used. Thin foils with the zone axis along the [210] direction were polished as explained elsewhere [2]. The high resolution TEM images were taken with a Philips CM200 ultra twin microscope, operating at 200 kV, with an objective lens spherical aberration of 0.52 mm. The suitable underfocus to obtain the image of the projected structure was estimated to be around 63 nm, following the procedure outlined in [7]. An objective aperture with a diameter of 14 nm^{-1} was used.

RESULTS

A typical high resolution TEM image containing part of a type I defect is shown in Fig. 1. The basal plane stacking sequences at both sides of the defect are indicated. A careful observation shows that the (0 0 18) and the (1 $\overline{2}$ 10) planes are shifted upon crossing the defect, giving rise to an additional displacement vector in agreement with the previous finding [2].

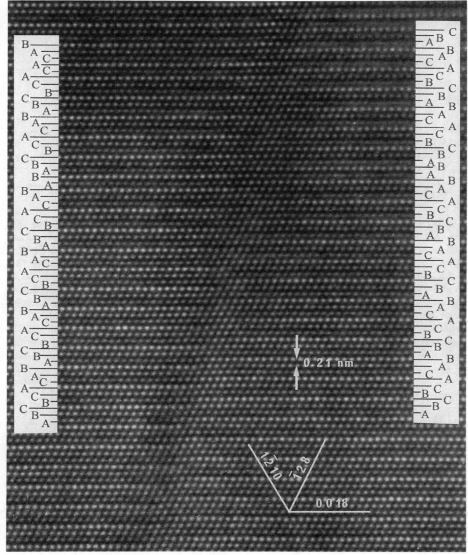

Figure 1.
High resolution image of a type I fault, from [$\overline{2}\,\overline{1}$ 0] zone axis. The stacking sequences at both sides of the defect are indicated. A shift of the (0 0 18) and (1 $\overline{2}$ 10) planes upon crossing the defect can be noted. Negligible shift along the ($\overline{1}$ 2 8) plane is observed.

In order to study in more detail the nature of this relaxation, an area centered at the defect, in Fig. 1, was analyzed using the Fourier transform method [8]. The corresponding diffractogram is shown in Fig. 2a. Figure 2b corresponds to the perfect 18R structure. It can be noted that the modulation along the c^* axis (normal to the basal plane) becomes weaker in the diffractogram corresponding to the defective area. This means that the corrugated planes of the 18R martensite become flatter at the defect, as can be noted in the image of Fig. 1. In addition, local rotation of the atomic planes and small changes in the interplanar distances can be measured by direct inspection of the high resolution images.

The following results can be emphasized:

- Negligible shift of the $(\bar{1}28)$ planes is observed upon crossing the defect.

  Both the (0 0 18) and the (1 $\bar{2}$ 10) planes are clearly shifted from one to the other side of the defect. The basal planes on the right-hand side of the defect, in Fig. 1, are shifted towards the top of the image in relation with the basal planes coming from the left, by an amount of $\delta_{0\,0\,18} = (0.25 \pm 0.10)\,d_{0\,0\,18}$, where $d_{0\,0\,18} = 0.213$ nm is the basal plane interplanar distance. On the other hand, the (1 $\bar{2}$ 10) planes on the right-hand side of the defect are shifted in the same way by $\delta_{1\,\bar{2}\,10} = (0.30 \pm 0.10)\,d_{1\,\bar{2}\,10}$.

- Both the (0 0 18) and the (1 $\bar{2}$ 10) planes are rotated by nearly the same angle θ, about (2 ± 0.5) degrees, in the defective area, as schematically indicated in Fig. 3. The rotation of the planes is clearly visible along a segment consisting of approximately 6 atomic columns in the defect. The distance between the columns on the basal plane is known to be 0.22 nm.

- The value of the overall area where the image of the perfect crystal is affected by the presence of the defect, I_T, measured in the image of Fig. 1, lies around 10 atomic columns.

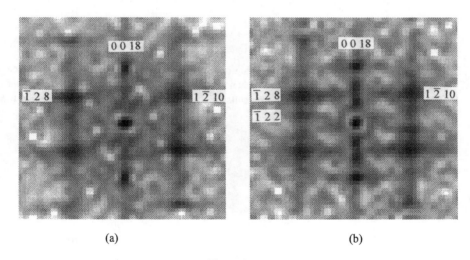

(a)   (b)

Figure 2.

Fast Fourier transforms (FFT) of different square zones (side = 1.5 nm) in Fig 1. a) Centered in the type I defect. b) From the 18R perfect zone. Note that the $\bar{1}22$ spot is much weaker in the defective zone.

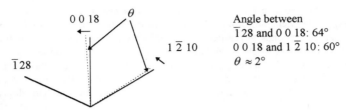

Angle between
$\overline{1}\,28$ and $0\,0\,18$: $64°$
$0\,0\,18$ and $1\,\overline{2}\,10$: $60°$
$\theta \approx 2°$

Figure 3.

Relative rotations of the spots in the defect (dotted lines) with respect to the 18R structure (full lines) as measured in Fig. 1.

- The value of the shift $\delta_{0\,0\,18}$ given above should be related to the local rotation of the basal planes and the actual defect width, w, by (see Fig 4):

$$w = \delta_{0\,0\,18}\Big/\tan\theta \tag{1}$$

using the measured results we obtain $w \approx 1.5$ nm.
The actual defect width cannot be directly obtained from the image because it is inclined by 7.7 degrees with respect to the normal of the foil [2]. In Fig. 4b the projection of the inclined defect is schematically shown. Three zones would be observed, i_1 and i_2 correspond to the overlapped area and I_P is the image of the pure defect. It is assumed, in Fig.4b, that if one of the components has a local thickness larger than 25 % then it dominates the image, thus:

$$i_1 = i_2 = \frac{1}{2}t\tan 7.7; \qquad I_p = w - i_1 \qquad \text{and} \qquad I_T = w + i_1 \tag{2}$$

where t is the foil thickness. Using the value of w given above and taking I_P as the distance between 6 atomic columns, this is 1.1 nm, then $i_1 \approx 0.4$ nm and $I_T \approx 1.9$ nm in good agreement with the observations. The foil thickness results $t \approx 6$ nm, which is a reasonable value owing to the good atomic resolution of the image.

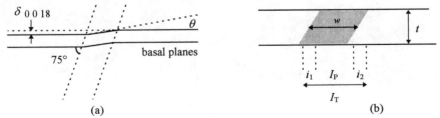

Figure 4.

a) Schematic drawing of Fig. 1 showing the rotation and the shift of the basal planes through the defective area. b) Schematic drawing of the foil sample with the defect edge-on. Due to the width of the defect an image of it can be observed in the inner zone I_P, while i_1 and i_2 correspond to the overlapped regions.

- The trace of the defect is (75 ± 2) degrees from the basal plane.
- The interplanar distance between the $(\bar{1}28)$ planes does not change.
- The distance between the $(0\ 0\ 18)$ planes becomes shorter by about $(1 \pm 0.6)\%$ while the $(1\ \bar{2}\ 10)$ lattice spacing increases by about $(2 \pm 0.8)\%$ at the defect. These values are consistent with the trace of the defect and the rotation of the planes.

DISCUSSION AND CONCLUSIONS

The observed shift of the basal plane, upon crossing the type I defects, is in good agreement with the relaxation vector determined previously from the image analysis using two-beam conditions [2]. On the other hand, it can be deduced from the previous work [2] that the component of the relaxation vector in the direction normal to the $(1\ \bar{2}\ 10)$ plane is $0.35\,d_{1\ \bar{2}\ 10}$, also in good agreement with the present results. Finally the very small shift of the $(\bar{1}28)$ planes corresponds quite well with the fact that the type I defects are almost invisible when observed in two-beam conditions with the $\bar{1}28$ diffraction spot [2].

It is interesting to compare the relaxation at the defects, given in terms of the plane rotations and lattice spacing variations, with the crystallographic changes in the martensitic transformation from β to 18R. This relationship can be directly obtained from the high resolution image of the habit plane, taken along the $[2\bar{1}0]$ zone axis in a Cu-Zn-Al alloy, given in the work of Lovey et al. [9]. From that image the following results can be measured.

- The rotations between the $(0\ 0\ 18)$ and the $(1\ 2\ 10)$ planes with respect to their corresponding $\{110\}_\beta$ planes are (3.8 ± 0.2) degrees.
- The distance between the $\{110\}_\beta$ type plane is shorter by $(2.5 \pm 0.5)\%$ than the $d_{0\ 0\ 18}$, and larger by $(5 \pm 0.5)\%$ than the $d_{1\ 2\ 10}$.
- No noticeable changes are observed neither in the interplanar distances nor in the relative orientation between the $(12\bar{8})$ planes and the corresponding $\{110\}_\beta$ type planes.

Since the type I defects were observed along the $[210]$ direction, which is equivalent to the $[2\bar{1}0]$, it can be deduced that the relaxation at the defects goes towards the second variant of β, being related to the original one by a 18R martensite symmetry operation [10]. The fact that the atomic planes become less corrugated at the defects is also consistent with this assumption. This relaxation towards the second variant of β is also consistent with the type I fault plane. The type I defects lie on a plane crystallographically equivalent to the habit plane and the presence of a thin layer of the second variant of β would create a low energy interface.

REFERENCES

1. M. Andrade, L. Delaey and M. Chandrasekaran, J. Physique **43**, C4, pp. 673-678 (1982).
2. A.M. Condó and F.C. Lovey, J. Physique IV, **5**, pp. 811-816 (1995).
3. F.C. Lovey, G. Van Tendeloo, J. Van Landuyt, L. Delaey and S. Amelinckx, Phys. Stat. Sol. (a), **86**, pp. 553-564 (1984).
4. M. Zhu, D.Z. Yang and C.L. Jia, Metall. Trans., **2**, pp. 1631-1636 (1989).
5. P.L. Rodriguez, A. Cuniberti, R. Romero and F.C. Lovey, Scripta metall. et mat., **27**, pp. 1133-1138 (1992).
6. T. Ohba, T. Finlayson and K. Otsuka, J. Physique IV, **5**, C8, pp. 1083-1086 (1995).
7. F.C. Lovey, W. Coene, D. Van Dyck, G. Van Tendeloo, J. Van Landuyt and S. Amelinckx, Ultramicroscopy, **15**, pp. 345-356 (1984).

8. Image Processing System developed by A.E. Scuri and S. Paciornik (private communication).
9. F.C. Lovey, G. Van Tendeloo and J. Van landuyt, Scripta Metall., **21**, pp. 1627-1631 (1987).
10. F.C. Lovey, A. Amengual and V. Torra, Phil. Mag. A, **64**, pp. 787-796 (1991).

STRESS EVOLUTION DURING CRYSTALLIZATION AND ISOTHERMAL ANNEALING OF TITANIUM-NICKEL ON (100) Si

Jinping Zhang, and D. S. Grummon, Department of Materials Science and Mechanics
Michigan State University, East Lansing, MI 48824

ABSTRACT

Thin films of near-equiatomic titanium-nickel are capable of thermally induced shape-strains and have recently been applied to silicon-based micromachined reversible actuators in force-production devices requiring energy-density levels substantially exceeding those available from piezoelectric, electromagnetic, electrostatic, or bimetal systems. Reversible actuation requires the presence of a resetting agent, or 'bias' force, capable of deforming the martensite phase during the exothermic transformation on cooling. Here, we show that both the initial martensite 'programming' force, and the bias force for subsequent reversible actuation, can readily be provided by manipulating intrinsic and extrinsic stresses developed during low-pressure sputter deposition of TiNi, and subsequent cooldown from either the deposition process temperature or from the crystallization annealing temperature. The stability of both intrinsic and extrinsic stresses at temperatures below the deposition temperature allows considerable flexibility in the design of the force-producing system, and reversible stress-production to levels beyond 0.7 GPa are shown to be readily achieved, corresponding to energy-density levels approaching 10 MJ-m^{-2}. The present paper will review recent results from our group on the development of residual stresses in TiNi sputtered on (100) Si, using the wafer-curvature method applied during isochronal and isothermal vacuum annealing of both amorphous and crystalline thin films, with an emphasis on the combined influence of deposition temperature and annealing temperature on stress-relaxation rates, and on the response of the system to temperature cycling in the displacive transformation range.

INTRODUCTION

Metals which undergo reversible displacive phase transformations underlying shape-memory and superelasticity effects can produce a large force-displacement product, and are thus applicable to silicon-based microelectromechanical systems (MEMS). Of the variety of intermetallic alloys that display thermotractive behavior, near-equiatomic TiNi has been the most intensively studied, its technical significance deriving mainly from its ductility and relatively high work-production capability [1]. On cooling, the ordered cubic B2 (CsCl) parent structure undergoes a 1st-order invariant-plane strain displacive transformation which forms a distribution of 24 possible crystallographic variants[1] of the monoclinic martensite phase. In the absence of stress fields this process is 'self-accommodated': the large transformational shear is nulled on a local scale by the formation of multiple variants of the daughter phase, and no macroscopic shape strain devolves. However, plastic deformation of the martensite by external forces is mediated by intervariant boundary motion, which occurs at stresses well below the critical shear stress for slip, with the result that the variant distribution is altered by an increase in the volume fraction of variants having shear displacements aligned with the applied stress. Subsequent heating the deformed martensite then reverse-transforms the altered variant population, unshearing each remaining variant to its prior orientation in the B2 parent lattice. This produces the displacements associated with the shape-memory by reversing the previous martensite deformation [2]. If constrained, the displacements that occur during the heating transformation can generate very high

---

[1] For the B2 cubic to B19'monoclinic transformation, there exist twelve lattice-correspondance variants which combine in twin-related pairs to satisfy the undistorted habit-plane requirements of the athermal displacive transformation, giving a total of 24 possible habit-plane variants.

Mat. Res. Soc. Symp. Proc. Vol. 459 © 1997 Materials Research Society

recovery stresses, which have been reported to approach 500 MPa in thin films [3, 4], theoretically corresponding to *one gram force* being produced by a film 10 μm wide and 3 μm thick. Sputtered thin films have been shown to display the full range of thermoelastic transformation characteristics [5], including transformational superelasticity [6, 7], and micromachined thin film elements of TiNi have been fabricated by conventional sputtering and etching processes [8-11].

The one-time shape strain produced by this mechanism is useful, but the reversible displacement required of actuators necessitates biasing arrangements to deform the martensite phase during the *cooling* transformation, and thus impart the latent strain to be recovered on heating. These forces may originate from coupled external elastic elements, or from internal stress-fields associated with defects or special precipitate distributions. In the case of titanium-nickel films applied to silicon substrates, the initial martensite biasing force can be conveniently obtained as an extrinsic stress associated with differential thermal contraction on cooling from an elevated process temperature. The latter is a ubiquitous feature of titanium-nickel thin film deposition processing, since such films are amorphous unless deposited at temperatures above approximately 650 K, or unless they are annealed at temperatures above ~700K [12], to obtain the required crystalline parent phase. Below we present some initial results on experiments which explore the development and relaxation of extrinsic and intrinsic stresses in both amorphous and crystalline phases of TiNi on (100) silicon wafers, and show how such stresses may be exploited for the purpose of biasing the martensite in ways applicable to high-energy-density, two-way, reversible actuators

EXPERIMENTAL METHODS

Ti-Ni thin films of thicknesses ranging from 1.4 to 2.8 μm were deposited on 0.3 mm thick (100) Si wafers by D.C. magnetron sputtering in a diffusion-pumped system which achieved base pressures better than 10^{-6} Pa, using ultra-high purity argon working gas at sputtering pressures below ~0.8 Pa. Sputter cathodes were spark-cut from an ingot of commercial NiTi with nominal composition Ti-51.6 at% Ni. Films were made both with and without external heating, allowing the production of both crystalline and amorphous films, with substrate temperature monitored by two thermocouples located on either side of the wafer. After deposition at elevated temperature, the films were allowed to furnace cool at 5 to 10 K/min.

Isothermal and isochronal annealing was conducted in a specially constructed turbo-pumped quartz bell-jar high-vacuum furnace integrated with a custom scanning laser wafer curvature measurement system described previously [5]. Pure titanium getters were used to minimize residual oxygen and nitrogen levels. Biaxial thin film stresses were calculated by the well-known Stoney equation. The films adhered well to silicon, and development of tensile stresses in thicker deposits was sometimes sufficient to fracture the wafer.

X-Ray diffraction to determine phase identities was conducted on a Scintag-2000 diffractometer using Cu-K$_\alpha$ radiation. Thin foils for transmission electron microscopic (TEM) study were prepared by electropolishing in a methanol solution containing 33% nitric acid at 243 K. TEM observation and selected-area electron diffraction (SADP) were conducted on a Hitachi H-800 machine operated at 200 kV.

RESULTS AND DISCUSSION

Intrinsic Stress, Thermal Expansion, and Stress Relaxation   Figures 1a and 1b, respectively, show the results of stress evolution experiments on amorphous TiNi films which were sputtered at (a) the process ambient temperature (estimated to be approximately 425K) and (b) at a slightly elevated temperature of 548 K (insufficient to induce *in situ* crystallization). Both films display a modest compressive stress at room temperature, and both respond to heating with an increase in the magnitude of the compressive stress over a temperature interval up to the approximate deposition temperature, whereafter the stress begins to relax toward lower compressive values. In Figure 1b, the linear increase in compressive stress between 400 and 500 K can be rationalized on the basis of differential thermal expansion, in which the expected rate of biaxial stress-evolution would be given as $d\sigma/dT = E_f(\alpha_f - \alpha_s) / (1-\nu)$, where E_f, α_f, α_s, and ν are the film modulus, the film CTE, the substrate CTE and the film Poisson's ratio. Taking the latter as 83 GPa [13], 15.4 x10^{-6} K^{-1} [14], 3x10^{-6} K^{-1} [15] and 0.33 respectively, gives an estimate of the extrinsic stress-evolution rate of about -1.5 MPa-K^{-1}, in agreement with the observed result.

The magnitude of the extrinsic stress induced on heating reaches a maximum at approximately the temperature that prevailed during deposition. Assuming that the evolution of extrinsic (thermal) stress is elastic and reversible, it may be concluded that this compressive stress maximum in fact represents the residual intrinsic stress achieved during deposition. As such, the intrinsic stress level is seen to increase with increasing process temperature. This is not unreasonable considering the general effect of working gas pressure on stress-evolution in sputtered films. At low pressures, the mean free path of particles in the process environment is relatively long, and few collisions occur between process gas atoms and species emanating from the sputter cathode. This results in a general increase in the kinetic energy of both the arriving deposit atoms and high-energy reflected charge-exchange neutrals elastically scattered from the cathode. The result is an augmentation of the atomic 'peening' process generally thought to underlie the development of compressive stress in sputtered films [16]. Although process pressures in the present study were nominally constant, an increase in process temperature still has the effect of rarefying the working gas in the vicinity of the cathode and substrate, thus increasing the compressive film stress by increasing mean free path for energetic species striking the growing film.

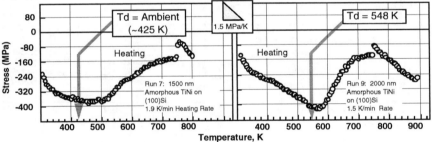

Figure 1. Stress evolution in amorphous TiNi films deposited (left) at ambient process temperature (estimated to be ~ 425 K); and (right) at a substrate temperature of 548 K.

Referring again to Figure 1, as the temperature increases past the deposition temperature, the film stress is seen to relax smoothly toward zero stress until the amorphous-to-crystallization transformation temperature[2] is approached at around 750 K. At this point, a transient stress-spike is observed which drives the film stress approximately 50 to 100 MPa in the tensile direction. This stress-transient is presumably associated with a general densification of the film as it adopts the more closely packed B2 crystalline structure. As temperature is increased beyond the crystallization event, the stress again develops in the compressive-going direction at the ~1.5 MPa/K rate dictated by the film/substrate differential expansion coefficient.

The relationship between stress-relaxation rate, annealing temperature and the deposition temperature is also observable in isothermal annealing, and is characteristic of both amorphous

---

[2] Crystallization temperature is a function of heating rate in isochronal anneals, and may be as low as 713K for isothermal conditions.

and crystalline films deposited at higher temperatures. Figure 2 shows isothermal stress-relaxation in anneals conducted at above and below the film deposition temperature, for an amorphous film deposited at 573 K, and for a film crystallized during deposition at a temperature of 673 K. In both cases, the degree of stress relaxation occurring during a one-hour isothermal anneal is negligible if the annealing temperature is below the deposition temperature, even though the compressive stress may initially be higher.

The same trends in stress-evolution during isochronal and isothermal annealing of amorphous films are generally observed in crystalline films in which the ordered B2 structure is formed in situ during the deposition process. In this case, the deposition temperature must be maintained above approximately 623 K, although films displaying classical displacive transformation characteristics require deposition at temperatures above about 673 K [6]. Figures 3 (a)

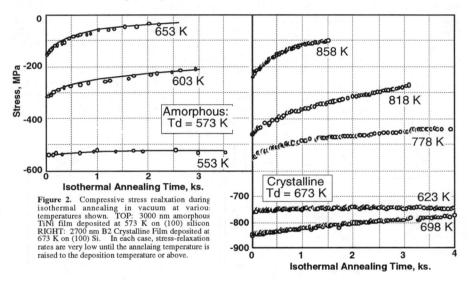

Figure 2. Compressive stress realxation during isothermal annealing in vacuum at variou: temperatures shown. TOP: 3000 nm amorphous TiNi film deposited at 573 K on (100) silicon RIGHT: 2700 nm B2 Crystalline Film deposited at 673 K on (100) Si. In each case, stress-relaxation rates are very low until the annelaing temperature is raised to the deposition temperature or above.

and (b) show results of isochronal annealing on crystalline films deposited at 693 and 723 K respectively. Here, as in the case for amorphous films, heating from room temperature initially induces an increase in compressive stress at the rate of ~1.5 MPa/K until the deposition temperature is reached, whereafter rapid relaxation to zero stress is observed. (Since the film is crystalline at the outset, no tensile spike occurs at 750 K.) In the 693 K deposition temperature data, it is apparent that differential thermal expansion is unable to compete with dynamic stress relaxation, such that the film stress remains at nominally zero as the temperature increase beyond 900 K. Cooling from 1100 K however does introduce some extrinsic stress, which becomes strongly tensile as the temperature is further decreased. It is therefore seen to be possible to control the extrinsic tensile stress at room temperature to a substantial degree by selection of the maximum temperature attained during the stress-relaxation process. Cooling directly from the process temperature results in near-zero or slightly compressive stresses, whereas virtually any level of tensile stress can, in principal, be obtained by relaxing the intrinsic compressive stress at elevated temperature, and then cooling to ambient. In this case the maximum tensile stress attainable would be limited by (a) the maximum permissible annealing temperature (constrained in most cases by film-substrate chemical reaction rates), and (b) the maximum force supportable by the substrate and film/substrate interface before substrate fracture or film separation would occur.

Effect of Film Stress on Displacive Transformations   Figure 4 illustrates the dramatic interaction between extrinsic thin film stress and the austenite-to-martensite transformation in titanium-nickel. In this plot, the biaxial film stress is plotted as a function of temperature for a 2800 nm thick crystalline film deposited at process-ambient temperature and fully stress-relieved at elevated temperature during the crystallization annealing process. The film was subsequently cooled in the stress-measurement apparatus to approximately 220 K, and then monitored as the temperature was scanned at ~0.5K/min from low temperature through an apparent martensite-to-austenite transformation at ~300 K. Heating was continued to slightly above 400 K, at which point the temperature was lowered to ambient at roughly 0.3 K/min.

     The behavior on the cooling scan may be understood in terms of the ability of titanium-nickel to undergo martensite transformation on cooling beginning at a temperature which is dependent on stress in the sense of the Clausius-Clapeyron equation, which predicts a monotonic

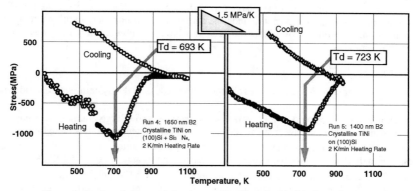

Figure 3. Isochronal stress evolution in crystalline films deposited at elevated temperature.

rise in the equilibrium transformation temperature with stress. This is typically characterized by a 'stress rate', $d\tau/dT$, giving the martensite-induction stress as a function of temperature as:

$$d\tau/dT = |\Delta S / V \cdot \gamma| \qquad (1)$$

where τ is resolved shear stress, T is the equilibrium transformation temperature at stress τ, V is the molar volume, and ΔS and γ are the entropy and the shear strain, respectively, associated with the displacive transformation. We may approximate ΔS as $\Delta H/T_O$, where T_O is the zero-stress equilibrium transformation temperature, taken as $(M_s + A_s)/2 = (220 + \sim 300 \text{ K})/2$ (i.e., ignoring hysteresis) and ΔH is the transformational enthalpy, with a

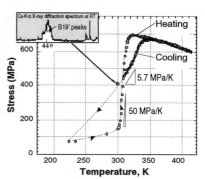

Figure 4. Stress relaxation and recovery due to displacive phase transformation in a 2800 nm thick crystalline TiNi film on (100) Si.

typical experimentally determined value of 17 J/g [6]. Then taking the theoretical transformational shear of 0.13 and a material density of 6.54 Mg/m², yields an estimate for the shear stress rate of approximately 2.85 MPa/K, for an equivalent biaxial stress rate, $d\sigma/dT$, of ~5.7 MPa/K.

This is quite close to the observed dependence of thin film stress on temperature during the cooling scan from 660 K, and it may thus be postulated that on cooling, a film with a nominal zero-stress M_s temperature of ~220 K begins to transform on cooling and in the presence of a 680 MPa tensile stress at 220 K + (680 MPa / 5.7 MPa-K^{-1}) = ~340 K. In the presence of stress, the martensite transformational shears effectively relax the tensile stress as the specimen cools further, forming a biased martensite variant distribution capable of expressing shape-strains on a subsequent heating excursion, as reported qualitatively by Lee. et al. [17]. X-Ray diffraction (see inset in Fig. 4) and TEM observation confirmed the existence of B19' martensite in the specimen at 300 K following cooling from 680 MPa.

The stress-recovery effect is strikingly apparent in the rapid rise in stress during the heating portion of the temperature scan in Figure 4. The reasons for the very large stress/temperature slope on heating are not yet clear, but it is obvious from the general result that extrinsic stresses in the (100)Si/TiNi system can very effectively bias the martensite transformation so as to allow development of very large forces during the martensite to austenite transformation on heating. Though we do not include the data in the present paper, additional temperature cycling of this specimen showed that the general behavior exhibited in Figure 4 was repeatable for a number of cycles.

CONCLUSIONS

From the present experiments on stress evolution in the (100)Si/TiNi system, we conclude the following:

1. Thin films sputter-deposited at low working gas pressure show large compressive intrinsic stresses which tend to increase in magnitude as the deposition temperature is increased.

2. In the absence of rapid stress-relaxation or displacive transformations, the accrual of extrinsic stress due to thermal expansion coefficient differences occurs at a rate of ~1.5 MPa/K.

3. In TiNi films deposited at temperatures below approximately 573 K, crystallization occurs at approximately 700 K, at which point the film stress relaxes rapidly and then experiences a rapid tensile spike of 50 to 100 MPa as the film densifies.

4. Stress-relaxation by diffusional mechanisms occurs at significant rates only when isothermal or isochronal annealing temperatures exceed the thin film deposition temperature.

5. Stress-relaxation by displacive phase transformations occurs on cooling at a temperature well above the nominal stress-free martensite start temperature, in general agreement with predictions of the Clausius-Clapeyron equation.

6. Stress-relaxation induced by extrinsic-stress-assisted austenite-to-martensite transformation is fully reversible on heating, and can generate a stress recovery with heating at a rate approaching 50 MPa/K, forming the basis for reversible cyclic actuation of TiNi/(100)Si composite structures applicable to microelectromechanical systems.

ACKNOWLEDGMENTS

The authors are grateful to the National Science Foundation for their support under grant #MSS9302270, the Michigan State University Composite Materials and Structures Center, and Ford Motor Company.

REFERENCES

[1] A. Keely, D. Stockel and T. W. Duerig in *Engineering Aspects of Shape Memory Alloys*, T. W. Duerig, K. N. Melton, D. Stockel and C. M. Wayman, Eds, Butterworth-Heinemann, p 181 (1990).

[2] K. Otsuka and K. Shimizu (1986), *Int. Met. Rev. 1986* **31**, 93.

[3] J.D. Busch, A.D. Johnson, D.E. Hodgson, C.H. Lee, D.A. Stevenson, *Proc Int'l. Conf. on Martensitic Trans.* (ICOMAT-89) (1989).

[4] J.D. Busch and A.D. Johnson, *J. Appl. Phys.* **68** (12), 6224-6228 (1990).

[5] D. S. Grummon, Li Hou, Z. Zhao and T. J. Pence, *J. de Physique* **IV**, Colloque C8 665-670 (1995).

[6] Li Hou and D. S. Grummon, *Scripta Metallurgica* **33**, 989-995 (1995).

[7] S. Miyazaki, T. Hashinaga, K. Yumikura, H. Horikawa, T. Ueki and A. Ishida, Proc. 1995 North American Conf. On Smart Structures and Materials, (1995).

[8] Li Hou, T. J. Pence and D. S. Grummon, Mat. Res. Soc. Proc. **360**, pp 369-374 (1995).

[9] J. A. Walker and J. Gabriel, *Proc. 5th Int. Conf. on Solid State Sensors and Actuators* (ext. abstr. B8), June 1989, Montreaux, Switzerland, 123 (1989).

[10] T. Minemura, H. Andoh, M. Nagai, R. Watanabe, S, Shimizu and I. Ikuta (1987), *J. Mat. Sci. Lett.* **6**.

[11] A.D. Johnson, *J. Micromech. Microeng.* **1**, 34-41 (1991).

[12] L. Chang, Ph.D. Thesis, Michigan State University, 1994

[13] *ASM Metals Handbook*, 10th ed., Vol 2, p 899; The Materials Information Society (1990).

[14] Z. Zhao, M.S. Thesis, Michigan State University (1995).

[15] *Handbook of Thermophysical Properties of Solid Materials*, A. Goldsmith, T. Waterman and H. Hirschorn, Armor Research Foundation, Pergamon (1961).

[16] D. W. Hoffmann and J. A. Thornton, J. Vac. Sci. and Tech. **20**, p 335 (1982).

[17] A. P. Lee, D. R. Ciarlo, P. A. Krulevitch, S. Lehew, J. Trevino and M. A. Northrup, Proc. 8th Int. Conf on Sol. State Sensors and Actuators, and Eurosensors IX, Stockholm, June 1995, pp 368-371 (1995).

THE ELECTRICAL RESISTANCE PROPERTIES OF SHAPE MEMORY ALLOYS

G.Airoldi[§], M.Pozzi[§], G.Riva[#]
§ Dipartimento di Fisica, Università di Milano, Via Celoria 16, 20133 Milano, I
airoldi@mvmidi.mi.infn.it
Pirelli Coordinamento Pneumatici, Viale Sarca 222, 20126 Milano, I

ABSTRACT

Shape memory alloys are known for their ability to build up large deformations under an applied stress, either in martensitic phase or in the pseudoelastic region.
The electrical resistance of shape memory alloys, traditionally used to define the transformation temperatures, shows interesting features when considering its modification as a function of strain.
The electrical resistance in NiTiCu ribbons, obtained by melt spinning, is here examined, either in martensite or in the pseudoelastic range or under a constant applied stress across the transformation range. In each examined case, the results obtained show that the electrical resistance follows a linear relationship as a function of the imprinted strain.

INTRODUCTION

Shape memory alloys (SMAs) are known for their functional properties related to their ability to recover a deformation and/or to develop a stress state during a constrained recovery .
These properties can actually be exploited in a monolithic sensor-actuator only if a physical property can be adopted to drive the deformation: this appears feasible, at least in well defined cases, when the electrical resistance (ER) shows a linear dependence from deformation, e.g. during the deformation recovery in specimens trained for the Two Way Memory Effect (TWME) [1].
The electrical resistance of SMAs, generally used to define the transformation temperatures, shows interesting features when considering its modification as a function of strain. In martensitic phase[2-4], under an applied increasing stress and in absence of a phase transformation, a linear or quasi linear increase with deformation is found for different SMAs. In the transformation range parent phase↔martensite, either during training under applied load or after training for the TWME[1], a linear behaviour is similarly found, at least when the electrical resistivity change related to the modification of electronic scattering processes is negligible.
The electrical resistance variation, either in martensite or in the pseudoelastic range, is here examined in $Ni_{25}Ti_{50}Cu_{25}$ ribbons, obtained by melt spinning. In this case one single thermoelastic martensitic transformation, from the cubic B2 to the orthorhombic B19 phase, is present with a rather small hysteresis cycle. It seemed appealing also to investigate the ER variation as a function of deformation during a thermal cycle between B2 and B19 under a constant applied stress, a condition that takes place in the extrinsic TWME.

EXPERIMENTAL

$Ni_{25}Ti_{50}Cu_{25}$ ribbons obtained by melt spinning (width=1.76±0.02 mm; thickness=32.7±1.3 μm), with typical specimen gauge lengths $\approx$ 120 mm, were used. The specimens were submitted to a thermal treatment of 500 °C for 300 s in air, in order to recrystallize the amorphous ribbons.
Tensile stress was applied by a Zwick 1445 testing machine (load cell 100N) at constant crosshead speed (0.5 mm/min). The specimen elongation was measured *via* the crosshead travel monitor with 1 μm resolution during tensile tests and with an external extensometer with 0.2 μm resolution during constant load tests. The constant stress states investigated were 3.3, 25, 50, 75, 100, 125, 150, 175 and 200 MPa.

The electrical resistance was synchronously detected during the mechanical tests, by the conventional four terminal DC method, using a microohmmeter Burster Resistomat type 2305.
The test temperature ranged from 10 °C to 90 °C under constant stress tests; during isothermal tests the temperature was kept constant using a circulating liquid thermostatic chamber equipped with a Lauda RC20 thermostat.

RESULTS

The Electrical Resistance (ER) vs. Temperature (T) is plotted in fig. 1, where one single transformation appears: the B2↔B19 transformation shows a small hysteresis cycle, approximately 5°C, with an appreciable variation of electrical resistance induced by the band structure modifications.
The transformation temperatures from fig. 1 are: M_f=52.1, M_s=66.3, A_s=56.2, A_f=69.5 °C. It can be seen that at T=50°C, on cooling, the martensitic orthorhombic phase is settled.
Open circles on the picture show

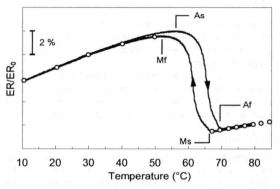

Figure 1: ER/ER_0 vs. T with no stress applied.

the temperatures selected to perform the tests, either in martensite or in parent phase.

Results in martensite (10 °C < T < 50 °C)
The stress(σ) - strain(ε) curves, in the temperature range 10 ÷ 50 °C, with the strain confined to the variant reorientation range, are plotted in figure 2a): the critical reorientation stress lowers with increasing temperature, as expected. All the pseudoplastic deformation was entirely recovered on heating to 75°C.
The ER/ER_0 vs. ε curves, detected at the same time as σ vs. ε, are given in figure 2b), where ER_0 is the ER at start when ε=0. A nice linear relationship can be appreciated all over the applied load region; the slope slightly modifies during unloading.

Results in parent phase (70 °C < T < 85 °C)
The σ vs. ε curves, for T>A_f in the temperature range 70 ÷ 85 °C, where martensite can be stress induced, are given in figure 3a): the pseudoelastic behaviour clearly appears with a hysteresis cycle width which modifies with the test temperature. The transformation strain is completely recovered on unloading.
The correspondent ER/ER_0 vs. ε curves, detected at the same time, are given in figure 3b), where, here also, ER_0 is the ER at ε=0. All the curves exhibit in the elastic deformation range a very small dependence upon deformation; when the stress induced transformation sets in, a linear law with a higher slope is found both on loading and unloading, though shifted on the deformation scale of approximately $\Delta\varepsilon$=0.1%. The shift seems to be due to a different elastic strain contribution during the direct P→M transformation in comparison to the reverse M→P.

Results under constant load
As above specified, the specimens were submitted to thermal cycles between 30°C and 90°C across the P↔M transformation under the following stress states: 3.3, 25, 50, 75, 100, 125, 150, 175 and 200 MPa.

Both deformation and ER, detected as a function of temperature, are plotted in fig. 4; here also the ER is given normalised to the ER_0. It can be appreciated how both ER and ε show a quite similar trend, chiefly in the transformation temperature range.

As a consequence, when considering the ER/ER_0 as a function of ε, it is not surprising to find out a linear relationship all over the transformation range with practically no hysteresis, as shown in fig. 5, except for stresses lower than 50 MPa.

DISCUSSION

Results in martensite

The stress-strain behaviour (see fig. 2a) follows the expected trend: the critical stress for reorientation increases with decreasing temperatures. The linear relationship between ER and ε is impressive (see fig. 2b); its slope is quite higher than expected from the deformation process itself with a fairly small dependence upon temperature. This behaviour proves that the rearrangement of martensite variants takes place as soon as the load is applied, at the very start of the deformation. A similar behaviour has already been found in literature[2] in NiTi alloys in martensite with a slope coefficient of the same order. In the case here examined less than half of the total ER variation is connected with the mere geometric factor which can be evaluated from the Poisson's ratio. A considerable contribution to the ER variation, higher than half of the total, is connected with the electronic contribution which modifies with the fraction of oriented

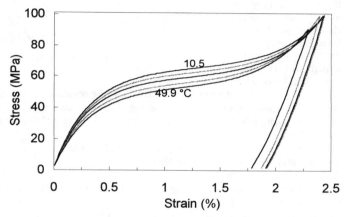

Figure 2a: σ vs. ε in martensite.

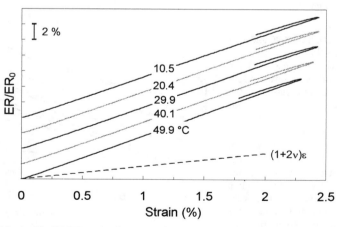

Figure 2b: ER/ER_0 vs. ε in martensite. Each curve is shifted on the vertical axis. The dashed line represents the geometrical contribution.

variants. The expected ER variation is:

$$\frac{\Delta ER}{ER_0} = \frac{\Delta\rho}{\rho_0} + (1+2\nu)\varepsilon$$

The experimental data suggest $\Delta\rho/\rho_0 = \alpha\,\varepsilon$, with $\alpha=3.9$; the reoriented variants exhibit a higher resistivity than the accommodated variants.

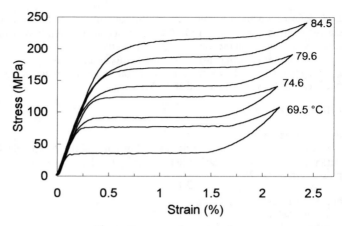

Figure 3a: σ vs. ε in parent phase.

Results in parent phase

Stress-strain curves show the pseudoelastic hysteresis cycle (fig. 3a), as expected, though with an elastic range which increases with increasing temperature both in the direct and the reverse transformation.

Beyond the elastic strain range, the ER/ER_0 increases linearly with ε with a rate of the order of 10.3, meaning that the variation of the normalized electrical resistance is ten times the deformation. Here also:

$$\Delta\,ER/ER_0 = \Delta\rho/\rho_0 + (1+2\nu)\varepsilon = (\rho_M - \rho_P)/\rho_P + (1+2\nu)\varepsilon$$

Experimental results point out here also:

$$\Delta\rho/\rho_0 = (\rho_M - \rho_P)/\rho_P = \beta\varepsilon$$

with $\beta=8.4$, supporting the reasonable hypothesis that the transformed fraction is linearly related to the transformation strain.

The hysteresis shown on the ER/ER_0 vs. ε curves between the direct and the reverse transformation stems from the difference of $\Delta\varepsilon = 0.1\%$ seen in the σ-ε curves in the elastic deformation range.

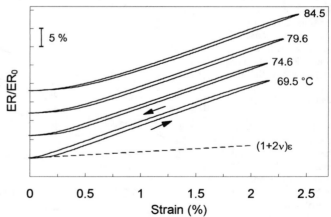

Results under constant load

The ER/ER_0 vs. ε curves at low stresses are different from the curves for $\sigma > 50$ MPa: that is especially true for the curve detected at $\sigma = 3.3$ MPa, a low stress level needed just to keep the ribbon straight. In this case one can appreciate the great ER variation due to the band structure

Figure 3b: ER/ER_0 vs. ε in parent phase.
The dashed line represents the geometrical contribution.

modification as well shown in fig. 3b. For σ=25 MPa and partially for σ=50 MPa, the applied stress does not succeed on cooling to obtain the martensite fully oriented, which arrives for higher stress level. The ER/ER$_0$ vs. ε curves are strictly linear up to a strain of approximately 1.45% and for higher stresses the slope slightly increase. The most astonishing feature is the constant slope shown through the transformation range, although an overall temperature shift is present induced by stress, as shown in fig. 4. Here also the slope is of the order of 10, in agreement with the value found in the pseudoelastic region.

CONCLUSIONS

In the NiTiCu specimens here investigated, the ER variation with ε either in martensite under increasing stress, or in the pseudoelastic region at constant temperature or under a constant stress state while crossing the transformation range, follow linear relationships with strain though with very different slope rates.

In martensitic phase a constant slope value around a value of 5.4 is found for all investigated temperatures. This result points out the relevant role of oriented variants which exhibit a higher resis-

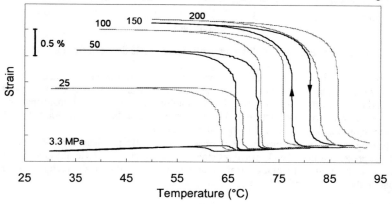

Figure 4a: ε vs. T under different applied stresses.

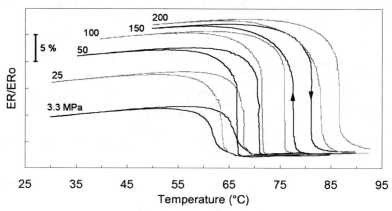

Figure 4b: ER/ER$_0$ vs. T under different applied stresses.

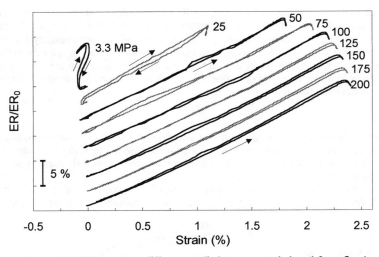

Figure 5: ER/ER$_0$ vs. ε at different applied stresses as deduced from fig. 4

tivity contribution with respect to the average contribution of accommodated variants.

As it concerns the pseudoelastic range, the ER variation with strain is connected both with the resistivity variation due to the electronic contribution modification and to the transformation strain which linearly increases with the fraction of the transformed phase: both contributions linearly add with the same sign to induce a high total rate with a figure of 10.

Similar appealing results are obtained in the extrinsic two way effect, at least for $\sigma > 50$ MPa.

At the light of present results, the electrical resistance variation in the NiTiCu alloys looks particularly appealing to be used as a strain control variable in actuators, due to the high rate of modification found in the transformation range.

ACKNOWLEDGEMENTS

The research has partially been supported by Istituto Nazionale per la Fisica della Materia (I.N.F.M.). NiTiCu specimens have been kindly supplied by Prof. A. Shelyakov.

REFERENCES

1) G. Airoldi, T. Ranucci, G. Riva, A. Sciacca, J. Phys.:Condens. Matter **7**, 3709 (1995).
2) G. Airoldi, T. Ranucci, G. Riva, J.de Physique IV, **C4**-439, 439 (1991)
3) G. Airoldi, T. Ranucci, G. Riva, in Shape-Memory Materials and Phenomena-Fundamental Aspects and Applications, edited by C.T. Liu, H. Kunsmann, K. Otsuka, M. Wuttig (Mater.Res.Soc.Proc. 246, Pittsburgh, PA, 1992), pp.277-291
4) G. Airoldi, T. Ranucci, G. Riva, Trans. Mat. Res. Soc. Jpn. **18B**,1109 (1994)
5) N.M. Matveeva, Yu.K. Kovneristyi, Yu.A. Bykovskii, A.V. Shelyakov, O.V. Kostyanaya, Russ. Metall. **4,** 166 (1989)

SHAPE MEMORY THIN FILMS OF THE SYSTEM Ti-(Ni-Pd-Cu)

E. QUANDT, H. HOLLECK
Forschungszentrum Karlsruhe GmbH, Institut für Materialforschung I, D-76021 Karlsruhe, Germany

ABSTRACT

Free-standing shape memory thin films of the system Ti-Ni-Pd-Cu exhibiting the two-way-effect have been fabricated by d.c. magnetron sputtering onto unheated substrates followed by annealing and training processes. Their transformation temperatures were investigated by differential scanning calorimetry and electrical resistivity measurements. By the Ni-Pd substitution the transformation temperatures (austenite/martinsite finish temperature; A_f/M_f) could be varied between 32°C/-38°C for the binary TiNi films to 570°C/498°C for the binary TiPd films. By the Ni-Cu substitution the transformation hysteresis (A_f/M_f) could be reduced from 70°C for the binary film to 20°C for films with 10 at % Cu. A similar behavior was observed for TiNiPdCu films.

INTRODUCTION

Shape memory thin films present a promising approach to realize new actuators that are in particular important for the advanced development of improved microsystems. As the traditional concept of down-scaling of successful macroscopic actuators is limited in size-reduction as their physical properties do not scale linearly with the dimension, smart materials like shape memory, piezoelectric, and magnetostrictive materials, which directly transduce electrical into mechanical energy, are an attractive alternative. Realizing actuators based on these smart materials, approaches using thin film techniques [1] are of special importance due to their high compatibility to microsystem process technology. Comparing shape memory with piezoelectric and magnetostrictive thin films, shape memory alloys show the highest work output resulting in high output forces combined with large motions, but due to their thermal actuation their applications are limited to the low frequency range. Using the binary TiNi films a further limitation of the maximum ambient temperature is given by the low maximum transformation temperature lower 60°C/140°C (M_f, A_f, respectively) [2] being obtained for Ti-rich films.

Therefore the development of new TiNi-based shape memory films is concentrated on an increase of the transformation temperatures and a decrease of the hysteresis as both features effect the cooling time which is responsible for the maximum frequency of these actuators. Within this work the influence of the Ni/Pd- and Ni/Cu-substitution on the transformation temperatures and their hysteresis was investigated.

EXPERIMENTAL

The shape memory thin films were d.c. magnetron sputtered (target diameter: 75 mm, sputtering power: 300..500 W, Ar sputtering pressure: 0,4 Pa) onto unheated SiO_2 substrates using hot pressed TiNiPdCu targets with a Ti content of 54 or 55 at % while the Ni and Pd contents were varied between 0 and 46 at % and the Cu content between 0 and 25 at %. The

Mat. Res. Soc. Symp. Proc. Vol. 459 © 1997 Materials Research Society

amorphous films were peeled-off the substrates, then crystallized at 750°C for 1 h and trained for the two-way-effect at 450°C for 1 h in a high vacuum furnace [3]. The composition of the films was determined by wave-length dispersive X-ray microanalysis (WDX), depth profiles of the composition were measured by Auger electron spectroscopy (AES). The crystallographic structure of the films were investigated by X-ray diffraction. The transformation temperatures were determined by differential scanning calorimetry (DSC) or electric resistance 4-point measurements. Double beam thin film actuator devices were realized by a shadow-mask sputtering and were used for beam-bending experiments as a function of the resistive heating current.

RESULTS AND DISCUSSION

Composition and microstructure

By X-ray diffraction the phase distribution of and lattice constants at room temperature of the Ti-Ni-Pd shape memory films, were determined using the Rietveld method. By TEM investigations the main grain size was found to be between 1 and 2 μm with a high number of precipitates both inside the grain and along the grain boundaries, whereas the type of the precipitates in the films depend on their Pd content. Binary Ni-rich TiNi films showed a high number of Ti_3Ni_4 precipitates (X-phase) which work due to their orientation as an inner bias spring [4]. These precipitates could not be found in films with Ni/Pd substitution although these films could be trained for the two-way-effect by the same stress/temperature process. The main precipitates in the Pd-containing shape memory films were found to be of the type (Ti_2Ni) for the lower Pd contents and of the type ($PdTi_2$) for the higher Pd contents. A more detailed description of the microstructure of shape memory films of the system Ti-Ni-Pd is given in ref. [5]. In the case of films of the system Ti-Ni-Cu again precipitates of the type (Ti_2Ni) and additionally of the type ($CuTi_3$) and (Cu_2Ti) were identified (fig. 1).

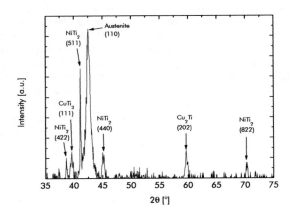

Fig. 1. X-ray diffraction peaks of a TiNiCu film exhibiting the two-way shape memory effect.

Transformation temperature and hysteresis

Thermal characterization of the binary TiNi films revealed the presence of an intermediate R-phase transformation (fig. 2) which was found to be the dominating phase transformation upon cooling. Films exhibiting these R-Phase transformations are of special interest due to their small hysteresis, which FWHM was found to be only 1 K [5]. Unfortunately, the Ni-rich composition required for the existence of the R-phase transformation limits the transformation temperature A_f to about 70°C. Attempts to increase this temperature by Ni/Pd substitution failed as samples containing Pd did not show this intermediate R-phase transformation at all (fig. 3).

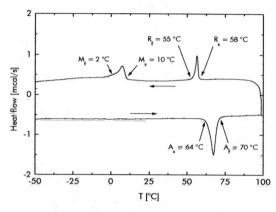

Fig. 2: Cyclic thermal characterization (A_s, A_f, R_s, R_f, M_s, M_f: austenite, rhombohedral, martensite start and finish temperature, respectively) of a TiNi film by DSC.

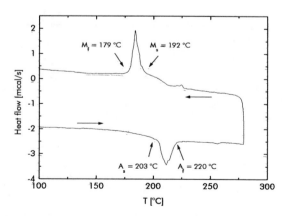

Fig. 3: Cyclic thermal characterization (A_s, A_f, M_s, M_f: austenite and martensite start and finish temperature, respectively) of a TiNiPd film by DSC.

Investigating the effect of the Ni/Pd substitution, two different NiNiPd compositions were compared to the binary TiNi and TiPd samples. Figure 4 shows the measured austenite finish temperatures as a function of the Pd content in comparison to the reported data of bulk materials [6,7]. These results indicate that for $Ti(Ni_{1-x}Pd_x)$ shape memory films a austenite finish temperature range between approx. 0 and 600 °C can be covered by a suitable adjustment of the Pd content x. Furthermore these thin film results show a good correspondence to bulk materials and, in the case of the TiPd sample, represent the highest transformation temperature obtained for shape memory thin films so far. Looking on the transformation hysteresis, in this case the difference of austenite and martensite finish temperatures (A_f-M_f), partly Ni/Pd substitutions lower the transformation temperature hysteresis (fig. 5) resulting in a minimum hysteresis of about 30 K at a at about 20 % Ni/Pd substitution.

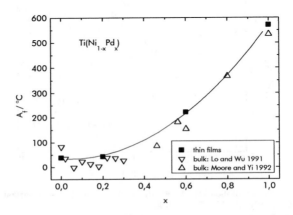

Fig. 4: Comparison of the austenite finish temperature A_f of bulk [6,7] and thin film $Ti(Ni_{1-x}Pd_x)$-materials.

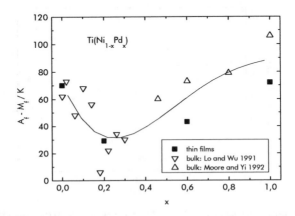

Fig. 4: Comparison of the transformation hysteresis (difference of austenite and martensite finish temperatures) (A_f-M_f) of bulk [6,7] and thin film $Ti(Ni_{1-x}Pd_x)$-materials.

By Ni/Cu substitution it was possible to reduce the transformation hysteresis. In case of the ternary system Ti-Ni-Cu this substitution almost effects only the martensite finish temperatures which are raised by increasing the Cu-content. Again the ternary films show no intermediate R-phase transition as the DSC data clearly indicate (fig. 6). The martensite and austenite finish temperature M_f, A_f, respectively as well as the hysteresis A_f-M_f are shown in figure 7 as a function of the Cu-content while the Ti content was kept constant. The results show that Ni/Cu substitution of about 20 % leads to a reduction of the hysteresis from 70 K to approx. 20 K, which is not further reduced by higher Cu contents. It was found that the Ni/Cu substitution mainly affects the martensite finish temperature which is significantly increased. First results on

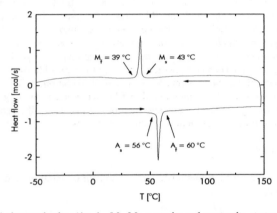

Fig. 6: Cyclic thermal characterization (A_s, A_f, M_s, M_f: austenite and martensite start and finish temperature, respectively) of a TiNiCu film by DSC.

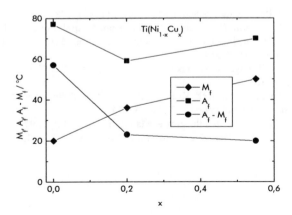

Fig. 7: Martensite finish (M_f) and austenite finish (A_f) temperatures as well as the transformation hysteresis (A_f-M_f) as a function of the Cu content in Ti(Ni$_{1-x}$Cu$_x$) shape memory thin films exhibiting the two-way-effect.

the TiNiPdCu films indicate that for these quaternary films the Ni/Cu substitution has the same effect on the hysteresis.

CONCLUSIONS

By Ni/Pd and/or Ni/Cu substitutions it was possible to adjust both the transformation temperature and the hysteresis. Intermediate R-phase transitions that show a very narrow hysteresis could only be obtained in binary TiNi films. Nevertheless, it was possible to reduce the hysteresis (A_f-M_f) by Ni/Cu substitution from 70 to 20 K, which almost equals the minimum R-phase hysteresis of 15 K. It was found that the Cu content affects the hysteresis by increasing the martensite finish temperature and the Pd content is responsible for the transformation temperature in general. As a result, it is possible to engineer shape memory thin films that meet a wide variety of requirements concerning transformation temperatures and hysteresis.

Future developments may focus on the optimization of deposition and annealing techniques, adjusting the materials composition (incorporation of additional alloying elements), to reduce the sensitivity of the shape memory films against oxidation, defining and optimization of the long-time behaviour of shape memory thin film actuators, modelling shape and size of these actuators, and finally to combine shape memory films with other smart materials and protective coatings resulting in intelligent thin film components.

ACKNOWLEDGEMENTS

The authors wish to thank Annette Müller for the film preparation and Dr. Klaus Feit for the DSC measurements.

REFERENCES

[1]   E. Quandt, H. Holleck, Microsystem Technology 1 (1995), 178.
[2]   J.D. Busch, A.D. Johnson, C.H. Lee, D.A. Stevenson, J. Appl. Phys. 68 (1990), 6224.
[3]   H. Holleck, S. Kirchner, E. Quandt, P. Schloßmacher, Proc. Actuator 94, Bremen, Germany, 15.-17.6.1994, p. 361.
[4]   K. Kuribayashi, T. Tanaguchi, O. Yositake, S. Ogawa, Mat. Res. Soc. Symp. Proc., Vol. 276, 1990, p. 167.
[5]   E. Quandt, C. Halene, H. Holleck, K. Feit, M. Kohl, P. Schloßmacher, A. Skokan, K.D. Skrobanek, Sensors and Actuators A53 (1996), 434.
[6]   Y.C. Lo, S.K. Wu, Scripta Metallurgica et Materialia 26 (1992), 1097.
[7]   J.J. Moore, H.C. Yi, Mat. Res. Soc. Symp. Proc. Vol. 246, 1992, p. 331.

FREQUENCY EFFECTS ON THE DAMPING CAPACITY OF SUPERELASTIC NITINOL

H. N. JONES
Physical Metallurgy Branch
Materials Science and Technology Division
Naval Research Laboratory, Washington DC 20375

ABSTRACT

The evolution of the stress-strain behavior of a superelastic NiTi alloy over a range of cyclic frequencies up to 10 Hz is presented. Due to the stress-strain hysteresis typical of these alloys work is dissipated as heat during every cycle. Also, the heat release and absorption induced by the cyclic phase transformation responsible for the superelastic effect causes the material to undergo a cyclic temperature history which is superimposed on the average temperature resulting from the dissipative nature of the stress-strain cycle. As the cyclic frequency is increased the imposed strain rate is also necessarily increased. The stress-strain loops at very low frequencies are under isothermal conditions. As the cyclic rate is increased nearly adiabatic conditions are achieved at the higher frequencies. This affects the hysteretic cyclic behavior and the damping capacity of the material.

INTRODUCTION

Of the various uses of shape memory alloys either proposed or implemented it is the employment of these materials for shock mitigation and vibration damping that has received some attention lately [1]. The potential advantages of these materials over others used for these purposes is the capacity for damping large displacements, resistance to ionizing radiation environments, and fire resistance. These advantages are important in some applications when compared to the properties of elastomeric dampers. The cyclic behavior of NiTi has been investigated extensively with most of the work done at relatively low frequencies under essentially isothermal conditions [2]. Some work at significant cyclic strains and imposed strain rates approaching $0.067 \ s^{-1}$ has been done at frequencies up to 1.6 Hz as reported in [3]. Exploring the higher frequencies at sufficiently large strains where the frictional aspect of the cyclic phase transformation is the predominant energy loss mechanism entails considerably more difficult experiments where the fatigue behavior of the test specimen becomes important [4,5]. This paper will be concerned primarily with the measurements of the mechanical behavior of NiTi in tension at cyclic frequencies in the range of 0.1 Hz to 20 Hz at sufficiently high cycle counts that steady state conditions have been achieved.

MATERIAL

Annealed polycrystalline NiTi having an austenite finish temperature (A_f) of 24 deg C was obtained in the form of 2.62 mm diameter wire from Shape Memory Applications, Inc. To condition the material so that a significant hysteresis loop could

Mat. Res. Soc. Symp. Proc. Vol. 459 © 1997 Materials Research Society

develop under cyclic loading the wire was first deformed in tension to a strain of nearly 0.06 to induce a significant volume fraction of the martensitic phase. A typical first cycle stress-strain curve for this material is shown in Fig. 1. It is this large first cycle hysteresis loop and the almost complete recovery of the strain after removal of the load that makes this material attractive for shock mitigation purposes. This behavior is strongly dependent on the ambient temperature and to a lesser extent on the imposed strain rate. After this initial straining the alloy will retain some residual martensite even after release of the load and, as a consequence, will exhibit a nonlinear stress-strain behavior on subsequent cycles. Without this conditioning no measurable hysteresis, with the exception of a very small thermoelastic component [6], can be observed as the material will behave in a completely reversible linear elastic manner at stress levels lower than 500 MPa.

EXPERIMENTAL TECHNIQUES

A servohydraulic load frame (Instron Model 1332) having a force capacity of 250 kN was utilized for the experiments. Since the maximum load that could be sustained by the wire was only around 4 kN this necessitated the insertion of a lower capacity (22.2 kN) load cell into the load train to provide an adequate load signal as shown in Fig. 2. The wire test specimen was gripped with a pair of 22 kN capacity side entry hydraulic wedge grips. In experiments where the test specimen accumulates a large number of cycles as in the tests reported here the conditions at the gripped ends become an important

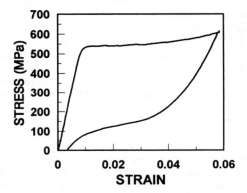

Figure 1. First cycle hysteresis loop for material used in the experiments at 25 C at an average imposed strain rate of $1.3 \times 10^{-3} s^{-1}$.

issue due to the possibility of fatigue failure at the stress concentration induced by the grip. Attempting to grip the wire directly would lead to the expected fatigue failure within a few hundred cycles. To alleviate this problem the gripped ends of the wire were encased within copper tubes. The grip pressure was then adjusted to crimp the tubes onto the wire without the wedge serrations cutting through and contacting the shape memory alloy. This technique provides a smooth transition within the grip that, for the most part, prevented early fatigue failures. Fig. 3 shows a test specimen with a copper tube crimped onto the gripped end along side a button headed test specimen design that had to be abandoned due to fatigue failures in the button shoulder at low cycle counts.

For measurement of strain a 12.7 mm gage length extensometer was attached to wire as shown in Fig. 4. This allows direct and accurate measurement of the hysteresis loops associated with the energy dissipation process responsible for the damping properties of these materials. In this configuration, which is near the minimum at which an extensometer could be mounted on the wire, there is about 50 mm of wire between grips. The maximum cyclic strain that could be applied to the specimen under load control with a 10 Hz sinusoidal wave form under these circumstances was 0.0086. This limitation was dictated by the dynamic characteristics of the

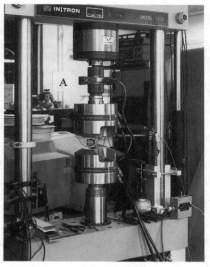

Figure 2. Photograph showing the overall arrangement of the servohydraulic load frame used to conduct the experiments. The 22 kN load cell is shown at "A".

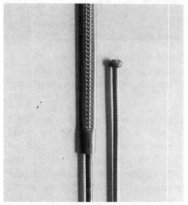

Figure 3. Photograph showing test specimen designs. The left hand example has a crimped copper tube and the right hand example has a button head end which had poor fatigue properties.

load frame at this frequency and the requirement that the wire not be put under compressive loads when the function generator was halted. The span and set point on the servovalve controller were adjusted to maximize the cyclic strain but still avoid buckling the wire at the end of a test.

Temperature measurements were made using a Type K thermocouple fabricated from 0.127 mm wires spot welded to the test specimen. This thermocouple was not beaded, but used the shape memory alloy as an intermediate metal as was done in [2]. This technique provides a very responsive temperature signal and is necessary to avoid the temperature lag that occurs when a separately beaded junction is used.

Data collection during an experiment was accomplished using two separate systems. The thermocouple signal was conditioned and recorded using a DEC-MINC-23 which incorporates chopper isolated thermocouple amplifiers using 5 Hz low pass filters for noise control since the temperature changes in these experiments are normally only a few degrees C different from ambient. To capture the hysteresis loop a MASSCOMP MC-5500 was utilized. This high speed data acquisition was operated at sweep rates of up to 5000 s^{-1} depending on the cyclic frequency. A sweep to sample a hysteresis loop was triggered by the sync signal from the function generator.

OBSERVATIONS

To evaluate the stress-strain behavior of this material under the nearly adiabatic conditions that would be encountered at high frequencies or impact conditions an experiment was conducted under strain control where a total strain of 0.06 was applied to the specimen over a time period of 0.16 s. The strain history

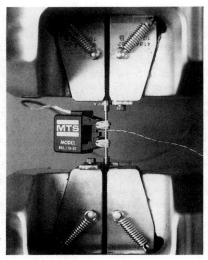

Figure 4. Photograph showing a test specimen clamped in the grips with the extensometer and thermocouple attached.

during this test showed that the peak rate attained during the latter part of the stress-strain curve where the hydraulic ram had accelerated to its peak velocity reached 1.2 s^{-1}. The heat released during this virtually adiabatic deformation raised the temperature of the material within the gage length of the extensometer 28 deg C above ambient. Assuming that the heat capacity of both the austenite and the martensite is 0.5 J/g [7] this implies that 14 J/g of heat energy was generated within the material during this deformation. It is interesting to note that the total "plastic" work done on the material in this test was only 4.5 J/g. As is typical, but not often commented on, the heat evolved during adiabatic deformation of a shape memory alloy at ambient temperatures just above A_f exceeds the amount of work done on it.

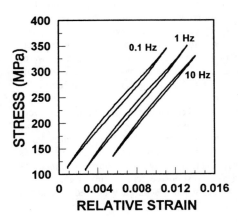

Figure 5. Hysteresis loops at 0.1, 1 and 10 Hz.

This is due to the fact that the imposed strain is being accommodated in part by an exothermic phase transformation.

While the first cycle hysteresis loop shown in Fig. 1 is large the strains associated with this cannot be sustained for a large number cycles without fatigue failure occurring at the grips or at the spot welded thermocouple. Fig. 5 shows the hysteresis loops observed at frequencies of 0.1, 1, and 10 Hz respectively. These hysteresis loops were generated by cycling the material between nearly the same stress levels that were achievable at 10 Hz with the setup shown in Fig. 4. These are relatively small compared to the strain range employed in other work, but are sustainable for large cycle counts without fatigue failure at the grips (>10,000 cycles). Table I lists some of the properties of the hysteresis loops shown in Fig. 5. An example of the temperature and strain histories that are observed in NiTi when cycled at 2 Hz under load control with a sinusoidal wave form is shown in Fig. 6. The temperature oscillates in phase with the imposed load cycles. The initial rise in the average temperature on the first cycle is primarily due to the phase transformation while afterwards it is maintained by the stress-strain hysteresis. At low frequencies, such as the 0.1 Hz case described in Table I, the average temperature tends to drift down with cycle count as the heat generated by the phase transformation during the initial loading to 115 MPa is lost.

Table I. Properties of hysteresis loops shown in Fig. 5 and observed temperatures.

Cyclic Frequency (Hz)	0.1	1	10
Dissipated Energy/Cycle (J/g)	0.0129	0.0117	0.0067
Temperature Oscillation During Cycling and Average($°C$)	23.6-24.2 23.9 Avg.	23.7-24.4 24 Avg.	could not be measured 26 Avg.
Difference From Ambient at Cycling Halt ($°C$)	-0.2	0	1.6
Stress Range (MPa)	115-344	108-350	135-330
Cyclic Strain	0.0102	0.0104	0.0086
Average Strain Rate/Cycle (s^{-1})	0.00204	0.0208	0.172

To achieve a 20 Hz frequency the length of wire between grips had to be reduced to minimize the actuator displacements. This required removal of the extensometer and the reduction of the effective length between grips to 25 mm with 13 mm of this length exposed outside the copper tubes. Without the extensometer the hysteresis loop cannot be measured accurately. In this configuration the predominate heat loss path was into the grips. Temperature rise for 20 Hz cycling between 110-350 MPa at an ambient temperature of 20 deg C was only about 3 deg C.

CONCLUSIONS

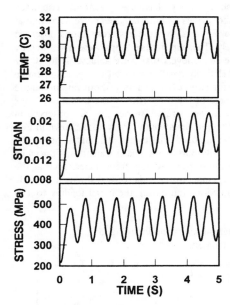

Figure 6. Typical temperature record observed at 2 Hz.

At high cyclic frequencies the area of the hysteresis loop tends to quickly minimize itself as the cycle count increases. As the average temperature of the shape memory alloy wire increases from the dissipation of work into heat the material's apparent stiffness also increases. This reduces the strain range under load control which also results in the loop area being reduced. This is a feedback mechanism which causes a steady state condition to be attained very quickly as the amount of work dissipated as heat is decreased until a balance with the rate of heat loss is achieved. An important consideration in the use of this material in damping applications is this effect and the possibility of fatigue failures at the gripped ends as this was one of the primary problems experienced in the work reported here.

ACKNOWLEDGMENTS

This work was supported by ONR.

REFERENCES

1. J. V. Humbeeck, J. Stoiber, L. Delaey and Rolf Gotthardt, Z. Metallkd. **86**, 176 (1995).

2. S. Miyazaki, T. Imai, Y. Igo and K. Otsuka, Met. Trans. **17A**, 115 (1986).

3. P. G. McCormick, Yinong Liu and S. Miyazaki, Mater. Sci. Eng. **A167**, 51 (1993).

4. R. Abeyaratne, C, Chu and R. D. James, Phil. Mag. A. **73**, 457 (1996).

5. J. L. McNichols, P. C. Brookes and J. S. Cory, J. Appl. Phys. **52**, 744 (1981).

6. V. K. Kinra and K. B. Milligan, J. Appl. Mech. **61**, 71 (1994).

7. P. H. Leo, T. W. Shield and O. P. Bruno, Acta metall. mater. **41**, 2477 (1993).

CORROSION KINETICS OF LASER TREATED NiTi SHAPE MEMORY ALLOY BIOMATERIALS

F. VILLERMAUX *, I. NAKATSUGAWA**, M. TABRIZIAN *, D.L. PIRON ***, M. MEUNIER ****, L'H. YAHIA *

* Biomaterials-Biomechanics Research Group, Institute of Biomedical Engineering, Ecole Polytechnique of Montreal, C.P. 6079, Succ centre-ville, Montreal, Quebec, Canada, H3C 3A7
** Institute of Magnesium Technology, 357 rue Franquet, Ste-Foy, Quebec, Canada, G1P 4N7
*** Department of Metallurgy, Ecole Polytechnique of Montreal, C.P. 6079, Succ centre-ville, Montreal, Quebec, Canada, H3C 3A7
**** Department of Engineering Physics, Ecole Polytechnique of Montreal, C.P. 6079, Succ centre-ville, Montreal, Quebec, Canada, H3C 3A7

ABSTRACT

NiTi shape memory alloy presents interesting mechanical properties as surgical implants. However, due to its high amount of Ni which may dissolve and release toxic ions in human fluids, the medical use of this material is a great concern. We have developed a laser treatment which modifies the oxide layer and enhances uniform and localised corrosion resistance of NiTi alloy.

In this paper we further analysed the effect of this treatment with potentiostatic and AC impedance measurements in physiological Hank's solution. We conclude that the laser treatment creates a stable passive film which results in improved corrosion resistance of this alloy.

INTRODUCTION

The use of NiTi shape memory alloy (SMA) as biomaterial is steadily increasing because of its properties of superelasticity and shape memory. NiTi implants in orthopaedics can be used as a substitute to complicated traditional devices. Screws, bolts and cements are replaced by simple apparatus such as staples and nails providing, after shape recovery, compressive stresses which results in acceleration of the healing process[1,2].

In North America, medical application of such an alloy is not recommended by Canadian health organisations. However, scientists have begun to evaluate the biocompatibility of these alloys in simulating human body media and also *in vivo* biocompatibility by animal implantation. Some *in vitro* studies showed that NiTi biocompatibility was close to the one of either Ti, Co-Cr, or stainless steel[3,4]. In addition, some authors also found that NiTi was not toxic for bone contact[5]. However, Assad et al.[6] and Berger-Gorbet et al.[7] revealed that NiTi plates induced *in vitro* a decrease of cell proliferation, and that some NiTi screws presented *in vivo* cytotoxic effect on bone cells as compared to the effect of stainless steel or Ti6Al4V.

Because of the high concentration of Ni in NiTi alloys, clinicians hesitate to use them. Ni could be involved in allergic, toxic and even carcinogenic reactions due to its release by corrosion process in the human fluids[4,8,9]. Even if NiTi generally presents a corrosion level between that of Ti6Al4V and stainless steel[10], it is characterized by a very unreproducible pitting resistance[11,12], which could render this alloy less safe for medical applications.

As the corrosion resistance, and thus the biocompatibility, depend on the surface properties of the alloy, many surface treatments have been considered. Covering the surface with a bioceramic (TiN and CTiN) thin film failed due to the cracking of the coating upon the large deformation due to the superelasticity and memory effect[13]. Oxide films obtained by heating at $900^{0}C$ for 10 minutes produced encouraging results[14] but induced changes in the bulk properties, modifying the transition temperatures of the alloy. Plasma polymerized tetrafluoroethylene coatings enhanced localized corrosion resistance because of the induced surface homogeneity improvement but this coating lacked cohesiveness and could not resist surgeon manipulations[15]. The enhancement of pitting and crevice corrosion resistance was also noticed with N^{+} implantation[16], while this treatment had no effect on the bulk properties[17]. We have developed a laser treatment able to improve uniform and pitting corrosion resistance of NiTi alloy[12]. This improvement was related to the oxide film thickening, N incorporation, and superficial Ni concentration decrease. All these modifications should be also favorable to a biocompatibility improvement.

The aim of this paper is to evaluate and understand the corrosion kinetics changes induced by this laser treatment, before performing cytotoxic tests, in order to assess its reliability for orthopaedic implants. The corrosion behavior of untreated and laser treated NiTi alloy is analyzed by AC impedance spectroscopy[18]. Potentiostatic measurements[19,20] examine the stability of the surface film and the repassivation speed capacity when the surrounding conditions become more aggressive.

EXPERIMENTALS

NiTi (50at% Ni - 50at% Ti) plates specifications and excimer laser treatment set-up and parameters have been described in a previous paper[12]. A 5 cm focal distance cylindrical lens giving a rectangular spot of 0.01x2 cm^2 on the substrate was used. This set-up allowed us to perform the treatment by using a simple scan at an intensity I=1200 mJ/cm^2/pulse with a scan speed V=0.2mm/s.

Corrosion measurements have been carried out at room temperature in aerated Hank's physiological solution of the following composition: NaCl 8 g/l, KCl 0.4 g/l, NaHCO$_3$ 0.35 g/l, KH$_2$PO$_4$ 0.06 g/l, Na$_2$HPO$_4$ 0.0475 g/l and C$_6$H$_{12}$O$_6$ (glucose) 1 g/l. A standard three electrodes set-up was used consisting of 4 cm^2 platinum plates, a saturated calomel electrode (SCE) and the NiTi sample, of which only 1 cm^2 was exposed to the solution.

AC impedance measurements have been performed with potentiostat/frequency response analyzer (Gramry Instrument PC3-150 and CMS-300) after 7 hours and 24 hours of immersion. The frequency ranged from 5.10^3 Hz to 1.10^{-3} Hz and the variations of the potential were controlled within ±5 mV around its corrosion potential E_{cor}.

Potentiostatic measurements have also been performed with the above equipment. After one hour immersion of the sample, its potential was changed from E_{cor} to E=0.1 V (SCE) in stepwise and the anodic current output was measured for 10 minutes.

RESULTS

AC Impedance Tests

Figure 1 shows the Bode plots for laser treated and untreated plates after 24 hours of immersion. Similar behavior was also observed after 7 hours immersion. The spectra suggest that the electrochemical equivalent circuit of this system can be expressed by a model with one relaxation time constant such as Randle's circuit.

Solution resistance R_s, corresponding to the Z modulus plateau at high frequencies, is negligible compared to the order of magnitude of the corrosion resistance of both kinds of samples.

In the case of untreated sample, Z modulus seems to attain a steady value around 2.5 $M\Omega.cm^2$ at 1 mHz, indicating a polarization resistance R_{cor} of 2.5 $M\Omega.cm^2$.

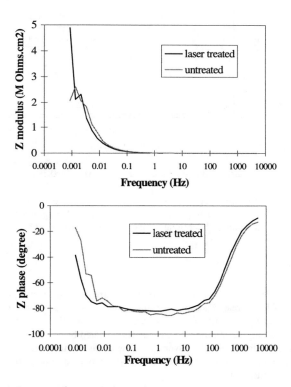

Figure 1. Bode diagram of the laser treated and untreated plates after 24 hours of immersion. Variations of Z modulus (a) and Z phase (b) as a function of the frequency.

With laser treatment, Z modulus reaches more than twice at this frequency, and further increase is expected at lower frequencies. The larger absolute value of Z phase supports this assumption. The measurement below 1 mHz was not conducted because of the small current out put ($\approx nA/cm^2$) and high signal/noise ratio. Then, the exact polarization resistance R_{cor} cannot be evaluated in this analysis, but the improvement by laser treatment is evident.

<u>Potentiostatic Tests</u>

Figure 2 shows the current-time curves of laser treated and untreated specimens observed in response to the potential step of 0.1 V (SCE). This polarization potential is around +200 mV more noble than E_{cor}, but still in the passivation range.

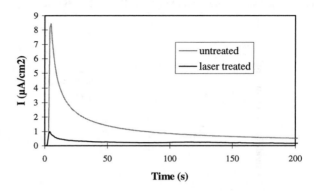

Figure 2. Potentiostatic measurements at E=0.1 V (SCE)

By applying laser treatment, the maximum of initial current decreases to a 10-fold smaller value. Besides, the laser treated sample repassivates 2 orders of magnitude faster than the untreated sample. The laser treatment is beneficial to passivation behavior, which contributes to a higher corrosion resistance of NiTi alloy.

CONCLUSIONS

Laser surface treatment is an effective tool to improve corrosion resistance of a NiTi alloy. The impedance of laser treated sample in Hank's solution increases more than twice compared to untreated sample. The potentiostatic measurements reveals smaller maximum of current and higher repassivation speed, suggesting the stable oxide film created by this treatment. This stability can be related to the thickening of the oxide layer and to the change of the oxide composition[12]. Indeed, the oxide layer promoted by such a treatment consists mainly of titanium oxide which induces a decrease of Ni concentration on the surface.

These results complete our previous study which indicates an improvement of the uniform and localized corrosion resistance of a NiTi alloy[12].

Finally, this laser treatment should enhance the reliability of NiTi alloy for medical use. Cytotoxic tests are being done to validate the corrosion data and these results will be published in a future paper.

ACKNOWLEDGMENTS

The authors would like to thank AMP Développement (France) for providing the NiTi samples.

REFERENCES

1. R.G. Tang, K.R. Dai, Y.Q. Chen and D.W. Shi in The First International Conference on Shape Memory and Superelastic Technologies, edited by A.R. Pelton, D. Hodgson and T. Duerig (SMST Proc., Pacific Grove, CA, 1994) pp.499-503.

2. K.R. Dai, X.H. Sun, R.G. Tang, S.J. Qiu and C. Ni, Injury **24**(10), 651 (1993).

3. J.L.M. Putters, D.M.K.S. Kaulesar Sukul, G.R.de Zeeuv, A. Bijma and P.A. Besselink, Eur. Surg. Res. **24**, 378 (1992).

4. R.S. Dutta, K. Madangopal, H.S. Gadiyar and S. banerjee, Brit. Corr. J. **28**(3), 217 (1993).

5. S.J. Simske and R. Sachdeva, J. Biomed. Mat. Res. **29**, 527 (1995).

6. M. Assad, S. Lombardi, S. Berneche, E.A. Desrosier and L'H. Yahia in The First International Conference on Shape Memory and Superelastic Technologies, edited by A.R. Pelton, D. Hodgson and T. Duerig (SMST Proc., Pacific Grove, CA, 1994) pp 731-736.

7. M. Berger-Gorbet, B. Broxup, C. Rivard and L'H. Yahia, J. Biomed. Mat. Res. **32**, 243 (1996).

8. S. Shabalovskaya, J. Cunnuck, J. Anderegg and B. Sachdeva in The First International Conference on Shape Memory and Superelastic Technologies, edited by A.R. Pelton, D. Hodgson and T. Duerig (SMST Proc., Pacific Grove, CA, 1994) pp 209-214.

9. J.K. Bass, H. Fine and G.J. Cisneros, Am. J. Orthod. Dentof. Orthop. **103**(3), 280 (1993).

10. G. Rondelli, B. Vincentini and A. Cigada, Corr. Sci. **30**(8-9), 805 (1990).

11. Y. Nakayama, T. Yamamuro, Y. Kotoura and M. Oka, Biomaterials **10**, 420 (1989).

12. F. Villermaux, M. Tabrizian, L'H. Yahia, M. Meunier and D. Piron, Appl. Surf. Sci. (in press).

13. K. Endo, R. Sachdeva, Y. Araki and H. Ohno, Dent. Mat. J. **13**(2), 228 (1994).

14. T. Sohmura in The Third World Biomaterial Congress (3rd WBC Proc., Kyoto, Japan, 1988), p574.

15. F. Villermaux, M. Tabrizian, L'H. Yahia, M. Meunier and D. Piron, Bio-Med. Mat. Eng. (in press).

16. S.M. Green, D.M. Grant, J.V. Wood, A. Johanson, E. Johnson, L. Sarholt-Kristyensen, J. Mat. Sci. Lett. **12**, 618 (1993).

17. D.M. Grant, S.M. Green, J.V. Wood, Acta Metall. Mater. **43**(3), 1045 (1995).

18. K.J. Bundy, J. Dillard and R. Luedemann, Biomaterials **14**(7), 529 (1993).

19. P. Sung, A.C. Fraker, in The third World Biomaterials Congress, (3rd WBC Proc., Kyoto, Japan, 1988), p 201.

20. A. Cigada, G. De Santis, A.M. Gatti, G. Rondelli, B. Vicentini, D. Zaffe in Clinical implant Materials, Advances in Biomaterials, Volume 9, edited by G. Heimke, U. Soltész and A.J.C. Lee (Elservier Science Publishers, Amsterdam, 1990) p. 51.

DETERMINATION OF THE UPPER LIMIT FOR THE CLUSTER MOMENT CONTRIBUTING TO THE GIANT MAGNETO RESISTANCE IN LASER DEPOSITED GRANULAR Cu-rich THIN FILMS

V. MADURGA*, R.J. ORTEGA *#, J. VERGARA*# K.V. RAO #

*Departamento de Física, Universidad Pública de Navarra, E-31006 Pamplona, Spain.
#Department of Condensed Matter Physics. Royal Institute of Technology, S-10044 Stockholm, Sweden.

ABSTRACT

We have fabricated granular $Cu_{95}Co_5$ thin films by laser ablation-deposition. Within a regime of annealing temperatures, these samples exhibit Giant Magneto Resistance (GMR), typically 5% in 0.5 Tesla at 5 K. The magnetic hysteresis loops are found to show finite coercive fields in the whole temperature range 2 K - 300 K. Below 9 K, the field dependence of the MR shows a split maximum. We interpret the data in terms of coercivity arising from blocking phenomenon of single domain superparamagnetic Co clusters. A quantitative determination of the upper limit for the cluster moment contributing to GMR is estimated to be 17000 μ_B (a cluster size of 5 nm).

INTRODUCTION

Granular solids exhibiting giant magnetoresistance (GMR) are fabricated by a variety of techniques, both in the form of thin films and bulk materials, including several microns thick melt-spun ductile long ribbons[1-5].

Post fabrication annealing of the as quenched materials produces a fine dispersion of magnetic precipitates in the metallic host matrix. The overall magnetic behaviour of such a material is superparamagnetic, concomitant with the phenomenon of GMR[6-8].

It is thus of considerable interest to study the evolution of the magnetic behaviour of such magnetic granules: the size dependence of the total spin per cluster, the intra- intercluster interactions, as well as the magnetic properties of the metallic host the clusters are embedded into. Clearly, a technique of fabricating materials that enables good control of the size and separation distance of magnetic particles will provide us an insight into the GMR phenomenon which has become a fundamental problem in condensed matter physics.

This work is an attempt to demonstrate the feasibility of producing metallic thin films by pulsed laser ablation of the immiscible elements, wherein a fine dispersion of magnetic particles is made to precipitate by subsequent annealing. A correlation between GMR and the magnetic properties of $Cu_{95}Co_5$ is presented by estimating the distribution of magnetic moments attained by this deposition technique.

EXPERIMENTAL

Thin films, with a nominal composition of $Cu_{95}Co_5$, have been fabricated by pulsed laser deposition technique (Nd:YAG laser, l=1024 nm, 20 Hz repetition rate, 4 ns pulse duration, energy 1 mJ per pulse, 2 mm in diameter beam spot size on target). The circular target used consisted of one ($\pi/10$) circular sector of pure Co, and one ($19\pi/10$) circular sector of pure Cu, and was rotated at an angular speed of ($\pi/3$) rad/s. The films were deposited onto ordinary glass substrates at room temperature, under a vacuum of 10^{-6} mbar, and had a total thickness of 100 nm. The samples have been subsequently annealed at different temperatures between 300°C and 400°C for 10 minutes in an inert atmosphere.

Microstructural measurements have been carried out in a X-ray diffractometer, using Cu-K_α radiation, as well as investigating the topography in a Scanning Tunneling Microscope. The magnetic properties have been measured in a Quantum MPMS2 SQUID magnetometer, in the temperature range 2 K - 300 K, in fields up to 1 Tesla. MR measurements have been made by means of an Automatic Resistance Bridge from A.S.L. model F26. 0.3 mA current was flowing

Mat. Res. Soc. Symp. Proc. Vol. 459 © 1997 Materials Research Society

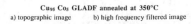

Cu₉₅ Co₅ GLADF annealed at 350°C
a) topographic image   b) high frequency filtered image

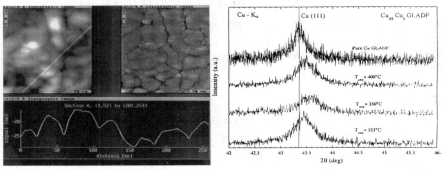

Fig. 1. a): 1. STM topographic image of the 350 C annealed sample. A pyramidal growth 200 nm lateral dimension and 20 nm height; 2. grains 10-50 nm rage of sizes from the h.f.f.i. 3. Profile of a section. b) X-Ray diffraction pattern: a shift of the peaks respect to pure Cu angle is observed.

along the sample 12 mm. long, 1 mm width, and the magnetic field was applied perpendicular to the current and parallel to the sample plane. An electromagnet was used to provide the magnetic field up to 2 Tesla. Measurements at 77 K were made in a maximum field of 1 Tesla. Measurement at temperatures lower than 77 K were made in a Janis cryostate and an Oxford Instrument superconductor magnet provides magnetic field up to 7 Tesla.

RESULTS AND DISCUSSION

As mentioned above, the films were grown at room temperature, thus it is not surprising that we obtain a pyramidal growth, as evidenced by topographic analyses of one the films (annealed at 350°C), in our scanning tunneling microscopy (STM). The pyramids have a basis rather irregular in shape, which can be as large as 200 nm in lateral dimension and 20 nm high. By scanning a flatter, narrower region of the film, we observed smaller ones. A high frequency filtered image of the latter topographic scan enhances the contrast at the boundaries, evidencing grains whose lateral size is in the range 10 - 50 nm, as shown in Fig. 1.a

We performed X-ray diffraction analyses on the annealed thin film samples as well as on a pure Cu thin film, deposited under the same conditions. Fig. 1.b show the patterns, in which only the (111) reflection of Cu could be detected, though in the case of the Cu-Co thin films, it is apparent that at least part of the Co is dissolved in the Cu grains, since the peaks are shifted to a higher Bragg angle. A measure of the FMHW for the sample annealed at 350°C gives, according to the Debye-Scherrer formula, an average grain size of 20 nm, in agreement with the STM observations. No reflections from Co could be detected, suggesting that the size of the Co particles are smaller than the resolution capability of our X-ray diffractometer.

Magnetization loops have been measured at temperatures between 2 K and 300 K as it is displayed in Fig. 2. For all of our samples, these curves show a ferromagnetic contribution as is evidenced by the presence of a remanence and coercitivity that decrease as the temperature of measurement is increased. It is also remarkable that the technical magnetization for all the samples is saturated in magnetic fields smaller than 1 Tesla.

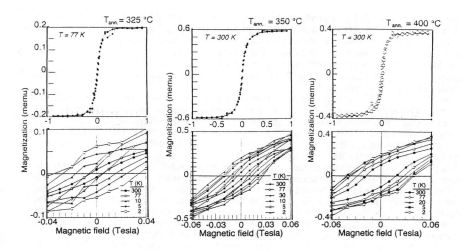

Fig. 2. Magnetization curves for three differently annealed samples. Note that for high temperatures the magnetization is saturated for all of our samples which exhibit coercitivity. The coercive field increases with annealing temperature and, for each sample, its magnitude increases when temperature decreases.

The Magneto-Resistance loops obtained for the same samples are shown in Fig. 3.

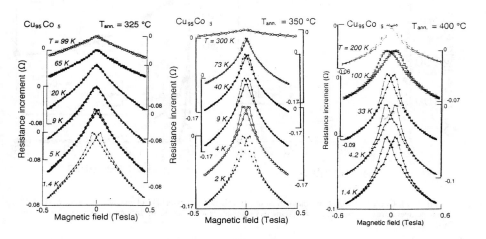

Fig. 3. Magneto-Resistance curves for three differently annealed samples. Note that even at low temperatures the M-R is not saturated for all of our samples. A reversible behaviour is observed for the 325 C annealed sample over 20 K. The coercive field increases with annealing temperature and, for each sample, its magnitude increases when temperature decreases.

For the three annealed samples the MR, defined as $(R(H)-R(0))/R(0)$ is negative and isotropic and reaches 5% for the 400°C annealed sample at 2 K in a magnetic field of 0.5 Tesla. In order to estimate the size of the magnetic clusters, we have focused our attention on the irreversible processes observed in the transport measurements in these samples, which give rise to a coercivity in the MR. On the basis of superparamagnetism (note the non saturated MR phenomenon in magnetic fields up to 7 Tesla. See Fig.5) it is possible to obtain the above parameters for the 325°C annealed sample as described below.

The MR of the 325°C annealed sample, at low temperatures, shows a splitting of the curve into two peaks, characteristic of a blocking process in superparamagnetic particles. The resistance reaches its maximum value in a certain magnetic field, the Magneto-Resistive Coercive Field, MR-Hc. As we increase the temperature, the MR-Hc decreases its value and eventually vanishes at 10 K, which we consider as the blocking temperature of the largest superparamagnetic particles that participate in the Magneto-Resistive process.

For the 325°C annealed sample, the evolution with temperature of both coercivities, the one obtained from the MR loops and the one from the magnetization measurement, is plotted in Fig. 4.a. The sharp increase in both coercivities at low temperatures can be atributed to the blocking process of the superparamagnetic particles. Above this temperature, the MR-Hc is zero, meanwhile the coercive field, measured from the hysteretic magnetic loops, remains more or less a constant. This weak temperature dependent part of the coercivity seems to arise from the larger ferromagnetic particles which do not participate in the Magneto-Resistance phenomenon.

In Fig. 4.b, we show the influence of the heat treatments in the values of the MR-Hc. All cases show an increase of this parameter when temperature decreases, which can be explaned by the presence of superparamagnetic particles in the three samples; this fact is in agreement with the non saturated MR in a 7 Tesla magnetic field. The increase of the size and number of the ferromagnetic particles, produced by annealing, gives rise to an increase in the MR-Hc values.

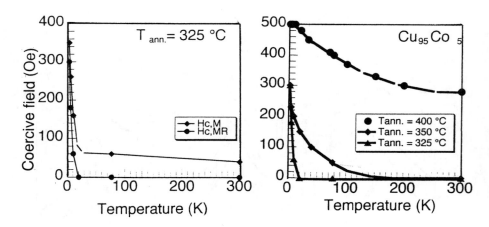

Fig.4. a). Coercive field of the magnetization, Hc,M, and coercive field of magneto resistance, Hc,MR, versus temperature. A split of this two parameters takes place above 20 K. b). Influence of the heat treatments in the Hc,MR.

From the data represented in Fig. 4.a is possible to estimate the magnetic moment of the largest superparamagnetic particles that enter in the Magneto-Resistance process. We have taken 20 K to be the maximum blocking temperature of the superparamagnetic particles in the 325°C annealed sample. Assuming both, an uniaxial anisotropy for the samples and a relaxation process according to the Arrhenius Law, the following relation holds[9]: s:

$$K_u V = 25 k_B T_b \qquad (1)$$

where K_u is the uniaxial anisotropy energy density, V is the volume of the largest magnetic cluster, k_B is the Boltmmann constant and T_b the blocking temperature for the previous clusters.

On the same basis, the temperature dependence of the coercive field for a system of single domain, single size, uniaxial particles is[10]:

$$Hc(T) = (0.96 K_u / Ms)[1 - (T/T_b)^{0.77}] \qquad (2)$$

Taking into account (1) and also the extrapolated value of MR-Hc at 0 K, which has been taken as 300 Oe, and from (2), the magnetic moment of the largest superparamagnetic clusters may be obtained from the expression:

$$m = 0.96 \cdot 25 k_B T_b / Hc(T=0K) \qquad (3)$$

For the 325°C annealed sample this value is found to be about 17000 m$_B$, which corresponds to a 5 nm size Co cluster (assuming fcc structure and 3.54 Å as lattice parameter) within the superparamagnetic regime.

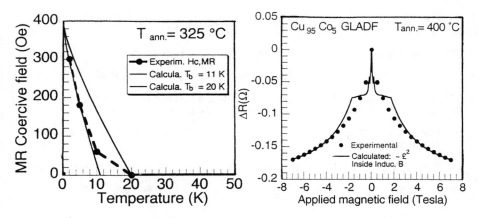

Fig.5. Experimental and calculated (using eq. 2) values of the Hc,MR, versus temperature. Two blocking temperatures limits for the clusters magnetic moments can be assumed.

Fig. 6. Experimental and calculated values of MR using the Langevin function for a magnetic moment of 14000 μ B and the inner ferromagnetic induction from the values of Fig. 2.

The 400°C annealed sample shows a maximum value of the MR although it is not saturated in a 7 Tesla magnetic field; this behaviour suggests the presence of superparamagnetic particles in this sample. These superparamagnetic particles involved in this process can not produce the observed high value of the MR-Hc because they would be much bigger than the observed sizes from structural measurements. One explanation of such behaviour can be given if one assume that the superparamagnetic particles and the ferromagnetic ones act coupled, giving rise to: a) high values of MR for low applied magnetic field -high ferromagnetic inner induction-; b) high coercive field for MR, for the same reason and c) high saturating magnetic field due to the superparamagnetic entities. In Fig. 6 we have represented the result of this assumption.

A complete comprehensive discussion of the results should take into account the mechanisms that give rise to the MR, as well as the consequence of interactions in the intermediate regime between superparamagnetism and local ferromagnetic character of the larger multidomain entities.

CONCLUSION

Granular thin films of $Cu_{95} Co_5$, fabricated by pulsed laser ablation, are shown to exhibit GMR. At low annealing temperatures, the magnetic loops show hysteresis, with finite coercivities at all temperatures, while the MR loops are reversible only at temperatures above 9 K. In this case, the maximum magnetic moment of Co particles contributing to GMR is estimated to be 17000 m_B, from the temperature dependence of the coercivity. In contrast, the samples annealed at higher temperatures, which exhibit even larger GMR, show hysteresis in both magnetic and MR loops even at 300 K, suggesting contributions from superparamagnetic as well as possible complex multidomain magnetic entities.

ACKNOWLEDGEMENTS

This work has been supported by the Comisión Interministerial para la Ciencia y la Tecnología, MAT94-0964, and by the Gobierno de Navarra. We gratefully acknowledge A. Gromov, RIT, for STM measurements and E.Azociti and S.Martínez for help to MR measurements.

REFERENCES

1. A.E. Berkowitz, J.R. Mitchell, M.J. Carey, A.P. Young, S. Zhang, F.E. Spada, F.T. Parker, A. Hutten and G. Thomas, *Phys. Rev. Lett.* **68** (1992) 3745

2. J.Q. Xiao, J.S. Chien and C.L. Chien, *Phys. Rev. Lett.* **68** (1992) 3749

3. J. Wecker, R. von Helmolt, L. Schultz and K. Samwer, *Appl. Phys. Lett.* **62** (1993) 1985

4. B. Dieny, A. Chamberod, J.B. Genin, B. Rodmacq, S.R. Teixeira, S. Auffret, P. Gerard, O. Redon, J. Pierre, R. Ferrer and B. Barbara. *J. Magn. Magn. Mater.* **126** (1993) 433

5. V. Madurga, R.J. Ortega, V. Korenivski and K.V. Rao. *March Meeting of the American Physical Society.* Pittsburgh 1994.

6. H. Wan, T. Tsoukatos, G.C. Hadjipanayis, Z.G. Li, J. Liu. *Phys. Rev.* **B49** (1994) 1524

7. B. Dieny, S.R. Teixeira, B. Rodmacq, C. Cowache, S. Auffret, O. Redon, J. Pierre. *J. Magn. Magn. Mater.* **130** (1994) 197

8. V. Madurga, R.J. Ortega, V. Korenivski, H. Medelius and K.V. Rao. *J. Magn. Magn. Mater.* **140-144** (1995)

9. C.P. Bean and J.K. Livingston, *J. Appl. Phys.* **30 Suppl** (1959) 120S
10. B.D. Cullity. *Introduction to Magnetism and Magnetic Materials* (Adison-Wesley, 1972), p. 338.

RADIAL, PLANAR AND HELICAL ANISOTROPIES INDUCED IN CYLINDRICAL AMORPHOUS CoP MULTILAYERS

C.FAVIERES*, C.AROCA* , E.LÓPEZ* , M.C.SÁNCHEZ *, P.SÁNCHEZ** AND V.MADURGA ***.
* Department of Physics of Materials.Universidad Complutense de Madrid.E-28040 Madrid, Spain.
** Department of Physics. ETSIT.Universidad Politécnica de Madrid.E-28040 Madrid, Spain.
*** Department of Physics. Universidad Pública de Navarra. E-31006.Pamplona, Spain.

ABSTRACT

Electrolytic cylindrical amorphous CoP multilayers, grown on copper wires, with controlled continuos variation of the magnetic anisotropy, in both, direction and magnitude, have been produced. The inner zone of the samples has been grown to exhibit radial anisotropy and the overlayer has been obtained with planar longitudinal anisotropy. Our studies have revealed that the magnetization at the surface of the samples is strongly coupled with the bulk anisotropy for thickness of the overlayer of $\approx$ 1.5 µm. Furthermore, it has been found that an angular deformation applied to the samples, when the copper wire is subjected to a torsion, modifies the direction of the longitudinal anisotropy to the helical direction. Also, a large spontaneous Matteucci effect is induced by means of this angular deformation.

INTRODUCTION

There are extensive efforts to control the magnetic properties of samples appropriated as active elements in different kind of microelectronics devices: e.i control of the surfaces anisotropies[1], modification of the giant magnetoresistance[2.] Investigations in this area are being performed in the last years.

The electrolytic technique[3] is process to fabricate, in a not expensive way, samples to be used as nuclei of these devices. In the present work, we show a study of the magnetic anisotropy in electrolytic cylindrical amorphous CoP multilayers, obtained varying the magnitude and the direction of the magnetic anisotropy: radial, planar or helical. The objective to produce these kind of samples was to obtain, in the same sample, two distinguished zones with different easy axis for the magnetization: the underlayer, with radial anisotropy and the overlayer, with in plane anisotropy, and study the magnetization processes monitored as a function of the multilayer thickness. Similar samples were obtained by us in a previous work, but the substrates were copper films [4]. The possibility to control the magnetization processes and the anisotropy in these samples, make them suitable for many applications in microelectronics devices.

Principally we have focus our studies in investigations on the evolution of the magnetic anisotropy direction of these samples and the Matteucci effect, this is, the appearance of a component of the magnetization in the circular direction in a magnetostrictive sample when it is subjected to a torsion and simultaneously a longitudinal magnetic field is applied. Some authors have reported results about spontaneous Matteucci effect [5] and helical induced anisotropies on amorphous wires [6]. We have also studied this effect in a previous work [7].

Mat. Res. Soc. Symp. Proc. Vol. 459 © 1997 Materials Research Society

EXPERIMENTAL

Amorphous CoP cylindrical samples were obtained by electrodeposition technique. The electrolyte was prepared according to reference [3]. During the electrolytic process, its temperature was maintained at 83 °C. A Co foil was used as anode and the substrate was a copper wire, rotating around its cylindrical symmetry axis during the deposition. We report here results obtained in wires of 0.6 mm diameter. Before the deposition, the surface of the wires were electrolytically polished to avoid possible inhomogeneous zones. Three types of samples were produced. In all the cases, the length of the samples was 4.5 cm and the substrates were 10 cm length.

First, over copper wires, we have deposited samples using a constant electrical current density of 200 mA·cm^{-2} during 30 minutes. As expected, the samples show a strong perpendicular magnetic anisotropy [8]. The thickness of these single layers samples was estimated observing the cross section of the wire in a microscope, where it can be appreciated the inner copper core and the CoP shell. We estimate a thickness of 25 μm for the amorphous sample.

The second type of samples are multilayers grown over copper wires, using an electrical current density of pulses, +500 mA·cm^{-2} during 65 ms plus -100 mA·cm^{-2} during 60 ms. The use of this electrical current density breaks the columnar growth, a composition modulated multilayer is produced and it appears an in-plane anisotropy [9]. The number of pulses was fixed at 8000. The thickness was measured as in the prior case, obtaining 25 μm.

The third type of samples were obtained in two successive steps: in the first one, a single layer as the first one described before was obtained and, in a second step, a composition modulated multilayer, using the electrical current density of pulses mentioned above, was grown over the surface of this single layer. We produced several samples, varying the number of pulses of the multilayer: 100, 300, 500, 800 and 1000, thus obtaining different thickness. Those were estimated too as in the other cases.

Studies about the domain structure at the surface of the samples and the magnetization processes were performed. We have measured, the inductive hysteresis loops, M_z-H_z, of all the samples and the magnetic domain-configuration by the Bitter technique.

The anisotropy constant, k, of the multilayers, second type of samples, is determined measuring the induced e.m.f in a pick up coil around the sample when it is magnetized by a longitudinal magnetic field, H_z, produced by a Helmholtz system. Simultaneously an alternating electrical current is through the copper wire[10.] In this situation, the magnetization rotates reversibly a small angle out of the axial direction, and the same angle should be rotated for different axial magnetic field when the sample is submitted to different circular magnetic field, H_ϕ produced by the electrical current. By minimizing the total free energy, and keeping constant the magnitude:

$$(\mu_o M_s H_z + 2k)/H_\phi = \text{constant} \tag{1}$$

RESULTS

Observing in Fig.1 the switching curve, M_z-H_z, at 20 Hz, of the first type of samples, it can be appreciated a radial anisotropy. Due to the columnar growth, magnetization rotation are observed, although there are initially irreversible wall displacements corresponding to the small closure domains. The induced anisotropy constant, k, is determined through magnetization work, The value for this radial anisotropy k is found to be ≈ 2200 Jm^{-3}.

The Bitter technique reveals that exist this radial anisotropy and small undulated closure domains with ramifications to reduce the magnetostatic energy. The domain pattern can be seen in Fig.2.

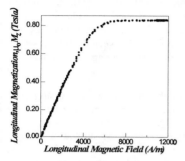

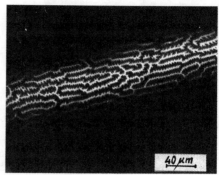

Fig.1. Switching curve of the first type of sample. The value of the anisotropy constant is 2200 J·m^{-3}

Fig.2. Domain structure of the first type of samples, showing stripes domain with small closure domains.

In Fig.3 we show one of the inductive hysteresis loop of the as-obtained multilayers, where it can be appreciated a longitudinal anisotropy, confirming the columnar growth is broken and there is no radial anisotropy. The domain pattern of one of these multilayer is seen in Fig.4, (a) and (b), showing, as expected, longitudinal magnetic domain walls. The value for the k, the anisotropy constant, measured as described above, is found to be ≈ 250 J·m^{-3}.

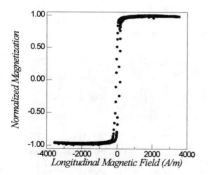

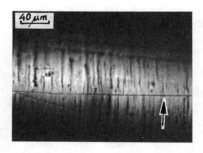

Fig.3.Hysteresis loop of a type two sample.

Fig.4.a.Domain pattern of a type two sample Without magnetic field.

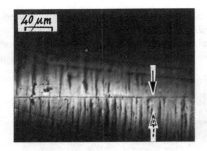

When a multilayer of 100 pulses, ≈ 0.3μm thickness, is deposited over the surface of a single layer, the magnetic structure observed in the surface varies, respect to the monolayer one, reducing the closure domains, although it exists a strong coupling between the magnetization at the surface and the underlayer radial anisotropy. No remarkable differences have been found between the hysteresis loops of the monolayer and this sample.

Fig.4.b.Under the action of a magnetic field ≈ 300 Gauss.

Similar results are obtained when increases the thickness of the multilayer, of ≈ 0.8μm (300 pulses), although, as expected, the effect of the surface anisotropy is here stronger that in the prior case. The closure domain are almost disappeared and the stripes are not clearly noted.

When a multilayer of 100 pulses, ≈ 0.3μm thickness, is deposited over the surface of a single layer, the magnetic structure observed in the surface varies, respect to the monolayer one, reducing the closure domains, although it exists a strong coupling between the magnetization at the surface and the underlayer radial anisotropy. No remarkable differences have been found between the hysteresis loops of the monolayer and this sample. Similar results are obtained when increases the thickness of the multilayer, of ≈ 0.8μm (300 pulses), although, as expected, the effect of the surface anisotropy is here stronger that in the prior case. The closure domain are almost disappeared and the stripes are not clearly noted. When the number of layers deposited over the single layer increases to 500, no perpendicular anisotropy is found at the surface of the sample. Magnetic domains walls can be observed, confirming the decoupling between the magnetization at the surface and the underlayer anisotropy. This decoupling is produced when the thickness of the multilayer is ≈ 1.5 μm. More clear qualitative results are obtained when the multilayer is ≈ 3 μm, corresponding to a 1000 pulses deposited over the monolayer. Fig.5 shows the domain structure at the surface of the sample. As expected, the magnetic bulk behavior is dominated by radial anisotropy; it can be appreciated when measuring the hysteresis loop, M_z-H_z of these samples, due to the small thickness of the multilayers compared with the underlayer.

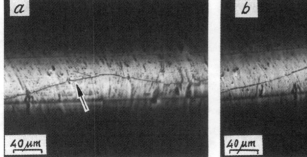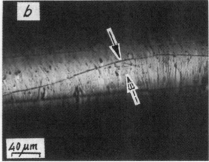

Fig.5. Domain pattern of a sample with a monolayer and a multilayer of 1000 pulses grown over its surface. (a) with zero applied magnetic field (b) With a magnetic field of ≈ 300 Gauss.

The application of an external torsion to the multilayers substrates, changes the direction for the easy axis for the magnetization: due to the magnetostrictive character of the sample, the in plane longitudinal anisotropy evolves to a helical anisotropy, as it is shown in Fig.6.

After a plastic deformation of the copper matrix, the samples exhibit a spontaneous Matteucci effect. Then, applying an ac longitudinal magnetic field with a solenoid that produces $\mu_0H=5,94$ mT/A, it appears an ac voltage generated at the ends of the CoP samples, directly proportional to the circular magnetization. As the value of the torsion applied to the matrix increases, the direction of the anisotropy changes monotonically, from the initial longitudinal one, to the helical one. In Fig.7. we show some of the magnetization curves obtained when the copper wire was twisted in successive steps. When the torsion is 1,6 turns counterclockwise, the ac voltage is maximum, indicating the anisotropy direction is in the helix at 45 deg. If the torsion is applied then clockwise, the ac voltage at the ends of the sample decreases, and finally, increasing the value in this direction, the anisotropy appears at the contrary helix, as shown in Fig.8.

Fig.6. Domain pattern of a twisted sample.(a) magnetic domain wall with zero applied magnetic field (b) With a magnetic field of $\approx$ 300 Gauss.

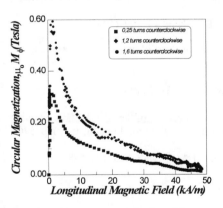
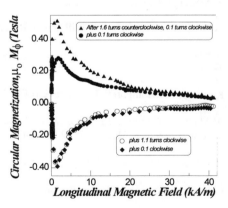

Fig.7. Switching curves, M_ϕ-H_Z, of a multilayer grown over a copper wire. The copper core has been has been twisted in succesive steps, counterclockwise.

Fig.8.Switching curves, M_ϕ-H_Z, of the same sample. The matrix has been twisted in sucessive steps clockwise.

CONCLUSIONS

CoP monolayers and multilayers have been electrolytically grown on copper wire substrates. A radial anisotropy, $k \approx 2200$ J·m^{-3}, is observed on the monolayers. A set of multilayers, with a modulated composition, has been grown over the single layer. The radial anisotropy of the monolayer evolves monotonically to a planar anisotropy as a function of the number of multilayers deposited. Planar, longitudinal anisotropy, $k \approx 250$ Jm^{-3} has been measured. An angular deformation applied to this multilayer, when the copper wire is subjected to a torsion, modifies the direction of the longitudinal anisotropy. The easy direction for the magnetization changes, continuously, from the initial longitudinal one to the helical direction, depending on the angular deformation. If the copper wire matrix undergoes a plastic deformation, a spontaneous Matteucci effect is observed in these samples: the maximum value of this voltage changes from zero to a maximum, simultaneously that the easy magnetization direction changes from zero to 45 deg. due to the torsion. This behavior makes these cylindrical films suitable for active elements in sensors like magnetic flux gate [11]. No thermal treatments or current annealing is needed to induce permanent helical anisotropies.

ACKNOWLEDGMENTS

This work has been partially supported by the Spanish CICYT, MAT93-0322, PB94-0288, and MAT94-0964 and by the Gobierno de Navarra. One of us (CF) wish to thank the University Alfonso X El Sabio for financial support.

REFERENCES

1. K.Saito, Y.Yanagida, Y.Obi, H.Itoho and H.Fujimori. Mat. Res.Soc.Proc. **384**, 403 (1995)

2. V.Madurga,R.J.Ortega,J.Vergara,K.V.Rao. Presented at the 1996 Fall Meeting, Boston, MA.

3. A.Brenner, D.E.Couch, E.K.Williams. J. Res. Natl Bur. Stand.**44**, 109 (1950)

4. C.Favieres, M.C.Sánchez, C.Aroca, E.López and P.Sánchez J.Mag.Mag.Mat.**140-144**, 591 (1995)

5. K.Mohri, F.B.Humphrey, J.Yamasaki. IEEE Trans.Mag **20**, 2107 (1985)

6. V.Madurga, C.Ortega, J.L.Costa and K.V.Rao in: Basic Features of the Glassy State. edited by J.Comenero and A.Alegría. (World Scientific Publishing Co.Pte.Ltd., London, 1990), p. 570

7. C.Favieres,M.C.Sánchez,E.López,P.Sánchez,C.Aroca.J. Mag. Mag. Mat. **157-158**, 329 (1996)

8. G.C.Chi and G.S.Cargill III.AIP. Conf. Proc. 147 (1975)

9. J.M.Riveiro and G.Rivero.IEEE Trans. Mag **17**, 30824 (1982)

10. A.Hernando, M.Vázquez, V.Madurga and H.Kronmüller. J. Mag. Mag. Mat. **37**, 161 (1983)

11. O.V.Nielsen, J.Gutiérrez, B.Hernando and H.T.Savage.IEEE Trans. Mag .**26**, 276 (1990)

A QUANTITATIVE MICRO-DEFORMATION FIELD STUDY OF SHAPE MEMORY ALLOYS BY HIGH SENSITIVITY MOIRÉ

Q. P. SUN, Terry T. XU, X. Y. ZHANG and P. TONG
Department of Mechanical Engineering, The Hong Kong University of Science & Technology, Clear Water Bay, Kowloon, Hong Kong, meqpsun@usthk.ust.hk

ABSTRACT

Quantitative micro-macro combined experimental research on the deformation field of single crystal CuAlNi shape memory alloy (SMA) is performed by using high sensitivity Moiré technique. The study is focused on the micro-macro correspondence of the deformation behavior of single crystal uniaxial tensile specimen during stress induced forward and reverse transformations. The aim of the experiment is to quantitatively relate the macroscopic applied stress with the deformation field in the mesoscale. The large deformation due to the lattice distortion during transformation was first successfully recorded by Moiré interferometry. Some important microstructure-related deformation features of single crystal SMA under uniaxial tension are first reported.

INTRODUCTION

Shape memory alloys (SMA), as both structural and functional materials, have many important applications in industry and daily life. Many intelligent structures (or devices) and control units are made of polycrystalline SMAs. In the design and reliability analysis of such kind of structures, a comprehensive knowledge on the constitutive behavior and fracture properties of the material is needed. From the fundamental and microscopic point of view, to obtain such knowledge requires a micro-macro combined constitutive and fracture study of single crystals. For single crystal SMAs, so far the experiments performed can be divided into two kinds: The first kind simply performs the uniaxial tension test at ambient temperature and records the stress-strain curves and temperature-strain curves [1-3]. Sometimes the stress rate-effect and micrograph of the specimen surface with microstructure observations are included. This kind of experiment is useful to illustrate qualitatively some typical features of the stress-strain curves. The second kind of test is more concerned with microstructure details and their effects on the macroscopic deformation behavior [4,5]. Such experiments represent an important improvement to the first kind. However, to have a quantitative picture of the macro-micro relationships and microstructure evolution pattern, a whole deformation field measurement is required and is necessary for the constitutive modeling. To achieve this aim, a whole displacement field measurement by Moiré interferometry is desired.

Moiré interferometry has been used widely in experimental mechanics to perform various kinds of in-plane displacement measurement [6]. It has several excellent properties like real-time, high displacement measurement sensitivity, high spatial resolution and high quality fringe patterns. These unique features make it an exceptionally attractive tool for measuring the deformation field of either single or polycrystalline shape memory alloys during transformation process. As a first step towards such a goal, in this paper the deformation field of single crystal CuAlNi shape memory alloy under uniaxial tension was measured by using high sensitivity Moiré interferometry. Some important deformation features of the material at different stages of transformation were first quantitatively revealed.

495

Mat. Res. Soc. Symp. Proc. Vol. 459 © 1997 Materials Research Society

EXPERIMENT

Specimen Preparation

The tensile specimen was cut from the Cu-Al-Ni (Cu-13.7%Al-4.18%Ni (wt%)) shape memory alloy single crystal boule (27mm diameter by 60mm long). This single crystal rod was produced by Prof. Tan Shusong (Central South University of Technology, China) using the improved Bridgeman method. The orientation of the rod is unknown originally. The specimen was cut from the boule using a wire-cutting electrical discharge machine (EDM). The shape and size of the specimen is shown in Fig.1 with 14 mm long and 3.5 mm wide gauge section and 2 mm thickness. The specimen was heat treated at $850^{\circ}C$ for 5 minutes and then quenched into 10% NaOH at room temperature. After this the specimen was hand-polished using 600 grit silicon carbide paper. The transformation temperatures of the specimen was measured by Differential Scanning Calorimetry (DSC) ($M_s=-37^{\circ}C$, $M_f=-48^{\circ}C$, $A_s=-37^{\circ}C$, $A_f=-20^{\circ}C$) to make sure that the initial state of the specimen is in the stable austenite so the tensile stress strain curve is pseudoelastic. The orientation of the tensile specimen is determined by the Laue X-ray diffraction. The orientation of the tensile loading axis in the orthonormal cubic basis is (0.087, -0.796, -0.605) and that of the normal of the specimen face is (0.975, -0.070, 0.217). A crossed line grating of 1200 lines/mm was replicated on the specimen by epoxy cement and was cured at temperature of $80^{\circ}C$ for six hours. By using this kind of grating, the displacement resolution is at least $0.1\mu m$.

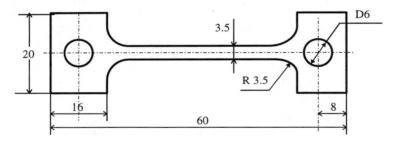

Fig.1(a) The shape and size of the tensile specimen

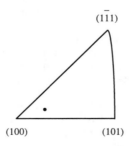

Fig.1(b) The orientation of the specimen

<u>Experimental Methods</u>

The basic principal of Moiré interferometry is described in detail in the book of Post, D., Han. B. and Ifju, P.[6]. In this method, a high frequency crossed-line diffraction grating is replicated on the surface of the specimen and it deforms together with the underlying specimen. Uniaxial loading is performed by a tensile loading frame with a load sensor. During the tensile loading, both elastic deformation due to stress and deformation due to transformation will contribute to the whole displacements. The resulting fringe patterns represent contours of constant in-plane U and V displacements, which are displacements in x (transverse) and y (axial) directions, respectively. The displacement can then be determined from the fringe patterns by

$$U = \frac{1}{f} N_x, \qquad V = \frac{1}{f} N_y \qquad (1)$$

where f (=2400 lines/mm) is the frequency of the reference grating, and N_x and N_y are fringe orders in the U and V field Moiré patterns, respectively. At each load the corresponding fringe patterns are first recorded on a holographic plate and then developed into the photograph in the dark room. In the case of small strain the in-plane strain components ($\varepsilon_x, \varepsilon_y, \gamma_{xy}$) can be simply calculated from the fringe pattern by

$$\varepsilon_x = \frac{\partial U}{\partial x} = \frac{1}{f}\left[\frac{\partial N_x}{\partial x}\right], \quad \varepsilon_y = \frac{\partial V}{\partial y} = \frac{1}{f}\left[\frac{\partial N_y}{\partial y}\right], \quad \gamma_{xy} = \frac{\partial U}{\partial y} + \frac{\partial V}{\partial x} = \frac{1}{f}\left[\frac{\partial N_x}{\partial y} + \frac{\partial N_y}{\partial x}\right] \qquad (2)$$

For single crystal CuAlNi considered here with large lattice distortion, the strain components should be calculated by the finite strain formula:

$$\varepsilon_x = \frac{\partial U}{\partial x} + \frac{1}{2}\left(\frac{\partial U}{\partial x}\right)^2 + \frac{1}{2}\left(\frac{\partial V}{\partial x}\right)^2 = \frac{1}{f}\left[\frac{\partial N_x}{\partial x}\right] + \frac{1}{2}\left[\frac{1}{f}\left(\frac{\partial N_x}{\partial x}\right)\right]^2 + \frac{1}{2}\left[\frac{1}{f}\left(\frac{\partial N_y}{\partial x}\right)\right]^2$$

$$\varepsilon_y = \frac{\partial V}{\partial y} + \frac{1}{2}\left(\frac{\partial U}{\partial y}\right)^2 + \frac{1}{2}\left(\frac{\partial V}{\partial y}\right)^2 = \frac{1}{f}\left[\frac{\partial N_y}{\partial y}\right] + \frac{1}{2}\left[\frac{1}{f}\left(\frac{\partial N_x}{\partial y}\right)\right]^2 + \frac{1}{2}\left[\frac{1}{f}\left(\frac{\partial N_y}{\partial y}\right)\right]^2 \qquad (3)$$

$$\gamma_{xy} = \frac{\partial U}{\partial y} + \frac{\partial V}{\partial x} + \frac{\partial U}{\partial x}\frac{\partial U}{\partial y} + \frac{\partial V}{\partial x}\frac{\partial V}{\partial y} = \frac{1}{f}\left[\frac{\partial N_x}{\partial y} + \frac{\partial N_y}{\partial x} + \frac{1}{f}\frac{\partial N_x}{\partial x}\frac{\partial N_x}{\partial y} + \frac{1}{f}\frac{\partial N_y}{\partial x}\frac{\partial N_y}{\partial y}\right]$$

RESULTS AND DISCUSSION

Fig.2 (a) and (b) respectively shows the fringe patterns of the U and V elastic displacement fields before the transformation happens. It is seen that the strain (and so the stress) is quite uniform except the regions near the two ends of the specimen.

With increase of the applied stress, the martensite variant with the most favorable orientation will nucleate and grow. The angle between the external load axis and the longitudinal direction of the martensite plates is about 50 degree . Fig.3 (a) and (b) show the U and V displacement fringe patterns at the initial stage of the transformation, from which the very thin martensite plates can be clearly identified by the deflection of the fringe patterns (or strain jump) across the martensite-parent interface (habit plane). By the way, due to the incompatibility between the martensite and the parent phase, a strain disturbance, relative to the original uniform strain field, can be seen from the fringe pattern inside the parent phase near the martensite plate. As will be seen below, this incompatibility effect will increase with increase in

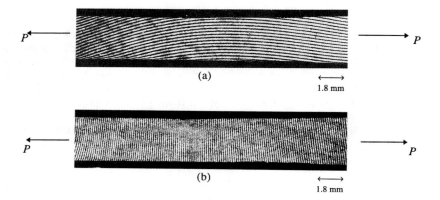

(a)

←——→
1.8 mm

(b)

←——→
1.8 mm

Fig.2 The elastic U (a) and V (b) displacement fringe patterns before the transformation.

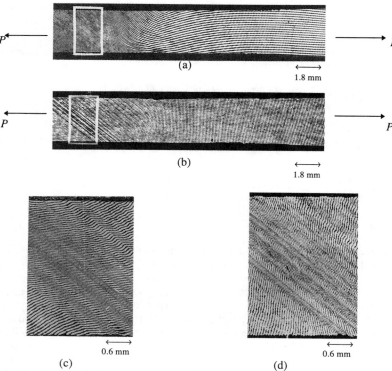

(a)

←——→
1.8 mm

(b)

←——→
1.8 mm

←——→
0.6 mm

←——→
0.6 mm

(c)   (d)

Fig.3 The U and V displacement fringe patterns at the initial stage of the transformation: (a) and (b) fringe patterns of the whole specimen; (c) and (d) amplified fringe patterns across the martensite and parent interface in (a) and (b).

the amount of martensite. The amplified U and V displacement fringe patterns are shown in Fig. 3 (c) and (d).

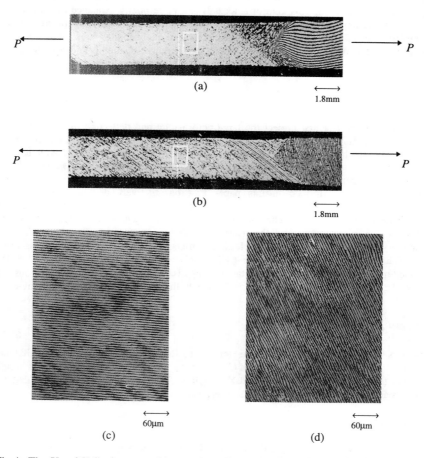

P — — — ← P

(a)

←—→
1.8mm

P ← — — P

(b)

←—→
1.8mm

←—→
60μm

(c)

←—→
60μm

(d)

Fig.4 The U and V displacement fringe patterns at the stage of the transformation where most part of the specimen has been transformed: (a) and (b) fringe patterns of the whole specimen; (c) and (d) enlarged fringe patterns inside the fully transformed zone.

Further increase of the loading causes the transformation bands to grow and coalesce and finally form a large band (see Fig.4(a) and (b)). The magnified fringe patterns (150x) inside the fully transformed band are shown in Fig.4(c) and (d), from which it is seen that the strain inside the band is uniform with the strain magnitude as high as $\varepsilon_x = -7.3\%, \varepsilon_y = 7.39\%, \gamma_{xy} = -3.28\%$. This calculated value is in agreement with the predictions based on the lattice parameters and the orientation of the habit plane normal and shear directions. A strong disturbance to the original uniform fringe pattern inside the parent phase is clearly seen from the fringe patterns. This disturbance will produce additional internal stress inside the austenite and so influence the transformation process. This internal stress effect is macroscopically demonstrated by the recorded softening stress strain curve during transformation, *i.e.*, the load decreases with the progress of transformation. This kind of autocatalysis can be briefly explained as follows: the internal stress is energetically favorable to the formation of new martensite, so macroscopically the transformation can proceeds without increase of the externally applied stress, thus leading to the observed stress-strain softening phenomenon. A detailed investigation is under progress.

CONCLUSIONS

In this on going research project the large deformation and deflection due to the lattice distortion during transformation was first successfully recorded by using Moiré interferometry technique. The results obtained have demonstrated that Moiré interferometry, with its high quality fringe image and high resolution, is a promising technique for the microstructure-based deformation measurement of SMA.

The experiment performed in this paper first quantitatively demonstrated a strong interaction between the deformation inside the martensite plates and in the parent phase even under uniaxial tensile stress state. The obtained data provide a quantitative basis for the analysis of the effect of microstructure on the macroscopic behavior of material such as effect of specimen orientation; internal stress and elastic strain energy on the macroscopic stress-strain curves. The experimental results will be used to check the theoretical predictions of the established models and be used as a quantitative comparison with the numerical simulations. By the way, the present work provides a basis for the further study on the fracture behavior of SMAs.

ACKNOWLEDGMENTS

The authors would like to acknowledge the Hong Kong Research Grant Committee (RGC) and the National Natural Science Foundation of China for supporting this work. The authors would also like to thank Professors F. L. Dai and X. L. Qin of Tsinghua University for many helpful discussions and assistance in performing the experiments.

REFERENCES

1. Y. Huo and I. Muller, Continuum Mechanics and Thermodynamics, **5**, 163 (1993).
2. I. Muller, H. B. Xu, Acta Metall. Mater., **39**, 263 (1991).
3. K. Otsuka and K. Shimizu, Int. Metals Reviews, **31**, 93 (1986).
4. T. W. Shield, , J. Mech. Phys. Solids, **43**, 869 (1995).
5. C. Chu, PhD Thesis, University of Minnesota, (1993).
6. Post, D., Han, B. and Ifju, P., <u>High Sensitivity Moiré: Experimental Analysis for Mechanics and Materials</u>, Springer-Verlag, New York, 1994.

FABRICATION OF PIEZOELECTRIC SENSORS FOR BIOMEDICAL APPLICATIONS

IQBAL HUSSAIN, ASHOK KUMAR, A. MANGIARACINA, S. C. PERLAKY*, C. C. MCCOMBS* , F. ZHONG*, J.J. WEIMER**, L. SANDERSON**
Dept. of Electrical Engineering, University of South Alabama, Mobile, AL 36688
*Dept. of Medicine, University of South Alabama, Mobile, AL 36688
**Dept. of Chemistry & Materials Engr., University of Alabama in Huntsville, AL 35899

ABSTRACT

Biosensors are a special class of chemical sensors that take advantage of the high selectivity and sensitivity of biologically active material. We are currently investigating the characteristics of various deposited electrode coatings (Au and Pt) on 10 MHz quartz crystals using the sputtering method. We are also investigating the effect of magnetic behavior (by intermixing Fe and Ni with electrodes) on the binding nature of antigen with the substrate. A change in mass occurs due to the binding of antigens and antibodies on the surface of the thin film coating. The frequency change as a result of a change in mass makes it possible to use these crystals as biological sensor devices. This paper describes the construction of antibody-based piezoelectric crystals capable of detecting mycobacterial antigens in diluted cultures of attenuated M. tuberculosis. The microstructural features of these crystals have been studied using scanning electron microscopy (SEM) and atomic force microscopy (AFM) techniques. The crystallographic properties have been characterized using the X-ray diffraction (XRD) method. The long term objective of this research is to develop a rapid quantitative method of analysis for the diagnosis of tuberculosis (TB) and other infections caused by mycobacteria, using biosensor technology.

INTRODUCTION

Piezoelectric immunosensors are based upon the measurement of small changes in mass. Piezoelectricity may be defined as the generation of electric charge in a substance by a mechanical stress that changes its shape, and a proportional change in the shape of a substance as a voltage is applied [1]. Some crystalline substances, such as quartz, possess this property. Certain crystals show positive and negative charges when they are compressed and these charges are proportional to the pressure on certain portions of the crystal's surface. The resonant frequency of a piezoelectric crystal can be affected by a change in mass at the crystal surface. The modification of the surface of the crystal with organic compounds or enzymes when bound to a particular substrate provides enhanced detection specificity.

Piezoelectric immunosensors are a special class of the biosensor family. A biosensor is a device that recognizes an analyte in an appropriate sample and interprets its concentration as an electrical signal via a suitable combination of a biological recognition system and an electrochemical transducer [2, 3]. Potential applications for biosensors range from mining and industrial to military and security, including agricultural and the clinical and veterinary biomedical areas.

Biosensors consist of a biocomponent that specifically reacts or interacts with the analyte of interest resulting in a detectable chemical or physical change. A biosensor has the following components: a) a receptor, which is responsible for the selectivity of the sensor. Examples are enzymes, antibodies, lipid layers etc.; and b) a detector, which plays the role of the transducer, detects the physical or chemical change by recognizing the analyte and relaying it through an electrical signal. Examples are a pH-electrode, an oxygen electrode, a piezoelectric crystal etc. [4]. The different component parts and the roles they play in a biological sensing device is shown in Figure 1.

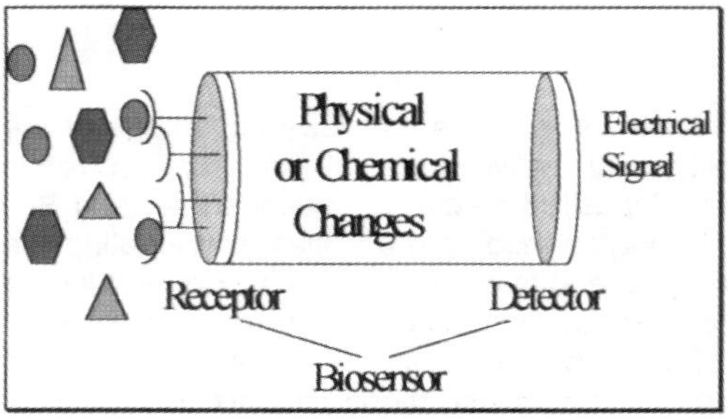

Figure 1: Schematic of a Biosensor Device

The interaction of antibodies with their corresponding antigens is an attractive proposition to the development of antibody based chemical sensors, otherwise known as immunosensors. An immunosensor is a biological sensing device that utilizes the specific molecular recognition qualities of antibodies [5]. The piezoelectric effect in various crystalline substances is a useful property that leads to the detection of specific analytes. We are investigating the antibody-antigen binding in the liquid phase using the piezoelectric properties of quartz crystals.

The relationship between the resonant frequency of an oscillating piezoelectric crystal and the mass deposited on the crystal's surface is formulated by the Sauerbrey Equation [6] :

$$\Delta F = -2.3 \times 10^6 \times F^2 \times (\Delta W/ A) \qquad (1)$$

where ΔF is the change in oscillating frequency of crystal in Hz, F is the oscillating frequency of the crystal, ΔW is the mass of the deposited film and A is the area of the electrode surface. The change in frequency is directly related to a change in mass of the crystal. From the above equation, for a 9 MHz crystal with 5 mm diameter electrodes on both sides, the change in oscillating frequency is calculated to be - 475 Hz. The detection limit is estimated to be 10^{-12} g [6]. This paper studies the response of piezoelectric crystals, under various conditions, for the determination of M. tuberculosis antigen concentration in gaseous state. The main focus of this research is to detect the TB antigen using piezoelectric sensor technology. TB remains the leading cause of death in the world from an infectious disease [7]. The airborne spread of this disease means that the principal risk behavior for acquiring TB infection is breathing [8]. The present methods of diagnosing TB all have drawbacks. They are either nonspecific [9] or too time consuming [10] . The biosensor approach discussed in this text would make it possible to diagnose the disease within hours and at a very low cost to the consumer.

EXPERIMENT

AT–cut quartz crystals (14 mm diameter), with approximate intrinsic resonance of 10 MHz, was used and gold electrodes were deposited on both sides of the quartz substrate using the sputtering method. The gold electrodes (8 mm diameter) were

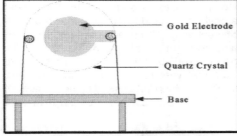

Figure 2: Schematic of a Quartz Crystal Biosensor

deposited on both sides using shadow masking. The vacuum deposition was done in an Ar environment at 50 mTorrs of pressure and 40–45 mA of current. The crystal was mounted on a holder with stainless steel wire leads. A silver composite was used to connect the electrode to the wire. Figure 2 shows the schematic diagram of the fabricated crystal attached to the base. The structural properties of the deposited crystal electrode was investigated using X-ray diffraction method (Rigaku DMAX 2200 with thin film attachment). A low frequency oscillator was used to electrically stimulate the crystals to oscillate at their inherent resonant frequency. A multi-function frequency counter was used to read the frequency. A detailed clinical process for detecting the TB antigen is discussed in the next section.

The characterization of the coated crystal was done using AFM and SEM techniques. The AFM analysis was done using a Nanoscope III (Digital Instruments, CA). Images were obtained in air, and no cleaning treatments were done on the sample surface before scanning. Scanning was done in tapping mode with standard Si tips of nominal radius 10-20 nm. The image was post-processed using a quadratic plane fit in both directions and a zero order line-by-line flattening routine. A low pass filter was run on the image before printing.

RESULTS AND DISCUSSIONS

The crystal electrodes were first modified with a thin (5 µl) coating of Protein A. This was done for better adhesion of the antibodies to the surface of the transducer [11]. Protein A is a polypeptide isolated from *Staphylococcus aureus* that binds specifically to the immunoglobulin molecules, especially IgG antibodies, without interacting at the antigen site. This property permits the formation of tertiary complexes consisting of Protein A, antibody and antigen. Prior to modification, the electrodes were anodically oxidized at constant current in 0.5M NaOH. They were dried in an incubator and the antibody (IgG) coating was then applied upon the Protein A coating. 10 µl of antibody was applied on both sides of the crystal using a pipette. After another drying period, 10 µl of antigens were coated onto the crystal and they were methodically dried again. The frequency of the crystal was recorded after each step. The crystal was washed after each step as a precaution against non-specific binding.

Control crystals and experimental crystals were coated with antibodies. The former were coated with an irrelevant antibody, one specific for an antigen not present in the solution containing the analyte. The experimental sample was coated with antibody specific for binding to the antigen. Both were exposed to the solution containing the analyte. Then the difference in frequency change between the control and experimental crystals were compared, reflecting the immunologically specific binding of analyte. This procedure was carried out first with crystal with gold substrate, and then using crystals coated with magnetic materials. A magnetic field was induced during the investigation of the latter.

The nature of the binding of antigen to antibody to the surface of the transducer is shown in Figure 3. The protein helps the antibody to bind to the gold electrode and the antigen in gaseous state binds to specific antibody. The change in frequency, ΔF, observed for the gold plated specimens were averaging at about 450 Hz for specific binding and at about 200 Hz for non-specific binding, but with a large standard deviation. This was possibly due to the fact that very few of the antibodies were binding strongly to the crystal surface, and the washing process eliminated most of them due to weak binding.

The use of magnetic materials underneath the gold coating have shown some innovative results to detect the antigen. A Cobalt supermagnet was introduced to create a magnetic field, and both control and experimental specimens were coated with antibodies cultured with magnetic beads for better adhesion to the surface of the crystal. The change in frequency in the induced magnetic crystal was 0.9 MHz, but

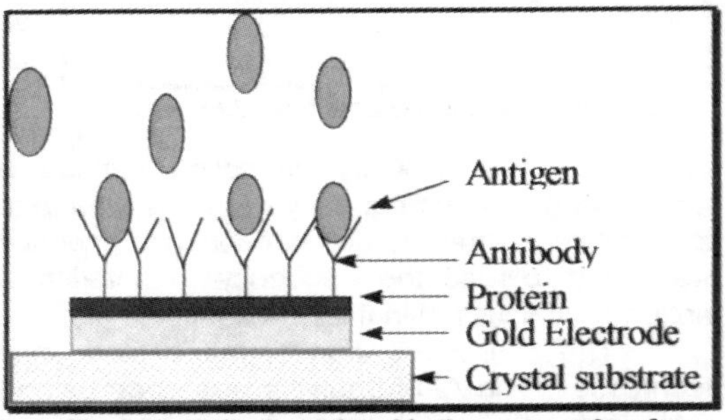

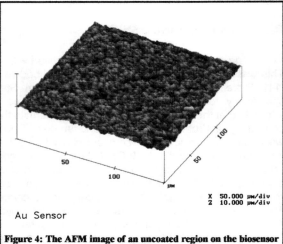

Figure 3: Antibody Antigen binding on crystal surface

only for a short period of time. This can be explained by the fact that the magnet used to induce a field might have dislodged the magnetic antibodies from the substrate and thus the frequency, in time, reverted back to the original frequency. More research is in progress to modify this experiment for improved results.

The AFM image of an uncoated biosensor surface is shown in Fig. 4. It shows a rough texture characteristic of protruding crystallites or grains deposited on the surface. The surface was difficult to scan without obvious convolution effects from tip-surface interactions

Figure 4: The AFM image of an uncoated region on the biosensor surface

because the edges of the protruding features were so sharp. In some cases, the tip was apparently not able to probe completely into the recesses between the grains. The RMS roughness from the region was calculated as 0.25 microns, and the mean roughness was 0.20 microns.

A protein coated region is shown in the image in Fig. 5 at the same scale as Fig. 4. The most obvious difference is the appearance of a film across a region of the surface. The size of this film is nearly macroscopic in proportion to the scan size. It also contained areas of particle-like projections. In an optical microscope, the film is clearly visible as a white, polymeric like, patchy layer distributed across the surface of the sensor. The protein coating was obviously not uniform in thickness even as viewed under an optical microscope. Interpretation of changes in microscopic features between the image in Fig. 4 and that in Fig. 5 at higher magnifications than those shown is therefore difficult.

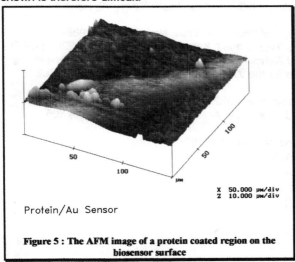

Protein/Au Sensor

X 50.000 µm/div
Z 10.000 µm/div

Figure 5 : The AFM image of a protein coated region on the biosensor surface

Further analysis of the sensors with and without proteins and antigens is ongoing [12]. The rough nature of the uncoated surface may limit the amount of information that can be obtained at higher magnifications than those shown in Figs. 1 and 2. In particular, we do not expect to obtain molecular scale resolution images on these surfaces until the nature of the adsorbed film is better understood at larger scales. The use of thinner coatings of protein on smoother surfaces would be extremely beneficial in this regard.

CONCLUSIONS

Piezoelectric quartz crystals have been successfully fabricated for the detection of mycobacterial antigens. The use of gold and gold with nickel underlayer electrodes have been used to exhibit the piezoelectric characteristics of the crystalline quartz substrate. The resonant frequency shifts of the piezoelectric immunosensors shows that specific binding of antibodies to antigens have been achieved. It has been demonstrated that the piezoelectric biosensor can be used to detect mycobacterial antigen in the gaseous phase. The AFM analysis exhibits the binding characteristics of the biomolecules to the transducer surface. This is our first work to implement the piezoelectric property of quartz crystals for the detection of micobacterial antigen. More analytical analysis is needed to shed more light on the optimum binding characteristics of the antigen and antibody to the piezoelectric transducer.

ACKNOWLEDGMENTS

This research was supported by the Alabama NSF Systematic Improvement Program and the NSF Academic Research Infrastructure (DMR # 9512324) grant.

REFERENCES

[1] M. D. Ward and D. A. Buttry, *Science*, **249**, 1000 (1990)

[2] H. Morgan and D. M. Taylor, *Biosensors and Bioelectronics*, **7**, 405(1992)

[3] R. S. Alberte, *Naval Res. Reviews*, **3**, 2, (1994)

[4] A.S. Dewa and W.H. Ko, *Semiconductor Sensors, Ed. S. M. Sze*, J. Wiley (New York), 415(1994)

[5] G. G. Guilbault and J. M. Jordan, *CRC Crit. Rev. Anal. Chem.*, **19**, issue 1, 1(1988)

[6] E. C. Hahn, Ph.D. Thesis, Dept. of Chemistry, Univ. New Orleans (1988)

[7] H. Muramatsu, J. M. Dicks, E. Taniya and I. Karube, Anal. Chem., **59**, 2760 (1987)

[8] C. J. L. Murray, K. Styblo and A. Rouillon, *Disease Control Priorities in Developing Countries,* Oxford Univ. Press for the World Bank (New York), 50 (1992); Bull Int Union Tuberc, 24 (1990)

[9] B. R. Bloom and C. J. L. Murray, *Science*, **257**, 1055 (1992)

[10] L. Arango, A.W. Brewin and J. F. Murray, *Amer. Rev. Respir. Dis.*, **108**, 805 (1978)

[11] J. C. Weissler, *Amer J. of Med. Sci*, **305**, 52 (1993)

[12] I. Hussain, A. Kumar et al. (Unpublished)

Part VIII

Magnetostrictive Materials and Applications

The Application of High Energy Density Transducer Materials to Smart Systems

J. F. LINDBERG
Naval Undersea Warfare Center Detachment New London, New London, CT 06320
janx@nuscxdcr.nl.nuwc.navy.mil

ABSTRACT

Recent NUWC research efforts in the field of high power sonar transducers designed to produce high acoustic outputs over significant bandwidths while being of minimal size and weight have been aided by advances in the continuing development of several new high energy density transducer drive materials. Both Terfenol-D, a rare earth magnetostrictive material, and lead magnesium niobate, a relaxor ferroelectric, have demonstrated a tenfold increase in field-limited energy density over a typical very hard lead zirconate titanate (i.e., Clevite PZT-8) piezoelectric ceramic. The Center's focus is to double the demonstrated performance of each material and to address such issues as hysteresis reduction in the magnetostrictive material and coupling coefficient improvements in the electrostrictive materials. Poly(vinylidene fluoride-trifluoroethylene) can also be considered a high energy density material because of its excellent energy density and its broad bandwidth possibilities. The application of these material technologies, either separately or as hybrid composites, to smart material design will be detailed.

INTRODUCTION

In the pursuit of higher performance but lighter active projector arrays for antisubmarine warfare sonar systems the Navy has taken the approach of exploiting new high energy density driver materials. If one then incorporates these materials into a compact transducer element this then facilitates the design of small and lightweight, high-powered active sonar arrays. These compact transducer source arrays will be able to achieve higher source levels and/or broader bandwidths while also reducing the number of transducers in the projector array. This will be achieved by replacing conventional piezoelectric ceramic driver material with new transduction materials capable of much greater energy density, and by developing new transduction mechanisms which better use these new driver materials. Development of mid-frequency and ultrasonic transducer arrays involve adaptation of new projector materials with innovative transducer array designs as a means to deliver high frequency, broadband, high-power transmissions for applications such as unmanned underwater vehicle (UUV) imaging sonars. This paper will address the pertinent attributes of several high energy density driver materials and then detail the transducer mechanisms that best exploit these materials.

Mat. Res. Soc. Symp. Proc. Vol. 459 © 1997 Materials Research Society

HIGH POWER PROJECTOR MATERIALS

High power projector materials have traditionally focused on the lead zirconate titanate compositions known by their copyrighted (by Clevite) names as PZT-4 and PZT-8. With the exception of a few barium titanate holdouts, all production underwater projectors in use in the U.S. Navy are of one of these two compositions. Recent trends have been to develop and introduce new high energy density drive materials into the Fleet and to tailor each material to its specific need or application. New drive materials include Terfenol-D [1-3], low temperature magnetostrictive materials [4], lead magnesium niobate (PMN) [5], and piezoelectric polymers [6]. Each of these materials has a potential use or niche and each has advantages and disadvantages. A useful way to rank these materials among themselves and to the PZT benchmark, is to calculate the field-limited energy density of the basic material. This metric is defined as one-half the product of the Young's modulus of the material times the square of the maximum strain obtainable at the electric or magnetic field limit of the material. This permits a comparison of materials on a per unit volume basis. Table 1 is an abridged summary of materials currently of interest. The field limit used for the PZT-8 benchmark is 10 V/mil which is the industry standard for reliability and acceptable losses. Lead magnesium niobate (PMN) is an electrostrictive ceramic which promises high strain while all of its other properties (notably cost) resemble that of PZT. The values given for PMN are representative of its current state of development. The energy density of polyvinylidene fluoride (PVDF) voided homopolymer is considerably less than that of PZT but its ρc makes for a very broadband device. Its copolymer, poly(vinylidene fluoride-trifluoroethylene) is also suited for broadband operation but is stiffer and at its full field limit becomes a true competitor to PZT if one can conquer the engineering challenge of reliably applying the field of 750 V/mil. The drive levels associated with the Terfenol-D tabulation are for the general case of limiting harmonic distortion to approximately 15% to permit broadband transducer operation and for the higher drive case in which it is presumed that operation will only be around resonance where the projector will act as a bandpass filter and suppress radiated harmonics. Terbium-dysprosium, used at cryogenic temperatures, yields the highest known magnetostriction of any material. The material cited is $Tb_{0.6}Dy_{0.4}$ at 77° K and represents data collected at a relatively low magnetic field.

The sonar system designer desires a transducer element which uses a completely characterized and tractable driver material which, if possible, possess electrical properties similar to existing materials so that an improved transducer can easily be incorporated as a simple upgrade to the system. At first look the incorporation of a high energy density relaxor material appears to be the most convenient of the choices since the material has mechanical properties similar to the PZTs, looks familiar, and reportedly should cost roughly the same as the PZT it replaces. But the need for a DC bias for linear operation complicates the situation and places new demands upon both the sonar system power amplifier and cable. The utilization of a magnetostrictive material such as Terfenol-D might be considered a radical departure from tradition but thanks to interspersed permanent magnets in the driver rod to provide magnetic bias, the magnetostrictive transducer can be considered a direct replacement except for a different value of the electrical impedance reflecting the driver type. The low temperature magnetostrictive alloys such as terbium-dysprosium or terbium-dysprosium-zinc exhibit giant strain levels but require cryocoolers (or liquid nitrogen) which at this time are both bulky and inefficient. Piezoelectric polymers are considered by many to be a radical departure from the established ceramic technology base but in practice can be configured to mimic ceramic stacks with quite different, and interesting, electromechanical properties.

At present relaxor ceramics are receiving considerable attention by material science investigators and transducer designers. The primary advantage of relaxors is the capability of producing high electrically-induced strains ($\varepsilon \sim 0.1\%$) while maintaining negligible hysteresis. The issues still requiring further development include (a) enhancements in ε to increase available acoustical power, (b) enhancements in electromechanical coupling coefficient to increase bandwidth, (c) aging of electromechanical properties, (d) materials characterization under realistic operating conditions as high-power transducers, and (e) synthesis and processing issues involved in scale-up towards multi-layer structures. To date, the measured coupling coefficients of lead magnesium niobate (PMN) have been disappointing and, if not improved, will shift a large burden (and cost) onto the power amplifier required to excite the transducer. Table 1 reports a recent coupling coefficient (k_{33}) measurement [7] of only 37% which is quite a bit lower than the 70% exhibited by the PZT family of piezoelectrics. The lack of a comprehensive data base on the material seriously restricts the transducer designer who is currently relegated to just one reference [5]. Recent difficulties during fabrication and testing of prototype transducers have highlighted the question of mechanical strength of the material. Initial testing of mechanical properties of candidate PMN compositions have revealed a mechanical flexural strength in PMN that is 2-2.5 times weaker than that measured in PZT-8. This effort is on-going [8] and still has to address the measurement of these properties under DC biased conditions.

High energy density magnetostrictive material research efforts are currently focusing on improvements to Terfenol-D, which is a commercial product. These efforts include single crystal studies and manufacturing techniques to produce thin laminations of the material.

It has been known for many years that the very high strain levels can be generated magnetostrictively by single crystal Terfenol ($TbFe_2$) using an excitation along the crystalline [111] direction [9]. Strains as high as 3.6×10^{-3} have been measured at room temperature in small samples. Single crystal [111] oriented crystals of Terfenol-D ($Tb_{0.3}Dy_{0.7}Fe_2$) possess theoretical magnetostrictions of ~2400 ppm. This is much higher than the commercially available Terfenol-D with a magnetostriction of ~1500 ppm. Single crystals of Terfenol-D have been prepared and examined as a function of applied compressive stress and field [10]. With an applied stress of 21 MPa, saturation magnetostrictions reached 2300 ppm with an applied field of only 100 kA/m (1.2 kOe), thus confirming the theoretical predictions. Because of the [112] growth morphology, it has been difficult to prepare large single crystal transduction elements of these giant magnetostrictive materials with the axis along [111]. Also, an ideal crucible material has not been identified. Some inroads have been made by a number of scientists at the Chinese Academy of Sciences in Beijing and at the Institute of Metals Research at Shendong (China) [11].

A major design constraint in the design of magnetostrictive transducers is the losses incurred in eddy currents. To alleviate this, the material must be laminated which places an added burden in the manufacturing process and associated costs. Because of the difficulty in producing thin laminations of Terfenol-D, the upper frequency bound for practical purposes has been approximately 10 kHz. A means to bypass this upper bound is being pursued using melt-spun methods. Rapid solidification by high speed rotating wheels and drums has produced thin samples of high magnetomechanical coupling ribbons [12]. These samples frequently contain glass former elements, such as boron and silicon. Recently it was shown that Terfenol could be solidified into a magnetostrictive amorphous alloy by both high speed sputtering and by rapid solidification processes [13]. Using the proper heat treatment these alloys are expected to revert to the lower energy cubic Laves phase crystal structure of Terfenol. Current efforts involve the manufacture of samples of rapidly spun amorphous Terfenol, prepared under various temperature gradients, pressures, and wheel speeds, which will be subjected to heat treatments in order to achieve oriented crystallites of the Terfenol compound. In order to yield the high strains characteristic of bulk samples, heat treatments will be made under stress and magnetic field. Upon achieving the proper crystallite size and orientation, saturation magnetostrictions, coupling factors, and elastic moduli will be measured using a unique experimental apparatus. Alloys prepared in this way can be expected to serve as ideal actuator/sensor components of transducers for frequencies as high as 500 kHz.

Table 1. High power projector materials.

Material	Field Limit	Young's Modulus	Strain	Coupling	Energy Density	Relative dB
	(rms)	Y (GPa)	(0-pk) (ppm)	k_{33}	$1/2\ Y\ S^2$ (J/m^3)	
PZT-8	10 V/mil	74	125	.64	578	0
PZT-4	10 V/mil	66	159	.70	830	1.6
PMN	15 V/mil	88	342	.37	5150	9.5
PVDF (voided)	750 V/mil	0.9	1020	.19	468	-9.2
P(VDF-TrFE)	750 V/mil	3	1023	.25	1570	4.3
Terfenol-D	64 kA/m	29	582	.67	5150	9.3
Terfenol-D	175 kA/m	59	810	-	19355	15.3
TbDy	22 kA/m	25	1980	.70	49005	19.3

Polyvinylidene fluoride (PVF_2) and its copolymer with trifluoroethylene (PVF_2EF_3) have been traditionally used in hydrophone constructions with recognized success. However, this family of materials has not been seriously considered for use as projectors, as several of the important material properties are incompatible for engineering requirements of high-power underwater projectors. In particular, the electric field required to produce the needed displacements are ~75 times that of traditional piezoelectric ceramics. A second problem is large dielectric loss factors, much higher than that of piezoelectric ceramics. Another problem is a low coupling factor. Despite these shortcomings, electroactive polymers have some promising attributes. They are more closely matched in mechanical impedance to sea water than their ceramic counterparts, and consequently can operate over broader bandwidths. The low mechanical impedance of the polymer proves to be of additional merit in that its mismatch to a metal backing plate is sufficient to preclude the need for an acoustic pressure release. Recent design efforts at NUWC have focused on the concept of a transparent half-wave projector. An initial design [6] consisted of 7 layers (~200 mm thick) of copolymer which were one inch square. (Materials were fabricated from powders and solvent cast, annealed, electroded, poled, and then consolidated.) The finished projector resonated at 542 kHz with a mechanical Q of 2.3, a transmitting voltage response of 156 dB/μPa-m/V, and an electroacoustic efficiency of 12%. Another approach (being pursued by Raytheon Co.) has concentrated on the development of a quarter-wave resonator, which can be used in closely packed planar arrays. A prototype had a resonance at 15.5 kHz and achieved a source level of 216 dB/μPa with a mechanical Q of 1.6. This source level was achieved under a drive level of 375 V_{rms}/mil. The mass and volume figures of merit of this prototype were 162 W/kg-kHz-Q and 633 W/m^3-Hz-Q, respectively. In a later design, the bandwidth was significantly increased, having a mechanical Q of 0.63 and an electroacoustic efficiency of 32%. The polymers possess a quality which can be of considerable advantage when designing and fabricating high frequency projectors. Ideally, for optimum source generation, a projector material in a transducer would be operated at its mechanical stress limit, at its electric field limit, under near-cavitating conditions, and while maintaining efficient electrical coupling to its drive electronics. If one takes a notional transducer and plots stress-limited and field-limited source generation as a function of bandwidth (Q_m), there is generally a maxima indicating an optimum Q_m. For materials in active use, this determines the optimum operational conditions of the transducer, in terms of its stress and field limits [1]. For example, Terfenol-D may be up to 8 dB superior over PZT-4 in radiated power capability in the low-Q, field-limited regime. In order for the projector to operate at its mechanical stress limit, the active driver material must be kept within its tensile limits. For piezoelectric and magnetostrictive transducers, a mechanical prestress means is required and in longitudinal vibrators this normally takes the form of a stress rod. A real problem occurs as the desired frequency of operation increases, since the scaling of components becomes hampered by size. For example, high powered vibrators working at 60 kHz have assemblies of such small size that practical execution of a stress rod design become very difficult. Herein lies a real advantage of polymer projector materials. Their inherent coercive strength results in a very high tensile strength and that permits design of vibrators which can produce very high strain levels without the need for externally applied prestress via mechanical means.

TRANSDUCER DESIGN

Transducers can be categorized as either body stress or surface force transducers. Body stress transducers are intrinsically linked to the active material, thus material science is of primary importance to the device whereas surface force transduction revolves more around the physical design of the transducer, rather than the properties of the driver material thus making transducer design innovation the primary issue. For high power acoustic sources the default transducer mechanism is of the body stress type (see Table 2). Array design is a means by which acoustic energy going to or coming from a specific direction can be focused, using a discrete number of sensors or sources. Because of complex array interactions both transducer design and driver material aspects are of importance for arrays. These concepts have been arrived at by judiciously combining either a new material or an anomalous materials property with a new transduction concept.

Table 2. Comparison of Transducer Mechanisms in Terms of Approximate Maximum Attainable Forces per unit Area [14]

TRANSDUCER TYPE		$(F/A)_{max}$ [N/M²]
Moving Coil		8,000
Electrostatic		16
Variable Reluctance		40,000
Magnetostrictive	Nickel	1,000,000
	Terfenol-D	39,500,000
Piezoelectric	PZT	8,000,000
	PMN	27,000,000
Hydroacoustic		7,000,000

The most common type of underwater transducer in use today is the longitudinal vibrator or tonpilz. These transducers are predominantly excited with PZT stacks and are ideally suited for a material upgrade of PMN. There have been prototype Terfenol-D longitudinal vibrators fabricated and tested and with proper design, they also lend themselves to the tonpilz shape and function. Less known to the materials community may be the variety of transduction mechanisms tailored to produce high power acoustic energy in a small volume. These types of transducers, coined 'miniature' by Woollett because of their small size in terms of acoustic wavelength in water, are ideal for the inclusion of high energy density drive materials since they are normally field-limited in terms of output power.

One of the most extensively studied of this type transducer for high-power low-frequency sonar applications is the flextensional projector. In a generic sense, flextensional transduction can be described as electrically-stimulated mechanical amplification. A specific example is the flexural excitation of a curved shell by an internal ferroelectric ceramic stack that, through the converse piezoelectric effect, is forced into extensional-compressional vibration by an applied electric field. The shell is contiguous to the acoustic medium and sound is radiated at wavelengths that are much larger than the dimensions of the transducer itself. Consequently, when operated at frequencies near its fundamental flexural mode, the 'miniature' flextensional transducer is omnidirectional. Further discussions concerning the operation and performance capabilities of flextensional transducers are given by Marshall et al. [15] and Oswin and Dunn [16]. In 1969, flextensional transducers were grouped into five classes by Brigham and Royster [17, 18] in order to simplify the identification of a growing number of flextensional designs at that time. Recently, their classification scheme was extended by Rynne [19] to include two additional classes. The criteria that distinguish the classes are based on specific performance enhancements and the shapes of the shells. The Class I transducers consist of eccentric convex shells that have rotational symmetry about their long central symmetry axes. Class II transducers have Class I shells but extended driving mechanisms for greater power capability. Class III transducers feature at least two contiguous Class I shells, each shell exhibiting its own flexural resonance frequency. If the resonances are in close proximity to each other, they can be combined to give large operating bandwidths. Class IV transducers are characterized by eccentric convex shells that have constant cross-sections when viewed along central orthogonal axes. Class V transducers have eccentric convex shells with rotational symmetries about their short symmetry axes. Classes VI and VII are the same as Classes V and IV respectively, but with concave shells.

In 1966 Howard Merchant, who was working at Honeywell at the time, was granted a flextensional transducer patent [20] which included a sketch of a peanut-shell-shaped flextensional. This Class VII flextensional, essentially a concave Class IV oval flextensional, apparently was never reduced-to-practice until 1995. The NUWC magnetostrictive barrel-stave projector development [21] identified many design and engineering problem areas of which two were stress failures encountered at the stave/end mass interface (i.e., the stave shears off) and difficulties in the design of a good magnetic return path for the magnetostrictive driver. Lockheed Sanders became involved in the design effort and what eventually became obvious was the idea that if one eliminated the joint between the barrel stave and the end mass by going to a continuous assembly and further discarded the circular symmetry of the barrel stave so that two Terfenol-D drivers in parallel can be used to simplify the magnetic circuit, one has designed a Class VII flextensional transducer. The basic Merchant design appears to suffer from a low coupling coefficient and this required design improvements and innovation to eliminate the deficiency. As a result of this evolution, two Class VII flextensionals were designed and built by Lockheed Sanders with Terfenol-D magneto-strictive drivers. The smaller of the transducers was designed for operation around 900 Hz and has a uniform wall thickness shell and Terfenol-D rods which are biased with DC imposed via the coils. The larger of the two transducers was designed for 500 Hz operation and utilizes a tapered shell and permanent magnet bias. The bias is achieved with neodymium-boron-iron magnets interspersed in the Terfenol-D drive rods. A significant advantage of a Class VII flextensional (over that of a Class IV) is that the imposition of hydrostatic pressure on the shell results in an increase in prestress on the driver. This is the exact opposite of the Class IV oval case, which must have a very stiff shell and has to be preloaded such that as the transducer is exposed to hydrostatic pressure (which subtracts out the preload), there is still sufficient preload in the shell at depth to prestress the driver and prevent it from going into tension during AC drive. Thus the Class VII (along with Classes I, II, III, & VI) has the advantage that initial high prestress is not required. This minimizes potential creep in both the shell and driver material over the life of the transducer. Unburdened from the requirement to statically prestress the driver, the designer is able to design a more compliant shell and better match it to the compliance of the driver. By combining this advantage with the low modulus of Terfenol-D, a transducer can be designed which maximizes coupling coefficient and bandwidth. Although a PZT or PMN driven Class VII flextensional will offer good performance and long-term advantage due to low static stress, the mating of the compliant Class VII shell with Terfenol-D results in a remarkable transducer. The DC-biased version of this transducer achieved a source level of 212 dB//1μPa-m at 930 Hz in a package the size of a lunch pail. The mass figure of merit (see Table 3 for a more comprehensive survey) measured was 210 W/kg-kHz-Q and the volume figure of merit was 710 W/m^3-Hz-Q. The permanent magnet biased version achieved a source level of 211.8 dB//1μPa-m at 490 Hz before the drive pulse length/duty cycle combination inadvertently heated the drivers to the point that the rare earth magnets started to demagnetize. This resulted in degraded performance and a mass figure of merit of only 113 W/kg-kHz-Q. It was reassuring that this transducer development demonstrated the ability to take a material energy density improvement and actually incorporate it into a device and retain the performance advantage. Table 1 indicates a 9.3 dB advantage of Terfenol-D over PZT-8 whereas the DC-biased Class VII prototype has an 8.5 dB edge over a Model 30 ceramic flextensional (see Table 3).

Table 3 is a compilation of transducer performance metrics for a variety of transducer types [22]. As can be seen those transducers which have been 'upgraded' to high energy density drive materials stand out in their superior performance.

A fairly new development in transduction technology is the hybrid transducer array which is an array of magnetostrictive/piezoelectric hybrid tonpilz transducer elements. The primary design is that of a double-ended tonpilz which takes advantage of both the inherent 90° phase difference between the magnetostrictive and piezoelectric velocities for the same driving voltages and an additional 90° phase shift due to time delays associated with the compressional wave speed within the transducer. Because of these phase differences, the transducer can be unidirectional out of either end of the vibrator [23, 24]. Another advantage of this transducer is that it is self-tuning. Additionally, as a result of the electrical and magnetic energy shared at the electrical resonance of the hybrid transducer, the coupling coefficient is increased by as much as 41%. The latest design [25] is configured as a unidirectional tonpilz, which as a result of mechanical design improvements, has a bandwidth exceeding 100%. Because of the 90° phase shift between the piezoelectric and magnetostrictive components, there is no intermediate cancellation between the two resonances in a three mass, two degree of freedom system and thus the resultant well-behaved transfer phase of the broadband hybrid permits its use for true broadband transmissions.

The decent performance achievable from piezoelectric copolymer combined with its mechanical impedance makes for a formidable material which can be used to develop acoustically transparent and multiplicative array designs. This will present new opportunities to array designers especially those working at high frequencies. One of the problems encountered in closely packed planar or conformal arrays is that of wiring. A new concept developed at NUWC is a design for a high frequency multi-element steerable array with simplified wiring [26]. A conventional n by n array requires electrical wiring to each element. Connections to the face of the array are accomplished by either wires in front of the array or feed-throughs that permit access to the rear of the array. For an array of considerable size, the amassment of (n-2)^2 wires in front of the array (or alternatively piercing through the array) becomes unmanageable and unworkable, especially if the array is a transmit array requiring high voltages. The new concept is based on the idea of a copolymer sheet with electrode strips configured so that a beam in one plane can be steered from broadside to an endfire. An identical layer is placed on top of the first askew at

Table3. Performance summary of various sonar projectors.

Projector ID	f_r (Hz)	Q_m	SL (dB)	η (%)	Mass (kg)	FOM_m (W/kg-kHz-Q)	FOM_v (W/m³-Hz-Q)
Sparton PZT Ring-Shell Model 34SA0400 Class V	400	3.5	213.2	83	225	55	81
Sanders Terfenol-D[a] Model 70 "dogbone" Class VII	490	4.5	211.8	45	54	113	250
Sparton PZT Ring-Shell Model 18SI0700 Class VI	700	3.9	205.2	90	43	25	71
DREA PZT Barrel-Stave large Class I	790	3.6	204.3	81	16.7	47	162
NUWC/Sanders Terfenol-D Model 8 "dogbone" Class VII	930	4.7	212.0	46	15.4	210	710
Sanders Model 30 PZT Class IV flextensional	1030	4.5	207.5	80	43	25	61
DREA Barrel-Stave[b] Terfenol-D Class I	1060	3.4	197.7	32	3.9	35	93
Sparton PZT Barrel-Stave Model 03BA1100 Class I	1100	5.2	194.1	74	2.0	19	50
DREA PZT Barrel-Stave high power Class II	1270	4.3	196.1	64	3.9	16	49
DREA PZT Barrel-Stave lightweight Class I	1450	4.1	194.4	50	1.5	26	63
NUWC Barrel-Stave[c] Terfenol-D Class I	1480	3.8	202.9	15	5.1	60	267
Edo PZT Barrel-Stave Model 6993 Class I	1560	3.0	194.7	50	4.1	14	49
Sparton Free-Flooded Ring[d] PZT Model 28FC1000	1600	1.4	217.7	57	270	43	96
DREA PZT Barrel-Stave broadband dual-shell Class III	1650	1.8	189.4	50	3.1	8	24
Allied Signal DX835E[e] PMN Class IV flextensional	2500	4.6	210.2	43	10	83	185
Allied Signal DX835 PZT Class IV flextensional	2500	5.7	206.6	>90	9.5	30	64
Lockheed Martin Class IV PMN flextensional	4600	2.8	204.9	61	0.88	244	374
Raytheon copolymer projector	11250	0.63	211.1	32	1.0	425	1570
Raytheon copolymer projector	15500	1.6	216.1	26	1.0	162	633
MSI 1:3 piezo-composite $\lambda/2$ resonator	84000	10.4	219.4	89	0.44	3	6

[a]Limited by internal heating problem.
[b]Limited by saturation in the magnetic circuit.
[c]Limited by stave attachment method.
[d]Limited by cable and amplifier ratings.
[e]Limited by cavitation and electrical difficulties.

a predetermined orientation, producing a steered beam at a different angle from the first. For example, if the desired steering resolution is 10°, then 18 layers are consolidated, with each layer at a 10° offset from the one below it. Because the impedance of the copolymer is near the ρc of water, the layer which is being excited to steer in a particular direction is not influenced by the other layers. The simplicity of the concept is that there are no wires in front of the array or behind it; connections are all made at the periphery of the array. Consequently, higher source levels can be manageable and workable.

CONCLUSION

The maturation of Terfenol-D and PMN as useable high drive, high strain materials has signaled a new design era in which performance gains can be made which will vastly improve the Navy's sonar systems and their warfighting ability. This will also prompt investigators to develop other materials which can offer improved performance over that now considered exemplary. The emergence of electroactive polymers as a active projector material opens up broad new vistas and offers the opportunity to transition to other technology areas such as medical ultrasonics.

REFERENCES

1. M. Moffett, A. Clark, M. Wun-Fogle, J. Lindberg, J. Teter, and E. McLaughlin, *J. Acoust. Soc. Am.* **89**, 1448-1455 (1991)
2. M. Moffett, J. Powers, and A. Clark, *J. Acoust. Soc. Am.* **90**, 1184-1185 (1991)
3. F. Claeyssen, D. Colombani, A. Tessereau, and B. Ducros, *IEEE Trans. on Magnetics* **27**, 5343-5345 (1991)
4. A. Clark, M. Wun-Fogle, J. Restorff, and J. Lindberg, *IEEE Trans. on Magnetics* **29**, 3511-3513 (1993)
5. K. Rittenmyer, *J. Acoust. Soc. Am.* **95**, 849-856 (1994)
6. G. Kavarnos and E. McLaughlin, NUWC Technical Report 10607, Naval Undersea Warfare Center (1994)
7. E. McLaughlin, J. Powers, M. Moffett, and R. Janus, 1996 ONR Transducer Materials and Transducer Workshop, State College, PA, 25-27 March 1996 (presentation)
8. L. Ewart-Paine, to be reported at American Ceramic Society meeting, May 1997
9. A. E. Clark, in Ferromagnetic Materials, ed. E. P. Wohlfarth, Vol. 1, Chapt. 7, page 531, North Holland Publishing Co. 1980.
10. A. E. Clark, M. Wun-Fogle, and J. B. Restorff, presented at the 1996 ONR Transducer Materials and Transducers Workshop, State College, PA, 25-27 March 1996.
11. Guang-Heng Wu, Xue-Gen Zhao, Jing-Hua Wang, Jing-Yuan Li, Ke-Chang Jia, and Wen-Shan Zhan, *Appl. Phys. Lett.* **67**, 2005 (1995).
12. A. E. Clark, L. T. Kabacoff, H. T. Savage, and C. Modzelewski, U. S. Patent #4,763,030, Aug. 9, 1986.
13. See, for example, Proceedings of the International Symposium on Giant Magnetostrictive Materials and Their Applications, Tokyo, Nov. 5-6, 1992.
14. C.H. Sherman, *IEEE Trans. on Sonics and Ultrasonics* **SU-22**, 281-290 (1975)
15. W.J. Marshall, J.A. Pagliarini, and R.P. White, *Proceedings of the IEEE Oceans '79*, (Institute of Electrical and Electronics Engineers, New York, 1979), pp. 124-129
16. J. Oswin and J. Dunn, Proceedings of the International Workshop on Power Sonic and Ultrasonic Transducers Design, edited by B. Hamonic and J.N. Decarpigny, (Springer-Verlag, Berlin, 1988), pp. 121-133
17. G.A. Brigham and L.H. Royster, *J. Acoust. Soc. Am.* **46**, 92 (abs) (1969)

18. L.H. Royster, *Appl. Acoust.* 3, 117-126 (1970)
19. E.F. Rynne, <u>Proceedings of the Third International Workshop on Transducers for Sonics and Ultrasonics,</u> edited by M.D. McCollum, B.F. Hamonic, and O.B. Wilson, (Technomic, Lancaster, PA, U.S.A., 1993), pp. 38-49
20. H.C. Merchant, "Underwater Transducer Apparatus," U.S. Patent Number 3,258,738 (June 1966)
21. M.B. Moffett and W.L. Clay, *J. Acoust. Soc. Am.* **93**, 1653-1654 (1993)
22. D.F. Jones and J.F. Lindberg, <u>Proceedings of the Institute of Acoustics; Sonar Transducers '95,</u> edited by J.R. Dunn, Volume 17 Pt 3 1995, 15-33.
23. J.L. Butler, and A.E.Clark, U.S. Patent Number 4,443,731 (April 1984).
24. J.L. Butler, S.C. Butler, and A.E. Clark, *J. Acoust. Soc. Am.* **88**, 7 (1990).
25. J.L. Butler, A.L. Butler, and S.C .Butler, *J. Acoust. Soc. Am.* **94**, 636 (1993).
26. J.F. Lindberg, U.S. Patent #5,530,683, "Steerable Acoustic Transducer", 25 June 1996 and U.S. Patent #5,511,043, "Multiple Frequency Steerable Acoustic Transducer", 23 April 1996

THEORY AND APPLICATIONS OF LARGE STROKE TERFENOL-D® ACTUATORS

M.J. GERVER*, J.R. BERRY*, D.M. DOZOR*, J.H. GOLDIE*, R.T. ILMONEN**, K. LEARY*, F.E. NIMBLETT*, R. RODERICK*, J.R. SWENBECK*
*SatCon Technology Corporation, 161 First Street, Cambridge, MA 02146, gerver@satc.com
**Present address: Foster-Miller, Inc., 350 Second Avenue, Waltham, MA 02154

ABSTRACT

Traditionally, control algorithms for Terfenol-D® magnetostrictive actuators have modelled the strain as a linear function of magnetic field and stress, but nonlinearity becomes important for strains of more than a few hundred parts per million (ppm), and for many applications even the maximum strain, about 1500 ppm, is inadequate. Larger strokes can be obtained by various types of stroke amplifiers, by resonant operation, or by inchworming. Previously, SatCon successfully used a 10:1 lever arm stroke amplifier in a helicopter flap actuator [1]. Current projects include a water pump using a hydraulic stroke amplifier, which potentially could be more compact and efficient than a lever arm amplifier, and linear and rotary inchworm motors for robotics. In all these designs, satisfactory performance requires careful attention to machining tolerances and to making the mechanisms and housing stiff enough or compliant enough. A model for inchworm motors has been developed, including finite load and resonant effects. Nonlinear control algorithms will be discussed, applicable to arbitrarily large Terfenol-D® strains, stresses, and magnetic fields.

INTRODUCTION

Terfenol-D® magnetostrictive actuators offer a number of advantages over Lorentz force or magnetic attraction force actuators, including much higher force per area, the ability to control stroke directly with high precision, and no fatigue. They also require much lower voltages, and can operate at much lower temperatures, than piezoelectric actuators. But the maximum strain that Terfenol-D® actuators can achieve, at least statically, is about 1500 ppm. Furthermore, the greatest strains require large fields, 1000 oersteds or more, and at these strains the response of the actuator is very nonlinear. This may not be acceptable for applications, such as vibration control, which have low tolerance for harmonic generation.

Over the past few years, SatCon and others have investigated a number of methods to overcome these limitations of Terfenol-D® actuators. Claeyssen et al [2] have shown that strains as high as 3500 ppm, with little harmonic distortion, can be achieved in sonar transducers if they are operated at resonance, although this requires high bias field and bias stress. A mechanical stroke amplifier, also making use of acoustic resonance, was used by ETREMA in an ultrasonic surgical tool [3], with a stroke amplification of 19, and SatCon recently used a lever arm mechanism to achieve a stroke amplification of 10 in a helicopter flap actuator [1]. A linear inchworm motor, invented and built by Kiesewetter [4], and a rotary inchworm motor by Vranish et al [5], can achieve unlimited stroke.

In this paper, we describe 1) a magnetostrictive water pump, built for NASA, which uses a 10:1 hydraulic stroke amplifier, 2) a linear inchworm motor built for robotic applications, and a planned rotary inchworm motor, differing from previous designs, and 3) a preliminary investigation of nonlinear behavior in Terfenol-D® and control algorithms for dealing with it.

MAGNETOSTRICTIVE WATER PUMP

The present portable life support system (PLSS) used by astronauts in extravehicular activity uses a pump, for circulating cooling water, powered by a motor which must be overhauled after 180 hours of use. In designing a new PLSS for future manned missions of long duration, including a possible Mars mission, NASA sought a pump using either piezoelectrics or magnetostrictives, which, having essentially no moving parts, could be expected to operate reliably for much longer periods. Piezoelectrics require

521

high voltage, a disadvantage in space where arcing is a problem, and they suffer from depoling, two problems not shared by magnetostrictives. A Terfenol-D® pump could potentially be used for cryogens. And unlike a pump using an attraction force actuator, it could still operate, though perhaps not as effectively, if there were a vacuum leak behind the piston. For all these reasons it was felt that Terfenol-D® could best meet NASA's requirement for a compact, reliable pump of reasonable efficiency.

Although Terfenol-D® pumps have been built before [6], they have been for applications in which high pressure (> 100 psi) and low but precisely controlled flow rate (< 2 ml/sec) were needed. Terfenol-D® actuators were able to work directly against high pressure, without the need for mechanisms to multiply force, and the low flow rates meant that the small stroke would not be a disadvantage, and it would not be necessary to operate at very high frequency, or to have a very large piston. The pump for the PLSS, in contrast, would have to produce a flow of 30 ml/sec, but only at 5 psi. The size of the pump was no more than 90 mm in diameter and 150 mm in length. The power was to be as low as possible, preferably under 10 or 15 W, and the pump cannot be excessively noisy.

In spite of the high flow rate required, it was hoped at first that the pump could be designed without a stroke amplifier. For a 75 mm diameter piston and a 100 mm long Terfenol-D® rod with strain of 1000 ppm (corresponding to H going from zero to 1000 Oe), the piston displacement would be 0.43 ml, and the pump would have to operate at 70 Hz to produce a flow rate of 30 ml/sec, even if the valves were 100% efficient (no backflow) and there were no compliance in the pump chamber. Preliminary tests showed that when the pump is first filled with water, there is typically between 0.5 and 1.5 ml of air trapped in the chamber, depending on the size and shape of the chamber and on how carefully it is filled. Although some of the air can be removed as the water circulates, it is difficult to remove all of it. Furthermore, the astronauts will frequently remove the intake and outflow tubing from the pump and replace them, bringing more air into the chamber. With a 15 psi intake pressure and 20 psi outflow pressure, and neglecting the pressure drop across the valves, the change in volume of 1.5 ml of trapped air during each cycle is 1.5 ml × (1 - 15/20) = 0.375 ml, and the net volume of water displaced per cycle is 0.43 - 0.375 = 0.055 ml. Then the pump must operate at 540 Hz to produce 30 ml/sec of flow. In fact, the assumptions of no backflow, and no pressure drop across the valves, are not realistic, especially at such high frequency, and such a pump would not be able to produce a flow rate anywhere close to 30 ml/sec. Even if the trapped air volume is smaller, and it is possible to achieve 30 ml/sec when the frequency is high enough, the trapped air can make the pump very noisy. The noise results from resonance of air bubbles. A spherical bubble of radius r_o, has a natural vibration frequency ω_{air} given by

$$\frac{\omega_{air}}{2\pi} = \left(\frac{r_0}{\pi\rho} \right)^{1/2} \left(\frac{3\gamma P_0}{4\pi r_0^3} + \frac{\partial P}{\partial V} \right)^{1/2} \tag{1}$$

where ρ is the density of water, P_o is the pressure, $\gamma = 7/5$ for a diatomic gas like air, and $\partial P/\partial V$ is the stiffness of the chamber. This frequency is typically 350 to 650 Hz for our pump, and can be observed as a ringing in the chamber pressure when a sudden impulse is applied to the piston by the actuator. Even when the drive frequency is much lower than this, bubble vibrations are driven by higher harmonics of the pump frequency, especially when the Terfenol-D® is driven hard and it becomes very nonlinear. For a given air bubble frequency ω_{air} and drive frequency ω, we have found that the noise level increases rather abruptly when the peak-to-peak ac magnetic field H_{max} in the Terfenol-D® exceeds a certain value (in oersteds),

$$H_{max} > 38\omega_{air}/\omega \tag{2}$$

In these tests, $H(t)$ varied sinusoidally between 0 and H_{max}, ω_{air} was either 350 Hz or 650 Hz, and ω_{air}/ω was between 5 and 25. It is likely that higher H_{max} could be used, without producing a lot of noise, if $H(t)$ were not sinusoidal, but were carefully tailored to produce a sinusoidal pressure variation in the chamber, taking into account the nonlinear response of the Terfenol-D®, and the valve dynamics.

To produce 30 ml/sec flow when there is trapped air, while avoiding this noisy regime, we must amplify the Terfenol-D® stroke. Greater stroke amplification leads to a pump with lower power dissipation. With stroke amplification factor $f = 10$, it is possible to design a pump with 75 mm piston diameter, dissipating only 15 W, while pumping 30 ml/sec at 5 psi. The design includes a bias H from permanent magnets, and a bias stress of about 1 ksi. The magnetic circuit efficiency is 52%.

The stroke amplifier should have a diameter no greater than 75 mm, so it fits within the pump housing, and a length no greater than 37.5 mm, an arbitrary figure that is 25% of the total length, which must also include the actuator, chamber, valves, and compensation bellows. Within this envelope, we first tried designing a lever-arm stroke amplifier, using flexures, with $f = 10$, subject to the constraint that the maximum stress encountered in the flexures should be less than 50% of the yield stress, so that the lifetime would be effectively infinite. Even with stainless steel, we were unable to find a design that was more than 65% efficient. Only 65% of the work done by the Terfenol-D® went into moving the piston, and the rest went into deforming the lever arms and flexures of the stroke amplifier.

A more efficient stroke amplifier, with the same envelope and $f = 10$, can be built using hydraulics, shown schematically in Fig. 1. A wide diameter bellows is attached to a narrow diameter bellows, and the space between them is filled with a nearly incompressible fluid (silicone oil), carefully avoiding any trapped air, and sealed. (Unlike the pump chamber, once the bellows are sealed there is never any need to open them and introduce more air.) The Terfenol-D® pushes on the end of the wide bellows, compressing them. If the Terfenol-D® stroke is d_1 and the cross-sectional area of the wide bellows is A_1, the volume of the fluid inside the wide bellows will be reduced by $A_1 d_1$. Ideally, if the fluid were incompressible and the bellows could not expand or contract radially, then the volume of the narrow bellows would have to decrease by the same amount, and the narrow bellows would contract axially by $d_2 = A_1 d_1 / A_2$, where A_2 is the cross-sectional area of the narrow bellows. Similarly, if the axial stiffness of the bellows were negligible, then the change in force F_2 on the piston would be related to the change in force F_1 on the Terfenol-D® by $F_1 / F_2 = A_1 / A_2$, so the work $F_1 d_1$ done by the Terfenol-D® would be equal to the work $F_2 d_2$ done by the piston, and the stroke amplifier would be 100% efficient. If the bellows have thick walls, then they will have little tendency to expand radially, but will have great axial stiffness, and the opposite will be true if the walls are thin. In either case the efficiency of the bellows, defined as $F_2 d_2 / F_1 d_1$, will be low. The maximum efficiency will occur when the bellows have walls of optimal thickness and optimal shape. For wide bellows of stiffness k_1, narrow bellows of stiffness k_2, Terfenol-D® stiffness k_T, and compliance $\partial V/\partial P$ (including radial expansion of the bellows, compressibility of the fluid, and any compliance due to air bubbles in the oil), we find

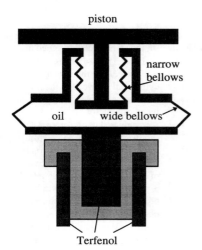

piston

narrow bellows

oil   wide bellows

Terfenol

Figure 1: Schematic of two-stage Terfenol actuator, hydraulic stroke amplifier, and piston.

$$d_2 = \left[d_0 \frac{A_1}{A_2} \left(1 + \frac{k_1}{k_T} \right)^{-1} - \frac{F_2 X}{k_T} \right] \left(1 + X \frac{k_2}{k_T} \right)^{-1} \quad (3)$$

where

$$X = \frac{k_T}{A_2^2} \frac{\partial V}{\partial p} + \frac{A_1^2}{A_2^2} \left(1 + \frac{k_1}{k_T} \right)^{-1} \quad (4)$$

and d_0 is the length of the Terfenol-D® times $H_{max} \partial \varepsilon / \partial H$, i.e. the stroke that it would have if the force were not changing. Its actual stroke d_1 is given by

$$d_1 = \left(d_0 - \frac{A_1 (k_2 d_2 + F_2)}{A_2 \ k_T} \right) \left(1 + \frac{k_1}{k_T} \right)^{-1} \quad (5)$$

and the change in force is $F_1 = k_T (d_0 - d_1)$.

Using finite element analysis, we have designed a

hydraulic stroke amplifier, within the required envelope, which allows the pump to produce a flow of 30 ml/sec at 5 psi, with the lowest possible power dissipation. Roughly speaking, this means maximizing d_2 /d_0, when F_2 is 5 psi times the piston area. Maximizing the piston stroke d_2 allows the drive frequency to be as small as possible, minimizing the hysteresis power dissipated in the Terfenol-D®. Minimizing d_0 minimizes H_{max}, and the coil resistive power. The optimal bellows design has an efficiency $F_2 d_2 / F_1 d_1$ of 75%, which is greater than the 65% efficiency of the best lever arm stroke amplifier.

An unusual feature of the actuator is the two-stage Terfenol-D® design, shown in Fig. 1. There are two pieces of Terfenol-D®, the outer one a hollow cylinder, and the inner one a solid cylinder nested in a piece of stainless steel nested inside the outer one, so that their strokes add up, like a telescope. This configuration, which was adopted to reduce the length of the pump and make better use of space, works fine with a few hundred lbs of force holding the pieces together. Without such force, the imperfect machining leaves gaps of up to 1.5 mils between the pieces, or between the Terfenol-D® and the bellows, which greatly reduce the stroke.

Due to an error in designing the bellows of the stroke amplifier, and possibly an error in manufacturing them, they are about 5 times stiffer than they should be, and their efficiency is only 50%. As a result, it would not be possible to reach the full 30 ml/sec flow rate at 5 psi. Although flow data for the pump was not available in time to include in this paper, the piston stroke was measured with no force on the piston ($F_2 = 0$), and the displaced volume was close to that in earlier tests, using a larger piston and no stroke amplifier, for which flow data was taken. The flow rate for the pump ought to be similar to what was seen in those tests, shown in Fig. 2. The valves are flap valves, opened and closed by the pressure difference. At frequencies greater than 35 Hz the inertia of the water behind the flaps prevents them from opening fully, and the flow rate stops going up with frequency. This might be avoided by using separate actuators to operate the valves, in phase with the actuator driving the piston, rather than relying on the pressure difference. However, the optimum drive frequency to minimize power would only be about 20 Hz (if the bellows had the correct stiffness), so the valve inertia will not adversely affect performance, and the pump can be simpler and more robust if the the valves are opened and closed by the pressure difference.

INCHWORM MOTORS

A linear inchworm motor was built and tested, and a rotary inchworm motor has been designed and is being built. Unlike the Kiesewetter motor [4], in which a cylindrical rod of Terfenol-D® fits snugly into a precisely machined cylindrical sleeve, and the windings encircle the sleeve, our inchworm motor consists of a rectangular slab of Terfenol-D® sandwiched between two spring-loaded flat plates. In addition to serving as a reaction surface, one of the plates contains a standard 3-phase winding. The winding provides the MMF which causes the Terfenol-D® to move. This configuration should avoid the problem of rapid wearing out seen by Kiesewetter, since it is only necessary for both surfaces to remain flat. An analytic expression was found for the inchworming speed v, when the

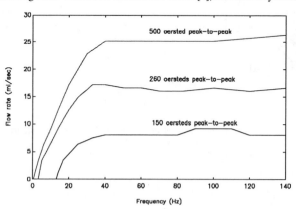

Figure 2: Flow rate as a function of actuator drive frequency and amplitude

windings induce a peristaltic wave of phase velocity ω/k in the Terfenol-D$^®$,

$$v = \left(1 - \frac{\omega^2}{k^2 c_s^2}\right)^{-1} \frac{\omega}{k}\left[\varepsilon_{mas} - \frac{1}{2}\left(\frac{\partial \varepsilon}{\partial \sigma}\right)\sigma_a\right] \qquad (6)$$

Here ε_{max} is the zero-to-peak magnetostrictive strain at constant stress, σ_a is the load stress on the Terfenol-D$^®$, $\partial \varepsilon/\partial \sigma$ is its compressibility, and c_s is the sound speed. The inchworming motion is in the opposite direction to the phase velocity of the excitation. The resonant denominator never leads to infinite speed, of course, but is limited by damping and nonlinearities, ignored here. In a larger motor the acoustic resonance could substantially increase the speed for a given current in the windings, but for our motor the resonance would occur at 42 kHz, where hysteresis losses would be much greater than power dissipated in the coils, so there would be no point in running close to resonance. Equation (6) is an overestimate of the inchworming speed, since it assumes that the regions over which the Terfenol-D$^®$ is locked against the plates are infinitesimally small. If that were true, then the perpendicular stress would be infinite at those points, and both the Terfenol-D$^®$ and plates would be deformed, so in practice they are in contact over a finite fraction of a wavelength, and the inchworming speed is correspondingly reduced. In our tests, the mechanical wavenumber k (which is twice the electrical wavenumber, since magnetostrictive strain is an even function of H) was $2\pi/(18$ mm), and mechanical frequency $\omega/2\pi$ (which is twice the electrical frequency) ranged from 40 to 2000 Hz. The peak current in the windings ranged from 20 to 85 amp-turns per mm, which we estimate (based on magnetic finite element analysis) produced a peak H equal to only 72% of that, due to leakage. At zero load, at the lowest frequency and current, the inchworming speed was 61% of what Eq. (6) predicts, which seems reasonable. The inchworming speed initially increases linearly with frequency, as expected, but starts to decrease due to skin effects above 800 Hz (400 Hz electrical), since neither the Terfenol-D$^®$ nor the poles were laminated in these preliminary tests. The inchworming speed was also lower, relative to what was predicted, at larger current, and it fell off much more rapidly than predicted with increasing load. It is likely that these effects are both due to the finite length of the locked regions, which may increase in these cases. The greatest speed seen, at 800 Hz and 85 A-turns/mm, was 2.8 mm/sec.

A rotary inchworm motor can be considered conceptually as a linear inchworm motor wrapped around into a circle. The plates can surround the Terfenol-D$^®$ either axially (like stacked washers) or radially; the former is likely to work better, because the surfaces will be planar, as in the linear motor. As long as there are an integer number of electrical wavelengths (an even integer number of mechanical wavelengths) in the full circle, Eq. (6) will still apply, and there is no need to put a cut in the ring of Terfenol-D$^®$, as we had earlier assumed. This kind of design should result in a motor that is simpler and more compact than the rotary inchworm motor of Vranish et al [5], which consists of four linear inchworm motors arranged in a circle.

NONLINEAR CONTROL ALGORITHMS

This work, which can only be described briefly here, included a study of the physical effects which cause nonlinear behavior in Terfenol-D$^®$, and an examination of control algorithms for reducing nonlinearity in actuators. At low amplitude, about 50 Oe peak-to-peak, it is sufficient to consider the nonlinearity of $\partial \varepsilon/\partial H$ and possibly $\partial \sigma/\partial \varepsilon$, and the amplitude dependent phase shift of the hysteresis in B vs. H, but the nonlinearity in $\partial B/\partial H$ can be neglected because the higher harmonics in B are filtered out by skin effects. At high amplitude, a few hundred Oe or more, a full nonlinear treatment is needed. Generally, even harmonics are generated by hysteresis, and odd harmonics by the nonlinearity of the anhysteretic $B(H,\sigma)$ and $\varepsilon(H,\sigma)$. We found that hysteresis in Terfenol-D$^®$ is well described by a classical Preisach model with two inputs (H and σ), with no need to invoke the nonlinear generalization of the Preisach model (where the permeability at a given H depends on past history), or models with reptation (where minor hysteresis loops drift toward the anhysteretic curves after many iterations) [7]. In particular, there was no detectable reptation in ε (less than 10 ppm) after 100 iterations of minor loops with $\Delta H = 150$ Oe or 250 Oe, and $\Delta\sigma = 300$ psi or 500 psi. Since it is unlikely that we were using the one ratio of ΔH to $\Delta\sigma$ for which reptation happens to cancel out, these conclusions probably apply separately to ε vs. H and ε vs. σ hysteresis, as well as to B vs. H and B vs. σ (since there is very little B vs. ε hysteresis).

Anhysteretic $B(H,\sigma)$ and $\varepsilon(H,\sigma)$ in a twinned Terfenol-D® rod with perfect [112] alignment can in principle be found by finding the angle of magnetization in each twin which minimizes the free energy of the rod. The important terms are the usual anisotropy, magnetoelastic, and $H\cdot M$ terms [8], averaged over both twins, and another term which is minimized when the angle of magnetization to the twinning plane is the same in both twins,

$$U_{twin} = (M_s^2/4\mu_0)(\alpha_x^+ - \alpha_x^-)^2 \tag{7}$$

Here M_s is the saturation magnetization, and α_x^+ and α_x^- are the sines of the angles of magnetization to the twinning plane for each of the twins. Having different angles of magnetization in the two twins also modifies the magnetoelastic term, which must now include the internal stress which develops so the strain tensors of the twins match each other. The free energy associated with domain walls, which generally requires finite element analysis to calculate, can be neglected in finding the anhysteretic curves, although it is important for hysteresis. The anhysteretic curves should thus be calculable by minimizing a simple analytic expression, without the need for finite element analysis.

ACKNOWLEDGMENTS

The work described in this paper was done under NASA Contracts No. NAS9-19114 and No. NAS9-19457, and under NSF Contract No. DMI-9561773. We are grateful to Scott Askew and Lee Willis of Johnson Space Center for their help and encouragement.

REFERENCES

1. D. A. Bushko, R. C. Fenn, M. J. Gerver, J. R. Berry, F. Phillips, and D. J. Merkley, in SPIE Symposium on Smart Structures and Materials, San Diego, CA, 26-29 February 1996, paper 2717-03.

2. F. Claeyssen, N. Lhermet, and R. Le Letty, in Proc. of 4th International Conference on New Actuators (ACTUATOR 94), June 15-17, 1994, Bremen, Germany, p. 203-209.

3. T. Toby Hansen and Solomon R. Ghorayeb, in Materials for Smart Systems, edited by E.P. George, S. Takahashi, S. Trolier-McKinstry, K. Uchino, and M. Wun-Fogle (Mater. Res. Soc. Proc. 360, Pittsburgh, PA, 1995), p. 259-264.

4. L. Kiesewetter, in Proc. of Second International Conference on Giant Magnetostrictive Alloys, edited by C. Tyren, Marbella, Spain, 1988.

5. J. M. Vranish, D. P. Naik, J. B. Restorff, and J. P. Teter, IEEE Trans. Mag. 27, 5355-5357 (1991).

6. Alfred Teves GmbH, a subsidiary of ITT, has a German patent (number 4,032,555) on a magnetostrictive pump for an anti-lock hydraulic brake system. ETREMA sells a pump which produces 1.7 ml/sec at 114 psi. Reference 2 describes a pump built by K. Suzuki for use in the chemical industry.

7. I. D. Mayergoyz, Mathematical Models of Hysteresis, Springer-Verlag, New York, 1991, Chapter II.

8. D. C. Jiles and J. B. Thoelke, IEEE Trans. Mag. 27, 5352-5354 (1991).

MAGNETOSTRICTIVE COMPOSITE MATERIAL SYSTEMS ANALYTICAL/EXPERIMENTAL

T.A. DUENAS†, L. HSU††, and G.P. CARMAN†††

†Mechanical and Aerospace Engineering Department, University of California,
Los Angeles, California 90095-1597
terrisa@cad.ucla.edu

††Hughes Space and Communications Company,
Los Angeles, California 90030-9009, Post Office Box 92919
00J5750@ccgate.hac.com

†††Mechanical and Aerospace Engineering Department, University of California,
Los Angeles, California 90095-1597
carman@seas.ucla.edu

ABSTRACT

Experimental and theoretical results are presented for a composite magnetostrictive material system. This material system contains Terfenol-D particles blended with a binder resin and cured in the presence of a magnetic field to form a 1-3 composite. Test data indicates that the magnetostrictive material can be preloaded in-situ with the binder matrix resulting in orientation of domains that facilitate strain responses comparable to monolithic Terfenol-D. Two constitutive equations for the monolithic material are described and a concentric cylinders model is used to predict the response of the composite structure. Experimental data obtained from the composite systems coincide with the analytical models within 10%. Particle size, resin system, and volume fraction are shown to significantly influence the response of the fabricated composite system.

INTRODUCTION

In the early 1960's scientists discovered that certain rare earth elements—i.e., terbium (Tb) and dysprosium (Dy)—exhibited giant magnetostriction effects on the order of 10,000 microstrain [1, 2]. These large effects, however, were only present in these rare earth materials at cryogenic temperatures. A decade later notable magnetostriction effects at room temperature as high as 2000 microstrains were discovered by combining Tb and Dy with magnetic transition materials [3]. Research on these magnetostrictive materials, like Terfenol-D, shows that they provide several advantages over their electro-mechanical counterparts including larger strains (in some instances) as well as freedom from depoling effects and breakdown strengths. These advantages have led investigators to explore the benefits of using magnetostrictive materials in various engineering structures, for example, using Terfenol-D in active control flaps for reducing the vibrations on a rotorcraft system [4].

While investigators are exploring applications for Terfenol-D, engineering models are still predominantly based on linear piezomagnetic behavior [5]; these linear models neglect critical coupling terms which must be considered to develop accurate predictive capabilities required in structural applications—including the effect of preload (prestress), magnetic field intensity, and temperature on the mechanical response as indicated by Clark [6]. Brown [7] describes constitutive approaches for magnetostrictive materials that include magnetic body moment coupling. Carman [8] uses a higher-order series expansion in thermodynamic relations to account for specific nonlinear coupling effects while Kannan and Dasgupta [9] suggest using magnetization as the dependent variable to circumvent nonlinear features. Notwithstanding, there does not exist a constitutive relation applicable to the wide range of operating scenarios.

Low fracture toughness is an additional limitation associated with magnetostrictive materials. In the case of piezoceramics which exhibit the same limitation, researchers partially solved the problem by embedding the material into a composite. Researchers also report (e.g. [10]) that by combining active materials into composite assemblages certain coefficients, such as the piezoelectric hydrostatic coefficient, could be enhanced. Bowen [11] and others report that 1-3 particulate piezocomposites can be manufactured with dielectric particles cured under a non-uniform electric field (dielectrophoresis); however, the alignment of the dielectric particles is difficult to achieve and the displacements provided by the composite are small. On the other hand, magnetophoresis can be used to align ferromagnetic particles in an epoxy resin in a similar 1-3 configuration. While a 1-3 magnetostrictive composite could provide an innovative active material because of its increased fracture toughness and ease of manufacturing, little published data is available describing its magneto-mechanical response. Research by Bi and Anjanappa [12] suggest possible enhancements for these materials in the small scale regime. The work by Roberts et al. [13] discusses potential benefits of this technology, such as the elimination of prestress requirements. The more detailed work by Sandlund et al [14] on 1-3 magnetostrictive composites indicates that the displacement output of the magnetostrictive composite rivals that of the monolithic.

In this paper we describe an innovative active material fabricated with Terfenol-D powder integrated into a resin system. The combined active composite is more durable than many monolithic active materials and can be easily manufactured into complex shapes. In addition to these advantages, the combined material system could increase the displacement output, have larger duty cycles, and facilitate hybrid actuation concepts.

CONSTITUTIVE RELATIONS

Constitutive relations can be derived in two fashions, that is, either phenomenological with models based on test data or from micromechanics with models based on physical evidence such as knowledge concerning domain distribution. The approach described in this paper attempts to use a phenomenological model to describe the response of a polycrystalline material subjected to a sufficiently large preload such that all the domains are oriented perpendicular to the load as well as a micromechanics formulation to describe the response for the material with a distributed domain structure. Two constitutive approaches are reviewed to understand the basis for this model. The first model suggested by Carman and Mitrovic [8] for Terfenol-D uses independent variables of stress, temperature, and magnetic field. With the appropriate free energy function expanded in a Taylor series the following relationship was proposed.

$$B_m = H_n \mu_{mn}^{T\sigma} + \Delta T P_m^\sigma + \sigma_{ij} H_n Q_{klnm}^T + \Delta T \sigma_{ij} H_n \alpha_{ijnm} + \frac{1}{2} \sigma_{ij} \sigma_{kl} H_n S_{ijklnm}^T$$

$$\varepsilon_{kl} = \sigma_{ij} S_{ijkl}^{TH} + \Delta T \alpha_{kl}^H + \frac{1}{2} H_n H_m Q_{klnm}^T + \frac{1}{2} H_n H_p \sigma_{ij} S_{klijnp}^T + \frac{1}{2} H_m H_n \Delta T \alpha_{klmn} \qquad (1)$$

where B is magnetic flux, H is magnetic field, T is temperature, ε is strain, and σ is stress. The constants μ is permeability, P is pyromagnetic, Q is magnetostrictive, α is coefficient of thermal expansion, and S represents compliance terms. Higher order tensor values of S represent magnetomechanical effects while higher order α's represent magnetothermal interactions. The superscripts on the constants define the physical conditions under which each constant was measured, i.e., constant stress, temperature, or magnetic field. While this equation does provide reasonable results for large preloads, where all domains are aligned, it does not provide information about saturation influences.

In an attempt to provide a better approximation, other forms of the constitutive relations were investigated. Experimental results indicate that the magnetization versus strain response for the material is independent of stress levels. Dasgupta and Kannon [9] also proposed to use magnetization as the independent variable rather than magnetic flux or magnetic field. In a similar development Hon and Shankar [15] developed an accurate constitutive relation for electrostrictive materials with polarization as the independent variable. Based on this information, the independent variables of magnetization, stress, and temperature were investigated. Using the appropriate free energy function the following constitutive relations can be derived. (See Hon and Shankar [15] for analogous development.)

$$\varepsilon_{ij} = S_{ijkl}^{M,T} \sigma_{kl} + Q_{ijkl}^T M_k M_l + \alpha_{ij}^M \Delta T$$

$$H_k = -2 Q_{ijkl}^T M_l \sigma_{ij} + \frac{M_k}{k|M|} \arctan h\left(\frac{|M|}{M_s}\right) + P_k^\sigma \Delta T \qquad (2)$$

Where k is a constant, M is the magnetization, and M_s represents the saturation magnetization.

Equation 2 has considerably fewer unknown constants when compared to Equation 1 and thus would represent a preferred relation, assuming that it provides reasonable agreement with experimental data. Both Equations 1 and 2 provide reasonable results for large preloads on the material, but both equally fail to provide an accurate prediction for low stress levels. To address this issue, assume when a polycrystal material is sufficiently preloaded it can be adequately characterized by a single homogenized domain state while for lower load levels a multi-domain state is a more appropriate description. To model the response of this single domain structure either Equation 1 or 2 could be used. Furthermore, assume that each domain in the multi-domain state can adequately be characterized by the equation used for the single domain structure. Therefore, by using the equation for a single domain coupled with knowledge about the domain orientation in the medium, the strains and the magnetic fields can be volume averaged over the structure and equated to a homogenous medium. This approach is mathematically expressed as follows.

$$\int_{vol} \varepsilon_{ij} dV = \sum_{domains} \int_{vol} a_{im} a_{jn} \varepsilon_{mn}^{90\,domian} dv$$

$$\int_{vol} H_i dV = \sum_{domains} \int a_{im} H_m^{90\,domain} dv$$

(3)

where a is the transformation matrix for each domain structure and the variable $\varepsilon_{mn}^{90domain}$ and $H_m^{90domain}$ are defined by Equation 2 or Equation 1 calculated for the bulk material with a sufficiently large preload applied. This equation can be further simplified by assuming that the distribution of domains can be adequately represented by two domains—one oriented parallel to the load and the other perpendicular. This reduces Equation 3 to the following.

$$\int_{vol} \varepsilon_{ij} dV = \int \varepsilon_{ij}^{90\,domian} dv^{90} + \int a_{im} a_{jn} \varepsilon_{mn}^{90\,domain} dv^0$$

$$\int_{vol} H_i dV = \int H_i^{90\,domain} dv^{90} + \int a_{im} H_m^{90\,domain} dv^0$$

(4)

where for a 2-dimensional problem the transformation matrix can be assumed to take the form

$$[a] = \begin{bmatrix} 1 & 1 & 0 \\ -1 & 1 & 0 \\ 0 & 0 & 1 \end{bmatrix}$$

MICROMECHANICS

A concentric cylinders approach was used to model the response of the 1-3 magnetostrictive composite [17]. An illustration of this model is provided in Figure 1. One of the principle assumptions is that the aligned particulate behave as continuous fiber elements. Additional assumptions include treating the particle as perfectly bonded to the epoxy; assuming the constituents are representative of a crystal with 6mm symmetry with the principal direction coincident with the axis of the cylinder; and considering the magnetic/thermal fields within the assemblage as invariant with respect to position. The equilibrium equations for this problem in cylindrical coordinates are

$$\frac{\partial \sigma_{rr}}{\partial r} + \frac{1}{r}\frac{\partial \sigma_{r\theta}}{\partial \theta} + \frac{\partial \sigma_{rz}}{\partial z} + \frac{\sigma_{rr} - \sigma_{\theta\theta}}{r} = 0$$

$$\frac{\partial \sigma_{r\theta}}{\partial r} + \frac{1}{r}\frac{\partial \sigma_{\theta\theta}}{\partial \theta} + \frac{\partial \sigma_{\theta z}}{\partial z} + \frac{2}{r}\sigma_{r\theta} = 0 \qquad (5)$$

$$\frac{\partial \sigma_{rz}}{\partial r} + \frac{1}{r}\frac{\partial \sigma_{\theta z}}{\partial \vartheta} + \frac{\partial \sigma_{zz}}{\partial z} + \frac{1}{r}\sigma_{rz} = 0$$

$$\frac{\partial B_r}{\partial r} + \frac{1}{r}\frac{\partial B_\theta}{\partial \theta} + \frac{B_r}{r} + \frac{\partial B_z}{\partial z} = 0$$

Using the assumption that the magnetic field is invariant with respect to position the last equation is not required. These equations reduce to a mechanical problem with stiffness properties functionally dependent upon the applied magnetic field. Using strain displacement relations in the constitutive equations (either Equation 1 or 2) and substituting these into the equilibrium equations, three partial differential equations are obtained as a function of three unknown displacements. General solutions to these partial differential equations have been developed for composite elements [16] or for electro-magnetic composite elements [17]. Applying the appropriate boundary conditions, the stresses, strains, magnetic fields, and magnetic flux on a point-wise basis can be calculated for both the resin and the particles. To calculate effective properties for the magnetostrictive composite the appropriate property is volume averaged over the entire element and compared with a homogeneous material.

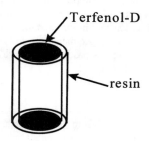

Figure 1:   Concentric cylinder model used to model 1-3 magnetostrictive composite

MANUFACTURING METHOD

A 1-3 Terfenol-D composite is manufactured by suspending Terfenol-D particulate in a resin system under the influence of a magnetic field. Two particulate sizes were studied: particles with dimensions less than 38 microns or particles less than 300 microns. These particulate were obtained from Etrema Products with sizing achieved with a sieve. The geometry of the particles are random in shape, neither tending to be ellipsoidal nor rod shaped.

Once manufactured, the particles are introduced into a resin system and cured in the presence of a magnetic field. When the magnetic field is applied, the particles find a microscopic "natural" orientation based on crystal structure and particle geometry where macroscopic chains form according to the magnetic flux lines. Because suspension of the particulate depends on both magnetic field and resin viscosity, a trade off exists, sometimes at the expense of evenly distributed particle chains. Seven resin systems with varying properties were studied in this research, including materials with different coefficient of thermal expansions CTE (30 to 50×10^{-6} /°C), Young's Modulus values (0.5 to 3 GPa), and viscosities (60 to 10,000 cps). For the resins studied, the low viscosity systems provided composites with larger displacement outputs due to the reduction in voids. Of the two low viscosity resins studied—Ciba-Geigy's modified bisphenol A epoxy resin and Polysciences Spurr epoxy—the Spurr appears to be a better candidate for mechanical actuators due to the relatively higher modulus of elasticity of 3 GPa compared to 0.5 GPa. In addition to varying resin systems and particle size, four different 1-3 magnetostrictive composite volume fractions (10, 20, 30, and 40%) were investigated.

The 1-3 magnetostrictive composites were cured in plexi-glass molds with dimensions of 17mm x 6mm x 6mm. The particulate and epoxy is mixed for approximately fifteen minutes followed by a pre-vacuum for two hours to remove entrapped gas introduced during the mixing process. Following this pre-cure process, a lid is placed over the cavity and fastened with a C-clamp and subsequently mixed again to redistribute the particulate that has settled in the mold. As mentioned, a combination of the correct resin viscosity and appropriate mixing provides an even distribution of particulate immediately prior to applying a magnetic field to the sample. When the magnetic field is applied to the sample, particles are attracted to the nearest magnetic flux line and join to form chains. This alignment is immediate and may not allow sufficient particle migration within the medium. In fact, the lower volume fraction samples (< 15%) typically produced specimens of questionable quality due to this lack of particulate homogeneity. Following this process, the entire assemblage is introduced into a vacuum oven for 8 hours and cured at 100°C.

EXPERIMENT

After a sample has been cured and retrieved from its mold the sample surface is prepared. Resistive strain gages (EA-06-062AP-120 by Measurements Group) are then attached to the surface in the longitudinal direction (i.e., fiber direction). To record strain response a high speed data acquisition system is used. Magnetic field is delivered from a laboratory fabricated solenoid connected to an HP 6268H 200 W DC power supply (0 - 40 V, 0 -30 A) while strain is recorded. Preloads are applied to the sample with a magnetic field by fixing the solenoid vertically along the compressive force path of an Instron 8516 as shown in Figure 2 below. This particular system is high-precision servo-hydraulic unit which can be run in displacement or load control; to simultaneously apply a preload during the generation of a magnetic field tests are conducted in the load control mode. Iron end pieces are fastened to the ends of the load plungers to reduce demagnetization and homogenize the magnetic fields generated by the solenoid. A compressive force is delivered to the upper and lower platen which is then transmitted to the load plungers. The composite sample is in direct contact with the iron end pieces but is secured in place with a nonmagnetic position mechanism to prevent slipping. When the load is applied the sample experiences a compressive force from the iron end pieces.

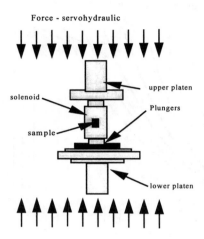

Figure 2:   Illustration of preload test set-up showing placement of solenoid and sample

ANALYTICAL RESULTS

In this section the analytical results for a monolithic material and a composite system are presented. In Figure 3 we present strain versus magnetic field results for the monolithic material using Equation 1 and experimental data obtained from Moffet et al. [18]. The constants used to generate these curves are called out in Carman and Mitrovic's paper [8]. The 8 curves in the figure correspond to different preloads applied to the material. The curves numbered 4-8 were constructed using the same coefficients while the curves numbered 1-3 required different coefficients in the model. This suggests that a physical phenomena is occurring at the lower preload levels which cannot be accounted for in the phenomenological model. Furthermore, while this constitutive relation produces reasonable correlation with experimental data it is unable to provide any indication of saturation influences at large magnetic field levels. In an effort to incorporate saturation influences we investigated the constitutive relation presented in Equation 2 of this paper. While accounting for saturation influences, Equation 2 was also unable to provide an accurate depiction of the materials response for both large and small preloads. These results were similar to the results obtained for Equation 1.

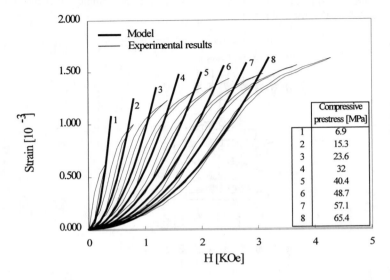

Figure 3:   Strain versus magnetic field results for the monolithic material using
              Equation 1 with experimental data [18]

In an attempt to remedy this dilemma, we used Equation 2 with Equation 4 to predict the materials response. In Figure 4, we present the analytical and experimental results for the same monolithic material. The material properties K, Q and Ms in Equation 2 play a key role in determining the strain/field response at different preloads. The analytical results were generated with K=0.0001[A/m], Q=0.002[1/Telsa^2] and Ms=0.8[Telsa]. Based on the experimental data the following hyperbolic function was used to approximate the domain distribution as a function of preload.

$$v_f^{90domain} = \tanh(\sigma / k1) \tag{6}$$

where σ is the preload and k1 is an experimentally determined constant; for this case it is 13MPa. While the results look promising, using an optimization algorithm to fit the data would have provided a better correlation than that presented in Figure 3. Nonetheless, results demonstrate good agreement over a wide range of loading scenarios including saturation influences using a single relationship.

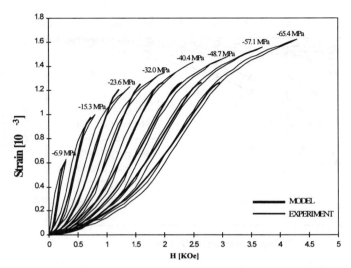

Figure 4:   Strain versus magnetic field results for the monolithic material using
             Equation 2 and 4 with experimental data [18]

Using the constitutive relations defined in Equation 1 and 4 with the micromechanics formulation described in the preceding section, the response of a 1-3 magnetostrictive composite is predicted. The calculations presented in this paper used isotropic properties for the stiffness values of Terfenol-D and the resin systems: $E^{TbDy} = 30GPa$ and $E^{epoxy} = 3GPa$ and $\nu = 0.3$. In Figure 5, the radial and axial stresses induced in the magnetostrictive material by the contraction of the resin during cure is presented. Stress values are normalized to the change in temperature during cure. Three different resin systems of varying coefficients of thermal expansions (CTE) are considered. The figure indicates that as the resin's CTE increases or the volume fraction of the Terfenol-D decreases the stresses imparted on the particulate increase. For a cure temperature of 100°C, axial stresses on the Terfenol-D particles range from 8 to 26 MPa (CTE=30×10^{-6} /°C and vf = 10 to 40%). This resin system is representative of the Spurr epoxy and cure temperature used in this study.

In Figure 6 we present the strain versus magnetic field for a 1-3 composite with different Terfenol-D volume fractions. The CTE of the epoxy is 30×10^{-6} [1/°C], comparable with the Spurr epoxy; the only preload applied to the sample is that which is induced during the cure process by the epoxy. All curves are truncated at saturation points and thus represent the largest strains achievable. These results indicate that the composite containing 20% volume particulate produced the largest strain of 600 microstrains. For composites containing 30 and 40%, the strain is approximately equal to 500 microstrain. The reason that the 20% volume fraction composite provides the largest strain output is attributed to the balance between in-situ preload by the resin and the force required to elongate the resin. However, as larger external preloads are applied the magnitude of the strain displacement will increase with the larger volume fraction composites providing larger displacements. In the

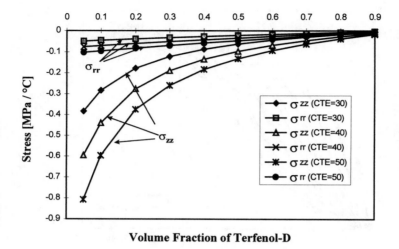

Volume Fraction of Terfenol-D

Figure 5:   Radial and axial stresses induced in the magnetostrictive composite material by the contraction of the resin during cure

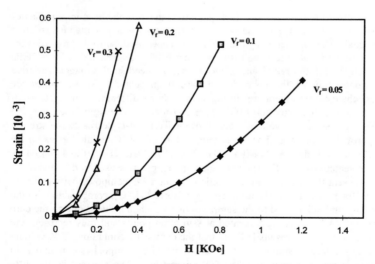

H [KOe]

Figure 6:   Extensional Strain of Terfenol-D Composites ($\alpha_{epoxy} = 30 \times 10^6$ [1/°C]) containing different volume fraction particles

Experimental Results section we show that the prediction presented in Figure 6 coincides with experimental data.

In Figure 7 we present the strain versus magnetic field for a 1-3 composite containing a resin with a CTE of 50×10^{-6} [1/°C]. Based on the results presented in Figure 5, this resin provides a larger in-situ preload than the 1-3 composite considered in Figure 6. In Figure 7, the strain output for an unloaded composite sample exceeds 900 microstrain, a 50% increase over the resin system with a CTE of 30×10^{-6} [1/°C]. The optimum volume fraction for this material system is 30% which is attributed to the larger in-situ preload applied on the larger volume fraction samples. As CTE values for the epoxy are increased the optimum volume fractions are shifted to higher volume fraction composites and the displacement capabilities increase approaching that of the monolithic. If the particles are aligned in a pseudo-single crystal format the response of the composite should exceed that of the monolithic material by as much as 30%. While experimental data is indicative of this pseudo-single crystal behavior, there is still insufficient evidence to conclude that this is occurring.

EXPERIMENTAL RESULTS

In Figure 8 experimental data generated for several 1-3 composite samples are presented. The magnetic field versus normalized strain (longitudinal) curves were generated on 4 composites manufactured with different resin systems. The magnetic field in the solenoid was not directly measured but calculations imply that for a 5 amp current the field should be approximately 0.7 KOe. This value is believed to be fairly accurate based on experimental data obtained for the monolithic material and comparisons with other published test data. In this figure all samples contained 40% volume fraction of particles less than 300 microns in size. The strains are normalized to the response of the monolithic material at a comparable prestress level. Results indicate that the composite samples fabricated with either Ciba-Geigy's or Polyscience's Spurr epoxy provided the largest strain response. The relatively larger response in these two samples are attributed to fewer voids when compared to the larger viscosity systems. The two preferred resins had viscosities ranging between 60 to 1000 cps while the larger viscosity resins ranged in value from 5000 to 10,000 cps. In all subsequent results presented in this paper the Polyscience resin is presented because of its relatively larger Young's modulus of 3 MPa compared to 0.5 MPa.

In Figure 9 magnetic field versus microstrain (longitudinal) is plotted for two samples. Both samples contain 30% volume fraction particulate less than 38 microns in size and were manufactured with the Polyscience's Spurr epoxy. The only difference between these two samples is that one sample was not cured in a magnetic field and thus had randomly oriented particles. As the results demonstrate the sample with random particle orientation yields a lower response than does its 1-3 composite counterpart cured under a magnetic field. This result was predicted but the strain magnitude for the aligned sample was less than expected. In general, all composites manufactured with the 38 micron size particulate produced only marginal strains.

In Figure 10 magnetic field versus microstrain (longitudinal) is plotted for two samples containing particles whose sizes are either less than 300 micron or less than 38 microns. Both samples contain 40% volume fraction and Polyscience's Spurr epoxy cured at the same temperature. While both samples were cured under a magnetic field to align particles, the

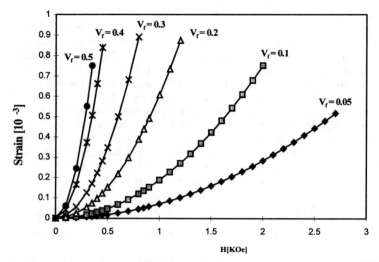

Figure 7: Extensional Strain of Terfenol-D Composites ($\alpha_{epoxy} = 50 \times 10^6$ [1/°C]) containing different volume fractions of particulate

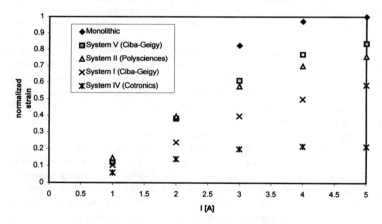

Figure 8: Experimental data for 40% volume fraction (less than 300 micron particle size) particulate magnetostrictive composites fabricated with different resins.

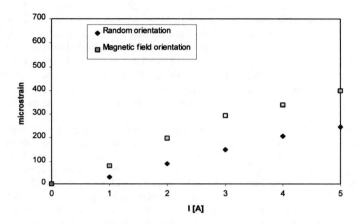

Figure 9:   Experimental data for 30 % volume fraction (less than 38 micron particle
             size) particulate magnetostrictive cured with and without a magnetic field

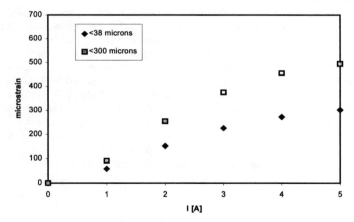

Figure 10:   Experimental data for 40% volume fraction magnetostrictive composites
             containing particles whose sizes are either less than 38 micron or less than
             300 microns

response of the sample with particles containing less than 38 micron particulate was substantially lower than the 300 micron particulate. This is a phenomena which has been reproduced in the 10, 20, and 30% volume fraction composites. Reasons contributing to the lower response could be attributed to oxidation influences, particle geometry, demagnetizing fields, or micro-voids around the smaller particles due to poor infiltration. Based on our experience with these materials, we believe that it is caused by micro-voids around the smaller particles produced by geometric influences which effect particle wetting. This would influence the preload applied to the particulate as well as provide a cavity in which the particle could expand. However, at this time there is no physical evidence supporting this contention.

In Figure 11 magnetic field versus microstrain (longitudinal) is plotted for samples varying only in volume fraction all cured at the same temperature using Polyscience's Spurr epoxy. All specimens were fabricated with less than 300 micron particulate. The 20% volume fraction sample yielded the largest strain response and corresponds with theory (See Figure 6). Theoretically the strain response of the 20% sample should be approximately 600 microstrain while the 30% and 40% volume fraction samples should provide about 500 microstrain. When reviewing Figure 10, these predicted values are within 10% of the experimentally determined values indicating that the analytical model reasonably depicts the responses of the composite. In regards to the 10% volume fraction sample, it produced the lowest strain response and was lower than predicted by the theory. This is attributed to the fact that the lower volume fraction samples had particle chains which were incomplete as described in the Manufacturing Method section of this paper.

In Figure 12 magnetic field versus microstrain (longitudinal) are plotted for a 40% volume fraction sample made with particles whose sizes are less than 300 microns and Polyscience's Spurr epoxy. Strain measurements were taken at different external preload values ranging from 5MPa to 10 MPa. While an unloaded sample produced 500 microstrains, a preload of 5 MPa produced a strain response of 700 microns at 0.7 Koe—a 40% increase in stroke. This response is comparable to the value reported by Sandlund [14]. However, Sandlund samples had a substantially larger volume fraction (75%), larger applied preload (15MPa), and larger magnetic field 1.5 KOe to achieve the same response. When the preload on the sample was increased to 10MPa in Figure 12, the magnetic field appears to be just at the beginning (or middle) of the magnetic "burst" effect suggesting that for a larger magnetic field a larger strain output would be achieved. While not presented here the experimental data presented in Figure 12 agrees well with the analytical predictions.

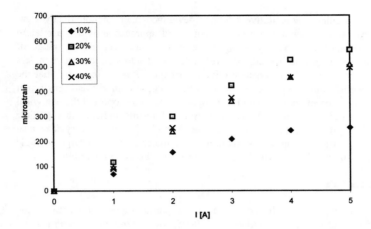

Figure 11:   Experimental data for (particles sizes less than 300 micron particles) magnetostrictive composites containing different volume fractions—i.e., 10, 20, 30 or 40%

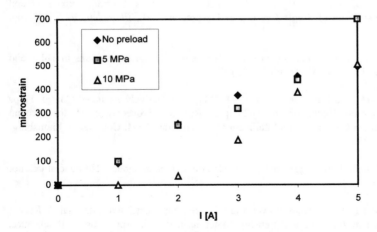

Figure 12:   Experimental data for a 40% volume fraction composite containing 300 micron particulate subjected to different external preloads (prestresses) with a mechanical load frame

CONCLUSIONS

In this paper two nonlinear constitutive relations developed phenomenologically were integrated with a micromechanics approach. The combined approach provided a relation which produced good agreement with existing experimental data over a wide range of operating conditions. The constitutive relations were introduced into a concentric cylinders model to predict the response of a magnetostrictive composite. Results indicated that the magnetostrictive material could be preloaded in-situ with the epoxy matrix and thus orient the domain structures. Experimental data on the composite systems supported the analytical modeling efforts in agreement within 10%. Experimental results indicate that particle alignment is critical for producing an actuation material. Particle size and resin systems also play an important role in producing a reasonable actuator material where a larger particle size, lower resin viscosity and large Young's modulus proved to be more favorable.

ACKNOWLEDGMENTS

The authors of this paper gratefully acknowledge the partial support provided by the Army Research Office under contract DAAH04-95-1-0095, contract monitor John Prater.

REFERENCES

1.   Legvold, S., Alstad., J., & Rhyne, J., "Giant Magnetostriction in Dysprosium and Holmium Single Crystals", *Phys. Rev. Lett.*, V. 10, pp. 509, 1963.

2.   Clark, A.E., Bozorth., R., and DeSavage, "Anomalous Thermal Expansion and Magnetostriction of Single Crystals of Dysprosium", B., *Phys. lett*, V. 5, pp. 100, 1963.

3.   Clark, A.E., and Belson, "Giant Room-Temperature Magnetostrictions in $TbFe_2$ and $DyFe_2$", *Phys. Rev.*, V. B5, pp. 3642, 1972.

4.   Millot, T.A. and P.P. Friedmann. 1994. "Magnetostrictively Actuated Control Flaps for Vibration Reduction in Helicopter Rotors", Proceedings of the Second International Conference on Intelligent Materials, June 1994, Colonial Williamsburg, VA., pp. 900-913.

5.   IEEE Standard on Magnetostrictive Materials: Piezomagnetic Nomenclature, no. 319, 1971.

6.   Clark, A.E. 1992. "High Power Rare Earth Magnetostrictive Materials," Recent Advances in Adaptive and Sensory Materials and their Applications, Blacksburg, VA., pp. 387-398.

7.   Brown, William Fuller, Magnetoelastic interactions. Springer-Verlang, 1966.

8. Carman, G.P. and M. Mitrovic. "Nonlinear Constitutive Relations for Magnetostrictive Materials with Applications to 1-D Problems", *Journal of Intelligent Material Systems and Structures,* V.6. no.5, Sept. 1995, pp. 673-684.

9. Kannan, K.S., and Dasgupta A., "Continuum Magnetoelastic Properties of Terfenol-D; What is available and what is needed", *Adaptive Materials Symposium,* Summer meeting of ASME-AMD-MD, UCLA 1995.

10. Cao, W., Zhang, Q.M., and Cross, L.E., 1992. "Theoretical Study on the Static Performance of Piezoelectric Ceramic-Polymer Composites with 1-3 Connectivity", *J. Appl. Phys.,* vol. 72 no. 12, Dec. 1992, pp. 5814-5821.

11. Bowen C.P., T.R. Shrout, C.A. Randall, of the Intercollege Materials Research Laboratory at Pennsylvania State University, University Park, Pennsylvania, "Intelligent Processing of Composite Materials", Ad-Vol. 35, Adaptive Structures and Material Systems, ASME 1993.

12. Bi, J. and Anjanappa, M., 1994. "Investigation of Active Vibration Damping Using Magnetostrictive Mini Actuators", *Smart Structures and Intelligent Syst.,* Orlando FL, V. 2190, pp. 171-180.

13. Roberts, M., Mitrovic, M., and Carman, G.P., "Nonlinear Behavior of Coupled Magnetostrictive Material Systems Analytical/Experimental", *Smart Materials,* San Diego 1995, pp.341-355, 1995.

14. Sandlund, L., Fahlander, M., Cedell, T., Clark, A.E., Restorff, J.B., and Wun-Fogle, M., "Magnetostriction, elastic moduli, and coupling factors of composite Terfenol-D", *J. Appl. Phys.,* May 1994.

15. Hom, C.L. and Shanker, N., "A Fully Coupled Constitutive Model for Electrostrictive Ceramic Materials", *Second International Conference on Intelligent Materials,* ICIM'94, pp. 623-634.

16. Hashin, Z., and Rosen, R.W., 1964 "The Elastic Moduli of Fiber-Reinforced Materials", *J. of Appl. Mech.,* V. 31, pp. 223-234.

17. Carman, G.P., K.S. Cheung and D. Wang. "Micro-Mechanical Model of a Composite Containing a Conservative Nonlinear Electro-Magneto-Thermo-Mechanical Material", *Journal of Intelligent Material Systems and Structures,* V.6, no. 5, Sept. 1995, pp. 691-700.

18. Moffett, M.B., A.E. Clark, M. Wun-Fogle, J. Linberg, J.P. Teter, E.A. McLaughlin. 1991. "Characterization of Terfenol-D for Magnetostrictive Transducers," J. Acoust. Soc. Am. 89, pp. 1448-1455.

MAINTAINING CHAOS

Mark L. Spano and Visarath In
Naval Surface Warfare Center
White Oak Laboratory
Silver Spring, MD 20903

and

William L. Ditto
Georgia Institute of Technology
School of Physics
Atlanta, GA 30332

ABSTRACT

The recognition of chaos as a new type of behavior for complex systems initially spurred efforts to avoid it and, later, to control it. Yet in many cases chaos may be beneficial. We present a method for maintaining chaos in physical systems and implement the method on a simple magnetomechanical system. Application to other systems is discussed briefly.

INTRODUCTION

A major thrust of the work in experimental chaos has been to convert the chaos found in various physical systems into periodic motion. Since the original theoretical exposition of the control of chaos[1] and its subsequent experimental demonstration in a mechanical system[2] (a magnetoelastic ribbon), the control of chaos has been implemented in lasers,[3] electronic circuits,[4] chemical reactions[5] and biological systems.[6,7]

Although chaos control may be very advantageous in many systems,[8] it has been suggested that the pathological destruction of chaotic behavior (possibly due to some underlying disease) may be implicated in heart failure[9] and some types of brain seizures.[7] For instance, some studies of heart rate variability suggest that losing complexity in the heart rate will increase the mortality rate of cardiac patients. Therefore some systems may require chaos and/or complexity in order to function properly. Another situation in which the maintenance of chaos might be useful is the mixing of fluids.[10] Maintenance of chaos could also be useful in machine tool chattering[11] applications for avoiding occurrences of low order periodic vibrational modes in the cutting bits during milling or cutting of parts. Another possible application is to apply maintenance of chaos to combustion systems in order to avoid flame-out.

Experimental work by Schiff et al.[7] demonstrated an ad hoc method for increasing the electrical complexity of an in vitro rat brain hippocampal slice preparation. Recent theoretical and computational work by Yang et al.[12] indicates that intermittent chaotic systems can be made to exhibit continuous chaotic behavior with no intermittent periodic episodes.

With this in mind we proceed to describe a general theoretical method for the maintenance of chaos, which is then implemented experimentally in a magnetoelastic ribbon demonstrating intermittency.[13] This intermittency appears as chaos interspersed with long periodic episodes. The method presented here is readily applicable to experiment and relies only on experimentally measured quantities for its implementation.

THEORY

An *ad hoc* method for increasing the system complexity (which they termed *anticontrol of chaos*) was implemented by Schiff *et al.*[7] Subsequent to this Yang *et al.*[12] proposed a method of anticontrol based on the observation that a map-based system in a regime of transient chaos, such as that near a transition from periodicity into chaos, has special regions in its phase space that they call "loss regions". If the system enters such a region, it immediately ceases its chaotic motion. Yang *et al.* identified these regions as well as n preiterates of each loss region. If the system enters a preiterate, they apply a small perturbation to an accessible system parameter in order to interrupt the progression of the system toward a loss region. This interruption makes use of the sensitivity of chaotic systems to small perturbations and places the system in a region of phase space that is neither a loss region nor a preiterate of one. Their method requires explicit knowledge of the map of the system and is accordingly difficult to accomplish experimentally. The lack of a systematic implementation of the Schiff method and the difficulty of experimental implementation of the Yang method are the motivation for the present work.

We describe a general anticontrol method that is more readily applicable to experiment and that relies only on experimentally measured quantities for its implementation. To start, only the following assumptions are made about the system: (1) the dynamics of the system can be represented as an n-dimensional nonlinear map (*e.g.*, by a surface of section or a return map) such that points or iterates on such a map are given by $\bar{\xi}_n = \bar{\mathbf{f}}(\bar{\xi}_{n-1}, p)$, where p is some accessible system parameter; (2) there is at least one specific region of the map (termed a "loss region") that lies on the attractor and into which the iterates will fall when making the transition from chaos to periodicity; and (3) the structure of the map does not change significantly with small changes $\delta p \equiv p - p_0$ in the control parameter p about some initial value p_0.

On the return map derived from such a system, the locations of loss regions are determined by observing immediate preiterates of points which correspond to periodic orbits. Clusters of these preiterates are then identified as the loss regions. The extent of each loss region is determined by the experimentally observed distribution of points in that region (Fig. 1). The time evolution of each region can be traced back through m preiterates, as desired.

Next, in a fashion similar to the OGY chaos control method, the parameter p is changed slightly; one then observes the resulting change in each loss region's location and estimates the local shift of the attractor $\bar{\mathbf{g}}$ for each loss region with respect to a change in p as:

$$\bar{\mathbf{g}} = \frac{\partial \bar{\mathbf{f}}(\bar{\xi}_n, p)}{\partial p} \approx \frac{\Delta \bar{\mathbf{f}}(\bar{\xi}_n, p)}{\Delta p}. \tag{1}$$

As an approximation, $\bar{\mathbf{g}}$ is taken to be constant for all loss regions on the attractor for sufficiently small parameter changes δp (otherwise calculation of $\bar{\mathbf{g}}$ for each loss region would be required). This is not strictly necessary in order to implement the method, but is simply a convenience that is approximately true for many systems (including our magnetoelastic ribbon) for small δp's.

Anticontrol can be applied once the system has entered the mth preiterate of the loss region. Since the map is constructed as a return map (or a delay coordinate embedding) with ξ_n versus ξ_{n-1}, the y-coordinate of the nth point becomes the x-coordinate of the $(n+1)^{st}$ point.

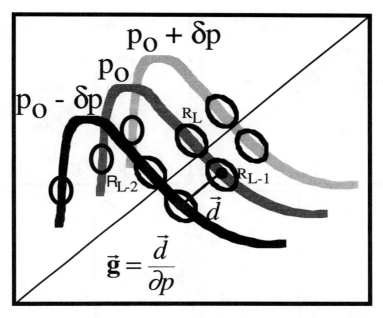

Fig. 1 Cartoon showing the attractor shift with different applied parameter values. The $\bar{\mathbf{g}}$ vector is calculated for each loss region by measuring the shifted distance of each loss region in response to the applied parameter change.

Since known values include the x-coordinate of the next point and the *size* of the region that this $(n+1)^{st}$ point would normally fall into, a minimum distance is calculated to move the attractor so that this next point falls outside of that region. This distance d is translated into the appropriate parameter change δp by

$$\delta p_n = \frac{d_{n+1}}{|\bar{\mathbf{g}}|}, \tag{2}$$

where the direction of the motion is along $\bar{\mathbf{g}}$.

If each of the m preiterates of the loss region is circumscribed by a circle of radius r_m, the worst case scenario gives $\delta p_n = 2\,r_m/|\bar{\mathbf{g}}|$, where it is understood that the $(n+1)^{st}$ point falls into the mth preiterate region. This is the maximum perturbation needed to achieve anticontrol and it guarantees that the next point will fall outside the mth preiterate region by moving the point one full diameter of the circle surrounding the loss region. This worst case can be improved upon. With a return map, the x-coordinate of the next point is known. Because there is a choice of whether to apply the perturbation in either the positive or the negative $\bar{\mathbf{g}}$ direction, the sign of the perturbation can be selected to move the next point to the left if this x-coordinate is in the left half of the preiterate region and vice versa. Thus, the minimum distance to move this x-coordinate is reduced to r_m and consequently $\delta p_n = r_m/|\bar{\mathbf{g}}|$, resulting in a significant reduction in the strength of the perturbation.

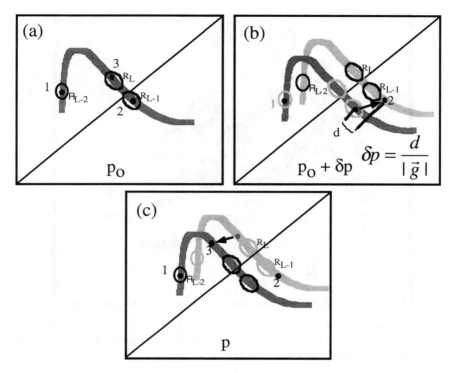

Fig. 2 Cartoon showing the sequence of the maintenance of chaos procedure. Fig. 2(a) shows the sequence of points without maintenance. The orbit first enters the loss region R_{L-2}. Then the next point will enter the R_{L-1} and proceed into R_L one iterate later. Fig. 2(b) shows the orbit entering loss region R_{L-2} and then perturbation was applied to disrupt the sequence. The resulting orbit is placed outside of the new loss region R'_{L-1}. After this, the perturbation is turned off and the subsequent orbit will fall outside of the loss region R_L as shown in Fig. 2(c).

Additionally, if the shape of the preiterate region of interest is approximately linear (line-like) and its slope is perpendicular to $\vec{g}$, then d is at most r_m and may even approach the thickness of this linear segment ($\delta p_n << r_m / |\vec{g}|$). Thus, while not necessary to achieve anticontrol, a detailed knowledge of the *shape* of the loss region and its preiterates can further reduce the size of the perturbation required to achieve anticontrol.

EXPERIMENT

The experimental system consists of a gravitationally buckled magnetoelastic ribbon driven parametrically by a sinusoidally varying magnetic field. The detailed experimental setup is discussed elsewhere.[14] An ac magnetic field of amplitude H_{ac} and frequency f added to a dc field of amplitude H_{dc} were applied such that $H_{applied}(t) = H_{dc} + H_{ac}\sin(2\pi ft)$. The values were chosen to be $f = 0.95\ Hz$, $H_{ac} = 0.961\ Oe$ and $H_{dc} = -1.221\ Oe$ in order to establish the system in a state of intermittent chaos. During the intermittent chaos, the system would switch between a periodic attractor and a chaotic attractor without any outside intervention. A return map was constructed by measuring the position, ξ_n, of a point on the ribbon once every driving period and then plotting the current position ξ_n versus ξ_{n-d}, where d is the delay. Here, where the data was strobed at the driving frequency, $d = 1$.

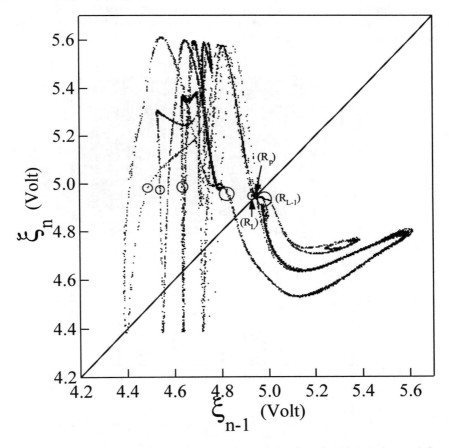

Fig. 3   Return map constructed from experimental data showing the loss region and its preiterates (circles). Region (R_L) denotes the loss region, region (R_{L-1}) denotes the first preiterate of the loss region going back in time and region (R_p) denotes the period-1 orbits clustering on the diagonal of the map. The other 4 circles denote the (R_{L-m}) preiterate regions. $(2 \leq m \leq 5)$ An orbit that falls into any one of the R_{L-m} regions will proceed quickly into the R_{L-1} region and then on to the loss region initiating the unwanted periodic orbit.

The loss region and its preiterates are identified on the return map (circles in Fig. 3). The loss region (R_L) is denoted by the circle immediately to the left of the diagonal. Its first preiterate (R_{L-1}) lies to the right of the diagonal. The other circles denote earlier preiterates (R_{L-m}). A point that enters any of these regions would eventually proceed to the region R_{L-1}. Once there, the point moves into the loss region R_L on the next iterate making the system periodic. This appears as a cluster of points (R_P) on the diagonal of the map. The points that enter the loss region mediate the intermittent transition from chaos to periodicity.[13] In this experiment during anticontrol, a perturbation was applied when an orbit entered the region R_{L-1}, so that the next point would fall outside of R_L.

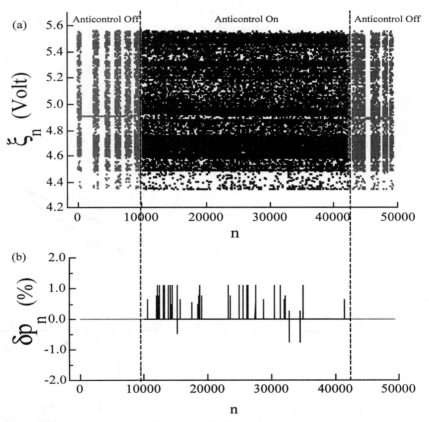

Fig. 4 (a) Experimental time series showing the system before, during and after anticontrol. Dashed lines separate the different regions. Before and after anticontrol, the system switches between chaos and periodicity (laminar phase). With anticontrol, the system lacks laminar phases. (b) Perturbations applied to achieve anticontrol, expressed as a % of the nominal DC magnetic field of $-1.221\ Oe$. $\delta p_{max} = 1.1\%$ and the fraction of time with $\delta p \neq 0$ is 0.0012.

The extent of the mth preiterate region was determined by observing the set of points that after m iterations fall into the loss region as well as neighboring points that do not fall into the loss region after m iterations. The boundary of the loss region lies between these points. The $\bar{\bar{g}}$ vector was determined by changing p_0 by $\pm 0.0068\,Oe$. As illustrated in Fig. 3, the two loss regions were very close to the cluster denoting the periodic orbit. Rarely during anticontrol the orbit was kicked into this period one region. This required the implementation of anticontrol on the succeeding iteration for the periodic region as well, in order to safeguard against the system remaining there. Of course when this happened a somewhat larger perturbation would be required to move the system away from this periodic orbit. A value of $0.0237\,Oe$ was adequate to control this problem in this experiment. A more elegant but computationally difficult solution would be to choose the original perturbation so that it avoided all of the loss regions and preiterates. Though this was possible, it was decided not to implement it here.

Fig. 4(a) shows the results of the anticontrol on the intermittency. The first 10,000 iterates were run with anticontrol turned off. During this time, the system intermittently switched between chaos and the period-1 motion. Once the anticontrol was turned on, the periodic behavior was eliminated for over 32,000 iterates. When the anticontrol was turned off, the periodic behavior reappeared. The corresponding anticontrol perturbations are shown in

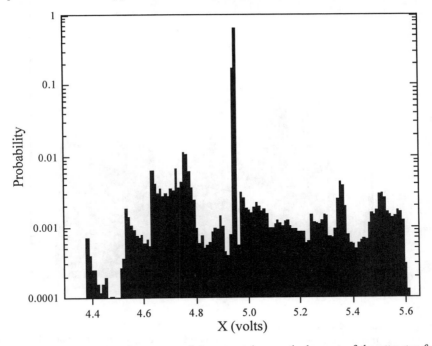

Fig. 5 Histogram showing the amount of time spent in a particular part of the attractor for the unperturbed time series data during intermittency. The system spent most of its time in period-1 motion, as indicated by the histogram peak at $x = 4.95$. Note that the vertical scale is logarithmic.

Fig. 4(b). The nominal dc magnetic field was $-1.221\,Oe$. During anticontrol the largest perturbation was 1.1% of this value ($\delta p_{max} = 0.020\,Oe$). It is also significant that the anticontrol signal needed to be applied *only 0.12%* of the time to keep the system chaotic.

The efficacy of the method is shown in Fig. 5. The figure is a probability histogram of the time series data for the unperturbed system. The large narrow peak near $x = 4.95$ reflects the fact that the system spent most of the time in a period-1 orbit. (Note the *logarithmic* vertical scale.) Contrast this with Fig. 6, where the anticontrol has been applied. The strong peak has been eliminated and its probability has been spread over the rest of the x-values with a distribution that approximates the original distribution of the intermittent chaotic data.

The effect of the anticontrol may be qualitatively appreciated by looking at Fig. 7. The 3-D histogram in Fig. 7 reflects the density of points over the attractor (from the return map of Fig. 3) of the unperturbed system. Most of the probability resides in the strong central peak that represents the period-1 orbit. The density of the points resulting from anticontrol is presented in Fig. 8. Here the probability is spread over the entire chaotic part of the attractor with a distribution that approximates that of the chaotic parts of the unperturbed chaotic system.

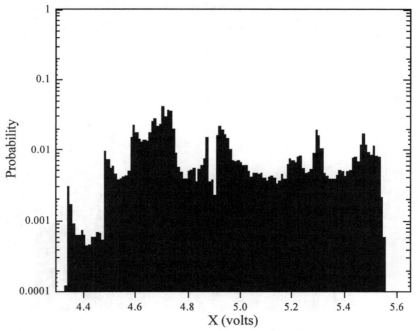

Fig. 6   Histogram of the anticontrolled time series data during intermittency. The large amount of time spent in period-1 at $x = 4.95$ is eliminated. What is left was the natural unstable period-1 of the chaotic attractor. Note that the vertical scale is logarithmic. Also note that the general shape of the histogram resembles that of Fig. 5 with the large spike removed, indicating that the orbit probabilities in the existing chaotic attractor are not significantly altered by the anticontrol.

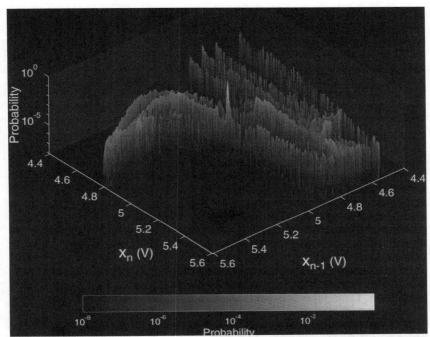

Fig. 7 3-dimensional histogram of the unperturbed data distributed over the attractor of Fig. 3. The vertical scale is normalized to the highest peak height. The period-1 peak (center peak) is where the system spent most of its time during the intermittency. Note that the vertical scale is logarithmic.

Note that there is still a period-1 peak in Fig. 8. This represents a period-1 motion that *does not destroy the chaos*. The anticontrol only prevents the period-1 motion that is initiated by following the sequence of preiterates which lead to the loss region and subsequently to the loss of chaoticity. However, the method did *not* interfere with sequences of points that enter the loss region by other routes and that did not destroy the chaos. This period-1 motion is *naturally* unstable and is properly one of the unstable periodic motions that comprise the chaos itself. Hence it is not removed. To reiterate, only the sequence of preiterates that leads to entrapment in a periodic motion is interrupted. It is because this approach was taken, rather than the approach of completely excluding the system from the region of phase space around the unstable periodic motion, that the method was able to maintain the chaos with only rare interventions (~0.12% of the time).

CONCLUSIONS

In summary, a general method for the anticontrol of chaotic systems has been presented which is straightforward to implement and needs to be applied very infrequently to keep a system chaotic. In addition, several methods of reducing the perturbation based on the nature of return maps and on the geometry of the experimentally measured attractor have been discussed. These methods are demonstrated to work in a magnetomechanical experiment with no failures.

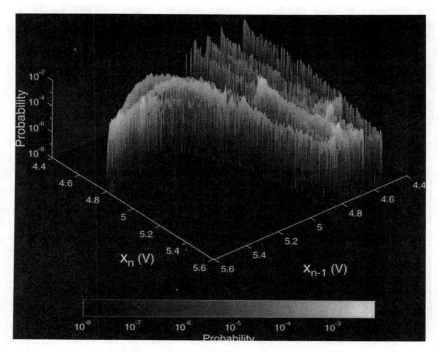

Fig. 8 3-dimensional histogram of the anticontrolled data distributed over the attractor of Fig. 3. The high center peak was eliminated by applying small perturbations during anticontrol. These perturbations were applied ~0.12% of the time to keep the system chaotic. The vertical scale is again logarithmic.

ACKNOWLEDGMENTS

Mark Spano wishes to acknowledge support from the NSWC Carderock Division ILIR Program and from the Office of Naval Research, Physical Sciences Division. William L. Ditto also acknowledges support from the Office of Naval Research, Physical Sciences Division. Visarath In acknowledges support from the ONR/ASEE Postdoctoral Fellowship Program

REFERENCES

[1]   E. Ott, C. Grebogi and J. A. Yorke, *Phys. Rev. Lett.* **64**, 1196 (1190).

[2]   W. L. Ditto, S. N. Rauseo and M. L. Spano, Phys. Rev. Lett. 65, 3211 (1990).

[3]   R. Roy, T. W. Murphy, T. D. Maier, Z. Gills and E. R. Hunt, *Phys. Rev. Lett.* **68**, 1259 (1992).

[4]   E. R. Hunt, *Phys. Rev. Lett.* **67**, 1953 (1991).

[5]   V. Petrov, V. Gaspar, J. Masere, and K. Showalter, *Nature* **361**, 240 (1993).

[6]   A. Garfinkel, M. L. Spano, W. L. Ditto and J. N. Weiss, *Science* **257**, 1230 (1992).

[7]   S. J. Schiff., K. Jerger, D. H. Duong, T. Chang, M. L. Spano, and W. L. Ditto, *Nature* **370**, 615 (1994).

[8]   For a nice review, see T. A. Shinbrot, C. Grebogi, E. Ott, and J. A. Yorke, *Nature* **363**, 411 (1993). Also of interest is W. L. Ditto and L. M. Pecora, *Scientific American* **269**, 78 (1993).

[9]   For instance, some studies of heart rate variability suggest that losing complexity in the heart rate will increase the mortality rate of cardiac patients. See, for example, M. A. Woo, W. G. Stevenson, D. K. Moser, R. M. Harper and R. Trelease, *Am. Heart Jour.* **123**, 704 (1992); A. L. Goldberger, *Ann. Biomed. Engin.* **18**, 195 (1990); and A. L. Goldberger, D. R. Rigney and B. J. West, *Scientific American* **262**, 42 (1990).

[10]   J. M. Ottino, *Scientific American* **260**, 40 (1989); J. M. Ottino, F. J. Muzzio, M. Tjahjadi, J. G. Franjione, S. C. Jana and H. A. Kusch, *Science* **257**, 754 (1992); J. M. Ottino, Guy Metcalfe and S. C. Jana, Proc. of the 2nd Exptl. Chaos Conf., p. 3-20, (World Scientific, Singapore, 1995).

[11]   Lee Hively, Oak Ridge National Laboratory, private communication.

[12]   W. Yang, M. Ding, A. Mandell and E. Ott, *Phys. Rev. E.* **51**, 102 (1995).

[13]   The statistics associated with the dwell times the system spends in the periodic motion are consistent with a type III intermittency. For in depth discussion of intermittency types see P. Berge, Y. Pomeau and C. Vidal, *Order Within Chaos* (John Wiley & Sons and Hermann, Paris, 1984).

[14]   W. L. Ditto, S. Rauseo, R. Cawley, C. Grebogi, G. H. Hsu, E. Kostelich, E. Ott, H. T. Savage, R. Segnan, M. L. Spano and J. A. Yorke, *Phys. Rev. Lett.* **63**, 923 (1989).

APPLICATION OF SMART MATERIALS
TO WIRELESS ID TAGS AND REMOTE SENSORS

RICHARD FLETCHER, JEREMY A. LEVITAN, JOEL ROSENBERG,
NEIL GERSHENFELD; The Media Laboratory, Massachusetts Institute of Technology,
Cambridge, MA 02139

ABSTRACT

Material structures having an electromagnetic or magnetomechanical resonance can be excited or detected remotely using an antenna. Incorporating smart materials into such structures provides new opportunities to encode ID and sensor information in the electromagnetic signature of the "tag." In this way, it is possible to create tags which not only have a unique ID but which can also respond to local changes in their environment (e. g. force, temperature, light, etc.). This principle forms the basis for a low-cost wireless ID and wireless sensor technology which has many potential applications in manufacturing, inventory control, security, surveillance, and new human-computer interfaces. As a means of illustrating this concept, two simple examples are given: a force sensor incorporating a piezoelectric polymer and a relative position sensor which incorporates a magnetoelastic amorphous metal ribbon.

INTRODUCTION

When probed by an electromagnetic field, magnetoelastic amorphous metal ribbons and planar inductor-capacitor (LC) structures exhibit a discrete resonant frequency and Q-factor. Operating at kilohertz and megahertz frequencies, respectively, these resonators can be engineered to encode identification or information about the local environment. Common approaches to electromagnetic tagging include RFID (radio frequency identification) or some type of IC chip with a sensor. The disadvantages of these schemes, however, is their inability to meet the needs of applications where the total cost of each sensor must fall below $0.10. The materials-based tagging technology described can meet this cost requirement and address a wide variety of applications, including disposable temperature probes, wireless force sensing devices, and small-scale remote identification.

Sensors based on LC resonators or magnetoelastic ribbons have been widely explored; however, most of these implementations are not wireless[1] or are performed as an inductive magnetic measurement in very close proximity to the sensing material[2]. To enable operation over non-trivial distances (> few centimeters), the material structures presented here employ a resonant mode of operation. As wireless sensors, the the sensor and ID

557

information is generally encoded in the resonant frequency and Q of the tag, which is then read from a distance (~2 meters or less) using a near-field antenna [Fig 1]. Object identification can be accomplished, for example, by using multiple magnetoelastic strips or multiple-layer planar tanks to yield many unique identities. If the magnetoelastic ribbons is used in conjunction with a small permanent magnet, the bias-field dependence of the resonant frequency can be used as a mechanism for sensing relative displacement or force; and in planar LC resonators, incorporating a smart material dielectric would provide a means of sensing an external stimulus (e.g. force, temp, light). Given a properly designed sensor package, a pyroelectric dielectric leads to a temperature sensor and a piezoelectric dielectric provides a pressure sensor, for example.

This paper focuses on the use of such structures as sensor tags. Two types of sensors are illustrated in the sections that follow, and the corresponding data is given.

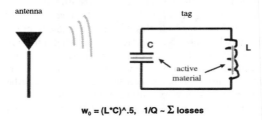

$$w_0 = (L{*}C)^{\wedge}.5, \quad 1/Q \sim \Sigma \text{ losses}$$

Figure 1. Schematic representation of a materials-based wireless sensor tag exhibiting an electromagnetic or magnetomechanical resonance.

SENSOR DESIGN AND EVALUATION

A simple displacement sensor was evaluated for the purpose of measuring the linear position of a piston in a small cylinder (10cm long). A strip of amorphous metal ribbon packaged in a plastic cavity was attached to the body of the cylinder. Additionally, a weak flat bias magnet made of Arnochrome 3 ^{TM} of ~ .5 Oe or so can be included with the ribbon to provide a small constant bias field; however, for simplicity this aditional magnet was not used for this paper. The magnetoelastic ribbons used were amorphous alloys manufactured by Allied Signal and prepared in a width of 1.2cm and length of 3.55 cm. A properly oriented permanent magnet was attached to the end of the piston, thus providing a bias field to the ribbon which varied with linear position of the piston but did not vary with azimuthal rotation of the piston shaft [Figure 2].

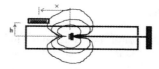

Figure 2. Schematic view of cylinder with position sensor.

Since the resonant behavior or the ribbon depends on the bias field as well as its material properties[3], the linear position of the piston could then be deduced by tracking the resonant frequency of the ribbon. The dependence of the local bias magnetic field presented to tag as a function of piston position could be varied by changing the mounting position of the tag on the cylinder; and the resulting resonant frequency shift resulting from this field could be also be tuned independently through annealing treatments of the amorphous metal ribbon.

In order to increase the linear distance over which the sensor could operate, a preliminary annealing study was carried out to investigate the optimum processing parameters for the ribbons that were tailored to this application. Since it was desirable to increase the usable range of bias fields, a slightly sheared M-H loop is desirable, so a transverse-field anneal was used. Samples of composition $Fe_{38}Ni_{39}Mo_{2.4}B_1Si_{0.2}$ were annealed at a temperature near 400 degrees Celsius using several different annealing fields. A second alloy used for this study was $Fe_{35}Ni_{33}Co_{19}B_8Si_5$, which was annealed by Sensormatic using another recipe. The resonant frequency shift as a function of an applied DC bias field was then measured using an Hewlett-Packard 8753D Network Analyzer. A representative sample of the measured data is shown in Figure 3. For $Fe_{38}Ni_{39}Mo_{2.4}B_1Si_{0.2}$, we suspect the annealing temperature was slightly higher than optimum, as exhibited by the extra degree of flattening in the curves likely due to partial recrystallization of the amorphous metal.

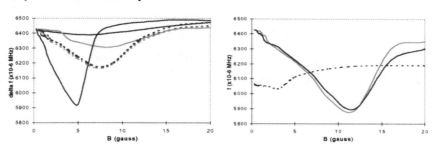

Figure 3. Bias-field dependence of the resonant frequency as a function of annealing treatment. On the left is data for $Fe_{38}Ni_{39}Mo_{2.4}B_1Si_{0.2}$ for transverse annealing fields of 0, 200, 300 Oe. (curves for higher fields are flatter) On the right, is the result for $Fe_{35}Ni_{33}Co_{19}B_8Si_5$ after the Sensormatic annealing treatment. Dotted lines denote pre-annealed as-cast result.

Force Sensor Design & Evaluation:

For characterizing the planar tank's response to pressure, a two-coil planar resonator was designed and etched from copper clad 1000K120 Kapton made by Rogers. The two coils were then folded upon each other, with a 25μm dielectric placed between the two layers. The dielectric region was comprised of either teflon sheet or the piezoelectric polymer polyvinyldifluoride (PVDF), supplied by AMP, Inc. The structure was then epoxied under vacuum or laminated to seal the dielectric between the layers.

The performance of the sensor tag was evaluated using an Instron 4411 mechanical tester and the HP network analyzer. The sensor tag was placed over a 2" diameter sapphire base containing a loop antenna. The whole assembly was placed onto the lower anvil of an Instron machine. On top of the coil assembly, a 3" x 3" x 3" cube of non-conducting foam was placed between the sensor tag and the top anvil. Both the network analyzer and Instron press were connected to a PC via a GPIB interface. The Instron was programmed to apply a load in increments and output both the applied load and displacement from the origin. The resonant frequency and Q-factor for of the sensor tag was simultaneously recorded. These numbers form the basis for the data. The specific force sensor demonstrated in this paper is designed to have a sensing range similar to a human finger, so an applied load range of 0-5 Newtons was chosen.

RESULTS AND DISCUSSION

Position Sensor: One sample of each type of amorphous metal sample was selected for use as the sensor tag and mounted on the cylinder. The resonant frequency of each tag was then recorded as a function of the piston position. The results are shown in Figure 4.

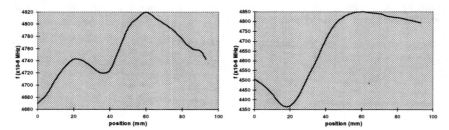

Figure 4. Plot of tag frequency vs. piston position for (left) and (right).

As shown in Figure 4, the usable sensing range was approximately 2 cm (from x = 40 mm to x = 60 mm) for $Fe_{38}Ni_{39}Mo_{2.4}B_1Si_{0.2}$, and was approximately 4 cm (from x = 20 mm to x = 60 mm) for $Fe_{35}Ni_{33}Co_{19}B_8Si_5$. By further optimizing the placement of the sensor tag as well as the strength of the permanent magnet used, we feel that the usable range of operation could be extended to 8 cm or more for a cylinder this size (10 cm long). For a larger size cylinder, the design can be scaled up using a larger size magnetoelastic ribbon; or for very long cylinders, separate tags can be used along the length of the cylinder to track the piston position along the entire length of its stroke.

Force Sensor: The resonant frequency response of the planar tank circuits with PVDF dielectric was compared to that of tags containing Teflon, a conventional high-frequency dielectric. Results are plotted in Fig 5.

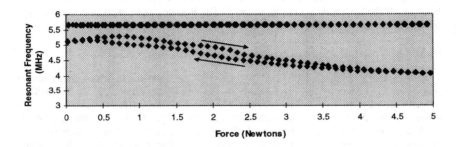

Figure 5. Data for wireless force sensor incorporating a piezoelectric polymer dielectric. The upper curve is the response of an identical resonant structure incorporating a normal dielectric material (teflon). Also plotted is the curve predicted by the linear elastic model.

For the resonator containing the normal dielectric, its response can be modeled as a simple LRC-circuit composed of an inductor, resistor, and plate capacitor with a dielectric material. By applying an elastic model to the deformation of the dielectric material under applied stress, the resonant frequency of the tag can be derived as a function of applied stress:

$$\omega_n = \omega_{n_0} \cdot \sqrt{\frac{E - \sigma}{E}} \qquad (1)$$

where ω_{n_0} is the resonant frequency of the tag absent any applied stress, E is the Young's Modulus of the dielectric material, and σ is the applied stress. Rearranging Equation (1) yields an expression relating the ratio of the change of resonant frequency versus initial resonant frequency and the induced strain, ε, in the dielectric material:

$$\frac{\Delta \omega}{\omega_{n_0}} = 1 - \sqrt{1 - \varepsilon} \qquad (2)$$

The curve predicted by this model is included in Figure 5 and very closely matched the measured data to within 0.1%.

In comparing the teflon response to the response produced using PVDF, this model indicates that in a typical dielectric material with Young's Modulus of about 3 Gpa (comparable to PVDF and clear teflon sheet), a 10% change in frequency would occur when there is a strain of 19%. Further manipulation of equation (2) shows that in order to produce in a 10% change in the resonant frequency of the tag, a force of 60000 Newtons would need to be applied to the tag. On the other hand, the smart material tag shows a significant response with an applied force of as little as 0.1 Newtons.

The primary advantages of the PVDF force sensor tag are its small thickness (< 0.5 mm) and good sensitivity to small forces. The main apparent disadvantage of this sensor is hysteresis; however, we feel this can be partially attributed to the packaging of the PVDF and thus can be improved with better packaging design. This type of sensor has also proved to be quite robust, and continued to perform with no noticeable degradation in sensitivity even after subjecting the tag to abuse, such as stepping on it or striking it with a hammer. If the hysteresis is intrinsic to the PVDF, it seems more likely due to some type of repeatable relaxation mechanism exhibiting no noticeable degradation.

SUMMARY

Two simple examples of wireless "tag sensors" were illustrated which make use of smart materials. Such sensors can function as conventional sensor devices but are made of simple material structures and are wireless. As a result, these structures can represent a low-cost robust alternative to semiconductor-based sensor technology. Although further work is certainly required to quantitatively design, analyze and model the behavior of such tag sensors, this application has great potential for further research given the ongoing commercial interest in low-cost wireless sensor technology.

ACKNOWLEDGEMENTS

The authors would like to acknowledge Dr. Robert O'Handley of the MIT Material Science Department for helpful technical discussions, and Sensormatic Inc., AMP Inc., and Festo Inc. for donation of materials.

REFERENCES

[1] Mohri, IEEE Transactions on Magnetics, vol. MAG-20, No. 5, Sep 1984.
[2] For example: Seekircher and Hoffmann, Sensors and Actuators, A21-A23 (1990), pp. 401-405.
[3] Mermelstein, IEEE Transactions on Magnetics, vol. 28, no. 1, Jan 1992.

Part IX

Actuator Materials

GIANT MAGNETOSTRICTIVE MULTILAYER THIN FILM TRANSDUCERS

E. QUANDT, A. LUDWIG,
Forschungszentrum Karlsruhe GmbH, Institut für Materialforschung I,
D-76021 Karlsruhe, Germany

ABSTRACT

Magnetostrictive multilayer films which combine exchange coupled giant magnetostrictive materials (amorphous $Tb_{0.4}Fe_{0.6}$) and materials with large polarizations (Fe or $Fe_{0.5}Co_{0.5}$) were prepared by dc or rf magnetron sputtering using a rotary turn-table technique in a stop-and-go mode. The magnetic properties of TbFe/Fe and TbFe/FeCo multilayers were investigated in relation to the layer thicknesses and the annealing temperatures. Giant magnetoelastic coupling coefficients (or magnetostrictions) are achieved at low fields, due to the magnetic polarization enhancement in such multilayers. Saturation magnetoelastic coupling coefficients of 20 MPa at 20 mT in the case of TbFe/Fe and of 28 MPa at 20 mT in the case of TbFe/FeCo were achieved. These high low-field magnetoelastic coupling coeffients and the possibility to engineer the material's properties by layer thickness variation are considered to be important features for applications of these films as thin film transducers in microsystems.

INTRODUCTION

Interest in giant magnetostrictive materials in thin film form has rapidly grown over the past few years due to their potential as a powerful transducer system for the realization of microactuators as they can easily be scaled down to small lateral dimensions. Special features of magnetostrictive in comparison to piezoelectric and shape memory thin films are their remote control and high frequency operation, their simple actuator lay-out, and the low process temperatures being compatible to microsystems and microelectronics fabrication technologies.

The development of magnetostrictive thin films is based on the rare earth - Fe_2 Laves phases which are well known to exhibit giant magnetostriction [1]. As the maximum applied magnetic fields in microsystems are limited, research on these materials concentrates in developing low field giant magnetostrictive properties. Attempts to reduce the magnetic saturation field are normally based around techniques for reducing the macroscopic anisotropy by using amorphous materials, either binary rare earth transition metals like Tb-Fe [e.g. 2] or for further anisotropy compensation even ternary alloys like Tb-Dy-Fe [e.g. 3]. Additionally, it was found to be extremely important to adjust an inplane magnetic easy axis [4], whereas the easy axis orientation is dependent on the film stress, itself being controlable by the fabrication conditions, especially by the applied rf bias voltage [5] or by stress annealing after deposition.

An alternative, very promising new approach uses the combination of two materials in a multilayer arrangement, which consists of an amorphous giant magnetostrictive and a soft magnetic material with very high magnetic polarization [6]. In order to achieve the required properties, the layers have to be magnetically coupled, i.e. they have to be thinner than the magnetic exchange length to avoid domain wall formation at the interfaces. In this case the magnetic properties of the multilayer will be defined by the mean value of the individual layers, leading to a remarkable reduction of the saturation magnetic field H_s - being proportional to K/M_s

- due to the decrease of the total anisotropy K and the increase of the saturation magnetization M_s obtained by the contribution of the soft magnetic, high magnetic polarization layers.

In this paper results on the magnetic properties of TbFe/Fe and TbFe/FeCo multilayer series with varying layer thicknesses are presented, compared to the state-of-the-art TbFe single layer films, and discussed in view of the described model for the exchange antiferromagnetic coupled multilayers.

EXPERIMENTAL

The multilayers were magnetron sputtered onto Si(100) substrates using composite-type targets (75 mm $\varnothing$) with a rotary turn-table technique in a stop-and-go mode. The TbFe films were fabricated using those conditions (150 W dc, 0.4 Pa Ar, 220 V rf bias) which in case of single layer films result in amorphous TbFe films with inplane magnetic easy axis after stress annealing, whereas the Fe or FeCo films were rf magnetron sputtered without bias voltage. The deposition rates for a target substrate distance of 50 mm were between 0.3 nm/s for FeCo and 2 nm/s for TbFe leading to typical deposition times between 1 and 30 s per layer. The total film thickness was in the range of 0.5 to 1.5 μm, the individual layer thicknesses were varied between 1.5 and 10 nm. After the deposition the multilayers were annealed in a high vacuum furnace at different temperatures with a hold-up time of 15 min.

The composition of the single layers was determined by wavelength dispersive x-ray microanalysis (WDX), and depth profiles of the multilayers were obtained using Auger electron spectroscopy (AES). The film stresses were calculated by measuring the difference of the curvature of uncoated and coated Si(100) substrates using a laser interferometer. The microstructures of the films were investigated by x-ray diffraction (XRD) and transmission electron microscopy (TEM). The magnetic polarization of the films was measured by a vibrating sample magnetometer (VSM). The magnetostriction was determined using the cantilevered substrate technique, in which the bending of the substrate due to the magnetostriction in the film was measured. This allows the magnetoelastic coupling coefficient of the film (b) and the magnetostriction (λ) to be directly determined [7] using:

$$ b = -\frac{\alpha}{L}\frac{h_s^2}{h_f}\frac{E_s}{6(1+\upsilon_s)} \quad \text{and} \quad \lambda = \frac{-b(1+\upsilon_f)}{E_f}, $$

where α is the deflection angle of the sample as a function of applied field, L is the sample length, E_s, E_f and υ_s, υ_f are the Young's modulus and Poisson's ratio for the substrate and the film, respectively; h_s, h_f are the thicknesses of the substrate and film, respectively. The saturation magnetoelastic coupling coefficient (b_s) and the saturation magnetostriction (λ_s) is obtained by

$$ b_s = \frac{2}{3}(b_\| - b_\perp) \quad \text{and} \quad \lambda_s = \frac{2}{3}(\lambda_\| - \lambda_\perp), $$

where the indices $\|$ and $\perp$ denote that the magnetic field is applied parallel or perpendicular to the measurement direction. The magnetostriction of the giant magnetostrictive multilayers and single layers was calculated using $E_f = 50$ GPa and $\upsilon_f = 0$.

RESULTS AND DISCUSSION

Microstructure

The giant magnetostrictive TbFe layers were deposited in the amorphous state in order to avoid high macroscopic anisotropy while the Fe and FeCo layers were found to be nanocrystalline with a mean grain size being equal to the layer thickness. Auger electron depth profiling revealed that the well-defined layer structure was not affected by annealing up to 300 °C, whereas e.g. annealing at 480 °C leads to a significant interdiffusion and alloy formation at the interfaces resulting in an important reduction of the magnetostriction [8]. The as-deposited multilayers are under compressive stress (typically 250...300 MPa). Preliminary results on the TbFe/FeCo films show that stress annealing of the multilayers at 300 °C results in almost stress-free films.

Magnetization

In spite of the compressive stress state, all multilayers exhibit an inplane magnetic easy axis but nevertheless stress annealing leads to a reduction in the inplane magnetic saturation fields. In figure 1 the magnetic polarization of the TbFe/Fe and the TbFe/FeCo multilayer series as a function of the transition metal (TM) layer thickness is compared to the theoretical behavior for spring-magnet type multilayers considering either parallel or antiparallel coupling of the TbFe and the TM layers. The experimental data clearly document the antiparallel coupling of the layers which can be related to parallel coupling of the Fe moments in the TbFe layer with the TM moments leading to antiparallel coupling due to the dominating Tb moments in the antiferromagnetic TbFe layer. The inplane magnetization loops of a $Tb_{0.4}Fe_{0.6}(4.5\ nm)/Fe(6.5\ nm)$ multilayer which was annealed at 280 °C in comparison to the corresponding loop of the giant magnetostrictive TbFe single layer film reveals the increase of polarization of the multilayer films combined with reduced hysteresis and saturation fields (fig. 2). The saturation magnetic

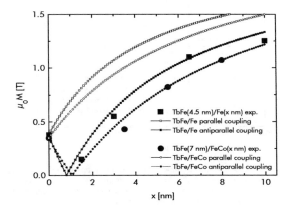

Fig. 1. Saturation polarization of TbFe(4.5nm)/Fe(x nm) and TbFe(7nm)/FeCo(x nm) multilayers as a function of the Fe or FeCo layer thickness in comparison to a simple model for exchange coupled layers considering either parallel or antiparallel coupling of the TbFe and the transition metal layers.

polarization of the Fe films ($\mu_0 M_s$ = 2.1 T) and of the FeCo films ($\mu_0 M_s$ = 2.3 T) are in good accordance to published data for bulk materials. Figure 3 compares different as-deposited inplane magnetization loops of a $Tb_{0.4}Fe_{0.6}$(7 nm)/$Fe_{0.5}Co_{0.5}$(x nm) multilayer series with the $Fe_{0.5}Co_{0.5}$ single layer film. These data show that an increase in the FeCo layer thickness leads both to an increase of the saturation magnetization and a decrease of the hysteresis.

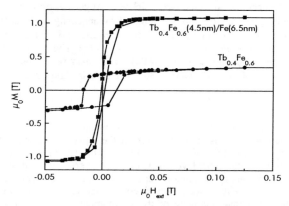

Fig. 2. In-plane magnetic polarization of a TbFe single layer and a TbFe(4.5nm)/Fe(6.5nm) multilayer showing the increased polarization in combination with the decreased hysteresis of the multilayer.

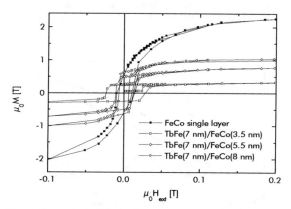

Fig. 3. Comparison of the inplane magnetization loops of as-deposited FeCo single layer and TbFe/FeCo multilayers for different FeCo layer thicknesses showing the increase of the polarization and decrease of hysteresis with increasing FeCo layer thickness

Magnetostriction

Figure 4 and 5 show the characteristic shape of the magnetostrictive hysteresis of TbFe/Fe and TbFe/FeCo multilayers respectively. Both multilayer systems can reach high saturation magnetoelastic coupling coefficients of 20 MPa (TbFe/Fe) and 28 MPa (TbFe/FeCo) at as low fields as 20mT. Also the hysteresis of the multilayers is relatively small compared to TbFe single layers.

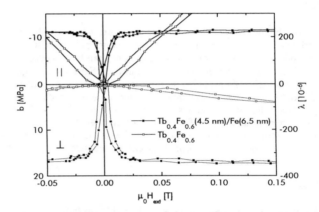

Fig. 4. Magnetostriction of a TbFe single layer and a TbFe(4.5nm)/Fe(6.5nm) multilayer film showing the increased magnetostriction at low fields.

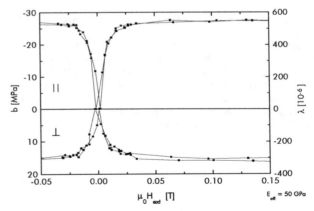

Fig. 5. Magnetostriction of a TbFe(7nm)/FeCo(8nm) multilayer film showing - in comparison to figure 4 - the increased saturation magnetostriction due to the magnetostrictive contribution of the FeCo layer.

The comparison of a TbFe/Fe multilayer and a TbFe single layer film sputtered under the same conditions as the TbFe in the multilayers clearly reveals a significant increase of magnetostriction at low fields but on the other hand a decrease of saturation magnetostriction in correspondence with the described model for the magnetic properties of exchange coupled multilayers. The improved magnetostriction in TbFe/FeCo compared to TbFe/Fe multilayers can be explained by the magnetostrictive contribution of the FeCo layer, which exhibited a saturation magnetostriction exceeding $100 * 10^{-6}$ [9].

CONCLUSIONS

The results on this novel type of multilayer structure, which combines exchange coupled giant magnetostrictive materials and materials with large polarizations, indicate that giant magnetoelastic coupling coefficients (or magnetostrictions) can be achieved at low fields, due to the magnetic polarization enhancement in such multilayers. Saturation magnetoelastic coupling coefficients of more than 28 MPa at 20 mT in the case of TbFe/FeCo multilayers are obtainable. These high saturation magnetoelastic coupling coefficients are not much lower than those of the best TbFe single layer film which shows saturation values of 45 MPa but exceeding the multilayers only for external fields higher 0.2 T [9]. Therefore such composite materials should be much more appropriate for low-field applications than the simple homogeneous alloy films studied up to now, especially, as the magnetic properties can be engineered by the choice of different materials and variation of the layer thicknesses.

ACKNOWLEDGMENTS

This work was carried out as part of the E.C. funded "MAGNIFIT" project.

REFERENCES

[1]   A.E. Clark, in: Ferromagnetic Materials, Vol.1, E.P. Wohlfarth (ed.), Amsterdam 1980, p. 531.
[2]   Y. Hayashi, T. Honda, K.I. Arai, K. Ishiyama, M. Yamaguchi, IEEE Trans. Magn. 29 (1993), 3129.
[3]   P.J. Grundy, D.G. Lord, P.I. Williams, J. Appl. Phys. 76 (1994), 7003.
[4]   F. Schatz, M. Hirscher, M. Schnell, G. Flik, H. Kronmüller, J. Appl. Phys. 76 (1994), 5380.
[5]   E. Quandt, B. Gerlach, K. Seemann, J. Appl. Phys. 76 (1994), 7000.
[6]   E. Quandt, A. Ludwig, J. Betz, K. Mackay, D. Givord, submitted to J. Appl. Phys.
[7]   E. du Trémolet de Lacheisserie, J. C. Peuzin, J. Magn. Magn. Mater. 136, 189 (1994).
[8]   E. Quandt, A. Ludwig, J. Mencik, E. Nold, submitted to J. of Alloys and Compounds.
[9]   E. Quandt, A. Ludwig, unpublished results.

GIANT MAGNETOSTRICTIVE, SPRING MAGNET TYPE MULTILAYERS AND TORSION BASED MICROACTUATORS

J. BETZ*, K. MACKAY*, J.-C. PEUZIN*, D. GIVORD*, B. HALSTRUP**
* Lab. de Magn. Louis Néel, CNRS, BP166, F-38042 Grenoble, France
** Inst. of Techn. Physics, Heinr.-Plett-Str. 40, D-34132 Kassel, Germany

ABSTRACT

There is a need for powerful active materials in microsystem actuators. Research on thin film magnetostrictive materials has concentrated on optimising the magnetostrictive response notably by reducing the required driving magnetic field. Here we present a novel type of multilayer, where 2 alloys having different properties are coupled together. The resulting composite system represents a new material with novel characteristics, which can not be fulfilled by simple alloys. In particular here we have investigated magnetostrictive multilayers, with remarkable low field performances (i.e. very large $\partial\lambda / \partial H$).

Most magnetostrictive microactuators are based on a bimorph structure. However, these simple structures are prone to thermal drift. We present here also some results on a torsion based magnetostrictive microactuator prototype being insensitive to thermal drift.

INTRODUCTION

The recent developments in the microsystem technologies have given rise to a growing interest of powerful transducer materials, which can be integrated into existing microelectronics production lines. Piezoelectric materials can be imagined as an adequate possibility for this kind of applications, but these ceramic based materials are very difficult to handle in micromachining technologies. Shape memory materials are, although powerful and simpler to prepare, rather slow in displacement. Magnetostrictive, metallic materials have a very large energy density and can be easily deposited by the well-established sputtering techniques. This is compatible with the Si micromaching processes and LIGA (LIthographie, Galvanik, Abformung [1]) techniques. The magnetostriction effect does not need any electrical contact as the magnetic fields can be applied from outside of the microsystem.

Recently many types of giant magnetostrictive thin films have been investigated [2, 3] to obtain high magnetoelastic effects in low fields. These films have been optimised in the structural, elastic, and chemical and magnetically point of view. Unfortunately the saturation fields of these films are still rather high with values of about 0.1 T. Here we present a composite material, based on a multilayer structure. The aim is to increase the saturation magnetisation while keeping the magnetostriction high enough.

A magnetostrictive material deposited on a passive substrate is a bimorph cantilever in which a torsion can be generated very easily [4]. In the second part of this paper, we present a new kind of microactuator based on a this torsional deformation. Such a deformation can be naturally generated using a magnetostrictive bimorph and this is used to separate the isotropic thermal deflection and the magnetostrictive mechanical output of a bimorph.

Mat. Res. Soc. Symp. Proc. Vol. 459 © 1997 Materials Research Society

THEORY

One possibilty to reduce the saturation fields while keeping the total magnetostriction sufficiently high, is to increase the saturation magnetisation M_s. Giant magnetostrictive materials used up to now have rather low magnetisation due to their ferrimagnetic nature (often even sperimagnetic). For the composition of interest Tb M_2 [1, 5], the Tb moments dominate and so an increase in the transition metal component will only further reduce the magnetisation at room temperature. The increase of the Tb concentration will decrease the ordering temperature and for this reason it is not possible to increase the magnetisation (at room temperature) with simple alloying.

However we can use multilayer systems to exchange couple two magnetic materials having different magnetic characteristics in a similar way to "spring magnets". There is one material (Tb-M) which shows a large room temperature magnetostriction but however a low magnetisation. The other one (Fe-Co) is magnetically soft with a very high magnetisation. When the individual layer thickness is smaller than the magnetic exchange length, domain walls can not be formed at the interfaces at room temperature. The whole system is entirely coupled to behave like a one layer system with its properties corresponding to an average of the characteristics of the two types of layers. In this case the saturation field is reduced while keeping relatively large values of magnetostriction.

For example, consider a very simple model having a multilayer system with layers of equal thickness. Also we suppose perfect coupling with high exchange stiffness so that we have a uniform magnetisation direction throughout the whole thickness. The soft material has a high magnetisation M_s^{hi}, and zero magnetostriction and no anisotropy $\lambda^{hi} = K^{hi} = 0$. The magnetostrictive layer has a magnetisation M_s^{MS}, a magnetostriction λ^{MS} and an anisotropy K^{MS}. Typically $M_s^{hi} \approx 5 M_s^{MS}$ so that the multilayer has got a magnetisation of $3 M_s^{MS}$. The anisotropy is $K^{MS}/2$ - so that the resulting saturation field of the mulitlayer is $H_s^{multi} = K^{MS}/3 M_s^{MS} \approx H_s^{MS}/6$. As for the magnetostriction of the multilayer we get $\lambda^{MS}/2$. Thus the a ratio (λ/H) is 3 times higher for the multilayer than for the simple magnetostrictive alloy.

The advantage of these composite multilayer materials over the normal homogenous alloys, is the fact that in each layer of the soft material, the magnetisations are very large and this can not be achieved in R based alloys at room temperature. In addition to this, large interface anisotropies can be developed and become very important increasing the magnetostriction performances as shown below. So the critical design criteria are the individual layer thickness, the interface quality and the composition of each layer.

RESULTS

We have studied the multilayer system ($Fe_{0.65} Co_{0.35}$ / $Tb_{1-x} (Fe_{1-y} Co_y)_x$) *n where we have varied the individual layer thicknesses t_{Fe-Co}, t_{Tb-Co}, the compositions x and y and the number of repetitions. The composition of $Fe_{0.65} Co_{0.35}$ was chosen because of its high saturation magnetisation of $M_s = 2200$ kAm^{-1} at room temperature.

For all multilayers, the Fe-Co layer is crystalline and the Tb-M layer is X-ray amorphous. The perpendicular grain size of the Fe-Co layer corresponds to the layer thickness. The total film thickness was measured by RBS and α-step profilometry.

The magnetic properties of the samples were measured using a VSM. A typical in-plane magnetisation loop is shown in figure 1 at 300 K where the multilayer is ferrimagnetically coupled. At 100 K a magnetic domain wall at the interfaces appears at about 2.5 T [6] increasing the total magnetisation. A parallel orientation of the magnetisation of the individual layer takes place with and progressif rotation of the Co-moments at the interfaces.

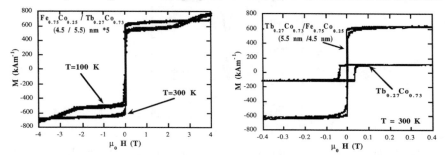

Figure 1 : Magnetisation loop of (Fe$_{0.65}$ Co$_{0.35}$ / Tb$_{0.27}$ Co$_{0.73}$) (4.5/5.5) nm at 300 K and 100 K (a) and its comparison with the single Tb$_{0.27}$ Co$_{0.73}$ film at 300 K (b).

The magnetostriction was measured using an optical deflectometer, in which the bending of the substrate due to the magnetostriction in the films was measured (see figure 4c). From the deflection angle we can determine the magnetoelastic coupling of the film b$^{\gamma,2}$ and then directly the magnetostriction coefficient λ$^{\gamma,2}$ [4, 9], using:

$$b = \frac{\alpha}{L} \frac{h_s^2}{h_f} \frac{E_s}{6(1+\nu_s)} \quad ; \quad \lambda = \frac{-b (1+\nu_f)}{E_f} \quad (1)$$

where α is the deflection angle of the sample as a function of applied field, L is the sample length, E$_s$ and ν$_s$ are the Young's modulus and Poisson's ratio for the substrate. h$_s$, h$_f$ are the thicknesses of the substrate and film respectively. b is proportional to the magnetostriction coefficient λ. These can not be reliably measured for thin films, however for comparison, we also give values of λ calculated using an effective Young's modulus E$_{eff}$ = E$_f$/(1+ν$_f$) of 50 GPa which is that normally used by other authors [7].

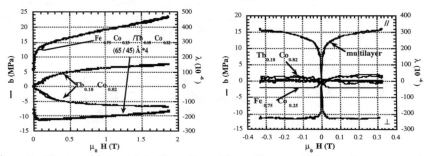

Figure 2 : *Magnetostriction of the Fe$_{0.75}$Co$_{0.25}$/Tb$_{0.18}$ Co$_{0.82}$ multilayerand a corresponding single films at high field (a) and low field (b)*

In the figures 2 we can see the magnetostriction behaviour of these multilayers as function of the applied magnetic field. A significant increase of the low-field slope is observed for the multilayer system. In addition, the total magnetostriction is higher which can be explained by interface anisotropy.

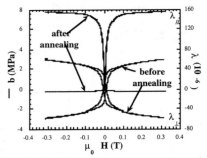

Figure 3 : Magnetostriction for the tri-layer ($Fe_{0.75}Co_{0.25}/Tb_{0.28}Co_{0.72}Fe_{0.75}Co_{0.2}$) before and after annealing under field of 2.4 T at 150°C for 1h

As has been already shown for simple, magnetostrictive films, an annealing under field can render the films magnetically softer due to a relaxation of the stresses and can define an uniaxial anisotropy in the plane of the film [2]. This fact can be observed in figure 3 where a clear transition occurs from an easy plane anisotropy before annealing to an easy axis anisotropy after annealing with an decrease of the saturation field. Thus the low-field slope increases strongly.

APPLICATION : A Drift-free Microactuator

Many microactuators are designed around a microbimorph (fig. 4a). This is a composite structure consisting of a Si-substrate and an active material deposited on one side. The mechanical output from such a structure uses normally a bending mode of deflection. However, these are particularly sensitive to thermal drift due to the difference of the thermal expansion coefficients. Magnetostrictive bimorphs can produce a torsional deformation as well as the bending deformation [4, 8, 9]. This is in strong contrast to the properties of the thermal and the usual piezoelectric bimorphs in which only bending deformations can be produced. This can be used to decouple the thermal and magnetic outputs. In addition, such torsional modes can be excited rather naturally in a magnetoelastic bimorph.

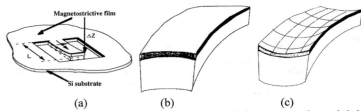

(a) (b) (c)

Figure 4 : *Magnetostrictive bimorph cantilever (a) and the isotropic thermal deformation (b) and a anticlastic magnetostrictive deformation (c) of a magnetostrictive bimorph*

The torsion is unaffected by differential thermal expansion. It is well known that the differential thermal expansion produces a simple isotropic curvature. This is in contrast with the magnetoelastic deformation which is anticlastic in nature (fig. 4b, c). Note that an anticlastic deformation consisting of two curvatures of opposite sign is equivalent to a torsion around an axis at 45° to the principal axes.

The torsion based micro-actuator, schematically shown in figure 5a, is composed of a square bimorph that is held at three of its corners by elastic hinges having "two degrees of freedom". The useful output motion of the actuator is that of the fourth free corner which can move perpendicularly to the bimorph plane.

If we assume that the hinges have negligible rotational stiffness for rotations about any axis lying in the bimorph plane and that the hinges have infinite stiffness for translational motion perpendicular to the bimorph plane then one can see that the differential thermal expansion produces no output, as illustrated in figure 5b. On the other hand, since the magnetoelastic effect can produce a torsion, a large vertical displacement of the fourth corner occurs (fig. 5c) having the same magnitude to that of comparable bimorph cantilever [10]. Therefore the thermal and the magnetoelastic effects are completely decoupled.

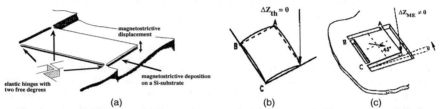

(a)   (b)   (c)

Figure 5 : *Torsion based, drift-free microactuator (a), compensation of thermal deflection (b) and magnetoelastic, torsional deflection (c)*

The validity of this new principle of actuator was tested by numerical simulation and by analytical calculation. In addition it was possible by these methods to take into account the effect of finite stiffness in the hinges. L-shaped hinges were used (fig. 5a) to get the required high perpendicular stiffness and low in plane stiffness with two easy rotation axes lying in the plane. A "residual drift ratio" was calculated, that is, the ratio of the thermally induced output to the useful magnetostrictive output of the actuator (for equal differential deformations of the active layer with respect to the substrate). This was 1:1 for the normal bimorph and 1:45 for the torsion actuator with L-hinges.

The actuators were fabricated using <100> and <110> Si-wafers by microlithography - and wet-etching techniques [11]. The dimensions of the hinges play a very important role for the functioning of the actuator. We have chosen for the L-shape hinges a length of $l = 100$ μm, plate- and hinge-thickness of $t = 50$ μm and widths of $w = 200$ μm and 20 μm, dimension being dictated by the fabrication techniques available.

To verify the principle of this actuator, 2 types of measurements were applied to obtain the curvature and the magnetoelastic behaviour of the actuator. With a Michelson Interferometer, it is possible to get the curvature of the whole structure out of the interference patterns. By this method we have observed isotropic deformations as well as pure anticlastic deformations. We observed high deformations on the acting plate due to strong deposition stresses bending the plate strongly. Work is continuing to relieve these stresses by appropriate sample preparation and annealing under field.

Magnetoelastic measurements under fields up to 2 T were made (fig. 6a) to compare the magnetoelastic output of these actuators compared to a simple bimorph cantilever having the same film (fig. 6b). The output measured experimentally is much smaller in this case (by a factor of 10), as the hinges of this actuator were not optimised (width of 200 μm). Analytical calculations based on the hinge dimensions led to a factor 9 between the output of a bimorph and the actuator. New actuators made out of <110> wafers were made in order to get better hinge dimensions with smaller widths. Characterisation of these devices is still in progress.

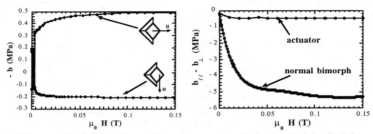

Figure 6 : Magnetoelastic behaviour of the microactuator in the 2 principle directions (a) and comparied with a bimorph cantilever (b)

CONCLUSION

A new kind of giant magnetostrictive multilayer structure has been presented which is adequate transducer material in microsystems. Due to the magnetic coupling between the layers, high magnetostrction is achieved in low fields. We also present a special drift-free, magnetostrictive actuator based on the torsion mode. In this case it is possible to produce a magnetoelastic displacement without having thermal drifts. Microactuators free from thermal drift should be feasable using the proposed solution.

ACKNOWLEDGEMENTS

The authors thank Dr. E. du Trémolet de Lacheisserie for helpful discussions. This work was supported by the European Union within the framework of the BRITE-EURAM Project "MAGNIFIT" (Contract No. BRE2-CT93-0536).

REFERENCES

1. R. Brück, K. Hahn and J. Stienecker, J. Micromech. Microeng. 5, (1995).
2. N. H. Duc, K. Mackay, J. Betz and D. Givord, J. Appl. Phys., 79, (2), 973-977 (1996).
3. E. Quandt, B. Gerlach and K. Seemann, J. Appl. Phys., 76, (10), 7000-7002 (1994).
4. E. du Trémolet de Lacheisserie and J. C. Peuzin, JMMM 136, 189-196 (1994).
5. J. Betz, K. Mackay and D. Givord, to be published.
6. S. Wüchner, J. Betz, D. Givord, K. Mackay, A. D. Santos, Y. Souche and J. Voiron, JMMM 126, 352-354 (1993).
7. F. Schatz, M. Hirscher, M. Schnell, G. Flik and H. Kronmüller, J. Appl. Phys. 76, (9), 5380-5382 (1994).
8. J.C. Peuzin and K. Mackay, J. Appl. Phys., accepted for publication, (1996).
9. J. Betz, E. du Trémolet de Lacheisserie and L. T. Baczewski, Appl. Phys. Lett. 68, (1), 132-133 (1996).
10. J. Betz, K. Mackay, J.-C. Peuzin, B. Halstrup and N. Lhermet, Actuator 96, edited by Axon (Conf. Proc. 5 , Bremen, Germany, Axon Technology Consult, 1996) pp. 283-286.
11. E. Bassous, IEEE Trans.on Elec. Devices 25, (10), 1178-1185 (1978).

SMART COMPOSITE BIOMATERIALS TiAlV/Al$_2$O$_3$/TiNi

Peter Filip*, Jaroslav Musialek**, Albert C. Kneissl*** and Karel Mazanec*

*Technical University of Ostrava, Institute of Materials Engineering, CZ-708 33 Ostrava-Poruba, Czech Republic,
**Municipal Hospital Ostrava-Fifejdy, Orthopedic Clinic, Nemocnicni 20, CZ-728 80 Ostrava-Fifejdy, Czech Republic,
*** University of Leoben, Institute of Physical Metallurgy and Materials Testing, A-8700 Leoben, Austria.

ABSTRACT
The TiNi shape memory alloy wires were plasma coated by α-Al$_2$O$_3$ ceramics and afterwards they were cast into Ti6Al4V or Ti6Al1.5V2.5Nb (Ti-alloy) matrix. The wire fibers melt and the alumina ceramics reacts with both TiNi and Ti-alloy if cooling rate is small. The interphases which are formed are Ti$_2$(Al, Ni) and Ti$_3$Al-type phases, respectively. The structure of TiNi wires changes to dendritic one, Ti-alloys contain 10 to 15% of primary α and transformed β phase. The optimal conditions for rolling are 900°C with reductions of 5%. At this conditions all the constituents of composite material can be deformed plastically. The adhesive strength of TiNi/Al$_2$O$_3$ is 60 MPa.

INTRODUCTION
Near equiatomic TiNi alloys are well known and widely used shape memory materials with excellent shape memory parameters, good mechanical as well as corrosion characteristics [1, 2]. The TiNi alloys inherently have sufficient corrosion resistance to the environment in which their are placed as implants [3, 4]. The alumina ceramic materials are typical with their low specific weight, high strength and stiffness as well as good corrosion resistance [5]. The α-Al$_2$O$_3$ represents a group of very frequently applied ceramics. Titanium alloys having a low specific weight (if compared to stainless steels and TiNi), adequate elastic modulus, reasonable high strength and good corrosion properties are well known too [6]. In this paper, we discuss new smart composite biomaterials. Following TiNi dynamic splints, fixation elements and TiNi/Ca10(PO4)$_6$OH$_2$ plasma coated (passive part) implants [7, 8, 9], these materials represent next step in development of smart biomaterials. The passive part of TiNi shape memory alloys is coated by alumina (thermal barrier) and cast into Ti-alloy. The optimization of coating process was described previously [10, 11], this paper concentrates on the technological parameters of composite processing and on the optimal treatment conditions. The attention is paid to the interfaces and interphases, observed between individual composite constituents.

EXPERIMENTAL
The TiNi wires with diameters ranging from 2.5 to 5mm were submitted by FIBRA Ltd. Their transformation temperatures were following: M$_f$ =-15°C, M$_s$=10°C, A$_s$=25°C, A$_f$=47°C. The wires were sand blasted, cleaned ultrasonically in an alcohol solution, dried at room temperature and only passive part was plasma coated with alumina ceramics (Ar-plasma, METCO System MN, U = 50 V, alumina powder with grain size ranging from 40 to 60μm). The thickness of coats was 400 μm and wires were put into a copper mold in such a way that the coated parts were cast into Ti6Al4V(Nb) alloys under vacuum. The Ti-alloys were melted in carbon crucibles, casting temperature was 1750°C. The composite materials were kept in a vacuum until their temperature decreased below 100°C (measured by Ni - NiCrAl thermocouple). The cooling rate was different depending on quantity of used molten Ti-alloy. A part of the ingots was further hot-rolled at different temperatures within the range 700 and 1050°C.
The specimens for scanning electron microscopy were prepared by water-beam cutting, grinding and mechanical polishing using a standardized routine. The scanning electron microscope Stereoscan Cambridge with Link analytical 360 EDX and WDX analyzers was used to investigate the structure and chemical composition of specimens. The pull out tests were performed using an Instron 1196 tensile test machine.

RESULTS AND DISCUSSION

The cross-sectional cut of specimen containing plasma-coated TiNi wire in Ti6Al4V alloy is shown in Fig. 1. From this Figure, it is visible that the Al_2O_3 coat reacts with melt and gas bubbles give rise. The TiNi wire (bright) with Al_2O_3 coat (dark layer) followed by transition layer in Ti6Al4V alloy matrix can be clearly seen. Analogous situation was observed in TiAlVNb alloys. The sources for creation of bubbles are different. The bubbles can form as a consequence of inappropriate flow of Ti-alloy melt during casting

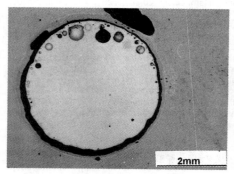
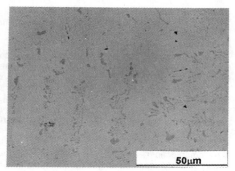

Fig.1. General view on TiNi/ Al_2O_3 embedded in Ti6Al4V matrix.

Fig.2. Dendritic structure of TiNi containing $Ti_4(Ni_xAl_y)_2(O,N)_z$-type particles (darker phase).

process, however, the plasma coats contain some quantity of gaseous phase in their pores, which can expand. Finally, the gaseous oxygen may originate during the reaction $12Ti + 2Al_2O_3 \rightarrow 4Ti_3Al + 3O_2$. It is apparent from Fig. 1 that the reaction intensity between alumina and individual metallic compounds (Ti-alloy or TiNi) differs depending on heat flow intensity. The plasma coating (dark layer surrounding TiNi wire) is thinner in direction towards the center of ingots and its thickness increases to the maximum value in direction where the maximum heat flow occurs (maximum cooling rate). Almost all bubbles are concentrated in region of greatest reaction intensity.

The typical structure of TiNi alloys is shown in Fig. 2. The casting process is connected with remelting of TiNi wire because the casting temperature is higher than the melting point of TiNi. The structure after this remelting consists of TiNi dendrites and the darker interdendritic phase. It was identified by EDX analysis that the content ratio Ti to Ni in this darker phase corresponds to the value 2:1. However, using WDX analysis, it has also been found that this phase contains more oxygen and nitrogen if compared to the matrix. Based on observed results we can conclude that the detected composition corresponds to $Ti_4(Ni_xAl_y)_2(O,N)_z$ phase [12].

The Ti6Al4V matrix consists of primary α-phase and transformed β-phase as can be seen in Fig. 3. The content of primary α-phase (darker phase in Fig. 3) ranges between 10 and 15%, and it has been found that this phase contains carbon. The source of carbon, found in α-phase, is carbon crucible. It is well known that carbon stabilizes α-phase, and its quantity depends on the carbon content.

The most important areas of interest are the boundaries between alumina ceramics and metal systems. The general view on the regularly observed good bonding between Al_2O_3 and TiNi or Ti6Al4V is shown in Fig. 4. The high temperature enables the aluminum atoms to diffuse into both TiNi and Ti6Al4V(Nb) alloys if the cooling rate is small enough and the interphases occur in the vicinity of different materials. Interphase which forms between Al_2O_3 and Ti6Al4V (Nb) contains approximately 75 at % of Ti, 22 at % of Al 2 to 3 at % of V, and approximately 0.2 at % of Nb in case of Nb alloyed variants. This composition corresponds to Ti_3Al phase containing vanadium and niobium atoms. Figure 5 (back scattered electrons) shows the transition between Ti_3Al and Ti6Al4V alloy.

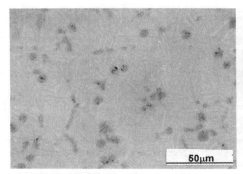

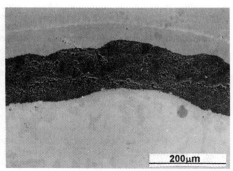

Fig.3. Structure of Ti6Al4V matrix consisting of transformed β-phase and primary α-phase (darker phase).

Fig.4. Connection between Al_2O_3 and metals (TiNi is in lower part of the micrograph).

It is evident that Ti_3Al phase is not homogeneous and contains bright particles with lower Al content. These bright particles contain more titanium (87 at % Ti, 11 at % Al and 2 at % V) if compared to Ti_3Al-type phase. No niobium was detected in similar bright particles, having very similar composition in case of TiAlVNb alloys.

On the other hand, the interphase which forms between TiNi and Al_2O_3 corresponds by its chemical composition to the $Ti_2(Ni_xAl_y)$ phase and it is identical with the interdendritic phase. It seems that the aluminum atoms dissolve in Ti_2Ni or more precisely in $Ti_4(Ni_xAl_y)_2(O,N)_z$ phase, which has lower melting point than intermetallic TiNi. The aluminum content in the $Ti_4(Ni_xAl_y)_2(O,N)_z$ phase near the $TiNi/Al_2O_3$ boundary was 6.5 at % and it decreases toward the center of TiNi wire to the value of 1,2 at % Al. No alloying of TiNi matrix with Al atoms was detected. Thus the transformation temperatures do not change significantly as a consequence of third element addition [12].

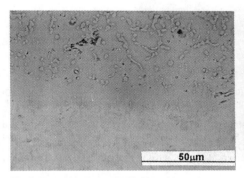

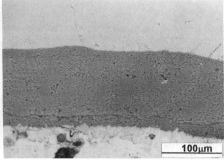

Fig.5. Transition between Ti_3Al-type interphase and Ti6Al4V matrix (lower part of micrograph).

Fig.6. The $Ti6Al4V/Al_2O_3/TiNi$ composite after rolling at 900°C (reduction of 5%).

As mentioned above, the formation of interphases is typical for small cooling rates. At higher cooling rates (ingots having smaller weight), the diffusion processes are limited and no interphase forms. Unfortunately, we are not able to specify the exact critical cooling rate.

The deformation of composites was performed at different conditions. The rolling temperature changed from 700 to 1050°C and the applied reductions varied from 3 to 10%. The optimal rolling temperature is 900°C and a maximum reduction should not be higher than 5%. Figure 6 characterizes the structure of specimen rolled with reduction of 5% after heating to 900°C. The total achieved deformation of specimens after several rolling passes was in this case 30%.

These optimized treatment conditions represent a reasonable compromise between the requirements of all structural constituents. The $Ti_4(Ni_xAl_y)_2(O,N)_z$ - type particles start to melt at 950°C and it means the rolling temperature must necessarily be well below 950°C. On the other hand, the plasticity of Al_2O_3 (if any) and Ti_3Al will be higher at elevated temperatures and these materials require higher rolling temperatures. In the case of Al_2O_3, however, the phase transitions occurs above 1100°C and further modification could be expected [9, 10]. It can be seen in Fig. 6, that some microcracks occur in Al_2O_3 ceramic as well as in Ti_3Al interphase even after optimized treatment. The analysis shows, however, that these microcraks are strictly located in these two structural constituents and do not propagate in the TiNi wires or Ti6Al4V(Nb) matrix. Even if the Al_2O_3 coating can not become apparently plastic at the proposed optimal rolling temperature (900°C), the hydrostatic pressure enables reasonable "deformability". The microcracks observed in Al_2O_3 and/or in Ti_3Al-type interphase grow further and become critical when the rolling conditions are not optimal as can be seen from Fig. 7 (reduction 30%, individual rolling passes approx. of 8% at temperature of 850°C).

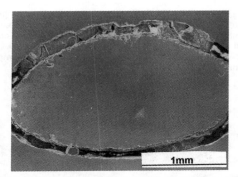

Fig.7. The cracks propagation in Al_2O_3 (rolling conditions were not optimal).

The composite materials without the interphases (cooled down with higher cooling rates during casting) are well formable at 900°C, too. It has been found in this case that the microcracks are dominantly localized in Al_2O_3 and good bonding between Al_2O_3 and TiNi or Al_2O_3 and Ti6Al4V(Nb) was found.

The pull-out tests results have shown that the strength of bonding between Ti6Al4V(Nb) matrix and plasma coated TiNi fibers is excellent. The shear strength of specimens with interphases is higher than 100 MPa and the failure occurs dominantly in Al_2O_3 ceramic. On the other hand the specimens without apparent formation of interphases have shear strength 60 MPa (Table 1) and failure occurs dominantly on TiNi/Al_2O_3 interface.

Table 1. Strength of TiNi shape memory wires/Al_2O_3/Ti6Al4V connection. The failure occurs in Al_2O_3 ceramics at lower cooling rates (CR) and on the Al_2O_3/TiNi interface at higher cooling rates (the interphases do not form).

PULL OUT TEST							
conditions	strength [MPa]						
Low CR	100	103	107	102	105	101	102
High CR	63	60	67	68	63	62	70

This behavior corresponds with previous results concerning the strength of TiNi/ Al_2O_3 adhesive bonding, however, the higher strength of Ti-alloy/ Al_2O_3 if compared to the TiNi/ Al_2O_3 signalize that the attractive forces between Ti-alloys and Al_2O_3 are higher even if the reorientation and stress induced martensite formation on the TiNi/ Al_2O_3 interface can eliminate the crack formation effectively [10, 11, 13]. The further study is necessary to explain these differences.

The transformation temperatures of active parts of TiNi shape memory wires (those without plasma coating which are not cast into Ti6Al4V alloy) do not change significantly if compared with as received conditions. The increase of M_s and decrease of A_s and A_f temperatures was detected. The extent of shape recovery (higher if compared to as received wires) and the generated forces during shape memory effect (lower than in as received conditions) are comparable with the TiNi wires annealed at 800°C. All the changes can be explained on the basis of modification of martensitic phase transformation and were discussed previously [1, 8, 12].

CONCLUSIONS

The Al_2O_3 plasma coatings react with TiNi wire as well as with Ti6Al4V(Nb) matrix during cooling of composite material after casting in vacuum if the cooling rate is small. The interphases which form at small cooling rate are $Ti_4(Ni_xAl_y)_2(O,N)_z$ and Ti_3Al-type, respectively. If the cooling rate is high enough, these interphases did not form. The optimal rolling temperature is 900°C and the applied reduction should not exceed 5% during individual rolling passes. At this conditions, total reduction of 30% can be reached.

The shear strength of Ti6Al4V(Nb)/plasma coated TiNi fiber is higher than 100 MPa if interphases form and 60 MPa in composites without interphase formation. The failure occurs dominantly in Al_2O_3 coating in the case of specimens with interphases and on the TiNi/Al_2O_3 interface in the case of specimens without interphases formation. The shape memory properties of "active parts" of TiNi wires (these which are not plasma coated and cast into Ti6Al4V matrix) are influenced if compared with properties of TiNi wires in as received conditions. This influence is comparable with annealing of TiNi at 800°C.

ACKNOWLEDGMENT

The authors are grateful to Grant Agency of Czech Republic for financial support (project 106/95/0480).

REFERENCES

1. P. Filip and K. Mazanec, Mater. Sci Eng., A 174, L41 (1994).
2. J. Van Humbeeck and Jan Cederstrom, ,The present state of shape memory materials and barriers to be overcome', Proceed. of the First Internat. Confer., Asilomar,CA, SMST Int. Comitee, 1 (1994).
3. Y.N.Zhuk in: Advanced Medical Applications of Shape Memory Implants in Russia, TETRA Consult, Moscow State University, Moscow, 1994.
4. G.Bensmann, F.Baumgart, J.Haasters, Technische Mitteilungen Krupp Forschungs-Berichte, 40, 1982, 123.
5. J.F. Shackefford, "Advanced eng. ceramics for biomedical applications", Key Eng. Materials, 56-57, 13 (1991).
6. F.H.Silver, Biomaterials, Medical Devices and Tissue Engineering, Chapman & Hall, 1994.

7. P.Filip, J.Pech, K.Mazanec: "Intelligent TiNi Shape Memory Alloy Applied for Dynamic Splints", Eighth Cimtec, Proc. Forum on New Materials-Intelligent Materials and Systems, ed.P.Vincenzini, 1995, 73.
8. P.Filip, J.Musialek, K.Mazanec, "Structure optimization of TiNi orthopedic implants", Jnl.de Physique IV, 5, 1995, C8-1211.
9. P.Filip, Progressive Biomaterials, TU Ostrava, 1995 (in Czech).
10. P.Filip, A.C.Kneissl, K.Mazanec, "Microstructure and the properties of hydroxyapatite coatings on TiNi shape memory alloys", Proceed.of Internat.Metallography Conference, ASM-Internat., Mater.Park (OH), 1996, p.397.
11. P.Filip, R.Melicharek, A.C.Kneissl, K.Mazanec, "Hydroxyapatite Coatings on TiNi SMA", Zeitschrift fuer Metallkunde (accepted for publication).
12. P.Filip, "Physical Metallurgy Parameters of Shape Memory Phenomena in TiNi Alloys and their Practical Use", Ph.D. thesis, TU Ostrava 1988 (in Czech).
13. P.Filip, M.Kaloc, M.Svicek, K.Mazanec, "C/TiNi composites- a prospective material for medical applications", Proc.of European Carbon Conference, The Royal Society of Chemistry, Newcastle, 1996, 106.

AUTHOR INDEX

SUBJECT INDEX